45-00

L

This

MW

Nuclear Power: Policy and Prospects

WORLD ENERGY OPTIONS

Series Editors

T. L. SHAW
Sir Robert McAlpine and Sons Ltd, London

D. E. LENNARD
D. E. Lennard and Associates, Orpington

P. M. S. JONES
Professorial Research Fellow, Surrey
University, Guildford.
U.K. Atomic Energy Authority, London

* * *

POLICY AND DEVELOPMENT OF
ENERGY RESOURCES

Edited by

T. L. SHAW, D. E. LENNARD and P. M. S.
JONES

BIOMASS

Edited by

D. C. HALL & R. P. OVEREND

NUCLEAR POWER: POLICY AND
PROSPECTS

Edited by

P. M. S. JONES

Nuclear Power: Policy and Prospects

Edited by

P. M. S. Jones

Professorial Research Fellow, Department of Economics, Surrey University, Guildford

and

Head, Economics and Energy Studies United Kingdom Atomic Anergy Authority, London

A Wiley–Interscience Publication

JOHN WILEY & SONS

Chichester · New York · Brisbane · Toronto · Singapore

Library of Congress Cataloging-in-Publication Data:

Nuclear Power.

(World energy options)
A Wiley–Interscience publication.
Includes index.
1. Nuclear industry. 2. Nuclear power plants—
Environmental aspects. 3. Nuclear power
plants—Waste disposal—Environmental aspects. 4.
Nuclear industry—Government policy—Case
studies. I. Jones, P. M. S.
II. Series.
HD9698.A2N79 1987 333.79'24 86–26714

ISBN 0 471 90732 4

British Library Cataloguing in Publication Data:

Nuclear power: policy and prospects.—
(World energy options)
1. Nuclear energy
I. Jones, P.M.S. II. Series
621.48 TK9145

ISBN 0 471 90732 4

Phototypeset by Input Typesetting Ltd., London SW19

Printed and bound in Great Britain

List of Contributors

J. Baumier	Commissariat a l'Energie Atomique, Paris.
Mme E. Bertel	Commissariat à l'Energie Atomique, Paris
K. Huttner	International Energy Forum, Tokyo
H. Inhaber	Risk Concepts Inc., Oak Ridge, USA
P. K. Iyengar	Director, Bhabha Atomic Research Centre, Trombay, India
P. M. S. Jones	Dept. of Economics, Surrey University, UK and UKAEA, London
A. N. Knowles	UKAEA, Risley, Warrington, UK
M. Mårtensson	Studsvitk Energiteknik, AB, Nykoping, Sweden.
P. A. H. Saunders	UKAEA, Harwell, Oxon, UK
D. Schmitt	Energie-BWL, Universitat-GH-Essen
I. D. Stancescu	Consultant, Hohenwaldeckstrasse 28/IV., 8000 München, FRG
T. Suzuki	International Energy Forum, Tokyo
J. J. Taylor	Vice President, Nuclear Power, Electric Power Research Inst., Palo Alto, USA
H. E. Thexton	Nuclear Energy Agency, OECD, Paris
N. G. Worley	Watt Committee on Energy Limited, London
T. Suzuki	International Energy Forum, Tokyo

Contents

Glossary

AGR	advanced gas reactor
ALARA	as low as reasonably achievable
ATR	advanced thermal reactor
AVM	Atelier de Vitrification de Marcoule
AVR	Arbeitsgemeinschaft Versuchsreaktor (HTR)
BHWR	boiling heavy water reactor
Bq	becquerel (unit of radiation)
bn	billion = 1000 million
BWR	boiling water reactor
d	days
ECCS	emergency core cooling system
ev	electron volt (energy unit)
FBR	fast breeder reactor
FR	fast reactor
G	giga = 1000 million
GCFR	gas cooled fast reactor
GCHWR	gas cooled heavy water reactor
GCR	gas cooled reactor
GWe	gigawatt electric
h	hour
HLW	high level waste (radioactive)
HPIS	high pressure injection system
HTR	high temperature reactor
HWR	heavy water reactor
ILW	intermediate level waste
k	kilo = 1000
kW	kilowatt (power unit)
kWh	kilowatt-hour (energy unit)
LLW	low level waste
LMFBR	liquid metal cooled fast reactor
LOCA	loss of coolant accident
LPIS	low pressure injection system
LWBR	light water breeder
LWR	light water reactor
m	metre
M	mega = 1 million
Mev	million electron volts
MOX	Mixed plutonium/uranium oxide fuel
mtce	million tonnes of coal equivalent (energy unit)
mtoe	million tonnes of oil equivalent
MW	megawatt = 1 million watts
MWd	megawatt-days (thermal energy unit)
MWe	MW electric
MWt	MW heat
NDT	non-destructive testing
NPT	non-Proliferation Treaty
OMR	organic moderated reactor
p.a.	per annum
PHWR	pressurized heavy water reactor
PNE	peaceful nuclear explosion
ppm	parts per million
PWR	pressurised water reactor
RBMK	Soviet water cooled graphite moderated reactor
SGHWR	steam generating heavy water reactor
SNG	substitute natural gas
Sv	sievert (unit of radiation dose)
SWU	seperative work unit (unit of enrichment)
t	metric tonne
TBq	terra Bq
TBP	tributyl phosphate

THTR	thorium high temperature reactor
W	watt (unit of power)
WOCA	world outside centrally planned economies

Series Preface

Each publication in this book series presents up-to-date technical, economic and environmental information about one energy source. The value of each source depends on the purposes for which it is used and where and how it is exploited. The wider consequences of its deployment, whether aesthetic or ecological, are also important when choices between options come to be made.

The Introductory Volume in this series, published in 1984 with the title 'Policy and Development of Energy Resources', emphasised the need for society to have access to a wide range of well proven energy supply options. Not only is this essential for choices to be made and flexibility secured through diversification, for example in the sources used to generate electricity, but it stimulates competition. Properly managed, this must be good for the consumer, both in terms of the cost of energy and the reliability of its supply.

Energy is a major political, economic and public issue. Few weeks pass without some reference being made to it at national and international level. Coal, oil and nuclear sources attract most attention for different reasons and their critics bring other sources including conservation into the headlines as alternatives. The choice is made to look deceptively easy, but it is not. The Introductory Volume emphasised the range of factors which lie behind sensible choice. Diversification is one of these. Another is the proper understanding of the full implications of each source available to meet a particular need.

The content of all these subsequent Volumes in the series seeks to provide the breadth of information and detail needed to judge each source in the context of each application. As the series title confirms, each and every energy source is dealt with in a world context. The order in which the sources are presented has been carefully chosen according to their technical and economic readiness for commercial use, as well as to the wider political need and for public education. Every precaution has been taken to ensure that factual information rather than personal opinion is given. Readers are strongly encouraged to translate this into the particular conditions for energy supply which they face. The fact that one source may be outstanding for one application in one country does not necessarily mean that it will be best for the same duty in another.

The days of limited choice in the energy market are passing. This is reflected in the scope of this book series. The further Volumes will be published as commercially important new developments bring more energy sources into the international market place. Likewise, earlier Volumes will be revised and re-issued as necessary.

The world has become increasingly aware of the importance of energy to a stable society and shows no sign of lessening its demands for plentiful and reliable supplies at the right price and with tolerable implications. The pressure is therefore on the suppliers and those who decide on investments in this sector, and it is to them that this series of books on the widening range of energy options is targetted.

The public also needs to be aware of the choices becoming available to meet their requirements and it is intended that the series will provide them with a better insight into the whole subject of their future energy supplies.

16 March 1987

Preface

This is the third volume in the World Energy Options series and the second to deal with a specific energy technology. As with the other volumes of the series the intention is to examine the technology and how it may develop in the light of its particular attractions and advantages and the constraints which it faces.

Although the world energy scene has changed considerably even since the first volume of the series was published, the fact remains that supplies of conventional fuels are not limitless and the world is approaching a time when continued extensive reliance on their use will become increasingly expensive. Exclusive reliance on fossil fuels also leaves the world vulnerable to problems associated with their environmental impacts, including their long term implications for world climate.

For these two reasons the development and selection of energy sources to take the world into the twenty-first century remains an important policy issue despite the apparent abundance and comparatively low cost of fossil fuels at this point in time.

This volume seeks to set out the basic nuclear technologies and the factors that have contributed to policy formation in different parts of the world. It looks at the advantages and problems associated with nuclear development and points to ways in which the nuclear industry might be expected to contribute in the coming decades.

Acknowledgement

I would like to record my gratitude to the United Kingdom Atomic Energy Authority whose support of my fellowship at Surrey University has enabled me to undertake the co-ordination, part authorship and editing of this work. I would also like to thank my secretaries, Miss Elsie Bailey and Mrs Wendy Marler, my colleagues in the UKAEA and the contributors, without whose assistance the task could not have been completed

The opinions expressed in the individual chapters are those of the separate authors and are not necessarily shared by the other contributors, by their companies or by their national authorities.

PMSJ
Guildford

March 1987

Nuclear Power: Policy and Prospects
Edited by P. M. S. Jones

1

INTRODUCTION

P. M. S. JONES
Department of Economics, Surrey University

One hundred years ago the idea that nuclear energy could, let alone would, contribute to meeting mans' energy requirements was unimagined. Even the prospects for using oil as a fuel were not clearly recognized, and at least one reputable economist (Jevons, 1863) saw a bleak future with almost total reliance on rapidly depleting coal resources.

The first speculations about the possibility of using atomic energy and its potential as a physically concentrated energy source began to appear in the early 1930s (Birkenhead, 1930), even before Chadwick's discovery of the neutron. However, the fission process itself, which turned such speculation into a practical prospect, was not discovered until the late 1930s (Meitner and Frisch, 1939: Hahn and Strassman, 1939). It was only after this that scientists realized that slow neutron fission of uranium could be exploited to provide a continuous source of heat and a substitute for fossil fuels (Gowing, 1964).

The priority development programmes which led to practical nuclear weapons during the Second World War were succeeded in most of the major industrial nations by intensive, largely government funded, R and D programmes aimed at early commercial exploitation of the newly available technology.

Within a remarkably short time radioactive isotopes produced in nuclear reactors were contributing to medical diagnosis and treatment as well as to the monitoring and automated control of industrial processes. They provided a new research and analytical tool which rapidly yielded new insights in such diverse fields as medicine, agriculture and hydrology, as well as in the physical sciences themselves. Of more direct relevance to this work is the fact that in 1953 the world's first commercial scale nuclear fuelled electricity generating plant, Calder Hall, was coupled to the United Kingdom's national electricity grid, whilst submarines powered by nuclear boilers were operating in the world's oceans by 1954.

In the atmosphere of the 1950s and 1960s, the enormous technological possibilities afforded by nuclear energy were themselves a sufficient justification for its development, often on a very wide front. Large numbers of alternative power reactor concepts were studied in parallel; many reached the experimental reactor stage before being dropped. Different countries, for good understandable reasons, pursued different designs; some favoured indigenous technology, others bought the technology in world markets; some changed their plans and abandoned their earlier preferences.

The choice of reactor and nuclear fuel cycle are closely linked but the bulk of fuel development has concentrated on uranium oxide fuels, which almost all countries initially planned to have reprocessed to recover plutonium and unburnt uranium, which could be used to fuel second generation reactors.

The concept of the breeder reactor, which produces more fuel than it consumes, evolved quite early on and was seen as immensely important in view of the very limited resources of uranium then known to exist.

As time has passed ideas have been modified, particularly on the timing of the need to move to plutonium fuelled reactors. Nevertheless development and deployment have continued so that nuclear power now contributes nearly 30 per cent of electricity consumed in Western Europe, about 18 per cent of electricity in North America and Japan and a rather smaller proportion (13 per cent) in the Comecon countries. Four basic types of nuclear reactor are currently being deployed; viz. the pressurized water reactor, the boiling water reactor, the advanced gas cooled reactor and the pressurized heavy water reactor; and three more are under active development, viz. the advanced thermal reactor, the high temperature gas cooled reactor and the liquid metal cooled fast reactor. Only two countries are reprocessing fuel on a commercial scale, though others plan to have plants on stream in the near future.

The object of this book is to set out simply and clearly where civil nuclear technology now stands, what options there are for the future, and what factors are important in making choices between them.

The chapters of Section I provide a general technical background to enable the non-technical reader to understand those aspects of the technology which are of particular relevance to policy choices. Chapter 2 introduces in a simplified way the physical processes underlying the production of nuclear energy. Chapters 3 and 4 set out the characteristics of the different reactor types and fuel cycles. Chapter 5 reviews non-electrical uses of nuclear power and its application in combined heat and power systems. The markets for these technologies seem unlikely to develop significantly within the general time-span covered by this book, and the technologies do not yet figure in the implementation plans of the countries covered in Section III. Nevertheless they are options that could begin to make worthwhile contributions to world energy supplies in the longer term and this volume would not be complete if it did not set out their attractions and disadvantages so that the reader can form a judgement on their likely future relevance. These chapters extend and update the very brief summary of the topics presented in the introductory volume to the series (Jones, 1984).

Section II looks in more detail at the sensitive issues of radiation and its effects and at reactor safety, to provide the reader with facts that will set these matters in a realistic perspective. The section then goes on to examine the question of spent nuclear fuel, nuclear wastes and the decommissioning of reactors and fuel cycle plant. All of the chapters in this section deal with matters that attract considerable attention in the media and which have been a focus of anti-nuclear pressures in recent years.

The individual chapters of Section III have been contributed by experts drawn from countries which were selected to span a range of choice and motivation which the reader can compare with his own country's position. They explain how different countries have reacted to the nuclear option, the policies they have chosen to pursue and the reasons for their choices. The authors go on to describe how they view the future and the technological options that they regard as having greatest significance in their own circumstances. There is no separate chapter covering those countries that have so far opted not to make use of nuclear power, like Austria and Australia, or who have put a strict limit on its use, like Sweden. Their position is discussed in the final section.

Section IV reviews the preceding material and seeks to bring out the main factors that have influenced and will continue to influence decisions in the civil nuclear field. Chapter 20 reviews the resource base; Chapter 21 includes a discussion of the economics of electricity generation, and on the environmental impacts of energy systems; two aspects that provide the major incentive for the adoption of nuclear power; Chapter 22 deals with the factors that have caused concern or have acted as constraints on nuclear development. Finally Chapter 23 discusses the prospects for the different nuclear systems, and the implications for these systems and nuclear power as a whole, of the widely different perspectives revealed in the earlier chapters.

REFERENCES

Birkenhead, The Earl of (1930). *2030 AD*, Hodder and Stoughton, London.

Gowing, M. (1964). *Britain and Atomic Energy 1939–45*, Macmillan, London, p.76.

Hahn, O., and Strassman, F. (1939). *Naturwiss.*, **27**, 11.

Jevons, W. S. (1863). 'The Coal Question', republished in part in *Environment and Change*, 1973, p.373.

Jones, P. M. S. (1984). 'The Nuclear Contribution', P. M. S. Jones, D. E. Lennard and T. Shaw (eds.), *Policy and Development of Energy Resources*, Wiley, London, pp. 151–170.

Meitner L., and Frisch, O. R. (1939). 'Disintegration of uranium by neutrons: a new type of nuclear reaction', *Nature*, **143**, 239.

Section I
TECHNICAL BACKGROUND

Nuclear Power: Policy and Prospects
Edited by P. M. S. Jones

2

The Underlying Physics

P. M. S. JONES
Department of Economics, Surrey University

2.1 Introduction

A lengthy exposition on the physics and detailed engineering of nuclear reactors would be out of place in a book that is mainly concerned with policy choices. Nevertheless some understanding of these matters is essential since different reactor types do have different resource requirements and implications, and different environmental and economic impacts; not only before and during their use, but also afterwards in terms of the wastes they produce.

This chapter and the one that follows describe in simple terms the process of fission which releases the energy trapped in the nucleus of heavy atoms, the conditions necessary to bring about fission in a controlled manner, and the characteristics of the different reactor systems designed to exploit the phenomenon. Readers wanting a fuller treatment should consult one of the many excellent source books (Marshall, 1983; Pederson, 1980).

2.2 The Physics of Nuclear Energy

2.2.1 *Atoms, Isotopes and Radioactive Decay*

For simplicity the atoms of which all materials are composed can be regarded as having a small nucleus carrying a positive electric charge, surrounded by a cloud of negatively charged electrons. The mass of the atom is concentrated in the nucleus which is itself made up of positively charged protons and electrically neutral neutrons; particles which have roughly equal masses, both being about 1800 times heavier than a single electron.

The nuclei of the different chemical elements are made up of varying numbers of protons and neutrons. The number of protons (*atomic number*) fixes the total electric charge on the nucleus and this determines the number and distribution of negatively charged electrons around the nucleus, which in turn determines the chemical properties of the element. All atoms with nuclei having a particular charge have identical chemical properties and are atoms of the same element. The numbers of neutrons in the nucleus of a given element can differ, however, and this gives rise to the so called *isotopes* which have the same chemical properties but different atomic masses and slightly different physical properties (see Figure 2.1). Most naturally occurring elements consist of a mixture of such isotopes.

In coming together to form a stable atomic nucleus, the agglomeration of neutrons and protons releases energy. Conversely, energy has to be supplied to break a stable nucleus up into its constituent neutrons and protons. The attractive force between the neutrons and protons has to be large enough to overcome the forces of electrical repulsion between the positively charged protons, otherwise the nucleus would fly apart.

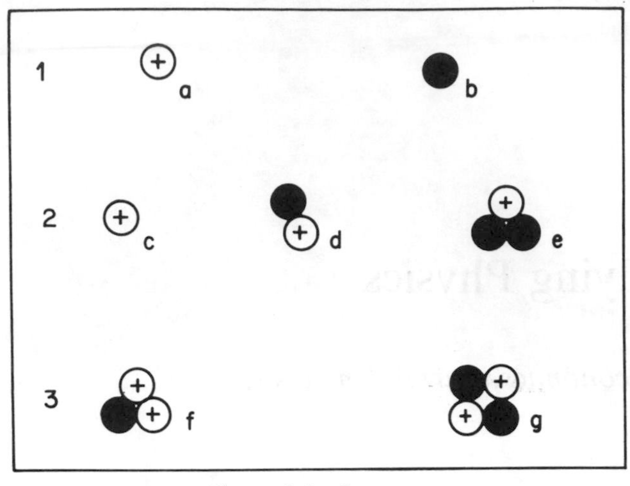

Figure 2.1 Isotopes.
Row 1 a = Proton; mass = 1, charge = + 1 :
b = neutron; mass = 1, charge = 0.
Row 2 Hydrogen isotopic nuclei: charge = + 1
c = hydrogen; mass = 1 : d = deuterium; mass = 2 : e = tritium; mass = 3.
Row 3 Helium isotopic nuclei: charge = + 2
f = helium-3; mass = 3: g = helium –4; mass = 4

In general, the larger the nucleus, i.e. the higher its atomic number, the higher the excess of neutrons over protons required within it to give it stability. For light elements (Figure 2.1) the numbers are roughly equal, whereas heavy elements like uranium have a large excess of neutrons: thus the uranium–238 atomic nuclei contain 146 neutrons and 92 protons and uranium–235 nuclei the same number of protons but three less neutrons.

With few exceptions the isotopes of chemical elements occurring in nature are those whose nuclei are stable—if they were not they would have disappeared long ago. The first exception is nuclei that are almost stable but decay so slowly that they have survived even over the 7 billion years or so since their formation (Dermott, 1978) which was long before the earth itself was formed 3 billion years ago (e.g. uranium–238). The second is nuclei that, although relatively short lived, are continuously produced in nature, either by the decay of the long lived isotopes in the first group (e.g. uranium decaying to produce shorter lived radon isotopes) or by the bombardment of elements such as oxygen or nitrogen by high energy cosmic rays entering the earth's atmosphere from outer space (e.g. radioactive carbon–14). The break up of nuclei (*radioactive decay*) occurs in general by the emission of electrons (also called β-particles or β-rays) or positively charged helium nuclei (also called α-particles or α-rays) together with short wavelength electromagnetic radiation called γ-rays. (There are other modes of radioactive decay and other products but these need not concern us here.)

2.2.2 *Fission and Fusion*

The stable nuclei formed by bringing together protons and neutrons in appropriate ratios have masses that are slightly less than the mass of their constituents and the difference is a measure of the *binding energy* for the nucleus. (The two are related by the well known Einstein equation $E = mc^2$). From Figure 2.2. it can be seen that the biggest mass deficits are for nuclei that are neither very light nor very heavy. Either joining (or

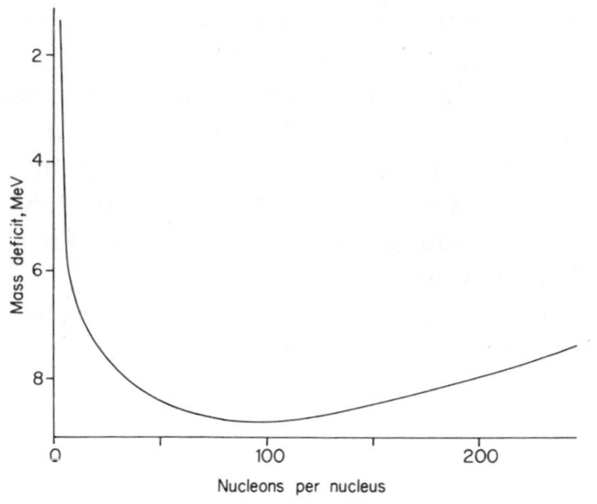

Figure 2.2 Mass deficit (MeV per nucleon)

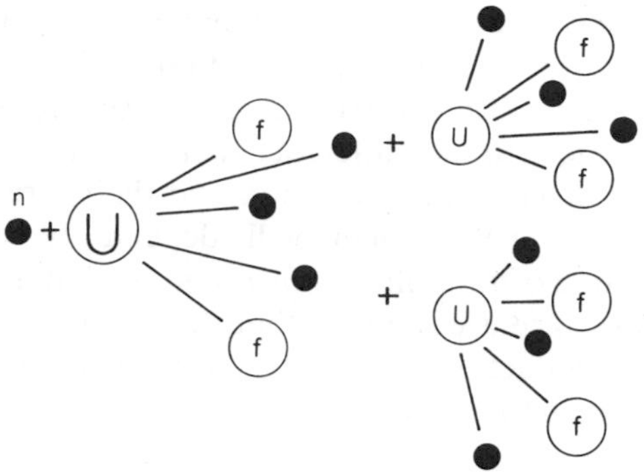

Figure 2.3 The chain reaction. Neutron (n) + uranium–235 (U) yields neutrons + fission products (f)

fusing) two light nuclei together or splitting (*fissioning*) a heavy nucleus to yield two lighter ones will release energy, and this is the basis of fusion or fission power.

To produce fusion one has to bring the light nuclei such as those of deuterium (hydrogen–2) and tritium (hydrogen–3) together, overcoming the high electric repulsive charges, and hold them together until they can rearrange their neutrons and protons and give up their excess energy. This will be looked at again in Chapter 22.

To produce fission one has to find a means of exciting the heavy nucleus so that it reorganizes its protons and neutrons and divides into smaller nuclei. This process can be accomplished by introducing an additional neutron into the nuclei of some isotopes of heavy elements such as uranium–233, uranium–235 or plutonium–239. The nuclei of these so called *fissile isotopes* split to yield two lighter nuclei called *fission products* and, importantly, around three free neutrons, which in their turn can produce fissions in other fissile nuclei. If the conditions are right a self-sustaining chain reaction can be established (Figure 2.3).

The majority of isotopes, even of the heavy elements, are not fissile and their nuclei may absorb neutrons to yield new heavier isotopes of the same element which will generally be unstable and undergo radioactive decay, or even a series of such decays, until, eventually, a stable nucleus of a totally different element is produced. This process of parasitic neutron capture is important to nuclear power for three reasons. It can interfere deleteriously with the establishment or maintenance of a controlled fission chain reaction; it can be used beneficially as a means of controlling the reaction; and it can lead to the production of new fissile isotopes (see Chapter 4), as well as to other radioactive isotopes which have uses in research, medicine, industry, etc.

The design of nuclear reactors is complicated by the fact that the ability of neutrons to induce fissions, and the ease with which they can be captured or absorbed by non-fissile nuclei, varies in a complex way with the speed at which the neutron is moving. In general lower speed neutrons both induce fissions and get captured more easily (Figures 2.4 and 2.5); but at some speed ranges small changes can alter the likelihood of fission or capture by as much as 1000–fold, depending on the specific nuclei involved.

2.2.3 *The Chain Reaction and Criticality*

The concept of *criticality* is most easily understood by considering a spherical mass of a fissile material, say uranium–238 with enhanced levels of uranium–235. Some neutrons are produced in this mass by spontaneous fission of uranium atoms and these induce further fissions in the surrounding uranium–235. A proportion of the neutrons are captured without fission by the uranium, however, and others will escape from the surface of the sphere. The net rate of increase in the number of neutrons in the sphere will depend on the difference between the net rate of neutron production and their rate of escape. The former will depend on the volume of the sphere $4/3\pi r^3$ and the latter on its surface area $4\pi r^2$ so that the relative ratio of production to losses will increase in proportion to the radius of the sphere, r. For small spheres the neutrons will escape as fast as they are produced and there will be no net multiplication. For large spheres the losses are less and the number of neutrons in the sphere will multiply rapidly. The point at which this multiplication commences defines what is known as the *critical mass* for the fissile material.

The critical mass differs for different fissile materials and for different geometries and

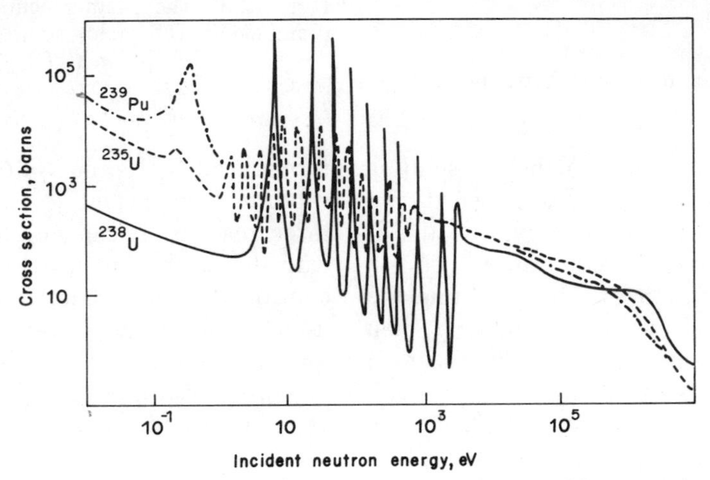

Figure 2.4 Neutron capture cross sections. Note: The central resonance cross sections for plutonium-239 are omitted for clarity *Reproduced by permission of Oxford University Press from Marshall (1983)*

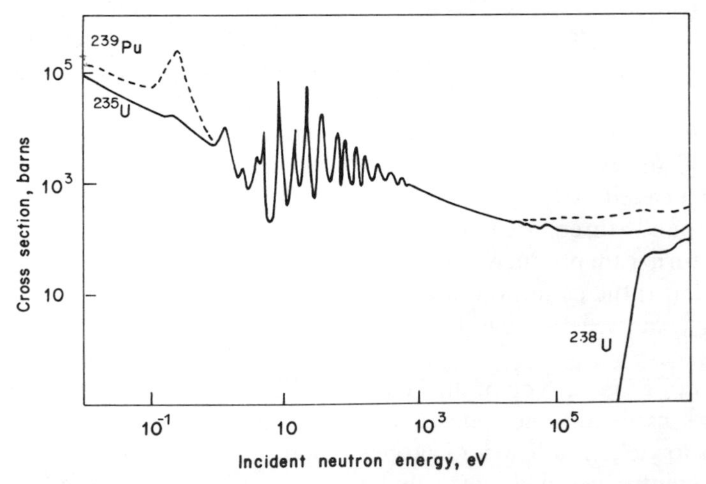

Figure 2.5 Fission cross sections. Note: The central resonance cross sections for plutonium-239 are omitted for clarity *Reproduced by permission of Oxford University Press from Marshall (1983)*

material densities. It will also depend on the presence or absence of a neutron reflector around the mass and be affected by the presence of neutron absorbing impurities.

Uranium nuclear reactor fuels produced using natural uranium, which contain 99.3 per cent of the non-fissile uranium–238 isotope and 0.7 per cent of the fissile uranium–235 isotope, or using uranium with slightly enhanced levels of uranium–235 (see Chapter 3) are incapable of sustaining a chain reaction with neutrons travelling at the high speeds with which they are emitted in the fission process. So many of the two to three high energy neutrons emitted per fission will be absorbed in the large excess of uranium–238 that the necessary condition for a chain reaction does not exist, i.e. that at least one neutron per fission produces further fissions.

There are two ways of overcoming this problem. Firstly the proportion of uranium–235 can be increased until parasitic capture by uranium–238 is reduced to acceptable levels. Alternatively, the neutrons can be slowed down to take advantage of the fact that the likelihood of uranium–235 fission increases more rapidly than the likelihood of neutron capture as the neutron speed decreases (Figures 2.4 and 2.5). Pursuing the former course leads to fast neutron reactors (or more concisely *fast reactors*) whereas the latter course leads to the so called *thermal reactor* in which neutrons move at speeds comparable to those of molecules in a gas.

It will also be apparent that control has to be exercised over any other materials introduced into a nuclear reactor, since they too may absorb neutrons and prevent the establishment of a chain reaction, or alternatively they may require higher enhancement of the uranium–235 content of the fuel than would otherwise be necessary.

Whilst fission neutrons are released instantaneously some further neutrons are released from fission products after delays ranging from fractions of a second to a few minutes (see section 4.2.2). These delayed neutrons account for only about 1 per cent of all the neutrons arising from the fission process but they are important because they simplify reactor control by increasing the time taken to establish a change in the fission rate. If an expanding chain reaction can maintain itself without relying on these delayed neutrons the nuclear material is said to be prompt critical. In such circumstances its energy output could increase by many orders of magnitude in small fractions of a second, making control extremely difficult. Reactors are designed and operated to avoid this condition.

REFERENCES

Dermott, S. F. (ed.) (1978). *The Origin of the Solar System*, Wiley, Chichester, pp.282–318.

Marshall, W. (ed.) (1983). *Nuclear Power Technology*, Vol. 1, Clarendon Press, Oxford.

Pedersen, E. S. (1980). *Nuclear Power; Volume 1, Nuclear Power Plant Design*, Ann Arbor Science, Ann Arbor, Michigan.

Nuclear Power: Policy and Prospects
Edited by P. M. S. Jones

3

Nuclear Reactor Types

P. M. S. Jones
Department of Economics, Surrey University

3.1 Introduction

A nuclear reactor is merely a device designed to induce and maintain a controlled fission reaction in the nuclear fuel whilst ensuring that the radioactivity and fission products produced are safely contained. Reactors can vary from low power research devices to high power devices specifically designed to produce heat, either for direct use or for use in producing steam to drive the turbines to produce electricity or propel ships.

For the reasons identified in Section 2.2.3 there are two basic types of reactor, classified by average neutron speed as either thermal or fast.

3.2 The Basic Reactor

3.2.1 *Thermal Reactors*

In a thermal reactor the neutrons emitted in the fission process are slowed down by allowing them to collide with materials called moderators. A good moderator should have the following characteristics:

1. Its atoms should not capture neutrons to any significant extent.
2. Its atoms should be light, because light atoms slow down neutrons more effectively.
3. It should be a dense material so that neutrons are slowed down within a short distance.
4. It should be readily available in pure form and cheap.

Materials matching these requirements reasonably well include carbon in the form of graphite, hydrogen as water or in suitably stable organic compounds and deuterium (hydrogen–2) as heavy water.

The most effective use of the moderator would be achieved if it and the fuel were intimately mixed. In the extreme case this would result in a homogenous reactor core in which the uranium fuel was dissolved or otherwise dispersed in the moderating medium, e.g. a uranium salt in water or in a mixture of fused salts, one of which acted as a moderator. Such systems have been developed and tested and have both advantages and disadvantages.

A reactor in which fuel is isolated from but surrounded by the moderator is less efficient in terms of neutron economy, but has the advantage that it is possible to put the fuel in containers (*cans*) which both retain the highly radioactive fission products and facilitate the replacement of fuel when its uranium–235 content has been reduced by fission to the minimum acceptable in the specific reactor design. This is the route that has been followed in almost all the world's existing power reactors.

If one were merely to construct a large matrix of moderator and insert containers of suitable uranium fuel, a point would be reached at which the system became critical and a self-sustaining chain reaction commenced. Alternatively introducing a moderator into a slightly subcritical mass of uranium could have the same effect. The ingress of water into the uranium mineral deposits at Oklo in Gabon produced a natural nuclear chain reaction in just this way many millions of years ago. Neither of these is a convenient means of control, however, and reactors are designed to contain a larger quantity of fuel than is required for criticality for the given geometrical configuration, structural materials and moderator. Control is then effected by introducing a neutron absorbing material, for example rods of boron steel. These, when in the fuel–moderator core, absorb so many neutrons that no chain reaction can begin. Gradual withdrawal of the rods will bring the system to the point of criticality. They can subsequently be used to shut down the system or to compensate for changes of fuel reactivity due to the gradual consumption of fissile uranium–235.

A further requirement in the basic reactor is a means of removing heat from the system and transferring it to where it is needed. This can be achieved using gaseous or liquid coolants passed through the fuel–moderator array. In systems where the moderator is fluid, e.g. water or heavy water, the moderator may itself be circulated. In most existing systems, as will be seen later, the coolant passes from the reactor into a heat exchanger where it boils water to provide the steam needed to drive conventional power generation turbines.

Because the fuel, moderator and coolant all become radioactive they have to be contained within a suitable pressure vessel and surrounded by materials that prevent neutrons escaping and absorb the penetrating γ-radiation produced within the core, in order to protect the workforce and the public.

A schematic layout for a simple reactor (Magnox) containing the necessary constituents is shown in Figure 3.1.

3.2.2 *Fast Reactors*

As has already been seen in section 2.2.3, reactors which do not have a moderator and rely on fissions induced by fast neutrons have to limit the amount of non-fissile material in the fuel to reduce parasitic neutron capture to an acceptable level. The fuel is therefore very concentrated and the heat output per unit volume (*power-density*) is potentially far higher than in a thermal reactor.

Control of fast reactors is no different to

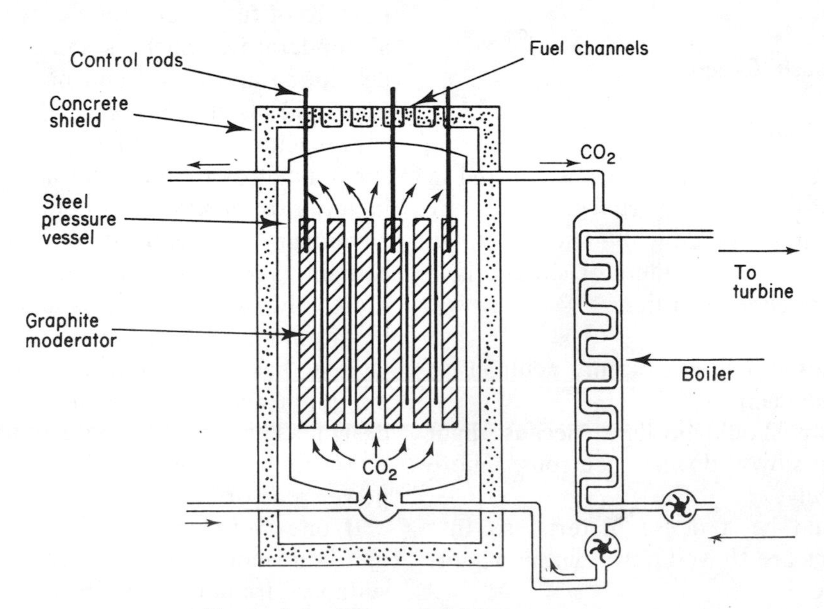

Figure 3.1 Gas cooled Magnox reactor

that of thermal reactors and can be accomplished using neutron absorbing control rods. Cooling, however, poses greater problems because the use of higher power densities to keep reactor size and costs down, demands enhanced heat transfer and very high rates of coolant flow. The greatest amount of development effort has been directed to the liquid metal cooled fast reactor, although gas cooling is also possible in principle. A schematic layout for a simple fast reactor is shown in Figure 3.2.

3.2.3 *General Design Considerations*

The reactor designer's objective will normally be to arrive at a least cost design, consistent with requirements for safety, reliability and ease of maintenance for the particular combination of fuel, moderator and coolant selected and the desired power output of the reactor.

He can vary the height and diameter of the reactor core, the design and spacing of the fuel elements, the composition of the fuel, the reactor operating temperature, etc., but he is constrained by the properties of the materials at his disposal. These must not melt, corrode, deform, crack, or otherwise fail under normal operation or any plausible fault conditions.

The pressure vessel or containment must be capable of withstanding the coolant pressures necessary for efficient heat removal from the core. These may be high for water cooled reactors, although lower for gas cooled systems and much lower for liquid metal cooling.

The designer must also take all possible modes of failure and their consequences into account, so that he can either eliminate them

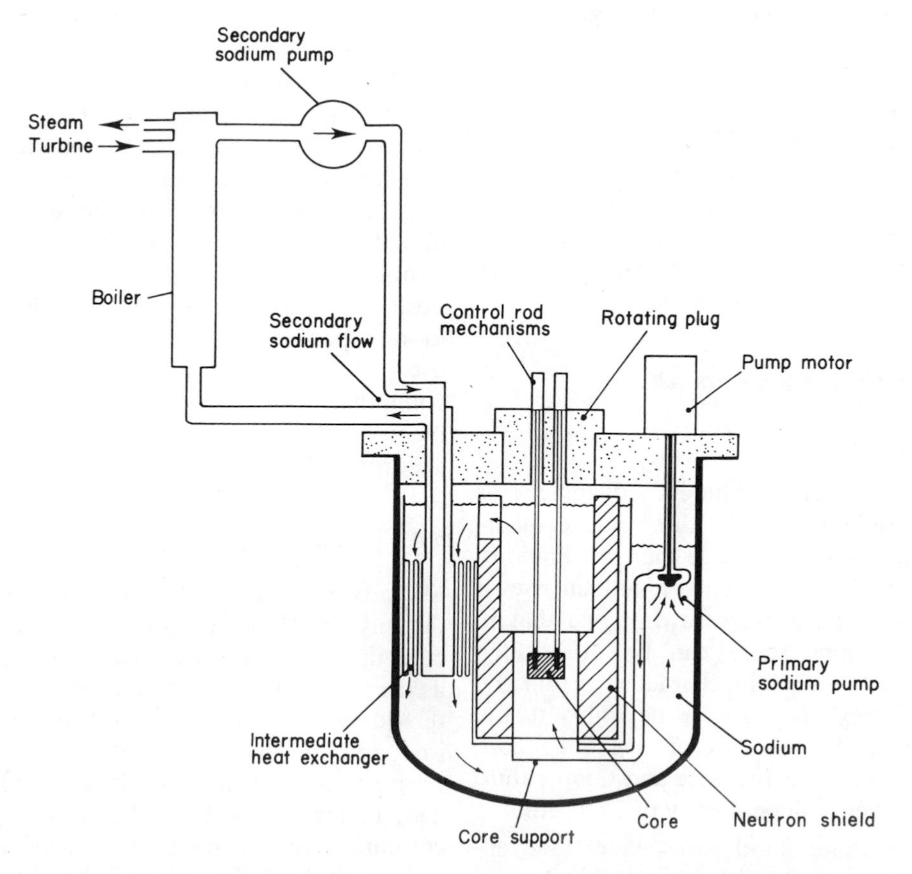

Figure 3.2 Pool type liquid metal cooled fact reactor. Source: UKAEA. *Reproduced by permission of Oxford University Press from Marshall (1983)*

or, by replication, ensure that the effects are controllable and kept to an acceptable level. Thus, in the event of coolant pump failure the reactor should shut down and auxiliary systems should come into play to remove the heat that continues to be produced by decay of the fission products in the reactor fuel. Based on past, sometimes unfortunate, experience, he will endeavour to avoid common mode failures by providing mechanical back-up to electrical systems, different power sources for main and back-up systems, and gravity operated fail-safe devices. He will also avoid routing all his services through single channels and will provide a number of coolant injection ducts, to guard against the remote risks of mechanical or other damage.

These design precautions inevitably add to the basic cost of nuclear reactors, not only because of the expense of the replicated components themselves, but also because of the effect they have on the size of the pressure vessel and containment buildings. However, as will be seen in Section III, it is extremely important to get the design settled at the outset. Most of the worst cost overruns and construction difficulties have arisen at sites where design changes have been introduced when construction was already well advanced or even, in some instances, after reactors have been completed.

3.3 Gas Cooled Reactors (GCR)

3.3.1 *Gas Cooling*

Gases have certain inherent advantages as reactor coolants. Unlike the liquid coolants they cannot undergo phase changes (liquid to gas) when they experience a sudden rise of temperature or a pressure drop, so that in fault conditions there can be no sudden change in the coolant characteristics in the reactor. They also provide flexibility to the designer in his choices of operating conditions, since their pressure and temperature can be varied independently of each other.

On the other hand gases have disadvantages. Compared with liquids their heat capacities and densities are low and they are less efficient at transferring heat from the fuel to the boilers. For any given fuel design, heat output per unit of fuel (fuel rating) has to be kept down and reactors tend to be larger with larger coolant circulating pumps, in order to get sufficient coolant flowing past the fuel.

There are, in principle, many different types of gas cooled reactor, depending on the choice of fuel, moderator and coolant gas. It is clearly desirable that the gas chosen as coolant should be stable chemically at reactor temperatures, even in high radiation fields, and it should not react with the moderator, fuel containers or other reactor component materials. It should have a low propensity to capture neutrons in order to minimize its own radioactivity, to aid over-all neutron economy, and, importantly, to avoid the risk of a rise in core reactivity if, for any reason, the reactor loses its coolant. Additionally the gas should have a high heat capacity and density so that it can remove heat efficiently from the fuel elements and transfer it to the boilers.

Although air was used as the coolant in some early small scale reactors, the almost universal adoption of graphite as the moderator for these systems has concentrated attention on carbon dioxide and helium. Carbon dioxide is cheap and has a high density but is not completely inert. Its use is therefore preferred at lower temperatures whereas the more expensive inert helium is preferred for use in high temperature reactors. Other common gases are either too chemically reactive or give relatively high neutron absorption.

3.3.2 *The Magnox Reactor*

Magnox reactors have one of the simplest designs of all power producing reactors in current use (Figure 3.1). They have graphite as the moderator and are cooled using carbon dioxide gas. They use natural uranium fuel, i.e. uranium with the naturally occurring proportions of uranium–235 and uranium–238, in the form of metallic rods sealed in containers (cans) made of magnesium alloy with small additions of berylium and aluminium. It is from this alloy, *Magnox*, that the reactors take their name.

The adoption of uranium metal as the fuel was linked to its simplicity and to the ease with which spent fuel could be dissolved and reprocessed to recover the plutonium it contained. However, it automatically imposes constraints on the operating temperature of the reactor, since to avoid physical damage the fuel has to have its central temperature kept below the metallurgical α-β phase transition. For a given fuel element thickness this limits its outer surface temperature and the fuel can temperature, and, depending on the rate at which coolant gas can remove heat, it sets the *heat rating* of the fuel, i.e. the rate at which heat can be produced within the fuel. The top temperature of the coolant carbon dioxide was initially kept to around 350 °C and gas pressures to around 7 bar (atmospheres).

The British Magnox reactors were constructed with a self supporting structure of dense graphite moderator with vertical channels to take fuel rods, neutron absorbing control rods and coolant gas flows. The designs allowed for dimensional changes in the graphite resulting from radiation damage. The whole was initially contained in a cylindrical or spherical steel pressure vessel and the reactors were controlled and refuelled from the top.

From the outset the Magnox reactors were designed for refuelling on-load. This is easier to accomplish with a gas coolant than with liquid cooling and has two great advantages. It avoids the need to build in a large initial excess reactivity to cater for loss of reactivity as the fuel ages, and it maximizes the period of time the reactor is in use and hence its output; always an important feature in reducing costs for capital intensive plant (see Chapter 21).

Table 3.1 Some Commercial Magnox Stations

Station	Commissioning date	Design output MWe net	Design net efficiency %	Reactor vessel	Coolant pressure bar
Berkeley, UK	1962	2 × 138	24.7	steel cylinder	9
Chinon-1, France	1962	68	22.7	steel cylinder	24
Bradwell, UK	1962	2 × 150	28	steel sphere	28
Chinon-2, France	1964	198	25.1	steel sphere	24
Hinkley A, UK	1965	2 × 250	26	steel sphere	13
Sizewell A, UK	1966	2 × 290	30.6	steel sphere	19
Chinon-3, France	1966	476	31	concrete with steel	26
Oldbury, UK	1968	2 × 300	33.6	concrete, integral design	24
St Laurent-1, France	1969	487	29.5	concrete, integral design	26
Wylfa, UK	1971	2 × 590	31.4	concrete, integral design	27
St Laurent-2, France	1971	516	30.5	concrete, integral design	26
Bugey-1, France	1972	547	28.4	concrete, integral design	41

Despite the constraints imposed by the choice of metallic fuel, experience enabled designers to move gradually to larger reactors, higher gas pressures and higher efficiencies (Table 3.1); a move facilitated by the introduction of prestressed concrete pressure vessels to replace the more expensive ones of steel. Oldbury, commissioned in 1968, was the first UK reactor to take advantage of prestressed concrete and was also the first with an integral design in which the reactor core, coolant circuit, boilers and pumps were all enclosed in the pressure vessel. The largest of the UK Magnox stations was Wylfa, commissioned in 1971 with a design output of 2 x 590 MW net, a design thermal efficiency of 31.4 per cent and a coolant pressure of 27 bar (atmospheres).

The French Magnox programme followed very similar lines and by the late 1960s they too were commissioning integral design reactors at St Laurent and Bugey. The adoption of hollow and later annular uranium fuel rods allowed them to move to higher fuel ratings, thus reducing the height of the core for a given power output. This and the adoption of downward flowing coolant with boilers below the core, yielded highly cost-effective reactor designs.

Some minor problems have inevitably arisen with Magnox reactors. The initial fuel can designs gave problems with vibration and creep and the UK reactors had an upper operating temperature limitation imposed (360 °C) to overcome corrosion by the coolant of some mild steel bolts within the reactor core. More recently the development of improved non-destructive testing techniques revealed minor cracks in parts of the cooling circuit which are believed to have been present since the reactors were built and of little significance. Nevertheless the reactors were shut down for the flaws to be eliminated. It is worth noting that none of these problems would affect new Magnox reactors. Lifetime availability factors of over 80 per cent could reasonably be expected (Table 3.2) from reactors based on the well tested and proven designs.

Uranium fuel in Magnox reactors has had its rating increased from below 3 MWt $tonne^{-1}$ in the early reactors to as much as 5.9 MWt $tonne^{-1}$ in France's Bugey 1. Fuel burn up (thermal energy obtained from a given weight of fuel) has also increased from initial targets

Table 3.2 Magnox Reactor Performance

Reactor	Annual load factor[1] %	Cumulative load factor %	MWe gross
Hunterston A	88.2	84.1	169[2]
Hinkley Point A	84.1	71.2	280[2]
Bradwell	75.7	61.9	174[2]
Trawsfynydd	74.1	61.2	292[2]
Dungeness A	70.7	57.5	286[2]
Vandellos	69.3	73.1	512
Wylfa	68.6	48.2	655[2]
Latina	67.7	59.7	210
Oldbury	62.9	56.1	313[2]
Tokai Mura	62.4	62.9	166
Bugey 1	58.5	61.3	560
Sizewell A	56.5	59.6	326[2]
St Laurent 2	49.3	51.8	530
St Laurent 1	44.3	52.0	500
Berkeley	43.8	66.6	160[2]
Chinon 3	0	32.5	500

[1]Year ending December 31, 1985
[2]Per reactor at twin reactor stations
Source: Nuclear Engineering International, June 1986

of 3000 Mwd tonne^{-1} to over 5000 MWd tonne^{-1} for current fuel.

Other graphite moderated, carbon dioxide, natural uranium fuelled reactors are in operation at Latina in Italy, Tokai Mura in Japan (both to British designs) and Vandellos in Spain.

3.3.3 *The Advanced Gas Reactor (AGR)*

One obvious route to reducing the costs of nuclear electricity is to increase the fuel rating in the reactor. This enables the designer to get more power out of a given volume of fuel and leads to more compact reactor designs. Two routes exist; improving the transfer of heat from the fuel to the coolant, or finding a fuel that is stable at higher temperatures than those acceptable with existing fuels (i.e. in the case of early gas cooled reactors, with metallic uranium). The former approach was intensively pursued in the Magnox reactor, with many designs of fuel can and fuel being developed. The latter approach has the added attraction that higher fuel, and hence higher outlet gas temperatures, will permit improvements in the steam conditions and hence in the thermal efficiency of the reactor.

The fuel chosen was uranium dioxide which was already being extensively employed in water cooled reactors. It has a very high melting point (2800 °C), it is chemically stable even in high radiation fields, and it retains its structure even in the presence of quite large quantities of fission products. This allows the achievement of higher fuel burn-up, currently up to 24 000 MWd tonne^{-1}.

Higher operating temperatures would lead to other problems, however. The Magnox fuel containers would be attacked by the coolant and a substitute was needed. The initial choice for canning material was beryllium, a material with low neutron capture characteristics which also acts as a neutron moderator. Difficulties with its fabrication into thin walled fuel cans, at reasonable cost, led to its abandonment and the substitution of stainless steel, whose less attractive nuclear properties forced a move to uranium oxide fuels enriched in the fissile uranium–235

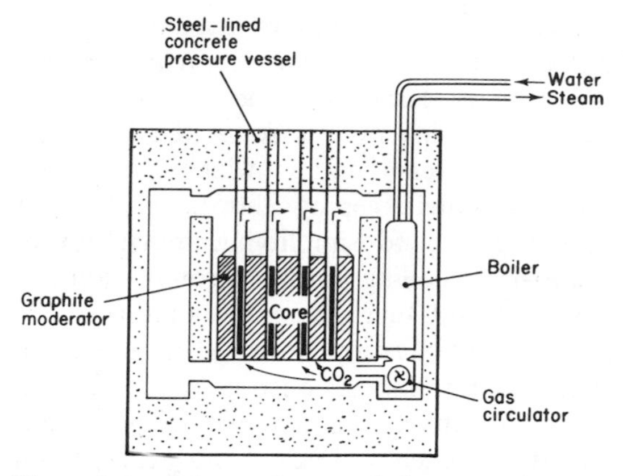

Figure 3.3 Advanced gas cooled reactor. Source: UKAEA. *Reproduced by permission of Oxford University Press from Marshall (1983)*

isotope. One of the attractions of the gas cooled reactor, its use of natural uranium, was therefore lost.

The higher coolant gas temperatures also led to a need to provide additional thermal protection for the prestressed concrete pressure vessel and, to inhibit its chemical reactions with the graphite moderator, the gas had to have trace additives and its purity had to be carefully controlled.

The British advanced gas reactors (Figure 3.3) were designed to have core outlet coolant temperatures of around 645 °C and coolant pressures of 42 bar (atmospheres) with an ultimate design pressure for the prestressed concrete pressure vessel of 115 bar. The steam conditions attainable (159 bar and 538 °C) are close to those for conventional fossil fuelled stations and lead to thermal efficiencies of around 37 per cent. The advanced gas reactors are designed for on-load refuelling, thus providing the ability to have high plant availability. In practice, operational problems with the lead reactors restricted the use of this facility initially, although refuelling on part load is now the norm and refuelling at progressively higher loads is planned.

The reactors are controlled and protected by gravity fed neutron absorbing control rods and have a back-up shutdown system relying on nitrogen gas injection into the coolant stream.

The AGR has many attractive safety features. It uses an *indirect cycle* in which the primary carbon dioxide coolant, which becomes radioactive, is separated from the secondary steam circuit which drives the turbines—thus keeping the latter free of radioactivity and simplifying maintenance. On-load refuelling permits prompt removal of any failed fuel elements and this keeps coolant radioactivity down. The use of a gaseous inert coolant ensures that there are no sudden changes of reactivity if the coolant circuit is breached. Even if the pressure fell from its design value of 40 bar to 1 bar through leakage, the remaining gas would be sufficient to remove the residual heat from the fuel elements, which falls to below 2.5 per cent of its operating level within minutes of the reactor being shut down. If the coolant circulating pumps fail natural convection is sufficient to remove the residual heat from the shut down core, provided gas pressure is maintained.

If the reactor should overheat the fuel expands and rapidly becomes less reactive, thus tending to stabilize the system. The graphite also expands, though much more slowly because of the large mass of it in the reactor and the resulting thermal inertia. This has the opposite effect, but the time delay makes this less important and facilitates reactor control.

The pressure of the system is low compared with water cooled systems, and the use of a prestressed concrete pressure vessel with a large number of separate steel tension cables ensures that the stresses on the vessel during operation are minimal and safety is not dependent on any single component.

One disadvantage of the AGRs was considered to be the fact that their construction required a great deal of work at the construction site itself compared to the 'factory' built pressurized water reactor with its steel pressure vessel. However, now the techniques are established differences are not as great as many imagined they would be (435 000 man weeks for an AGR compared with 355 000 man weeks for a PWR).

The characteristics of the UK AGRs are summarized in Table 3.3. Only Hinkley B and Hunterston B have reached the point where useful load factor data are available.

3.3.4 *The High Temperature Reactor (HTR)*

The HTR is a gas cooled graphite moderated reactor designed to operate at temperatures well above those of the AGR and at a high fuel rating with a small core, to reduce capital costs. The higher temperatures offer the prospect of higher thermal efficiency steam cycles for electricity production or, eventually, the use of coolant gas to drive turbines directly, thus eliminating the expense of boilers and the secondary coolant circuit. The HTR has also been regarded as a possible source of direct heat for use in steelmaking, synthetic natural gas manufacture and hydrogen production (see Chapter 5).

In order to achieve outlet coolant tempera-

Table 3.3 UK Advanced Gas Reactors

Station	Commissioning date	Design output MWe gross	Net efficiency %	Coolant pressure bar	Design steam conditions bar °C	Load Factor % Annual[1]	Load Factor % Cumulative
Dungeness B	1985	2 × 660	41.0	34	159 566		
Hinkley B	1976	2 × 660	41.5	42	159 538	65.5	51.1
Hunterston B	1976	2 × 660	41.5	42	159 538	77.0	47.4
Hartlepool	1985	2 × 660	41.5	41	159 538	14.4	—
Heysham I	1985	2 × 660	41.5	41	159 538	23.9	—
Heysham II	1986		41.5	43	159 538		
Torness	1986		41.5	43	159 538		

[1]To December 1985

tures of 700 °C to 1000 °C the fuel has to be made entirely of refractory ceramic materials and the coolant gas has to be completely inert. The fuel has therefore consisted of enriched uranium oxide or carbide and the coolant has been helium.

The fact that the core consists entirely of ceramic materials, which are able to withstand considerable local temperature variations, has simplified reactor design by reducing the importance of the close matching of coolant flows and power rating compared with those required for metal clad fuel assemblies. Spherical fuel pellets (*c.* 1 mm diameter) consisting of highly enriched uranium carbide are coated with concentric dense carbon and silicon carbide shells which retain the fission products even at the core fuel operating temperatures of up to 1200 °C or 1300 °C. These fuel pellets can be used in conjunction with pellets of fertile thorium oxide which produce fissile uranium–233 fuel *in situ* via neutron capture by thorium–232 and β-decay (see Chapter 4).

In the so called prismatic core HTR the graphite moderator is arranged in vertical columns, usually hexagonal, with spaces for fuel assemblies and coolant flows. The coated pellets of fuel and fertile thorium bonded in a carbonaceous matrix are encased in cylindrical or annular graphite tubes which are inserted into holes in the graphite core structure. If desired the fuel and fertile material can be mixed in the same coated particles. Uranium–238 can also be used as a fertile material in place of thorium.

The compact core, with a thermal power density of some 5–10 MWt m^{-3}, compared with 2.5 MWt m^{-3} for AGRs, is contained within a prestressed concrete pressure vessel which has to be carefully insulated. Additionally the helium coolant gas purity has to be carefully maintained since traces of moisture, air or other impurities can react with the graphite moderator at the reactor operating temperatures. Air or moisture ingress as a result of a failure in the pressure circuit has to be avoided.

The high thermal capacity of the graphite core and the stability of the fuel are attractive safety features which ensure that no safety problems are encountered for several hours even if the coolant pumps fail, although core temperatures could rise in this time to some 2000 °C. The uranium–thorium HTR also has the desirable characteristic that its reactivity declines as its temperature rises (*negative temperature coefficient*), thus providing a degree of self regulation.

An alternative HTR design developed in the Federal Republic of Germany is the so called pebble-bed reactor, in which large numbers of 6 cm diameter solid spheres of graphite containing dispersed coated particles of fuel and fertile thorium are loosely held in a vessel whose walls are also made of graphite to act as a neutron reflector. The fuel spheres are cycled through the reactor to remove those that are damaged or have reached the desired burn-up. The remainder are returned to the core together with fresh fuel spheres.

Control of HTRs can be accomplished using suitable neutron absorbing rods although the strong negative temperature coefficient allows core temperature and hence coolant flow to be used as a means of controlling reactor power.

Despite its evident attractions the HTR has not yet been brought to full commercialization. Small experimental reactors have been operated in the USA (40 MWe Peach Bottom), the Federal Republic of Germany (15 MWe Arbeitsgemeinschaft Versuchsreaktor), and the United Kingdom (20 MWt Dragon). These reactors have allowed development and demonstration of the viability of the fuel and they have led to the construction of the US prismatic core 330 MWe Fort St Vrain reactor and the Federal Republic of Germany's pebble-bed 300 MWe thorium high temperature reactor (THTR). Further details are given in Table 5.7.

Larger reactors have been designed and in the USA orders were placed for four such reactors based on Fort St Vrain, although these were subsequently cancelled in the mid–1970s. The main interest in HTRs at present is concentrated in FRG and Japan.

3.4 Light Water Cooled Reactors (LWRs)

3.4.1 *General*

Ordinary water, called light water to distinguish it from deuterium oxide or heavy water, has many attractions as a reactor coolant. It is cheap, it has a high specific heat, high density and low viscosity. The high specific heat means that a small mass can carry away a lot of heat; the high density that this mass only occupies a relatively small volume; and the low viscosity that the coolant can be pumped around the reactor with pumps of relatively modest size. This combination of properties gives it advantages over gas cooling.

To retain the full advantage of liquid phase cooling at temperatures above 100 °C, the water has to be kept under pressure, and the higher the desired temperature the higher the pressure. The pressurized water reactor (PWR) operates with its coolant at high pressures to ensure that this desired condition is achieved. The boiling water reactor (BWR) on the other hand, allows limited boiling to occur in the core, thus losing the simplicity of the single phase coolant. However, it too is kept under pressure so that the temperature at which boiling takes place is well above the normal boiling point of water.

Both the PWR and BWR take advantage of the fact that light water is a good neutron moderator, although it suffers from the disadvantage that it also captures neutrons so that the uranium–235 content of the fuel has to be increased above the natural ratio to about 3 per cent, in order to sustain a chain reaction.

3.4.2 *The Pressurized Water Reactor (PWR)*

The PWR is the reactor that has been adopted in most countries of the world to provide the basis of their thermal nuclear power programmes. As already indicated, the reactors are both cooled and moderated using light water under pressure. They use enriched uranium oxide pellets for fuel, contained in corrosion resistant zirconium alloy cans.

The pressurized liquid water coolant is circulated through the core and through an external heat exchanger, where it boils the water used to drive the turbines. This arrangement keeps the primary coolant, which becomes radioactive through neutron capture to produce tritium (hydrogen–3) and by dissolving small quantities of fission and activation products, away from the turbines.

The reactor core and its light water moderator and coolant are contained in a thick walled steel pressure vessel. The higher the reactor design temperature, the higher the pressure and the more expensive the pressure vessel becomes. A balance has to be struck between this cost penalty and the advantage, in terms of thermal efficiency, of operating at higher temperatures. In practice this balance has been struck at a temperature of around 300 °C with a corresponding pressure of 150 bar. This temperature is less than half that of the AGR and the pressure three and a half times larger.

The power density of the PWR core is around 100 MWt m^{-3} which is larger than that of the AGR (*c.* 2.5–3 MWt m^{-3}), or HTR (5–10 MWt m^{-3}). This high rating and the relatively small heat capacity of the core and its surrounding coolant-moderator means that considerable and rapid temperature and pressure rises could occur in the event of loss of coolant circulation through pump failure. Rapid temperature rises would also follow any loss of coolant following a break in the water circuit. In this respect the PWR is inherently more vulnerable than the gas cooled reactors and careful consideration has to be given to the design of reactors that have back-up systems that can continue to keep the fuel cooled in the event of any malfunction in the primary coolant circuit.

The PWR has evolved gradually from early designs developed in the USA for submarine propulsion. The first such reactor powered the Nautilus which was launched in 1954. The first PWR specifically constructed for electricity production, the 60 MWe Shippingport reactor which went critical in 1957, was designed to operate at 138 bar with fuel clad temperatures below 335 °C. Subsequent

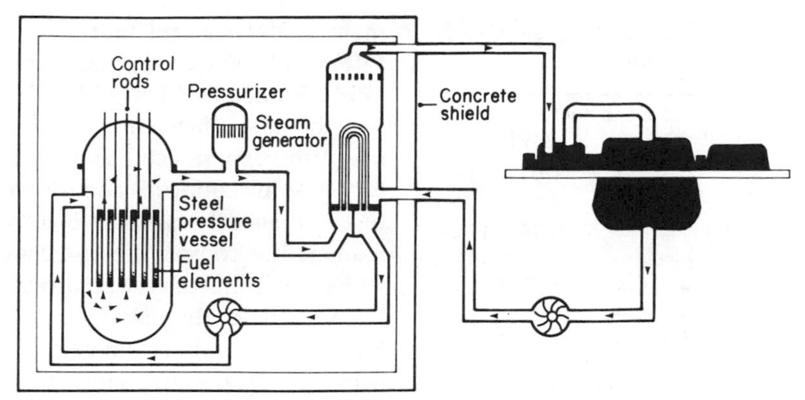

Figure 3.4 Pressurized water reactor. Source: *Nuclear Power Reactors*, UKAEA

development has been towards progressively larger reactors, but these have retained the same basic characteristics (Figure 3.4).

The concerns over safety have been met by a series of measures. The first of these employs a number of independent primary cooling circuits (two to four), each with its own heat exchanger, so that in the event of complete failure of any one circuit the reactor could continue to function. The pressure in the primary system is kept constant by a separate electrically heated pressurizer built into the water circuits which contains coolant and saturated steam. In the event of any pressure rise the vessel can be cooled to return the pressure to its design value. In the event of any coolant loss the emergency core coolant systems (ECCS) are brought into operation. The first of these systems consists of a number of reservoirs containing water under pressure which will be fed to the reactor automatically in the event of any loss of primary coolant pressure. The water contains boron compounds in solution which absorb neutrons and act as a complement to the solid control rods in shutting the reactor down. This system relies only on gravity and pressure operated non-return valves to function. It is supplemented by other high capacity systems which can inject and recirculate water from an outer containment sump where it would collect if there were a major failure of the primary coolant circuits.

The whole of the PWR primary coolant circuit and the pressure vessel surrounding the core are surrounded by a steel lined reinforced concrete containment building. This building is capable of withstanding modest pressure rises (of the order of 3.5 bar) and prevents any contaminated coolant or fission products released from the primary circuit getting out to the environment. This containment building can itself be spray cooled internally using chemical sprays to dissolve gaseous radioactive contaminants (except the inert gases) in the event that any major fission product release has occurred.

Refuelling the PWR is a more complex procedure than refuelling gas cooled reactors. The reactor has to be shut down and the pressure vessel head and its attachments removed. The reactor cavity is normally flooded so that the whole operation can be conducted under water and the spent fuel, which is still producing considerable quantities of heat from radioactive decay processes, transferred under water to cooling ponds where it can be stored. This operation has been refined and can now be accomplished in as little as eleven days, but it does represent a loss of reactor output.

Since a small number of the tens of thousands of zirconium alloy clad fuel elements may be leaking at any time, the coolant water becomes contaminated with traces of fission products in the course of normal reactor operation and it has to be chemically purified to keep this radioactive contamination at

Table 3.4 LWR Performance—Average Annual Load Factors for 1985

Country	Load factor % PWR	BWR
F.R. Germany	85.4 (9)	83.5 (7)
Belgium	83.3 (5)	—
Sweden	—	76.4 (7)
Taiwan	—	74.4 (4)
Japan	74.2 (14)	71.7 (15)
France	74.0 (34)	—
USA	65.1 (55)	53.0 (28)
Spain	54.5 (4)	—

Number of reactors in parentheses
Source: Howles (1986)

acceptable levels and to minimize operator exposure during refuelling and maintenance of the reactor.

Exposure of operatives to radiation during refuelling and maintenance also occurs for other types of reactor but the existing water cooled reactors are significantly worse in this respect than gas cooled or liquid metal cooled systems. Average operator exposure for PWRs is about 1 man-rem per MW year, of which 10 per cent is due to routine operation, 8 per cent due to refuelling, 30 per cent to routine maintenance, 40 per cent to special unplanned maintenance, and 12 per cent to other sources.

PWR performance has not fully lived up to initial expectations (Table 3.4). Few reactors have consistently reached the 75 per cent or more load factors that are generally regarded as desirable, and the average cumulative load factor for all reactor sizes and for all countries was 62.4 per cent by the end of 1985 on a capacity weighted basis (Howles, 1986). The main reason for this shortfall has been failures in the heat exchangers due to corrosion damage which accounts for about 25 per cent of all major outages (Pathania and Tatane, 1980). Other problems have arisen from cracking of pipework, vibration damage, etc. Considerable effort has been devoted to reducing the frequency of all failures, which not only affect reactor economics adversely but also contribute to the comparatively high operator radiation doses.

PWRs have been constructed in a great many countries and there has been competition between industrial suppliers in the USA, France and FRG. The USSR has also developed similar systems. Construction and operating experience has been more satisfactory in some countries and for some manufacturers than others. France and Japan have managed to complete construction within six years, a period which was also achieved in the USA in the early years of their programme. However a combination of factors, including licensing delays and labour problems have significantly extended this period, particularly in the USA, where construction periods of eight to ten years from ordering to commercial operation have become the norm.

Factors underlying reactor performance differences have been the subject of much academic debate (Komanoff, 1982, 1983; Surrey and Thomas, 1980, 1983; Lucas and Hall, 1980). In addition to the manufacturer and country, reactor size, reactor age, reactor vintage and operator experience have been considered significant (Table 3.5). Despite the comparatively large number of PWRs in operation, however, the processes of evolutionary change have meant that little confidence can be placed in broad brush statistical analyses (Jones, 1984) and more detailed analysis of causes of poor performance (such as Pathonia and Tatane, 1980) are of greater value in looking for improvement.

3.4.3 *The Boiling Water Reactor (BWR)*

This system was developed rather later than the PWR, since it had to await improved understanding of heat transfer under

Table 3.5 Factors Considered in Reactor Performance Analyses

Reactor	Location	Experience
Type	Country	Supplier
Size	Region of country	Operator
Vintage		Cumulative nuclear capacity
Age		Date

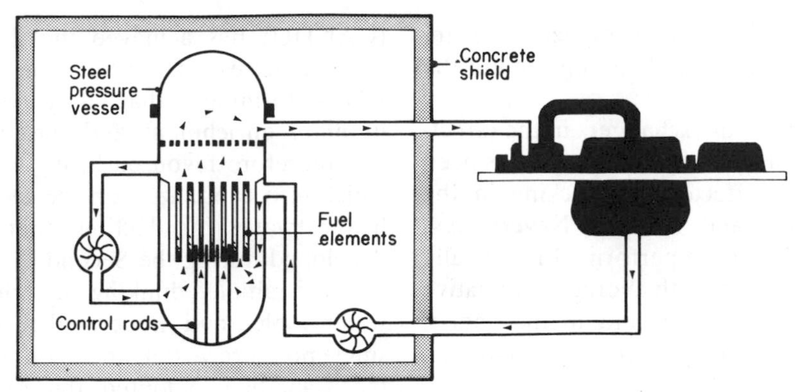

Figure 3.5 Direct cycle boiling water reactor. Source: 'Nuclear Power Reactors', UKAEA

conditions where the hot source (fuel element) was in contact with both liquid and vapourized coolant. Like the PWR it uses enriched uranium oxide fuel contained in zirconium alloy cans and is controlled with movable neutron absorbing rods.

Following initial experimental designs the first commercial power producing BWR was the 200 MWe Dresden–1 in the USA, which operated on a dual-cycle and went into operation in 1960. In this reactor the *direct steam cycle*, in which coolant is allowed to boil in the reactor and the steam used directly to drive the turbines, was complemented by an indirect-cycle similar to that in the PWR (Figure 3.5). This approach, whilst affording good control, was expensive and, although it was used in reactors in India, Italy and FRG as well as in the USA, it was subsequently abandoned in favour of the direct cycle.

As with the PWR the reactors were gradually scaled up and their power density increased, reaching 50 MWt m^{-3} in 1000 MW designs by the late 1960s. In the typical large scale BWR, the pressures are now maintained at about 70 bar and the water coolant at its corresponding boiling point of 290 °C. The power density in a typical modern BWR is about 50 MWt m^{-3}, half that of the PWR.

The BWR also has to have emergency cooling systems to cater for any failure in the main coolant circuit. These protective systems include three separate means of water injection from different sources linked with an automatic depressurization system. The reactor vessel in which the core is situated is itself supported in a steel lined reinforced concrete containment building designed to withstand the full pressure surges that could be associated with a loss of coolant accident.

In addition to the control rods for starting up or shutting down the reactor, its power output can be controlled by varying the water flow through the core using pumps. Higher water flows reduce the rate of steam production and, by introducing more water (and less steam) between the fuel elements, increase the neutron moderation and hence the rate of fissions in the fuel. This increases the fuel temperature which then causes boiling and a new equilibrium is established at a higher power rating and with more steam to drive the turbines. Decreasing the water flow has the opposite effect. The over-all system pressure is kept constant and control rods are moved only if absolutely necessary.

Steam passing to the turbines of the BWR contains some volatile fission products and hydrogen and oxygen produced by the action of radiation on the cooling water. The latter are recombined catalytically to avoid the risk of explosion. The fission products are separated from the steam and stored to allow decay of short-lived isotopes before discharge to the atmosphere. The average occupational exposure to BWR operators is 1.5 man-rem per MW year, roughly 50 per cent above that for the PWR. The causes of exposure are similar to those for the PWR reported in the previous section.

There are no heat exchangers in the direct-cycle BWR system and performance has been most seriously affected by cracking in the steel pipework and nozzles. Nevertheless BWRs as a class have performed marginally less well than PWRs with average cumulative load factors of 60.1 per cent to the end of 1985, when capacity weighted (Howles, 1986).

3.5 Heavy Water Reactors

3.5.1 *General*

Heavy water is chemically similar to ordinary (or light) water but contains the hydrogen isotope deuterium (atomic mass 2) in place of ordinary hydrogen (atomic mass 1). It has excellent moderating properties and a lower propensity to capture neutrons than ordinary water. A heavy water moderated reactor, unlike the LWR, is able to sustain a chain reaction using uranium fuel with the naturally occurring 0.7 per cent uranium–235 content.

Heavy water has many of the other characteristics that make light water attractive as a reactor coolant with one significant disadvantage—its price. Although it occurs naturally in ordinary water, its concentration is only one part in 5000, and its separation from the predominant light hydrogen is demanding in terms of capital plant and the energy inputs required.

In principle, heavy water could be used to moderate and cool reactors with similar designs to those of the PWR and BWR. In practice however, a different design philosophy has prevailed which uses pressure tubes to isolate fuel and coolant from the moderator. Such designs facilitate the use of alternative coolants such as gas or light water.

3.5.2 *The Pressurized Heavy Water Reactor (PHWR)*

Of all the heavy water reactor designs the Canadian Deuterium Uranium Reactor (CANDU) has achieved the greatest commercial success. The Canadians (see Section III) determined that they would avoid uranium enrichment and fuel reprocessing and therefore favoured heavy water systems which also capitalized on the technology for heavy water production that they had developed during the Second World War.

After initially thinking in terms of heavy water cooled and moderated reactors with a single pressure vessel, analogous to the PWR, the Canadians adopted the pressure tube design which offered considerable advantages. This design has a tank of low temperature, low pressure heavy water moderator through which horizontal pipes pass from face to face of the tank. Each pipe or channel contains a concentric zirconium alloy pressure tube into which fuel bundles are slid and through which heavy water coolant under pressure is circulated (see Figure 13.3). The coolant after flowing over the fuel passes to a heat exchanger where it boils ordinary water to feed steam turbines. The system operates at about 300 °C and 70 bar pressure.

The fuel for CANDU consists of short bundles of elements each of which has natural uranium oxide pellets enclosed in zirconium alloy fuel cans. Nine such bundles are contained in each pressure tube, eight of them within the active core region, during normal reactor operation. Burn-up of 7500 MWd tonne^{-1} is achieved. These bundles are fed in, one at a time, at one end of the pressure tube and removed at the other; a process that can be conducted with the reactor on load. This, combined with much happier Canadian experience with the heat exchangers than that observed in the USA with PWRs (despite inherently similar designs), has contributed to the achievement of excellent over-all reactor performance (see Table 3.6). Average load factors have been 68.8 per cent on a capacity weighted average (Howles, 1986). This average does not do full justice to the PHWR however. With only 8 per cent of world reactors being PHWRs, they nevertheless provide one third of the world's top 23 reactors ranked on a cumulative load factor basis.

Table 3.6 Heavy Water Reactor Performance

Station	Country	Annual load 1985	Cumulative load to end 1985	Capacity MWe gross
Pickering 1	Canada	0	68.4	542
		0	69.2	542
Pickering 2	Canada			
Pickering 3	Canada	61.8	77.9	542
Pickering 4	Canada	76.6	81.5	542
Pickering 5	Canada	77.5	74.3	540
Pickering 6	Canada	72.9	76.0	540
Pickering 7	Canada	92.7	89.3	540
Pt. Lepreau	Canada	97.4	83.6	680
Bruce 1	Canada	94.2	84.3	826
Bruce 2	Canada	71.8	75.8	904
Bruce 3	Canada	87.9	87.2	826
Bruce 4	Canada	72.4	84.4	904
Bruce 6	Canada	81.8	79.8	865
Gentilly 2	Canada	57.3	50.5	685
Atucha	Argentina	47.1	71.5	367
Embaise	Argentina	73.1	51.6	648
Rapp 1	India	13.5	27.4	220
Rapp 2	India	57.2	44.0	220
Mapps 1	India	46.3	47.2	235
Fugen	Japan	51.9	53.8	165
Wolsung	S. Korea	94.9	70.0	678

From Howles (1986)

The pressure tube design has several attractive features. It was believed that pressure tube failure, unlike that of the large thick walled PWR pressure vessel, should be preceded by slow leakage into the annular space between the pressure tube and the reactor tube surrounding it: leakage which could be detected by monitoring inert gas fed through the annulus for radioactivity. The pressure tubes can be isolated from the coolant flow and replaced individually if they develop defects, unlike the PWRs' large pressure vessel, and this replacement could be done without necessarily closing the whole system down. The pressure tube also limits spillage of valuable heavy water if leaks occur, and separates off the bulk moderator from the coolant which can be contaminated by fission products from leaking fuel bundles.

The first commercial scale CANDU was the 208 MWe Douglas Point reactor commissioned in 1967. Subsequent reactors have been scaled up to 600 MWe and 800 MWe and the Canadians have sought to get scale and operational economies by co-locating four identical reactors on suitable sites. This reduces the need for spares and cuts maintenance and operating costs through the provision of common services.

PHWRs have been built in Argentina, India, Korea and Pakistan as well as Canada and plans are well advanced to construct another in Rumania.

The CANDU has not, however, been entirely free from problems. The early reactors suffered from pump, valve and seal difficulties and from degredation of the heavy water coolant through light water in-leakage at the heat exchangers. Corrosion products circulating in the coolant circuit contributed to high radiation exposures for operatives. Most of these problems have been reduced to manageable proportions but operator doses still remain higher than for other systems, even those using light water coolant. There have been problems with cracks in pressure tubes that have necessitated significant programmes of replacement and, contrary to expectation, some pressure tubes have failed suddenly without prior warning, due to unforeseen corrosion problems, and they too have had to be replaced.

3.5.3 *The Boiling Light Water Reactor (BHWR) and Steam Generating Heavy Water Reactor (SGHWR)*

Despite the apparent difference in the names both concepts are similar and analogous to the BWR, but the designs adopt a vertical pressure tube array rather than the single reactor vessel of the BWR. These direct-cycle reactors employ heavy water as the moderator with light water coolant. A variety of designs have been produced notably by Canada (Gentilly–1), the United Kingdom (Winfrith SGHWR), Italy (Latina Cirene) and Japan (Fugen, Advanced Thermal Reactor).

The Canadian 200 MWe and Italian 35 MWe reactors are designed to use natural uranium oxide fuel similar to that in their PHWRs, whereas the UK (100 MWe) and Japanese (169 MWe) reactors are designed for enriched uranium fuel or also, in the latter case, mixed plutonium uranium oxide fuel (MOX).

The concept of the SGHWR developed from earlier consideration of gas cooled heavy water moderated reactors (GCHWR) which had been considered as a possible alternative to the AGR. The need to keep the gas separate from the liquid moderator led naturally to the pressure tube concept and it was rapidly appreciated that this lent itself to modular designs, provided the fuel was fed in on a vertical axis. A large tank of moderator could have any number of identical fuel channels containing fuel assemblies with their associated cooling, and the over-all power of the reactor would be proportional to the number of assemblies. The heat output of the reactor could also be controlled very simply by raising or lowering the heavy water moderator level, which in effect increases or decreases the size of the active core. Thus, conceptually, reactors ranging from a small 100 MWe to over 1000 MWe could be built from largely common components, and benefits of scale and replication could be made available even to users of the smaller sizes.

The UK SGHWR has a tank or *callandria* containing the heavy water moderator which is circulated through external heat exchangers to keep it at low temperature (below 80 °C) and pressure. Through the callandria run vertical pipes or channels within which are located the concentric pressure tubes to house the fuel and contain the coolant. The coolant, which is light water, is fed to each pressure tube separately from the bottom and each can also be fed directly with emergency cooling water in the event of any circuit failure. (The layout is shown in simplified form in Figure 3.6) The fuel consists of enriched uranium oxide pellets which are enclosed in zircalloy cans, several of which are linked together in a fuel assembly. Unlike CANDU only one long fuel assembly is used for each fuel channel.

The reactor can be shut down by draining out the heavy water moderator but quick shutdown is achieved using solutions of neutron absorbing materials (boric acid enriched in boron–10) introduced through vertical tubes located between the pressure tube channels in the callandria. Small power changes can be made by adjusting the moderator level and long term reactivity changes can be compensated by the introduction or removal of soluble neutron absorbing materials in the moderator itself.

The outlet coolant at about 280 °C and 65 bar pressure is allowed to boil and drive the turbines on direct cycle as in the BWR. Great care is taken in both the coolant and moderator circuits to maintain the water purity to minimize corrosion and reduce operator exposure during maintenance operations.

The SGHWR has all the attractive operational features of CANDU plus some of its own. Pressure tubes should not fail catastrophically and early warning of failure can be detected by monitoring the annular space around the tube for radioactivity. In the event of failures (none have so far occurred) individual pressure tubes could be replaced without draining the whole reactor or even removing fuel from other pressure tubes. The

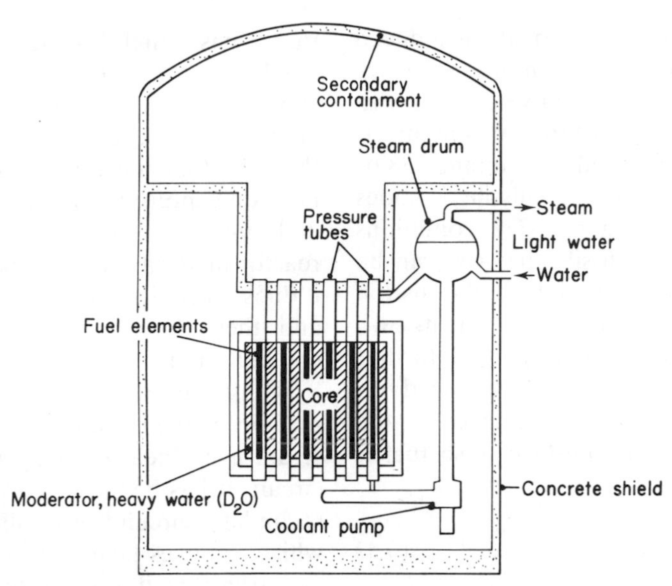

Figure 3.6 Steam generating heavy water reactor. Source: *UKAEA. Reproduced by permission of Oxford University Press from Marshall (1983)*

vertical channel layout avoids frictional wear of the pressure tubes during refuelling and, when fuel has been removed, the integrity and condition of the pressure tubes can be examined simply using non-destructive testing techniques.

The Japanese ATR is similar to, although larger than, the UK SGHWR. The Italian Cirene and Canadian boiling light water cooled heavy water moderated reactors also use vertical fuel channels but employ natural uranium oxide fuel in short bundles similar to those adopted in CANDU.

3.5.4 *Gas Cooled Heavy Water Reactors (GCHWR)*

Reactors of this type employ a heavy water moderator but cool the fuel with flowing gas rather than pressurized heavy water (CANDU, PHWR) or boiling light water (SGHWR). The use of gaseous coolant also favours the pressure tube design but allows greater flexibility over fuel and coolant temperature than the liquid cooled systems. Indirect cycles are required with intermediate heat exchangers to produce the steam for the turbines.

Small systems of this type have been constructed in France, the Federal Republic of Germany, Switzerland and Czechoslovakia, but all have now been shut down.

3.6 The Fast Reactor (FR)

3.6.1 *General*

The fast reactor differs from all the reactors dealt with in the preceding sections of this chapter in that it has no moderator. Fission is induced by fast neutrons travelling at the speeds similar to those with which they are produced in the fission process itself. As indicated in section 2.2.3, criticality can only be maintained by reducing the proportion of neutron absorbing materials present, and the fuel for fast reactors has to contain a much higher concentration of fissile isotopes; in general either plutonium–239 or uranium–235.

The absence of moderator means that the reactor core can be very compact and power densities can be of the order of 600 MWt m^{-3}. With such large quantities of heat being produced in a small volume the coolant must be very efficient if fuel temperatures are to be kept at acceptable levels.

A second characteristic of fast reactors is

their ability to breed. Fast neutron induced fissions produce, on average, more secondary neutrons than those induced by neutrons with thermal energies. Some of the extra neutrons can therefore be absorbed in uranium–238 or thorium–232 to produce the fissile isotopes plutonium–239 or uranium–233. If conditions are right, more fresh fissile material can be produced than is consumed in the fission process itself, and the reactor breeds its own fuel together with a surplus to fuel further reactors. If Ψ neutrons are produced on average per fission in a reactor core, the number C available to produce fresh fuel is given by

$$C = \Psi - 1 - L \qquad (3.1)$$

where one neutron is needed to sustain the chain reaction and L is the number lost on average by capture in other reactor materials. For breeding to occur C has to be larger than one and is called the breeding ratio. If C equals one the reactor is self-sustaining (neglecting material losses in fuel processing), and if C is less than one the reactor is a net consumer of fissile material. The latter is the case for thermal reactors, although useful quantities of fresh fissile material may be produced and some consumed *in situ*. For the uranium fuelled LWR C has a value of around 0.6 and for AGR and HWRs about 0.8.

3.6.2 *The Liquid Metal Cooled Fast Breeder Reactor (LMFBR)*

The fast reactor concept receiving the greatest attention has been the liquid metal cooled reactor. As indicated earlier (section 3.4.1) liquids have attractive heat transfer properties. Water and heavy water, however, which act as moderators can not be used in fast reactors, and other liquids with low neutron absorption, good thermal capacity, low viscosity, high density and comparative cheapness have to be sought. Sodium is a metal which turns liquid at a modest temperature and has the great advantage over water that it has a high boiling point and does not need high pressures to keep it in the liquid state at the coolant outlet temperature of about 560 °C. It suffers from the disadvantage that it does become radioactive through neutron capture by sodium–23 to produce β- and γ-active sodium–24. Given suitable reactor materials it is comparatively inert but it does react vigorously with air and moisture and these have to be excluded from the coolant circuit.

Whilst in principle the liquid sodium reactor coolant could be used to generate steam using heat exchangers in the primary circuit, it has been judged prudent to have a secondary circuit, also filled with sodium, which removes heat from the primary circuit and transfers it to further heat exchangers where the steam for the turbines is produced. In this way any failures in the water–sodium heat exchanger do not involve radioactive sodium and the hydrogen produced can be safely vented to the atmosphere. The introduction of the secondary sodium circuit adds considerably to the capital cost and offsets the savings made possible by the compact reactor core.

There are two alternative approaches to LMFBR design. The first, the *pool design* (Figure 3.2) has the reactor core, the primary coolant pumps and the intermediate sodium–sodium heat exchangers all contained within the reactor vessel. The second, the *loop design* removes everything but the core and controls from the vessel and has the intermediate heat exchangers and primary sodium pumps outside the vessel. In both designs the sodium coolant is held at a slight positive pressure under argon gas to prevent air ingress but the reactor vessels are not pressure vessels in the same sense as those used in the AGR and LWR.

The pool approach, which has been favoured in Europe, has the advantage of simplicity but it requires a larger pressure vessel whose volume is further increased by the need to include a neutron shield to prevent the sodium in the secondary circuit from becoming radioactive. The large volume requires a large roof which has to be strong enough to support the reactor components. The design is arranged to keep relatively cool sodium in contact with the pressure vessel to minimise thermal stressing.

The *loop design*, which has been favoured in the USA and India, has a smaller, cheaper pressure vessel surrounded externally by the neutron shield. The design of the vessel itself is, however, more complex and it is subject to thermal stress from the temperature gradients in the sodium circulating through it.

The fast reactor core is made up of an array of fuel elements loaded from the top. Each element consists of a number of fuel pins usually, at present, made from pellets of mixed uranium–plutonium dioxides (*MOX*) contained in thin walled stainless steel tubes (cans). The fuel composition can be varied depending on the desired burn-up and specific reactor design, but it usually contains around 15 per cent to 30 per cent of plutonium oxide. As in the case of thermal reactors the fast reactor fuel cans have to retain the fission products including those that are gaseous. Because of the higher burn-up demanded of fast reactor fuels, the quantities of fission products are larger than in thermal fuel and pressures within the fuel element cans can become quite high. For this reason a space is allowed at the top of the fuel can to contain this gas and keep the pressures moderate. The burn-up currently achievable with fast reactor oxide fuels is around 10 per cent (100 000 MWd t^{-1}) with 15 per cent or 20 per cent considered a reasonable development target.

The central core fuel in the breeder reactor is surrounded by a quantity of fertile material, such as natural uranium, uranium tails from enrichment plant or thorium, all in the form of oxide pellets. Some of this breeder blanket is included above and below the mixed oxide pellets in the fuel pins themselves; this is termed the *axial breeder blanket*. The remainder, the *radial blanket*, is in separate assemblies surrounding the core itself.

Like other reactors the fast reactor is controlled by neutron absorbing rods fed into the core from the top of the reactor vessel. It can also be designed so that if its temperature rises for any reason the reactivity of its fuel decreases, thus providing a degree of inherent stability. The pool design has a large volume of coolant with a large heat capacity which is at low pressure. If the reactor vessel were damaged the primary coolant would not vapourize and escape rapidly (as it would in a water cooled system) and in suitable designs the core and secondary heat exchangers can be kept immersed even after failure. Provided the secondary coolant circuit is operating, failure of the primary coolant pumps can be accommodated without emergency cooling. If the secondary circuit is lost the thermal inertia of the primary coolant is such that many hours would elapse in the absence of emergency cooling before fuel temperatures rose to levels where fuel can rupture was likely.

As with other reactor types, the fast reactor is protected by its reactor vessel and by its containment building. It has multiple automatic trips and cut-outs so that it would be shut down if anything untoward or abnormal were detected. To guard against the failure of its secondary coolant circuit an emergency cooling system is incorporated which operates by convection and conduction alone. The primary coolant gives its heat to an alloy of sodium and potassium which is liquid at room temperature, and this in turn is cooled by allowing it to flow through an air cooled 'radiator'.

The liquid metal cooled fast reactor is inherently very safe but designers have to consider all possible modes of failure no matter how unlikely. Precautions have to be taken to guard against improbable hot fuel–coolant interaction and fuel movement or failure that could lead to major increases in core reactivity.

Demonstration and experimental liquid metal cooled fast reactors have been operating in many countries, including France, the United Kingdom, the Federal Republic of Germany, Japan and the USSR. Operational experience goes back for more than 25 years (Dounreay Fast Reactor, 1959).

3.6.3 *The Gas Cooled Fast Reactor (GCFR)*

The use of gaseous coolants for fast reactors has both attractions and disadvantages. One of the principal advantages of the fast reactor is the fact that in the absence of a moderator its core can be compact and have very high power densities. The rate of heat removal from such a core has to be large if the fuel is to be kept at reasonable temperatures and this can only be achieved using gases if they are at high pressures. The alternative would be to use more dilute fuel in thinner fuel elements and hence have a larger core and larger fuel inventory which would be likely to increase capital costs. One of the main advantages of gas cooling would be the removal of the need for the intermediate heat exchanger circuit (and the internal neutron shield of the pool design LMFBR) and this would reduce capital costs significantly. Against this the use of a gaseous coolant at around 100 bar pressure demands considerable investment in protective and emergency cooling systems to cater for the possibility of loss of coolant in the event of any failure in the pressure circuits.

Design studies have considered the use of superheated steam, carbon dioxide and helium as coolants. The nuclear properties of helium make it particularly attractive but carbon dioxide is also a serious contender. The reactors, if built, should operate at temperatures similar to the LMFBR and achieve similar fuel burn-up.

At the present time, given the more advanced stage of development and demonstration of the LMFBR, it seems unlikely that much effort will be devoted to the GCFR.

3.6.4 *Fast Reactor Fuels*

The reactors described so far have planned to use mixed oxide fuels. The use of plutonium uranium carbide has attractions in that the material is also very stable at high temperatures and it has a higher thermal conductivity than the ceramic mixed oxide. A higher thermal conductivity means that fatter fuel pins or pins with a higher fissile content can be used without running into temperature limitations and both alternatives offer the prospects of lower costs, provided comparable fuel burn-up can be achieved. Many countries have experimented with carbide fuels but they are only being adopted by India for use in their 15 MW prototype loop design LMFBR. The United States is looking at metallic fuels which can be fabricated by powder metallurgical routes.

The very high fast neutron fluxes in fast reactors and the high burn-up sought from the fuel give rise to design problems, since the fuel and other reactor materials suffer radiation damage (and, in the case of fuel, accumulate gaseous fission products) which lead to significant dimensional changes. The problem is not solely confined to fast reactors (see section 4.2.3) but it is at its most acute in these systems and a lot of development effort has been aimed at identifying and overcoming the problems it could cause. The reader should consult specialist books on radiation damage if he needs more detail on this topic (IAEA, 1974).

3.7 Water Cooled Graphite Moderated Reactors (RBMK)

In addition to the gas cooled water moderated reactors described earlier there is another system which is, in effect, a hybrid between the more conventional lines of gas cooled graphite moderated reactors (Magnox, AGR, HTR) and water cooled water moderated reactors (PWR, BWR, PHWR, BHWR, SGHWR). This is the Soviet RBMK, a graphite moderated pressure tube boiling water cooled reactor.

This line was developed early in the USSR, mainly as a plutonium producing system. A small 5 MWe plant was commissioned in Obinsk in 1954 and followed by six 100 MWe units in Siberia in the late 1950s and early 1960s. Larger commercial demonstration systems (200 MWe) were brought into operation in 1964 and 1967 at Beloyarsk.

The commercial reactors have vertical fuel channels passing through the cylindrical graphite moderator. These channels consist of zirconium–niobium pressure tubes within which two joined assemblies of zircalloy clad enriched (2 per cent) uranium dioxide fuel pins are housed.

The fuel is cooled by boiling light water contained in the pressure tubes and operates on a direct steam cycle. The operating pressure is about 60 bar with a channel outlet temperature of 300 °C. The moderator temperatures are high (some 700 °C) and the core is contained in a steel vessel in an inert atmosphere of helium and nitrogen to prevent oxidation of the moderator.

Control is effected by conventional vertical control rods and the control rod channels are used to provide additional water cooling of the moderator.

The cooling system has built in redundancy with spare pumping capacity and emergency core cooling (ECC) is designed to deal with failures of the feed water or steam outlet pipework. This has a passive injection system from pressurized accumulators and pumped injection from large storage tanks. The design also allows for removal of decay heat from the shut down reactor using natural circulation alone.

The large commercial reactors and their cooling systems are contained in thick walled concrete buildings. 1000 MWe twin reactor plants have been in operation in Leningrad, Kursk, Chernobyl and Smolensk; four reactors at each of the first three sites and two at Smolensk. A further two are under construction at Chernobyl. Two larger 1500 MWe stations have also been commissioned at Ignalino (1984) with power densities in the fuel some 50 per cent above that in the 1000 MWe stations. The RBMK reactors have been providing (pre-Chernobyl, 1986) some 6 per cent of the USSR's electricity needs. No reactors of this type are in operation outside the Eastern Bloc.

The physics of a heterogeneous system like RBMK with a two phase coolant is complex and there is always some risk that failure of a pressure tube will bring steam into contact with very hot graphite, leading to chemical reactions which could not occur in the conventional gas cooled LWR or HWR systems.

Although it has been reported (Chapter 7) that USSR safety design targets are not as strict as those in the West they have had some 250 reactor years of safe operating experience with the RBMK prior to the Chernobyl disaster and the accident has been attributed to 'gross breaches of operational regulations by workers' who 'conducted experiments with turbogenerator regimes' which were not cleared by the safety authorities. The work was not properly supervised and in order to carry it out the automatic reactor shutdown and emergency core cooling systems were shut off. Due to a combination of circumstances the operators tried to run the plant at low power at a time when extensive xenon poisoning had occurred, a state in which the reactor was unstable and had a significant positive temperature coefficient of reactivity. These actions created conditions which would not have arisen in normal operation (see Chapter 7).

3.8 Other Reactor Types

There are many other combinations of moderator, coolant, fuel and other operating regimes which are not covered here since they have not been considered suitable for development into commercial systems. Some of these have been the subject of significant research, others not.

Fused salt reactors in which the fuel and moderator are a homogenous high temperature fluid (such as an eutectic mixture of sodium and uranium fluorides) were considered seriously at one time. They had the attraction that the fission products could, in theory, be separated off and the reactor refuelled continuously. Control would have been by normal control rods and heat would have been removed through a conventional water–steam system. Suitable salt mixtures were highly corrosive and the ideas were dropped in favour of the mainstream LWRs and gas reactors, although one prototype was run in the USA for some years.

REFERENCES

Frost, B. T. (1982). *Nuclear Fuel Elements*, Pergamon, London.

Howles, L. (1986). 'Nuclear station achievement; *1985 Annual Review', Nucl. Eng. Int.*, 1986 (June), pp.75–79.

IAEA, (1974). *Proceedings of a panel on the behaviour and chemical state of irradiated ceramic fuels*, IAEA, Vienna.

Jones, P. M. S. (1984). 'Statistics and Nuclear Energy', *The Statistician*, **33**, 91–102.

Jones, P. M. S. (1980). 'Comment on nuclear plant performance', *Futures*, **12**, 245.

Komanoff, C. (1982). *The Westinghouse PWR in the United States; Cost and Performance History*, Komanoff Energy Associates, New York.

Komanoff, C. (1983). *Proof of Evidence to Sizewell Inquiry, CPRE/P/1, and transcripts of cross examination*, July 1983.

Marshall, W. (ed.) (1983). *Nuclear Power Technology*, Vol. 1, Clarendon Press, Oxford.

Lucas, N. J. D., and Hall, J. A. (1980). *Age, size, learning and country effects in light water reactor operating data*, Department of Mechanical Engineering, Imperial College, London.

Pathania, R. S., and Tatane, O. S. (1980). *Steam generator tube performance*, Atomic Energy of Canada Ltd, Report No. AECL 6852.

Surrey, J., and Thomas, S. (1980). 'World-wide nuclear plant performance', *Futures*, **12 (No.1)**, 3–17.

Surrey, J., and Thomas, S. (1983). *World-wide nuclear plant performance revisited*, Science Policy Research Unit Occasional Paper No. 18, SPRU, Sussex University, Brighton.

Nuclear Power: Policy and Prospects
Edited by P. M. S. Jones

4

The Nuclear Fuel Cycle

P. M. S. JONES
Department of Economics, Surrey University

4.1 Introduction

Nuclear fuel differs from conventional fossil fuels in two respects. Firstly it is used in the form of an engineered product manufactured to close tolerances and a tight specification. Secondly, after it comes out of the reactor it will, in general, contain significant quantities of potentially useful uranium and plutonium and only a small part (*c.* 2 per cent for light water reactor fuels taken to 33 000 MWd t^{-1} burn-up) can properly be regarded as waste. Additionally, since the spent fuel removed from a reactor is highly radioactive and produces large amounts of heat for very long periods of time, there are significant costs incurred in its management as well as in fuel purchase and fabrication.

This chapter explains the distinction between fissile and fertile materials, briefly examines the processes involved in fuel manufacture and management, and describes the alternative nuclear fuel cycles with their merits and disadvantages. A discussion of the availability of nuclear fuel materials is deferred until Chapter 20, when the potential future contribution of nuclear power to world energy supplies is considered.

For convenience the management of nuclear fuel can be divided into three stages: the so called *front-end* stage comprising fuel production and fabrication; the *back-end* dealing with fuel after its removal from the reactor; and the stage in between when fuel is actually in the reactor. This division will be used in the relevant sections of this chapter. The over-all cycle is shown in Figure 4.1.

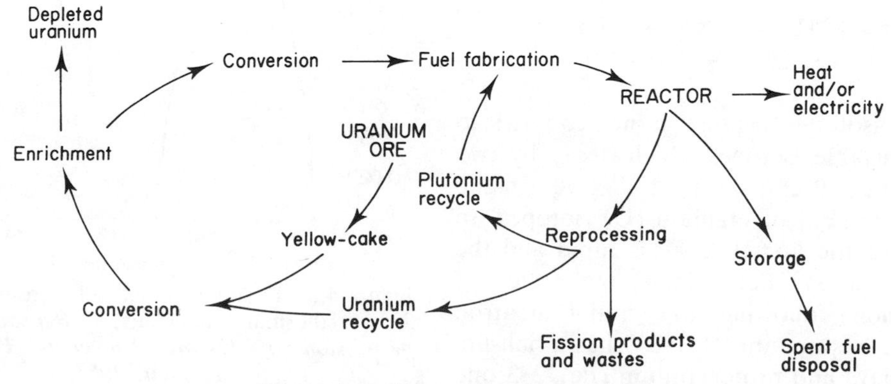

Figure 4.1 The nuclear fuel cycle

4.2 Nuclear Reactions

4.2.1 *Fissile and Fertile Materials*

The splitting or fission of heavy atomic nuclei by neutrons has been described in section 2.2.2. Only some isotopes of the heavy elements, the so called fissile isotopes, are readily split in this way at all neutron energies, however. The commonest naturally occurring *fissile isotope* is uranium–235 which occurs together with 'non-fissile' uranium–238 in the approximate ratio 0.7 per cent to 99.3 per cent. Other fissile isotopes include uranium–233, plutonium–239 and plutonium–241, and, in general, heavy nuclei containing an even number of protons and an odd number of neutrons are found to be fissile.

All isotopes of the heavy elements, including those with an even atomic number, can undergo fission. Indeed 1 kg of uranium–238 has about ten of its atoms splitting spontaneously in each second. However, most isotopes only fission with incident neutrons which have very high energies; uranium–238 requires 1 MeV neutrons and thorium–232 even greater energies. Whilst such fissions contribute something to the output of fast reactors they have little significance in thermal reactors.

Only the readily fissile isotopes are useful reactor fuels but some non-fissile isotopes can be converted to fuels by the processes of neutron capture and radioactive decay. Such isotopes are termed *fertile isotopes*. Uranium–238 and thorium–232 are the commonest fertile materials.

$$^{238}U + n \rightarrow {}^{239}U \rightarrow {}^{239}Np \rightarrow {}^{239}Pu$$
$$^{232}Th + n \rightarrow {}^{233}Th \rightarrow {}^{233}Pa \rightarrow {}^{233}U$$

Both isotopes capture a neutron (n) to yield unstable isotopes which decay by two successive β-emissions to give fissile plutonium (Pu) or uranium (U) isotopes. In each case the first β-decay is rapid and the second decay determines the rate of production following the initial neutron capture. Neptunium (Np)–239 has a half-life of 2.4 days and protoactinium (Pa)–233 one of 27 days. This difference is of some importance to the thorium (Th) based fuel cycle as will be seen later.

4.2.2 *Reactions in Fuel*

As indicated in section 2.2.3 a nuclear reactor has to contain above a minimum quantity of fuel if it is to become critical and produce sustained energy output. This quantity is dependent on the geometry of the fuel, the presence or absence of moderator and the presence or absence of other neutron absorbing materials. With an efficient moderator such as heavy water or graphite it is possible to sustain a chain reaction in fuel made from natural uranium either in the form of the metal or as its dioxide. However, most reactors now use uranium with its uranium–235 content raised above the natural 0.7 per cent to compensate for neutron absorption in the moderator (of LWRs) or canning materials and to give higher fuel burn-up.

The principal nuclear reactions taking place in the fuel when it is in a reactor are fission of the uranium–235 and parasitic neutron capture by uranium–238. In addition to the essential two to three fission neutrons, the former process yields a mixture of isotopes of lighter atoms (Figure 4.2) which tend to have too many neutrons in their nuclei for stability. They are therefore radioactive and decay by emitting β-particles (or

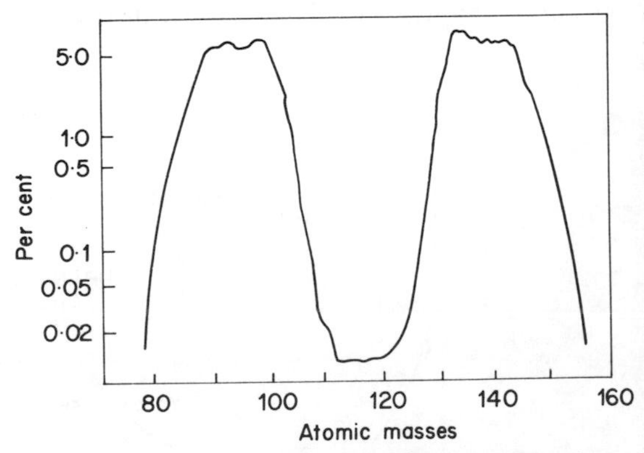

Figure 4.2 Fission product yields from Uranium-235. (thermal neutrons). *Reproduced by permission of Oxford University Press from Marshall (1983)*

electrons) from their nuclei; a process which produces an isotope of a different element with its nuclear charge (atomic number) one higher than its parent. The new nucleus is usually in an excited state and emits γ-rays before itself decaying by β-emission. The process continues until a stable atomic nucleus is produced. Each radioactive isotope has its own *half-life* (the period taken for half the isotope to decay), some very short and some much longer. All of the fission products and the elements resulting from their radioactive decay are themselves exposed to the neutrons in the reactor and some of the isotopes will react with the neutrons, thus reducing the number of the latter available to produce further fissions.

The capture of neutrons by uranium–238 produces, with an average delay of some days, plutonium–239 which itself has a relatively long half life of 24 000 years. This fissile plutonium isotope will therefore tend to accumulate in fuels containing fertile uranium–238. However a proportion of the ^{239}Pu will undergo fission itself to produce a slightly different mixture of fission products to those from uranium–235. (About 25 per cent of fissions in a PWR at equilibrium are in plutonium–239 produced *in situ* rather than in its feed uranium–235 fuel). The uranium–238 may also undergo a so called n,2n reaction in which the nucleus captures one neutron and immediately emits two, thus becoming uranium–237 which decays to long lived neptunium–237 by β-emission.

Both uranium–235 and plutonium–239 (once produced) can capture one or more neutrons to yield higher isotopes of uranium or plutonium some of which decay by α-emission and others by β-emission. In the latter case elements of higher nuclear charge than plutonium such as americium (Am) and curium (Cm) are produced. Thus

$$^{239}\text{Pu} + 2\text{n} \rightarrow {}^{241}\text{Pu} \rightarrow {}^{241}\text{Am}$$
$$\text{and } {}^{241}\text{Am} + \text{n} \rightarrow {}^{242}\text{Am} \rightarrow {}^{242}\text{Cm}$$

It will be evident from even this superficial examination that the fuel in the reactor changes quite rapidly from relatively pure uranium dioxide or uranium metal into a complex mixture of isotopes spanning the whole range of chemical elements including new unstable elements heavier than uranium that no longer occur naturally in the earth's crust.

The precise mixture will depend not only on the initial composition of the fuel and the number of fissions that have occurred, but also on the rate and duration of neutron irradiation and the energy spectrum of the neutrons in the reactor. At high irradiation rates there is greater likelihood of short lived intermediate products (such as uranium–239) capturing neutrons or undergoing fission before they decay. There is greater likelihood of multiple neutron capture in long lived products (such as plutonium–239) if irradiation continues over an extended period, and the relative likelihood of the different fission and capture reactions is influenced by the neutron spectrum.

The *reactivity* of nuclear fuel in the reactor changes with time, not only because the initial fissile material it contains is steadily used up, but also because of the new fissile isotopes produced *in situ*, and the changing mixture of neutron absorbing species present. This effect can be compensated for to some extent by incorporating neutron absorbing materials (*burnable poisons*) in the fresh fuel elements and increasing the initial fissile content. The poisons absorb neutrons and decrease the fuel's reactivity but in so doing they are consumed so that their effectiveness can be arranged to decline at a rate that matches the net decline of the fissile content of the fuel.

A further important nuclear reaction in fuel is the emission of *delayed neutrons* by fission products. Some neutron rich fission products decay by neutron emission from their nuclei rather than electron emission (e.g. bromine–87, iodine–137, isotopes of rubidium). The half lives of these neutron emitters range from a fraction of a second to a few minutes contrasting with the instantaneous emission of the fission neutrons themselves (see section 2.2.3).

4.2.3 *Effects of reactions*

Over 80 per cent of the energy released in the fission process is in the form of *kinetic energy* of the fission products. These fly apart at high speed giving up their energy in collisions with the other atoms in the fuel. After a sufficient number of collisions the fission products are halted and the orderly arrangement of atoms in the initial fuel will have been brutally disturbed, with atoms forced out of position and the foreign fission product atoms sitting uneasily where they have been stopped. The vibration and motion induced in the fuel by the energetic fission products (and to a lesser extent by other nuclear processes) manifests itself as heat which is removed by the reactor coolant.

The disturbed arrangement of atoms in the fuel has a larger volume than the initial orderly arrangement and the fuel swells. The more mobile fission products will diffuse slowly through the fuel to find more favourable sites such as grain boundaries, or, in the case of gases, the pores between fuel particles or the voidage around the fuel. The strength of the fuel itself is thereby affected and volatile fission products can escape from the fuel to collect in its surrounding container.

All of these factors contribute to limit the number of fissions that can be allowed to take place in a fuel element. It is preferable to remove it from the reactor before the fuel can or the structure of the fuel itself fail. The number of fissions occurring in the fuel before its removal from the reactor determine the thermal energy derived from the fuel and this energy, or *burn-up*, is measured in millions of watt days per unit mass of fuel (MWd t^{-1}). Fuels designed for high burn-up have to have a higher initial fissile content than low burn-up fuels and may make use of burnable poisons (above) to regulate their reactivity.

During the time spent in the reactor the fuel becomes highly radioactive due to the presence of the β- and γ-emitting products of fission and neutron capture. Its α-activity is also increased due to the production of less stable heavy elements through n,2n and neutron capture reactions. Around 7 per cent of the energy associated with fission appears as radioactivity, although the short life of many fission products means that they have decayed before the fuel has left the reactor. Indeed, because short lived products predominate, the radioactivity of fuel when it leaves the reactor core is closely related to the energy rating of the fuel and almost independent of the length of time the fuel has been in the core.

The total radioactivity of spent fuel removed from the reactor falls rapidly; by a factor of 20 over the first month. The heat output, which is initially mainly due to fission product radioactivity, also falls. A freshly discharged PWR fuel element gives out many hundreds of kW but this falls to about 4 kW after one year and 1 kW after five years. There is no direct correlation between total radioactivity and total heat however, since the heat output depends on both the number and energy of emitted particles.

The characteristics of spent fuel are such that it has to be contained within a radiation shield to protect the workforce and public. It also has to be cooled to prevent its temperature rising to levels where its containment could be breached and volatile fission products released to the environment. The methods of doing this are described later.

4.3 The Uranium Fuel Cycle

4.3.1 *The Front End*

4.3.1.1 General The so called front end of the uranium fuel cycle consists of all those stages from prospecting for uranium ores to the final supply of fabricated fuel assemblies to the reactor site (see Figure 4.3). In this section the separate stages are described in outline with an indication of the alternative processes available.

4.3.1.2 The Uranium Supply Industry Uranium is widely distributed in nature and occurs both in the world's oceans and land masses. The average abundance in the crust is only about two to three parts per million and higher concentrations have to be sought before it can be recovered economically.

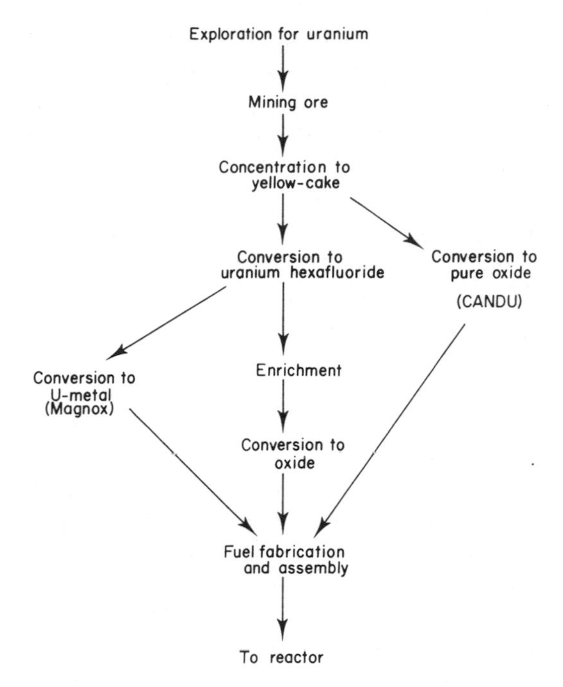

Figure 4.3 Front-end of uranium fuel cycle

Fortunately natural geological processes have produced such concentrations in a wide variety of mineral formations (Nininger, 1983), notably as vein type deposits where the uranium minerals have been forced into cracks in a host rock; sandstone deposits where uranium has been concentrated in sedimentary deposits by favourable chemical processes; and quartz pebble conglomerates. Although some deposits contain remarkably high concentrations of uranium amounting to well over 10 per cent, the majority of economically workable ores contain only around 0.2 per cent. Some lignite deposits contain nearly as much while some large shale deposits and phosphate rock deposits contain about a tenth of this concentration. None of these can be worked economically at present.

A wide variety of methods are used to locate promising ore bodies. Surface ores can be pinpointed using radiation detectors which, for wide area coverage, may be mounted in aircraft or road vehicles. Buried ore bodies are harder to locate since their radioactivity is screened off by the overlying strata. The slow diffusion of radioactive radon gas from such ores towards the surface provides a pointer, however, and it can be detected by radiation probes in shallow bore holes. The chemical analysis of the water and sediments in streams can also help to indicate areas with potential. The combination of geological understanding with even more specialized techniques relying on small variations in isotopic abundances, the concentration of elements in biological materials, etc., may also become increasingly important when the more readily located deposits are exhausted and the deeper more difficult deposits are sought.

Once a promising ore body is found its presence is confirmed by chemical analysis and its extent delineated by drilling and analysis. This process is progressive and serves to establish the recoverable uranium content of those ore bodies considered potentially exploitable. If initial results are discouraging, or if the economic climate is unfavourable, the ore body may not be fully investigated and this leads to different degrees of confidence over the uranium content of different deposits (see Chapter 20).

The uranium bearing minerals are usually worked by open-cast methods or by conventional deep mining. The former is suited to large deposits such as sedimentary sandstones which occur near the surface, and the latter to deeper deposits. The ores are crushed in a processing plant or mill and the uranium extracted by chemical methods appropriate to the form of mineral concerned. Some 85 per cent to 95 per cent of the uranium will normally be recovered as a solid concentrate called *yellow cake*. For each tonne of uranium extracted some 500 t of rock and ore will have been crushed and treated and the resuidues, called *mill tailings*, together with the treated waste liquors from the chemical processing, now have to be deposited in such a way that they minimize the transfer of uranium and its decay products to ground water. The tailings piles which contain most of the radioactivity from the original ore are also covered over with soil to help retain the radon and to screen off γ-radiation. Monitoring and management of mill tailings was not always satisfactory in the early days of the industry and a great deal of attention has been paid to this problem over recent years to ensure that environmental and potential health impacts are minimized (Nuclear Energy Agency, 1984; Uranium Institute, 1984).

In addition to the conventional mining methods, uranium is now being recovered by solution mining or *in-situ leaching*, in which chemicals in solution are pumped via bore-holes through the ore body and the uranium containing leachate extracted through other bore holes, prior to concentration and treatment to produce yellow-cake. This method has the advantage that it does not extract large quantities of unwanted rock and spoil but it is only suitable for special geological situations where the liquors can not escape. It is, so far, less efficient in its extraction of the uranium from the ores.

Uranium mining, in common with other underground mining operations (Saihs, 1974) exposes workers to radiological hazards. Uranium itself is only weakly radioactive and in massive form it is harmless. However the dusts containing uranium and its daughter products such as thorium–230 and radium–226 can lodge in the lungs, and the inert gas radon–222, released from the rock by the mining operations, whilst itself not a major hazard, can deposit its solid daughter products. These include the α-emitting polonium isotopes 218, 214 and 210, which can lead to excessive radiation doses to the lung tissue. Radon can also be a problem in open-cast mines and even in well-insulated houses, where radon from the soil and building materials can accumulate (O'Riordan *et al*, 1983). The mining risks can be overcome using suitable protective measures including air filtration and good ventilation.

4.3.1.3 Conversion The yellow-cake concentrate is an oxide form of uranium which is further purified by dissolution in nitric acid and chemical processing to obtain a pure uranyl nitrate. This compound can then be further processed to yield either metallic uranium for fuelling Magnox reactors, uranium dioxide for fuelling CANDU reactors, or uranium hexafluoride, a volatile compound of uranium, which is used as the feed to the most widely used enrichment processes.

The chemical processes involved are thermal decomposition of uranyl nitrate to produce uranium trioxide, followed by reduction with hydrogen to produce uranium dioxide. For uranium metal production the dioxide is converted to uranium tetrafluoride using hydrogen fluoride and this tetrafluoride is reduced using magnesium metal. Uranium hexafluoride feed for enrichment processes is produced by reacting the tetrafluoride with fluorine and the enriched hexafluoride can be reconverted to uranium dioxide using steam and hydrogen.

4.3.1.4 Enrichment Enrichment is the process in which a desired isotope of an element such as uranium–235 is preferentially concentrated at the expense of unwanted isotopes. Thus natural uranium which consists of 99.3 per cent uranium–238 and 0.7 per cent uranium–235 passes through the enrichment process to produce the material containing 3 per cent to 4 per cent of uranium–235 required for light water reactor fuel, or the more highly enriched uranium used in some research or advanced reactors.

The process of enrichment (or isotope separation) has in the past relied on the fact that although the isotopes of an element have virtually identical chemical properties, their masses differ. This difference can be exploited using a variety of techniques, some of which are described briefly below. For uranium the majority of the processes require the material in the gas phase and the only suitable compound which is gaseous at relatively low temperatures (> 60 °C) is uranium hexafluoride. This compound is extremely reactive and has to be kept away from moisture, air, organic materials and many metals. It has the advantage that fluorine has only one isotope, fluorine–19, so that all the mass differences between the uranium hexafluoride molecules is due to their uranium atoms.

London (1961) reviews the main techniques of isotope separation and the theory behind them. The following paragraphs briefly explain the basis of the processes that have been used for uranium isotope separation and their present status.

(a) *Magnetic Separation* This technique was used in the early stages of the Manhattan Project in which the United States and her allies developed the atomic bomb during the second World War. The principle is that of the mass spectrometer (London, 1961). Molecules of uranium hexafluoride or atoms of vapourized uranium are ionised and the ions accelerated in an electric field before being injected into an evacuated chamber. A strong magnetic field perpendicular to the direction of the uranium ion beam deflects the ions into circular paths with radii which are smaller for light isotopic ions than heavy ones. The isotopes can therefore be collected separately in suitably located containers.

The advantage of the method, which is still used for separation of some stable isotopes in small quantities, is that virtually complete separation is possible in a one stage process. The disadvantage is that it is expensive and not readily adaptable for large scale production.

(b) *Diffusion* The diffusion process (see for example. Massignon, 1979) relies on the fact that molecules of a gas at a given equilibrium temperature have on average the same energy and those with lower masses will therefore, on average, be travelling faster. (The kinetic energy of the molecules is $mv^2/2$ where m is their mass and v their velocity. For molecules of the same energy but different masses the velocity is inversely proportional to the square root of the mass.) The light molecules therefore strike the wall of a containing vessel more often and, if the walls are porous, more light molecules will enter the pores. If these pores are of molecular dimensions and pass right through the vessel walls, then light molecules will escape preferentially into any outer containment, leaving the gas remaining in the vessel depleted in the light isotope. In a closed system, with a chamber of gas separated from a vacuum by a suitable porous membrane, the vacuum chamber will initially receive material enriched in the light isotope, but as the pressure equalizes between the two chambers so would the isotopic composition. To avoid this the feed stream of fresh gas has to be passed across one side of the membrane and the diffusing gas has to be pumped away from the other. If this gas is itself passed over a second porous membrane a further enrichment can be achieved, and so on through as many stages as are required to reach the desired degree of enrichment. In practice a cascade of membranes is used in which natural uranium hexafluoride vapour is fed into the plant and two product streams are removed; one enriched in the ^{235}U isotope and the other depleted. The depleted product

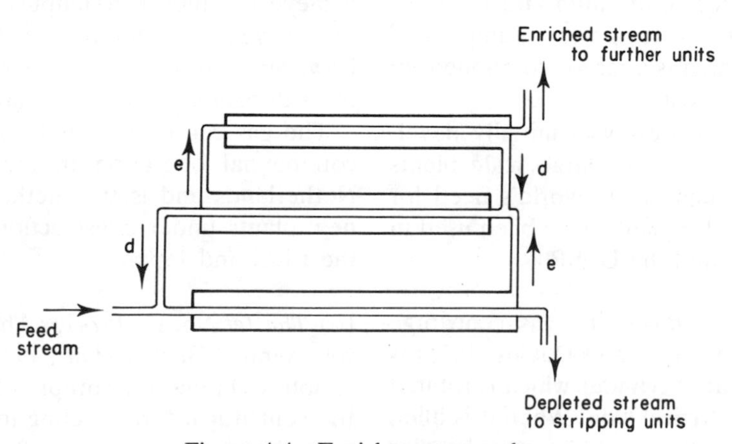

Figure 4.4 Enrichment cascade

from the individual stages is passed back and mixed with the feed to the preceding stage (Figure 4.4).

The enrichment achieved at each membrane is dependent on the ratio of the square root of the masses of the uranium hexafluoride molecules and the proportion of the feed allowed to pass through the membrane at each stage. The higher the enrichment required, the larger the number of stages needed and the smaller the quantity of product. Because the mass ratio of the uranium hexafluoride molecules is small (less than 1.01) the separation achievable at each stage is small (less than 0.5 per cent change in isotopic ratio) and the diffusion process requires a large number of stages. Vast quantities of feed uranium would be required if only a small proportion of the feed stream were removed and this would be prohibitively costly. The diffusion plant is therefore designed to strip as much ^{235}U as possible from the uranium feed before the latter is finally rejected. Typically the reject uranium or enrichment plant tails have had a ^{235}U content of 0.2 per cent to 0.3 per cent over recent years. This means that for each 1 t of enriched uranium product containing say 3.2 per cent of ^{235}U, the quantity of feed uranium has to be about 7 t. A typical diffusion plant might have some 1200 stages of which two-thirds are in the enrichment section and one-third in the stripping section. The mode of plant operation and the selected ^{235}U content of the tails can be adjusted to match movements in the price of uranium and the price of electricity used to power pumps and compressors, which is a major component of total enrichment costs.

The diffusion process was initially developed in the USA and its large scale plants have supplied much of the world's need for enrichment. Similar plants have been used in the UK, France and the USSR.

(c) *The Gas Centrifuge* The gas centrifuge (see for example Avery and Davies, 1973) is simply an evacuated cylinder which is rotated rapidly about a vertical axis. When it is filled with uranium hexafluoride vapour the heavier molecules are preferentially concentrated towards the outer wall of the cylinder and the lighter towards the central axis. A convective gas flow can be introduced by means of a temperature gradient along the tube, which moves the lighter isotope to the top of the centrifuge and the heavier to the bottom. By introducing the feed gas at the mid-point of the cylinder and withdrawing product and waste streams from the top and bottom, a continuous flow system can be set up and the individual centrifuges can be linked in a cascade like the diffusion stages discussed in the previous section.

The centrifuge has the great advantage that the isotopic separation is dependent on the difference of mass between the molecules and not the square root of the ratio. It is therefore as efficient at separating uranium–235 from uranium–238 as it would be separating light hydrogen isotopes. The separation also varies as the fourth power of the peripheral speed and the length of the centrifuge cylinder. High speeds are therefore important in achieving large separations and reducing the number of centrifuge stages needed. Materials limitations restricted development of the process for many years but it has now come into its own. A typical plant producing enriched uranium for LWR fuel at 0.25 per cent ^{235}U tails assay would require only about 25 stages compared with the 1200 for the diffusion plant, although size limitations demand that large numbers of centrifuges be used in parallel at each stage in order to achieve the high throughputs required. The power required to run such a plant is, however, only about 4 per cent of that needed for a diffusion plant of the same size.

The gas centrifuge has been brought into commercial operation in the UK and the Netherlands and is the method selected for new plants under construction in the USA, the FRG and Japan.

(d) *The Jet Nozzle Process* This method (see for example Becker *et al.*, 1975) is extremely simple and relies on isotopic separation under the centrifugal forces acting in a fast moving curved jet of uranium hexafluoride mixed in a suitable inert carrier gas such as hydrogen. The gas stream containing about 5 per cent uranium hexafluoride is fed into a jet located in proximity to a fixed hemispherical wall. A suitably placed barrier divides the emergent gas stream into two fractions, one enriched in ^{235}U and hydrogen and the other depleted.

The separation factor per stage, whilst not as high as in the centrifuge process, is nonetheless four times that for a diffusion plant, although the energy requirements for the jet nozzle and diffusion plants are of similar magnitude. The method can be used with gases at moderate pressure and the individual stages are therefore compact. Since fewer of them are required the jet-nozzle process should compete economically with diffusion. Research on this process has been in progress in FRG and South Africa and a demonstration plant is being built in Brazil.

(e) *Laser Separation* Laser separation (see for example Jensen *et al.*, 1983; Farrar and Smith, 1972) relies on a totally different principle to the methods described earlier. The atomic route exposes uranium vapour at over 2000 °C and at low pressure to intense light from a carefully tuned laser that selectively ionises the desired uranium–235 isotope. The ions can then be separated electromagnetically from the unionized uranium–238. In the molecular route a gaseous compound such as uranium hexafluoride is irradiated with tuned infra-red light that selectively excites or dissociates the molecules containing uranium–235 so that the light isotope can be removed preferentially by chemical means. Both routes are based upon the fact that the absorption of electromagnetic radiation by the different atomic or molecular species differs slightly with the mass of the isotope.

The atomic route should give almost complete separation, but processes of charge exchange between ^{235}U ions and ^{238}U atoms and the effects of the high temperatures on the energy absorption, reduce the over-all efficiency. Other difficulties arise from the corrosiveness of uranium vapour and the performance of the lasers themselves.

Selective molecular excitation is more difficult because there are a large number of overlapping vibrational energy levels. In consequence both ^{235}U and ^{238}U containing molecules are removed from the vapour and only a partial separation is achieved. The laser route based on currently envisaged technology therefore requires almost as many stages as the centrifuge process to achieve a given degree of enrichment.

The laser routes are under active investigation in a number of countries but it will be some years before the technology reaches the stage where it could be considered, economics permitting, as a competitor to the existing processes for uranium isotope separation.

(f) *Chemical Methods* Chemical methods are widely used for the separation of light isotopes such as those of hydrogen.

Molecules containing different isotopes have slightly different energy levels and bond strengths due to their mass differences and they react chemically at slightly different rates. These rate differences can affect the relative concentration of the isotopes in separate molecular species at equilibrium.

Although the separation factors for isotopes of light elements can be quite large, this is not the case for heavy elements such as uranium. Attention has concentrated on oxidation-reduction reactions with the uranium in hexavalent and tetravalent, or even metastable trivalent, forms. (Vanstrum and Levin, 1977; Commissarait a l'Energie Atomique, 1974.) The forms with different valencies can be kept in different phases (e.g. aqueous and organic solutions; solid adsorbate and solution) which when brought into contact will result in preferential concentration of the desired isotope in one or other phase. A cascade process similar in concept to that described for diffusion will then facilitate the achievement of efficient multi-stage enrichment.

Chemical methods for uranium are limited by the time taken to establish the exchange equilibria and the small separations achievable in each stage. No large scale commercial processes have yet been developed. One advantage claimed for this route is that the slowness of the stages makes the achievement of high enrichment a laborious and time consuming process. The method would therefore be unsuitable for producing material for weapons use although, economics permitting, acceptable for civil fuel enrichment requirements.

(g) *Other Methods* There is a wide variety of methods based on different physical processes that can be used to separate isotopes, particularly of the lighter elements (London, 1961). These include simple fractional distillation, thermal diffusion, chromatography, electrolysis or selective physical adsorption (Jones and Hutchinson, 1967), and more complex devices such as plasma centrifuges (Nathrath *et al.*, 1975).

4.3.1.5 Fuel fabrication To produce fuel pins, the fuel material itself, uranium metal or more commonly uranium dioxide, is sealed within gas tight containers called *cans* or *cladding*. A number of such fuel pins are held together in a lattice of fixed geometry to produce a *fuel element*.

The composition of the fuel including the enrichment, the material used for cladding, the dimensions of the fuel pins, and the number of pins in an element, are dependent on the reactor in which the fuel is to be used (Table 4.1). The enrichment and composition may also vary depending on the intended location of the fuel in the reactor core and the stage in the reactors life at which the fuel is to be introduced.

The reactor designer wants the maximum rate of heat production per unit of fuel, consistent with his ability to remove it with acceptable coolant flow and pressure conditions, in order to minimize the size and cost of the reactor. In practice he is limited by the temperatures that the fuel can tolerate without either melting or releasing its fission products too rapidly, and by the operational requirements to minimize the frequency of refuelling for those reactors that can only refuel off-load.

The design, manufacture and quality control of nuclear fuel is clearly extremely important to the safe and economic production of nuclear power. The principal requirements are that the fuel should be extremely pure and have the minimum of unwanted neutron absorbing impurities, other than any burnable poisons that may be introduced to control reactivity (section 4.2.3). It should be physically stable to the limit of its design life. The canning material

Table 4.1 Fuel for Different Reactor Types

Reactor type	Fuel	^{235}U concen-tration %	Clad	Peak coolant temp. (°C)	Average burn-up (MW d t^{-1})	Average fuel rating (W g^{-1} of U or Th)
Magnox	natural uranium metal	0.7	magnesium alloy	400	5 000	3
AGR	uranium oxide	2.5	stainless steel	650	21 000	15
PWR	uranium oxide	3.0	zirconium alloy	320	30 000	38
BWR	uranium oxide	2.2	zirconium alloy	285	28 000	20
SGHWR	uranium oxide	2.3	zirconium alloy	280	20 000	14
PHWR	natural uranium oxide	0.7	zirconium alloy	305	7 000	19
HTR	uranium oxide	10	carbon and silicon carbide	800	100 000	112
THTR	U–Th oxide	10	carbon	800	100 000	100

should be strong enough to retain fission product gases and should resist corrosion by the reactor coolant. The thermal conductivity of the fuel and canning material should be high to get the highest possible outlet coolant temperatures. Thermal contact between the fuel and its cladding should be good for the same reason. The thermal expansion of the fuel and canning material should be small and as close as possible to each other to minimize risks of damage to the fuel or the can.

These requirements will usually conflict with one another. Thus uranium dioxide has a high melting point but a poor thermal conductivity. Its thermal expansion is twice that of the corrosion resistant zirconium alloys used for PWR fuel cladding, but half that of the stainless steel used in the high temperature oxidizing environment of AGRs. Fuel design is therefore an interactive process of compromise.

(a) *Uranium Metal* The Magnox reactor has short (1 m), machined metallic fuel elements made from cast uranium rods with trace alloying additions. The metal has good thermal conductivity and good fission product retention but, due to phase transitions involving volumetric change, the temperature at which it can be used is limited to below about 660 °C at the fuel element centre. The cladding material is a magnesium alloy, Magnox, with a low propensity to capture neutrons. The cladding is bonded on to the fuel by by pressing it into machined grooves in the uranium. This prevents ratcheting due to differential thermal expansion.

The Magnox fuel pins or elements are used singly, not in assemblies, with some six to twelve stacked vertically in the fuel channels in the graphite moderator. The cans have external fins or helical spirals to provide extended heat transfer surfaces and improve contact with the coolant gas.

(b) *Uranium Dioxide* This is used to fuel AGRs, LWRs and HWRs. The fine powdered material produced by reduction of uranium hexafluoride is pressed into cylindrical pellets and sintered in a hydrogen atmosphere at about 1650 °C to yield a dense product with controlled closed porosity. In some processes a volatile binder is added to facilitate pressing. This is then driven off by heating to 800 °C before sintering. The pellets can be produced in a range of sizes as solid or hollow cylinders, and solid cylinders are generally dished at the ends to ensure that thermal expansion movement in the reactor is determined by the outer cooler regions of the fuel. The pellets are needed to close tolerances and are finally ground to the required dimensions and carefully checked to ensure that there are no chips or cracks.

The sintered pellets are loaded into fuel cans which are then sealed. In the case of the LWRs these cans are thick walled and are usually made of corrosion resistant zirconium alloy, although some reactors use stainless steel. The fuel pellets are not in immediate contact with the fuel can when at their operating temperature and the can is filled with helium gas under pressure to provide efficient heat transfer from the fuel to the can. For PWRs this pressure is around 20 to 30 bar whereas for the BWR it has been 5 bar or less, both with the fuel element cold. The LWR fuel cans have a space without fuel, at one or both ends, called a *plenum* which, together with internal closed voidage and the gaps between pellets and between the pellets and the can walls, provides a buffer volume to keep escaping fission product gas pressures within the can at acceptable levels.

The fuel elements for PWR consist of a fixed 14 x 14 to 17 x 17 array of pins compared with the BWR's 6 x 6 to 8 x 8 array of somewhat fatter pins. Both reactors have the fuel pins occupying the full core length unlike Magnox, AGR or CANDU. The assemblies are designed to keep the fuel pins in their proper geometrical array whilst they are in the reactor.

The unenriched uranium oxide fuel for the CANDU PHWR also consists of pressed pellets contained in zirconium alloy cans, but the fuel assemblies contain fewer pins (viz. 37) and are shorter (0.5 m) so that nine assemblies are used for each horizontal fuel channel.

The AGR, like the LWR, uses enriched uranium oxide fuel but the sintered pellets are annular and are contained in relatively thin walled stainless steel cans. Unlike the LWR however, the AGR fuel is segmented into separate assemblies about 1 m long, for ease of loading, and several assemblies, each containing 36 fuel pins, are loaded or unloaded as one unit held together by a tie-bar. The whole array is surrounded by a double skin graphite sleeve which helps to insulate the bulk moderator from the hottest coolant gas. It is changed with the fuel.

The thin walls of the AGR cans can not withstand high pressures and segmentation of the fuel precludes the inclusion of plenum volumes. The cans are therefore filled with helium at low pressure and the use of hollow (annular) fuel pellets provides the necessary volume to keep fission gas pressures down.

Fuel pins can be filled with vibration compacted oxide powders in place of sintered pellets if so desired, and good fuel densities can be achieved with this method using carefully graded materials. The process is not widely used for thermal reactor fuel.

(c) *Other Uranium Compounds* Uranium carbide and uranium nitride have attractions as fuel materials particularly in view of their good density and thermal conductivity; both being better than uranium dioxide. There has, however, been little interest in their use in thermal reactors because of their potential reactivity with the conventional coolants. They are considered further under plutonium and thorium fuel cycles.

(d) *Safety factors* Fresh uranium and its compounds are only weakly radioactive and workers and the general public have only to be protected from inhalation and ingestion of uranium containing dusts. This can be simply achieved by ensuring that working areas are adequately ventilated and that effluent air is filtered. The uranium recovered by reprocessing spent fuel can be purified so that its use in enrichment and fuel fabrication requires no additional precautions. A lesser degree of purification would necessitate the adoption of radiation shielding and, in the extreme, remote handling techniques.

Although not a major consideration for natural or slightly enriched uranium fuels, the possibility of criticality has to be borne in mind. The quantities of material in any place are kept well below those that could lead to criticality through their accidental movement or through the ingress of some moderating material such as water (see section 2.2.3).

4.3.2 *At Reactor*

The fabricated fuel elements are transported to the reactor site, suitably protected to prevent contamination and damage, and are stored in air at the site until they are needed. The operator will normally hold a sufficient stock of fresh fuel to guard against any supply disruption he considers possible.

The elements are transferred to the core of the reactor when required and they remain in the core until they have reached their planned burn-up. This will differ with the fuel, the reactor operating regime and the stage in the reactor's life. For the LWR, refuelled off-load, the period in the reactor will be from three to five years depending on the refuelling schedules selected.

At the end of this time the spent fuel is removed. It is highly radioactive and gives out a great deal of heat; typically hundreds of kilowatt from a single PWR fuel element. The fuel has therefore to be transferred to a cooled shielded facility, generally a water filled pond, where it is left for two years or more to allow its radioactivity to decay to more easily manageable levels. The cooling at the reactor can be continued for as long as the reactor operator wishes, subject to the corrosion resistance of the fuel and the availability of pond space. Although not designed for prolonged immersion in water, gas cooled reactor fuels have generally been placed in ponds for their initial cooling also.

4.3.3 *The Back-End*

4.3.3.1 General The so called *back-end* of

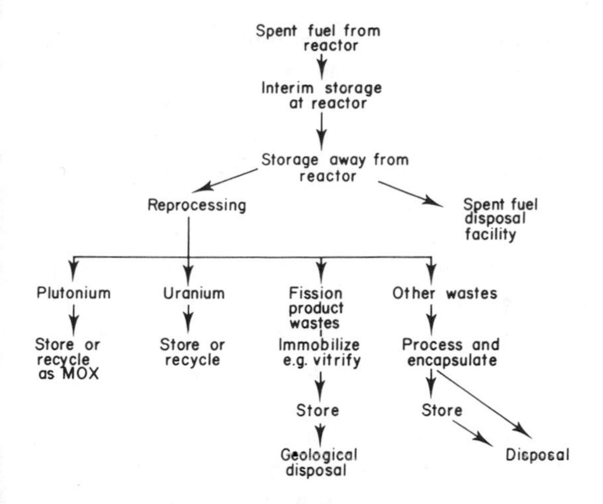

Figure 4.5 Back-end of uranium fuel cycle

the nuclear fuel cycle consists of all those stages from removal of the spent fuel from the cooling ponds at the reactor site to the final disposal of the spent fuel or the wastes separated from it by reprocessing (see Figure 4.5). The reactor operator has a number of options:

1. The spent fuel can be stored indefinitely in any one of a number of alternative ways and decisions on its ultimate fate can be left in abeyance.
2. It can be stored for a pre-selected time and then, with or without prior treatment, encapsulated and disposed of; again in any one of a number of ways.
3. Finally it can be stored for a pre-selected time then reprocessed to recover plutonium and unburnt uranium and the reprocessing wastes in turn can be either stored or encapsulated and stored or disposed of. The recovered uranium and plutonium may be either stored or recycled in fresh fuel for thermal or fast reactors.

The choice of policies for the 'back-end' will depend on a combination of technical, environmental, economic and strategic factors which will differ from country to country (Jones, 1985; Detilleux, 1985.)

4.3.3.2 Transport Spent PWR fuel assemblies still emit about 5 kW of heat one year after removal from the reactor and 1 kW after five years. The assemblies removed from the reactor cooling ponds after two or more years will therefore still need to be cooled and shielded during and after transport. Transport containers are massive and are designed to provide shielding to personnel and the public, to protect the spent fuel in the unlikely event of a transport accident and fire, and to radiate heat efficiently to keep the contents below temperatures at which the fuel cladding might fail and release fission products into the container.

The containers may be mounted on rail or road vehicles or shipped by sea and their design and use is controlled by national and international regulations.

4.3.3.3 Interim Storage In the early days of nuclear power the majority of spent fuel storage was expected to take place either in the reactor cooling ponds or in buffer ponds at a reprocessing plant, and neither was expected to be prolonged. Many reactor operators have made no final decision on the ultimate fate of their fuel, however, and storage has therefore become an extended phase of the fuel cycle.

Whilst it may be possible to extend reactor pond facilities this is not always practicable or cheap, and some fuels can not be left in water for long periods without suffering damage. Other forms of storage have therefore been explored.

(a) *Pond Storage* Special away from reactor pond stores using water cooling are perfectly satisfactory for water reactor fuels which have been designed for the aqueous environment. Water is cheap and provides a good radiation screen. It can be readily cycled through chemical purifiers to minimize corrosion of the fuel cans and through heat exchangers to reject waste heat from the fuel. The fuel assemblies have to be suitably spaced to avoid criticality but the spacing can be reduced if required by incorporating suitable neutron absorbers. This also allows the dismantling of fuel assemblies and closer packing of fuel pins if this appears economically worthwhile.

The ponds are usually steel lined reinforced concrete tanks and can have leakage detectors between the liner and the outer vessel. Pond water can become radioactive due to the presence of defective fuel elements, although it should be noted that there have been no instances of LWR or HWR fuel failure in over 20 years that could be attributed to corrosion by pond waters.

Gas cooled fuels were not designed for prolonged exposure to water and Magnox cladding in particular is susceptible to attack in acid solutions. Careful control of pond water acidity can minimize the problem but it is not planned to store Magnox fuels for prolonged periods in this medium. Where there is a risk of corrosion there are advantages in placing the fuel inside sealed water filled 'bottles' which are themselves placed in the main pond. This limits fission product concentrations in the main pond and reduces operator exposures.

(b) *Dry Stores* Dry stores keep the dried fuel in either air or an inert gas atmosphere rather than water. The cooling is less efficient and operators have therefore preferred to consider dry storage as a sequel to a period of pond storage of, say, five years, rather than as a complete substitute. Dry stores have the advantage that they are less susceptible to some hazards such as loss of coolant and seismic damage; they produce less radioactive wastes and reduce operator exposures; they are easier to maintain and less hostile to the fuel and fuel cladding. Some designs may also be cheaper to construct and operate.

There are three main types of dry store. The first, the *vault*, is merely a large concrete shielded building or cave with suitable fuel receipt and handling facilities. Fuel assemblies are stored either in a forced stream of cooling gas or allowed to cool by natural convection. The heated gas rejects its heat through a heat exchanger or can be filtered and vented to the atmosphere through a stack.

Casks are massive sealed metallic vessels designed to contain a limited number of fuel assemblies (for example, nine PWR assemblies) in a helium atmosphere. They can be self contained and used to transport as well as store fuel. They give up their heat directly to the atmosphere acting as their own heat exchanger.

Drywells are, as their name implies, boreholes in the ground. They are fitted with concrete or steel liners into which the fuel assemblies are placed with suitable overpacking. The geological medium provides radiation shielding and heat loss is directly to the geological surroundings. The assemblies have to be recoverable and the environment dry to avoid corrosion during the interim storage period.

All the dry storage methods can be employed for fuel assemblies or separated fuel pins. In cases where the fuel has been damaged in the reactor it may be necessary to place it within a new sealed container before placing it in the dry store. Considerations of safety, operator exposure and economics will determine the choice between the alternatives.

4.3.3.4 Reprocessing

(a) *General Considerations* The spent fuel emerging from the reactor contains a mixture of unused uranium (238 and 235); plutonium–239 and other plutonium isotopes and actinides produced by neutron capture; and intensely radioactive fission products. The object of reprocessing is to separate the useful uranium and plutonium from the unwanted fission products.

The plant and processes involved need to be kept as simple as possible to minimize the need for maintenance and, for preference, they should be suitable for continuous operation. The processes selected for commercial exploitation were dissolution of the fuel and the use of solvent extraction together with oxidation–reduction reactions.

The chemistry of reprocessing is basically simple but the high levels of radioactivity mean that all the plant has to be shielded and operated remotely, and the effects of radiation on solvents and materials have to be taken into account. Because the plant will have significant quantities of fissile materials in neutron moderating solvents, the plant has to be designed to prevent critical concentrations accumulating. Precautions have also to be built to cater for any possible plant failure and to ensure that the workforce and public are not exposed to undue risk. Good housekeeping and national regulations also require that radioactive effluent discharges to the environment be kept as low as possible.

Detailed consideration of reprocessing plant design and operation is beyond the scope of this work, particularly since there are so many design and operational variants, and practices differ depending on the nature of the fuel, the scale of the operation and the environmental protection standards which the plant is expected to meet. The following paragraphs outline the basis of reprocessing in current plant. Further information can be obtained from appropriate technical publications (Allardice *et al.*, 1983).

(b) *Dissolving the Fuel* The fuel assemblies arriving from reactor stores or interim stores are fed into the reprocessing plant buffer store from which they are removed and stripped of as much non-fuel material as possible or necessary. Magnox fuel cans are removed and the uranium fuel passed to the dissolver. The short AGR elements are dismantled and the pins topped and tailed before they are fed to the dissolver where only the oxide fuel is dissolved. The longer LWR fuel elements are chopped into smaller lengths to facilitate solution. The dissolver can be operated either batchwise or continuously.

The solvent used is strong boiling nitric acid which dissolves all the actinides and the majority of the fission products whilst leaving the short lengths of cladding behind for disposal as solid wastes. During the shearing and dissolution process the volatile fission products such as tritium, iodine–129, carbon–14 (as its dioxide) and krypton–85 are released. Some, such as radio-iodine, are removed by chemically scrubbing the effluent gas stream. Alternatively the gases can be passed through a suitable absorber bed such as charcoal. Even the inert gas krypton can be removed using fluorinated solvents or by freezing. In practice the more noxious elements are removed and the remaining gas is vented to the atmosphere through a suitable tall stack. Some fission products do not dissolve completely and these are separated off using a centrifuge before passing the clarified solution to the solvent extraction stages. The precise nature of the solid residues depends on the fuel and its radiation history but they are highly radioactive and produce significant quantities of heat which have to be taken into account in plant design.

(c) *Solvent Extraction* The acidity of the final solution of plutonium, uranium, other actinides and fission products is adjusted and suitable agents are added to ensure that the plutonium is in the most favourable tetravalent state. The aqueous acidic solution is then intimately contacted with a suitable organic solvent with which it is immissible. This extracts the uranium and plutonium compounds but leaves the fission product nitrates in the aqueous acidic phase. The two liquids are then allowed to separate and the uranium–plutonium rich organic phase removed.

The mixing can be done batchwise in mixter–settler tanks or continuously using vertical counterflow columns through which the less dense organic solvent flows upwards whilst the more dense aqueous phase flows downwards. A cascade of either type of unit allows highly efficient separation. Many organic solvents can be used. In the UK the Butex (dibutyl ether of diethylene glycol) originally used at Windscale has been replaced by TPB (tri-*n*-butyl phosphate) in kerosene solution in the so called Purex process.

In the Purex process some 99.9 per cent of the uranium and 99.98 per cent of the plutonium are extracted into the TBP in the first cascade stage leaving 99.5 per cent of the fission products in the nitric acid solution. The plutonium tetranitrate which is readily soluble in TBP is then reduced chemically to the trinitrate which is not, and a further extraction with fresh nitric acid then removes the plutonium into the aqueous phase whilst leaving the uranium in the organic (TBP) phase. The two product streams have some slight cross-contamination and small quantities of fission products remain which are removed in further solvent extraction stages.

The efficiency of the initial extraction process can be assisted by using additives to decrease the solubility of the uranium and plutonium species in the aqueous phase, and the counterflow columns can be packed with an inert filler (metal rings) to increase the intermixing of the liquid phases. This helps to reduce the necessary column length. Further reductions by up to a factor of five can be achieved by pulsing the fluid in the columns so that an up and down reciprocating motion is superimposed on the overall directions of flow of the two phases.

Mixer–settlers have both advantages and disadvantages compared with columns. Their capital costs may be lower, they are potentially more flexible in operation and they are easier to instrument. On the other hand they have a larger liquid hold-up, are slower in operation and have more moving parts. The hold up leads to greater radiation degradation in the solvent. The mixer–settler approach could be speeded-up and the volumes reduced using centrifugal contactors but these are mechanically complex and one would need to be sure of their long term reliability before they were adopted for use in a highly radioactive low-maintenance environment.

It is clearly desirable in the interests of economy to recycle the organic solvents. Since these are broken down by radiation in the plant they have to be re-purified by chemical washing before they are returned to the plant.

(d) *Criticality Control* Criticality control can be achieved by a number of methods. The mass of fissile materials in the different sections of the plant can be limited so that a critical mass can not arise; the volumes and/or shapes of vessels can be designed to ensure that criticality can not occur; the concentration of fissile materials in process streams can be regulated and closely monitored; or neutron poisons can be used, either incorporated into the plant fabric or dissolved in the process liquors.

(e) *The Products and their Re-cycling* The product uranyl nitrate stream containing traces of plutonium and fission products can be purified by solvent extraction and chemical means until it can be handled in exactly the same way as fresh uranium feed for fuel manufacture (section 4.3.1.3). A lesser degree of purification would lead to the need for additional handling precautions.

The uranium recovered from Magnox fuel (and CANDU fuel if it were reprocessed), has less than the natural 0.7 per cent of uranium–235 (*c*. 0.3 per cent)., whilst uranium recovered from AGR or LWR fuel is still slightly enriched (*c*. 0.8 per cent). In either case it can be fed back into the enrichment plant to produce fresh uranium fuel of any desired degree of enrichment. The recycled uranium, however, does contain higher quantities of non-fissile uranium–236 than the feed from newly mined uranium and, since this has a high neutron absorption, the feed made from recycled uranium has to have a somewhat higher ^{235}U enrichment to compensate. Recovered uranium also contains enhanced levels of uranium–232 as a result of neutron capture and radioactive decay sequences. Both ^{236}U and ^{232}U are concentrated in the re-enrichment process and would accumulate on successive recycles. To overcome this the recycled uranium could be mixed in with fresh uranium. The economic attractiveness of using recovered uranium in preference to fresh uranium depends on its residual ^{235}U content, its ^{236}U content, the price of fresh uranium and the prices of enrichment and fabrication services.

The recovered uranium can also be used without enrichment as feed to mixed oxide fuel manufacture (section 4.4). The plutonium solutions in nitric acid can be used, after purification, as feed to one of the plutonium fuel cycles, or the plutonium can be converted to oxide for storage (section 4.4.)

(f) *Other Reprocessing Methods* Other processes have been considered from time to time for the separation of uranium, plutonium and fission products (Detilleux, 1986). Some of these avoid the use of solvents and, in principle, might give rise to smaller volumes of radioactive effluent. The conversion of spent fuel to fluorides and the volatilization of uranium hexafluoride away from non-volatile fission product fluorides is one obvious route.

France and the USSR have been reported (Nuclear Fuel, 1984) to be studying the feasibility of melting steel clad fuels at 1500 °C, dissolving the fuels in molten fluorides and separating the plutonium and uranium hexafluoride by fractional crystallization and sublimation. The method is said to be simple and compact and avoids some of the difficulties arising from radiation induced degradation and heat release in the conventional wet processes. However corrosion and handling problems will have to be overcome before the process could become a commercial prospect.

4.3.3.5 Waste Treatment

(a) *The Nature of the Wastes* It will be evident from the earlier sections of this and the preceding chapter that radioactive wastes are produced throughout the nuclear fuel cycle and during the course of reactor operation (Figure 4.6). Apart from the uranium mining and milling wastes described in Section 4.3.1.2, uranium contaminated materials and process effluents will arise at the front-end of the fuel cycle in the course of conversion, enrichment and fabrication. In the reactor itself fission products released from damaged fuel and neutron activation products from the reactor materials enter the coolant and may be collected on filter beds or ion-exchange resins. Reactor core and coolant circuit components become contaminated and solid wastes can arise both in maintenance and ultimate decommissioning (Chapter 10).

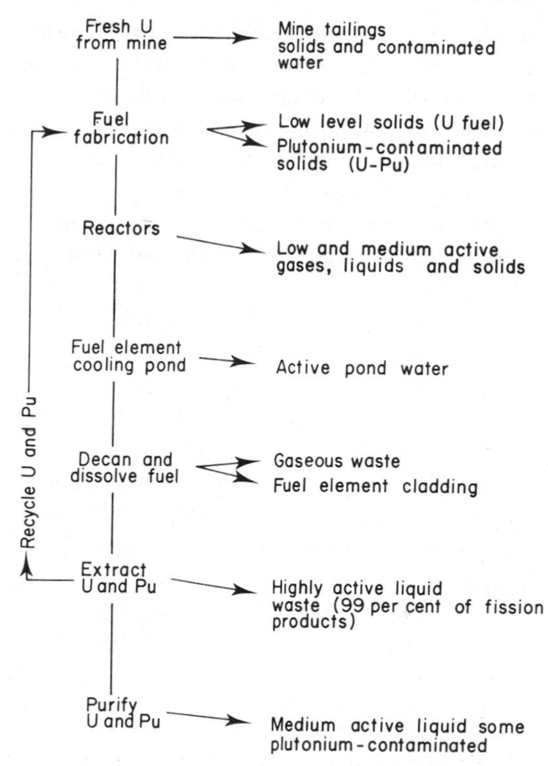

Figure 4.6 Wastes from the fuel cycle

The highest radioactivity is associated with the spent fuel itself and with the wastes arising from its handling and reprocessing. These include fuel assembly components and the fuel cladding, the gaseous fission products released when spent fuel is opened and dissolved, the separated fission product streams, and the other washing and waste liquors from the reprocessing plant. The components of the reprocessing plant replaced during maintenance or when the plant is finally decommissioned are also contaminated.

(b) *The Objectives of Waste Management* The main objective of radioactive waste management is to keep waste volumes and their contained activity to a minimum, consistent with the direct relationship between energy production in the reactor and the quantity of fission products produced (IAEA, 1984). This is a matter of good planning, good design and good housekeeping. Once wastes have been produced they have to be prevented from harming the public or workforce and suitable means of disposal have to be identified which then determine the treatment the wastes receive.

The two options for disposal are either to dilute and disperse the waste so that the population is not exposed to harmful radiation levels, or to contain it in some medium that isolates it from the biosphere until it has decayed to tolerable levels. In the former case dilution to levels well below the natural background levels to which everyone is exposed from birth might be considered satisfactory, although possible concentration of dispersed radioisotopes by physical or biological processes (e.g. food chains) has to be taken into account. In the latter case guaranteed isolation until the radiotoxicity of the wastes has fallen to levels comparable to those found in nature (e.g. uranium or thorium ore bodies) might be the criterion. There are as yet no internationally agreed standards and countries are formulating their own strategies in the light of informed scientific opinion and national and international public attitudes.

(c) *Categorization of Wastes* There are several ways of categorizing wastes; by phase (solid, liquid or gas), by origin (reactor operation, decommissioning, reprocessing, etc.), by the lifetime of its activity (short or long-lived), by its radioactivity level (high, medium, low) or by the routes favoured for its disposal. The whole question of radioactive waste management is dealt with more fully in Chapter 9, but it is appropriate here to indicate briefly the sources of the waste and the general options available for its treatment and disposal. For convenience they will be categorized by activity level and phase.

(d) *High Level Liquid Waste (HLW)* The aqueous acidic liquors emerging from the first cascade of the reprocessing plant contain over 99 per cent of the non-gaseous β- and γ-emitting fission products together with small but important traces of plutonium and other actinide elements. They continue to produce considerable quantities of heat and this is a major factor in determining how they should be dealt with. These waste solutions have customarily been concentrated by evaporation and stored in large screened water-cooled double walled stainless steel tanks, which can be monitored to ensure that there is no leakage from the inner container. In the event that a leak were detected the contents would be transferred to a new tank. There is no fundamental reason why such storage could not continue indefinitely with complete safety. However, the conversion of the wastes to a solid form would eliminate even the remotest risk of leakage and dispersion, reduce the level of monitoring required and facilitate transport of wastes to other stores or repositories. For this reason countries which reprocess their fuel plan to evaporate the high level wastes to dryness and incorporate the residues in a solid matrix. The selected matrix has to have a good capacity to accommodate the mixture of elements in the HLW, has to be resistant to radiation and mechanically strong, and has to be suitable for commercial production in remotely operated low maintenance plant.

With ultimate disposal in mind the medium chosen has been a borosilicate glass which would retain the fission products for prolonged periods even if the glass blocks were exposed to ground water. The calcined fission product residues are mixed into the molten glass at 900–1000 °C, which is then allowed to cool and solidify in thick walled metal containers which can be several metres long and about a metre in diameter. The vitrification process (AVM) has been employed on a commercial scale in France since 1979.

The quantity and age of waste incorporated into each block, its dimensions, and the necessary cooling are inter-related and the operational choice will depend on economic and technical factors. Thus the activity of the waste introduced should not be such as to heat the glass to its softening point and the surface temperature of the blocks has to comply with any constraints arising from the characteristics of the repository. The more waste incorporated, the hotter the blocks and the wider the spacing that may be needed between them in a store or repository. This may force up repository costs and outweigh any savings in material or manufacturing costs.

Whilst borosilicate glass has received the greatest attention and forms the perfectly satisfactory basis of existing high level waste encapsulation plans in France and the United Kingdom, research is still proceeding on other matrix materials such as ceramics, metals and synthetic rock compositions. Proponents of the latter argue that it is closer to the materials which have proved their stability and ability to retain relevant chemical elements in the earth's crust over long time periods. The ease of manufacture is, however, another major consideration which affects possible plant operator exposure to radiation and the economics of the overall waste management process.

The high level waste in its glass (or other) matrix and its steel container can be safely stored in surface or underground air cooled stores with either natural convective or forced draught cooling, depending on the radioactive content, the matrix and the store design. Ultimately the containers with, if considered necessary, additional physical barriers can be transferred to suitable geological repositories where they will remain isolated from the biosphere for tens of thousands of years. The initial supervised storage period is optional but may last for 50 years or more to allow the radioactivity to decay and reduce the heat output. This will reduce the size and cost of the final repository at the expense of more intermediate storage capacity.

The rock media suitable for disposal include salt domes, clay and granite, all of which, in suitable areas, are capable of providing a stable long term environment (see Chapter 9).

Other disposal routes considered have been burial in deep ocean sediments (Nuclear Energy Agency, 1984), burial in permafrost or deep ice in the Antarctic and despatch to the sun or deep space in rockets. The former of these appears the most practicable and has technological and environmental attractions.

(e) *Intermediate and Low Level Liquid Wastes* Intermediate level wastes containing low concentrations of fission products arising from reprocessing plant operations would normally be concentrated by evaporation and added to the high level waste stream. Low level liquid wastes arising from spent fuel storage ponds, from activated reactor coolant, from laundries handling protective clothing, and from the low activity washings from reprocessing plants are treated by appropriate means such as filtration, flocculation, ion-exchange or reverse osmosis to remove the bulk of the remaining radioactive material. Only when the radioactivity has been brought down to controlled and specified levels can remaining very dilute wastes be discharged to rivers or the sea, and even then it is carefully monitored to ensure that releases to the environment are kept within the limits laid down for the site by the regulatory agencies.

(f) *Low-level Solid Wastes* These are produced in large quantities and include waste paper, protective gloves and clothing, tools and other equipment that has been (or could have been) slightly contaminated in the course of operating and maintaining reactors, plant and laboratories. In the UK wastes containing less than 20 millicuries m^{-3} of α-activity and 60 millicuries m^{-3} of β-activity, with a surface dose of less than 0.75 rad h^{-1}, are contained in metal drums and can be buried in shallow trenches at suitable sites. The site has to have suitable geology so that rain water percolating through the wastes does not enter aquifers that might be used for drinking. The wastes are deposited in trenches some 10–15 ft deep, compacted and covered with at least 3 ft of soil.

The use of such a site has to be licensed and monitored to ensure that the quantities of long lived waste are strictly limited and that public access to the site is controlled for as long as is considered necessary.

In the past, solid wastes of somewhat

higher activity have been sealed in a cement matrix in steel drums which have then been deposited on the deep ocean floor in the Atlantic Ocean some 800 km off Lands End. This sea dump was conducted annually under the London Convention and disposed of some 300 t of solid wastes per annum. The steel of the drums would eventually corrode and the radioactive contents would be slowly released into the sea water where it would add negligibly to the quantities of natural radioisotopes already present.

International pressures and trade union opposition to the sea dump have led to a temporary moratorium and further scientific investigations to reassure objectors of the inherent safety of the practice. These wastes, if not consigned to the sea, will require land burial under more stringent conditions than the low level wastes described earlier.

(g) *Intermediate Level Solid Wastes* These arise when filters or ion-exchange resins are used to recover radioactive materials from gas or liquid waste streams. They also include items of contaminated plant and equipment and some decommissioning wastes.

Those that are combustible, including α-active wastes, can be incinerated under controlled conditions and the residues, together with non-combustibles, incorporated in a suitable matrix, such as cement or bitumen, within steel drums.

In the UK waste of relatively low β- and γ-activity not requiring thick shielding was expected to be packaged in drums with 3 in of concrete shielding and consigned to concrete lined land trenches 19 m deep. The waste was to be buried under 3 m of clay and the whole trench would be capped with a 2 m thick reinforced concrete shield to prevent inadvertent human access.

Intermediate level solid wastes of higher β- and γ-activity requiring thicker shielding were to be similarly immobilized but would be provided with re-usable shielding over-packs for transport to the waste site. The burial would be similar but the packages would be deposited using remote handling techniques in shielded cells, similar to those at Centre de la Manche in France, built within the trenches. A recent governmental decision means that these wastes will be consigned to deep geological repositories ultimately.

The hydrological and geological constraints on these burial sites are similar to those for the low level waste site at Drigg.

Wastes containing significant quantities of long-lived α-emitters like plutonium can be similarly incinerated to reduce their volume (if appropriate) and encased in cement or bitumen in steel drums. Because of the greater radiotoxicity of these wastes they have to be more completely isolated from the environment and plans have focused on deep geological repositories such as salt domes in Germany and under-sea granite tunnels in Sweden. The UK contemplated using the Billingham anhydrite mines. The conditions required are broadly comparable to those for disposing of high level vitrified wastes except that the wastes do not produce significant quantities of heat and this simplifies the repository design.

(h) *Gaseous Wastes* These are produced in the fission process and through neutron activation of reactor materials. Limits are laid down for the radioactivity allowed in gaseous emissions from reactors or plant to the atmosphere, to ensure that they are not a significant health hazard. In order to keep within the prescribed limits effluents are filtered to remove entrained particles and aerosols, and they can be scrubbed or passed through absorber beds to reduce levels of tritium or iodine–129 if necessary (see Section 4.3.3.4).

The controlled gaseous discharges are made through high stacks to ensure efficient dilution and dispersion of residual radioisotopes such as argon–41, arising from activiation of air used to cool early Magnox steel pressure vessels, and fission product krypton–85, released when spent fuel is dissolved.

4.3.3.6 Spent Fuel Disposal If the spent fuel arising from reactors is not reprocessed it has either to be stored indefinitely using the methods described in section 4.3.3.4 or it has to be encapsulated and isolated from the environment in much the same way as high level virtrified waste.

The direct disposal of spent fuel has the advantage that it retains the fission product gases and the fission products in the ceramic matrix of the fuel (for oxide fuels) and hence gives smaller short-term radioactive emissions to the environment than the reprocessing route. On the other hand it still contains large quantities of uranium and, more importantly, plutonium and other actinides, which add considerably to its toxic potential. Additionally the disposal of the plutonium and uranium which could be re-used in fuels adds to the need for fresh uranium and adds to the mining and milling wastes associated with that activity.

A country opting to dispose of spent fuel, despite these disadvantages, can elect to store it for some decades to allow its initial radioactivity and heat output to decay. The fuel assemblies can be dismantled for closer packing of the fuel pins in the disposal containers, or they can be disposed of intact.

The Swedish plans envisage the packaging of spent fuel in thick copper canisters and burial in deep hard rock repositories (Chapter 9).

4.4 The Plutonium Fuel Cycle

4.4.1 *General*

Plutonium does not occur naturally but is produced via neutron capture in uranium isotopes and subsequent β-decay reactions (sections 4.2.1, 4.2.2). The principal fissile isotope, and the one produced in greatest quantity in thermal reactors, is plutonium–239. Other plutonium isotopes are also present in spent fuel however and their relative concentrations depend on the radiation history of the fuel (Table 4.2.) High burn-up fuel contains higher proportions of non-fissile plutonium–240, and other isotopes of higher mass. (Non-fissile is used in the sense defined in Section 4.2.1. ^{240}Pu is fissioned by fast neutrons.)

The fissile behaviour of recovered plutonium in thermal or fast reactors depends on its isotopic composition and it is therefore customary to describe quantities of plutonium not in terms of its total mass but in terms of its ^{239}Pu *equivalent mass* (kg Pu(E)). This is based on the average neutron multiplication factor arising from the mixture of isotopes relative to that for plutonium–239 (Baker and Ross, 1963). For an oxide fuelled fast reactor plutonium equivalents for specific isotopes are: plutonium–239, 1.0 (by definition); plutonium–240, 0.15; plutonium–241, 1.5; plutonium–242, 0.15; uranium–235, 0.7. However, these factors will differ with the incident neutron energy spectrum since the number of neutrons produced per fission and the relative likelihood of neutron capture compared to fission both vary. Quite different factors are appropriate for thermal reactors where the fission neutron yields are lower and the ratio of capture to fission is much higher for all the plutonium isotopes. The plutonium–239 equivalent of plutonium reco-

Table 4.2 Isotopic Composition of Plutonium from Uranium Fuelled Reactors

Reactor	Fuel burn-up MWd t^{-1}	Percentage of isotopes when fuel discharged ^{238}Pu	^{239}PU	^{240}Pu	^{241}Pu	^{242}Pu
Magnox	3000	0.1	80.0	16.9	2.7	0.3
	5000	n.a.	68.5	25.0	5.3	1.2
PHWR	7500	n.a.	66.6	26.6	5.3	1.5
AGR	18 000	0.6	53.7	30.8	9.9	5.0
BWR	27 500	2.6	59.8	23.7	10.6	3.3
	30 400	n.a.	56.8	23.8	14.3	5.1
PWR	33 000	1.5	56.2	23.6	13.8	4.9

n.a. = not applicable

vered from spent LWR fuel (33 000 MWd t^{-1}) is only 70 per cent of the total plutonium content. Because plutonium produces significantly more neutrons per fission at high incident neutron energies than uranium–235 it is particularly well suited for use in fast breeder reactors. Its use as a substitute for uranium–235 in thermal reactor fuels does not capitalize on this advantage, and suffers from the relatively high neutron absorption of all the plutonium isotopes at low neutron energies, including a high resonance absorption in plutonium–239 at 1 ev.

4.4.2 *Safety*

Plutonium is a toxic material and depending on its irradiation history it and its decay products emit a mixture of α-, β- and γ-radiation. Plutonium produced from lightly irradiated fuels can be almost pure plutonium–239 which is a long-lived α-emitter which requires little shielding. High burn-up fuel on the other hand can contain as much as 15 per cent plutonium–241 (which decays to americium–241) and higher actinides which emit high energy γ-radiation, which requires much more shielding and the use of remote handling techniques. Plutonium from high burn-up fuels which has been held in store for any length of time may therefore need to be re-purified chemically to remove americium and other actinides before it is fed to a fuel fabrication plant.

Plutonium in its pure state is a highly concentrated fissile material and care has to be exercised in its separation, storage and use to ensure that a critical mass is not inadvertently created.

4.4.3 *Thermal Reactor Fuels*

4.4.3.1 Plutonium Concentration Plutonium can be used as a substitute for uranium–235 in thermal reactors and can be added either to fresh uranium, or to enrichment tails, or to uranium recovered from spent fuel, as an alternative to using enrichment (Chamberlain and Melches, 1977; Schenk and Riedel, 1984). Its use saves on uranium purchases and on the costs of enrichment, but against this higher costs are incurred in fuel fabrication due to the need to increase shielding and ventilation (Jones, 1985).

For the mixture of isotopes of plutonium recovered from LWR fuel the concentration of plutonium–239 equivalent needed in mixed plutonium–uranium oxide (MOX) fuel is slightly larger than the corresponding concentration of uranium–235 due to their different neutron production and absorption characteristics. Plutonium which had been repeatedly recycled would accumulate higher non-fissile isotopes and its use as a uranium–235 substitute becomes progressively less effective. (The first cycle fissile content falls from about 70 per cent of total plutonium produced in a PWR to under 60 per cent on repeated recycle in MOX fuel.)

At the present time it is considered that MOX fuel equivalent to 3 per cent to 4 per cent enriched uranium fuel could be used to provide up to 30 per cent of the core fuel in existing LWRs without upsetting the reactor's physics or control. There are advantages in using fuel pins enriched in either uranium or plutonium rather than both, since fabrication costs will be less if two-thirds of the fuel can be manufactured without the special precautions needed for MOX.

Thermal reactors designed to burn wholly plutonium enriched MOX fuel would not be self sustaining in fuel and would require the support of plutonium recovered from three to four ordinary uranium fuelled reactors of similar capacity.

Alternatively a single reactor or group of reactors could be operated in the self generation mode with MOX fuel produced from their own reprocessed fuel. If the plutonium is only used once (i.e. a once-through mode) then some 15 per cent to 20 per cent of a typical LWR's fuel might be in the form of MOX in a settled down system. On the other hand, if the plutonium fuels were themselves reprocessed and the plutonium repeatedly recycled an equilibrium would eventually be established in which the percentage of MOX fuel would be in the region of 30 per cent (Table 4.3), and, to compensate for the

Table 4.3 Self Generated Plutonium Fuelled PWR Cycle

	First reload	Second reload (first recycle)	Third reload (second recycle)	Fourth reload (third recycle)	Fifth reload (fourth recycle)	Sixth reload (fifth recycle)
% ^{235}U in UO_2	3.0	3.0	3.0	3.0	3.0	3.0
% Pu in MOX	–	4.72	5.83	6.89	7.51	8.05
% MOX in total fuel	0	18.4	23.4	26.5	27.8	28.8

degraded plutonium, 8 per cent plutonium (total) would be needed in the MOX.

4.4.3.2. Fuel Manufacture and Processing As pointed out in section 4.4.2 the toxicity and radioactivity associated with plutonium necessitate additional precautions at the front-end of the fuel cycle and this can add considerably to the costs. Because plutonium is used mixed with uranium, both as oxides, additional care is needed in the preparation of fuel pellets to ensure that they are homogeneous, otherwise differential heating could occur in the reactor and damage the fuel. The variation of plutonium isotopic content depending on its source necessitates careful analysis and control of fuel manufacture and fuel batches will be more variable in total composition than those using uranium alone. This need for close control also adds to costs. The mixed oxides can be prepared either directly from a solution of the mixed nitrates or by dry blending the separate oxide powders. The dry powder routes have disadvantages in terms of operator exposure and research has been in progress for several years on alternative routes (the sol-gel process) which could reduce the problem.

Once the fuel has entered the reactor it becomes more like uranium fuel. The back-end of the fuel cycle can be precisely the same except that care has to be taken in the reprocessing plant to allow for the higher concentrations of heavier plutonium isotopes and other actinides present.

4.4.4 *Fast Reactor Fuels*

4.4.4.1 Plutonium Concentration For the resons given in the preceding sections plutonium in the form of MOX is the preferred fuel for fast reactors (Marsham *et al.*, 1972). Plutonium contents above 5 per cent (239 Pu kg (E)) are needed to achieve criticality, even in very large masses of fuel, and to produce a conveniently small reactor core and reasonable energy ratings 15 per cent to 30 per cent plutonium content is acceptable. In practice at the high neutron energies in the fast reactor about 15 per cent of the fission occurring takes place in the 'non-fissile' uranium–238.

4.4.4.2 Fuel Manufacture and Processing This is essentially the same as for thermal reactor fuel except for the fact that the high plutonium concentrations require even greater attention to avoidance of criticality in both front-end and back-end plant.

Additionally fast reactor fuels are very highly rated and the pin diameters are reduced to keep the centre temperatures down to acceptable levels. This means that the costs of fabricating and assembling fast reactor fuel pins is higher per unit mass of fuel.

Another feature of fast reactors is the introduction of breeder material into and around the core. Axial breeder is included in the form of pellets of depleted uranium dioxide within the stainless steel fuel can. In current designs each 1m core length of MOX pellets has 0.4 m of uranium dioxide pellets above and below it. The elements also contain a plenum volume to retain fission gases.

Radial breeder elements consist entirely of uranium dioxide and it is possible to process this material separately from fast reactor fuel in a thermal fuel reprocessing plant. There are operational advantages however in reprocessing the radial breeder and fuel together and either course may be followed.

The breeder elements containing uranium–238 absorb the excess neutrons produced in the core to produce plutonium–239 which, when recovered by reprocessing, can be used to produce fresh fast reactor fuel.

The higher radiation fluxes to which they are exposed, and the high burn-up sought in fast reactor fuels (100 000 MWd t^{-1} compared with 33 000 MWd t^{-1} for current PWR fuel and 5000 MWd t^{-1} for Magnox fuel) causes swelling in fast reactor fuels and this is important in fuel design (Sections 3.6.4; 4.2.3).

4.4.4.3 Other Fuel Materials The interest in plutonium and uranium carbides or nitrides for use as fast reactor fuels has been mentioned earlier (Section 3.6.4). The carbides, whilst not compatible with carbon dioxide and water coolants, are stable in liquid sodium, the coolant favoured for fast reactors.

The higher thermal conductivity of carbide would permit use of fatter fuel pins and reduce fuel fabrication costs but this is offset by the greater difficulty of manufacturing carbide fuels through solid state reduction of the corresponding oxides with carbon at high temperature. Quality control is also more difficult in such processes. Fuel pellets can be produced using carbides in the same way as oxides. Further complications arise however in fuel reprocessing where the relatively simple dissolution procedures adopted for oxides are no longer suitable.

Nitrides can be produced by the reaction of nitrogen or ammonia with uranium–plutonium metals or their hydrides at high temperature. Fabrication is similar to that of MOX (described earlier) but again reprocessing would require the development of new chemical techniques. Nitride fuels might prove compatible with carbon dioxide coolant in gas cooled breeder reactors or the AGR, but the neutron absorption of nitrogen–14 would have adverse effects on neutron economy unless nitrogen–15 were separated and used in preference to natural nitrogen.

A great deal of work would be required to bring either carbide or nitride fuel to the same state of development as oxide fuels.

The USA is examining the feasibility of using plutonium enriched metallic fuels for fast reactors. Use of powder metallurgy and pyrochemical reprocessing is claimed to give low costs (Lindley and Studer, 1986).

4.5 The Thorium–Uranium–233 Fuel Cycle

4.5.1 *General*

Thorium, which is not itself fissile, is more abundant in nature than uranium. Its predominant isotope, thorium–232, is a fertile material which via neutron capture and two β-decay reactions produces fissile uranium–233. Thorium can therefore be used to replace uranium–238 in reactor fuels or breeder blankets and the resultant uranium–233 can be used as a fuel in place of plutonium.

Uranium–233 gives a higher neutron yield per fission than either uranium–235 or plutonium–239 at most energies below about 40 kev, and it is therefore a good thermal reactor fuel. It is less attractive than plutonium–239 in fast reactors, partly because of its lower neutron output per fission and partly because thorium–232, its precursor, is much less fissile than uranium–238 and has a higher capture cross section, both of which affect the fuel enrichment necessary and reduce the breeding gains achievable compared with those in MOX fuelled fast reactors.

The use of the thorium ^{233}U cycle has been examined for high conversion ratio thermal reactors, including high temperature reactors (Huddle, 1969; Walker, 1978; Shapiro *et al.*, 1977).

Whereas the standard LWR produces only about 0.6 atoms of fissile plutonium per atom of fissile uranium–235 consumed, due to the significant absorption of neutrons in the moderator, a reactor with less moderator and hence a fast neutron spectrum will give more absoprtion in any fertile materials present, such as uranium–238 or thorium–232. Careful optimization of a reactor using uranium–233 fuel, with its higher neutron output–fission ratio, in a thorium–232 matrix, can lead to conversion ratios (^{233}U produced/^{233}U consumed) close to or even, in theory, exceeding unity. High conversion reactors or light water breeder reactors (LWBR) have been considered which obtain their initial ^{233}U fuel from a pre-breeder reactor which employs thorium with uranium–235 or plutonium fuel. A uranium–233–thorium–232 LWBR core has been under test at Shippingport in the USA. The same can be done even more easily with heavy water moderated reactors using uranium–233 fuel as a result of the good moderating properties and lower parasitic neutron capture in heavy water itself.

Canadian research has concentrated on achieving high conversion ratios for once-through fuel (Cristoph *et al.*, 1975; Slater, 1977). The thorium–232 in the initial fuel with uranium–235 produces enough uranium–233 *in situ* to replace the consumed fissile material. Fuel burn-up might be limited to around 10 000 MWd $^{-1}$ and the reactor would consume thorium–232 rather than uranium or plutonium.

4.5.2 *Fuel Manufacture*

Thorium oxide can be produced directly from thorium minerals by chemical extraction and purification processes, although its occurrence with chemically similar rare earth elements makes for greater difficulty than uranium recovery and concentration. It can be pelletized for use in fuel assemblies in precisely the same way as uranium. The irradiated thorium oxide can be reprocessed to recover the uranium–233 chemically. The necessary processes are not well developed but would consist of similar solvent extraction processes to those used for recovery of uranium and plutonium. The problems of recycle are, however, more acute than those for plutonium, and fuel fabrication facilities and the fuel itself would require heavy shielding to cope with the radioactivity from the decay products growing into the uranium–233. The most important of these arise from the presence of uranium–232, which is produced from uranium–233 in the reactor by n,2n reactions, and which decays to yield high energy γ-emitting daughter products, particularly thallium–208. Uranium–232 produced can not be separated from the uranium–233 fuel by simple means (see too section 4.3.3.4).

Special fuels for the high temperature reactors have been mentioned in section 3.3.4. The German pebble bed reactor has thorium and enriched uranium oxides dispersed in 6 cm diameter carbon spheres. It is operated in the once-through mode with the thorium producing uranium–233 fuel *in situ*.

The more complex fuels for the Dragon HTR employed highly enriched (*c.* 90 per cent ^{235}U) uranium carbide in 1 mm diameter spheres coated with carbon deposited thermally on the fuel. This in turn is coated with a layer of high density pyro-carbon, an impervious layer of silicon carbide to retain fission products, and another pyrocarbon layer. The slightly larger fertile thorium oxide pellets are also coated with carbon layers. The fuel and breeder material can withstand temperatures of 1200 °C to 1300 °C. No routes have been developed for reprocessing this fuel although for maximum fuel economy such reprocessing is as necessary to the thorium–uranium cycle as to its plutonium–uranium counterpart.

REFERENCES

Allardice, R. H., Harris, D. W., and Mills, A. L. (1983). 'Nuclear Fuel Reprocessing in the U.K', in W. Marshall (ed.), *Nuclear Power Technology*, Clarendon Press, Oxford, pp. 209–281.

Avery, D. G., and Davies, E. (1973). *Uranium Enrichment by gas centrifuge*, Mills and Boon, London.

Baker, A. R., and Ross, R. W. (1963). 'Comparison of the value of plutonium and uranium isotopes in fast reactors', in *Proc. Conf. on breeding, economics and safety in large fast power reactors*, ANL 6792, National Technical Information Service, Springfield, Virginia.

Becker, E. W., Berkhahn, W., Bley, P., Ehrfeld, U., Ehrfeld, W., and Knapp, U. (1975).

'Physics and development potential of the jet nozzle separation process', in *Proc. Int. Conf. on Uranium Isotope Separation*, British Nuclear Energy Society, London.

Chamberlain, A., and Melches, A. (1977). 'Prospects for the establishment of plutonium recycle in thermal reactors', in *Proc. of IAEA Conf. on Nuclear Power and the fuel cycle*, IAEA, Vienna, Vol. 3, p.271.

Cristoph, E., Milgram, M. S., Veeder, J. I., Banerjee, S., Barday, F. W., and Hamel, D. (1975). *Prospects for self-sufficient thorium cycles in CANDU reactors*, AECL 5501, Atomic Energy of Canada Ltd, Chalk River, Ontario.

Commissariat a l'Energie Atomique (1974). *Isotopic enrichment of uranium with respect to an isotope*, French Patent Spec. 1467/74.

Detilleux, E. J. chmn. (1986). *Nuclear Spent Fuel Management*, Nuclear Energy Agency, NEA/OECD, Paris.

Farrar, R. L., and Smith, D. F. (1972). *Photochemical Isotope Separation as applied to uranium*, Oak Ridge Report K–3054 Rev.1., National Technical Information Service, Springfield.

Huddle, R. A. (1969). 'Fuel elements for high temperature reactors', in *Proc. Symp. on advanced and high temperature gas cooled reactors*, IAEA, Vienna, pp. 631–645.

IAEA (1984). *Radioactive Waste Management, Proc. Int. Conf.*, IAEA, Vienna.

Jensen, R. J., Judd, O'D. P., and Sullivan, J. A. (1983), Los Alamos Science.

Jones, P. M. S., chmn. (1985). *The Economics of the Nuclear Fuel Cycle*, OECD/NEA, Paris.

Jones, P. M. S., and Hutchinson, C. (1967), 'Adsorption of hydrogen isotopes on Charcoal', *Nature*, **213**, 490.

Lindley, R. A., and Studer, J. E. (1986). 'American designers achieve reduced costs', *Nucl. Eng. Int.*, **31 (July)**, 31–33.

London, H. (1961). *Isotope Separation*, Newnes, London.

Massignon, D. (1979). *Uranium enrichment—gaseous diffusion*, Springer-Verlag, Berlin.

Marsham, T. N., Bainbridge, G. R., Fell, J., Iliffe, C. E., and Johnson, A. (1972). 'The technical problems and economic prospects of using plutonium in thermal and fast reactor programmes', in *Proc. of 4th Int. Conf. on the Peaceful uses of atomic energy*, IAEA, Vienna, Vol. 9, p. 119.

Marshall, W. (ed.) (1983). *Nuclear Power Technology*, Vol. 1, Clarendon Press, Oxford.

Nathrath, N., Kress, H., McClure, J., Mück, G., Simon, M., and Dubbet, H. (1975). 'Isotope Separation in rotating plasmas', in *Proc. Uranium Isotope Conf.*, British Nuclear Energy Society, London.

Nininger, R. D., chmn. (1983). *Uranium Resources, Production and Demand*, OECD/NEA, Paris.

Nuclear Energy Agency (1984). *Seabed disposal of high level radioactive wastes*, OECD/NEA, Paris.

Nuclear Energy Agency (1985). *Technical Appraisal of the Current Situation in the Field of Radioactive Waste Management*, OECD/NEA, Paris.

Nuclear Fuel (1984). *France and USSR consider experimental dry reprocessing of breeder fuel*, June 4th, p.14.

O'Riordon, M. C., James, A. C., Rae, S., and Wrixon, A. D. (1983). *Human Exposure to radon decay products inside dwellings in the U.K.*, National Radiological Protection Board report NRPB-R152, HMSO, London.

Saihs, S. A. (1974). 'The approach to radon problems in non-uranium mines in Sweden', *Proc. Third Int. Congress of the International Radiation Protection Assoc.*, Washington.

Schenk, H., and Riedel, E. (1984). 'Plutonium and uranium recycling in light water reactors', paper to *World Nuclear Fuel Market and 11th Ann. Meeting and Int. Conf. on Nuclear Energy*, Florence.

Slater, J.B. (1977). *An overview of the potential of the CANDU reactor as a thermal breeder*, AECL–5679, Atomic Energy of Canada Ltd., Chalk River, Ontario.

Shapiro, N. L., Rae, J. R., and Matzie, R. A. (1977). *Assessment of thorium fuel cycles in PWRs*, EPRI NP–359, Electrical Power Research Institute, Palo Alto, California.

United Nations (1982). *Ionizing Radiation: Sources and Biological Effects*, UN Scientific Committee on the Effects of Atomic Radiation; United Nations., New York.

Uranium Institute (1984), *Principles of Uranium Mill Tailings Isolation and Containment*, U. Inst., London.

Vanstrum, P. R., and Levin, S. A. (1977). 'New processes for uranium isotope separation', IAEA-CN–36/12, in *Proc. Int. Conf. on nuclear power and its fuel cycle*, IAEA, Vienna, Vol.3.

Walker, R. F. (1978). 'Experience with the Fort Saint Vrain reactor', *Ann. Nucl. Energy*, **5**, 337–356.

Nuclear Power: Policy and Prospects
Edited by P. M. S. Jones

5

Nuclear Energy for uses other than Electricity Generation

N. G. WORLEY
Watt Committee on Energy Limited

5.1 Introduction

The first practical application of nuclear energy, without which it might never have been made available for other uses, was as a military explosive. Explosives can have civil as well as military uses however, and the second section of this chapter will review briefly the present situation on peaceful explosions.

As indicated in section 5.3 nuclear fission is accompanied by the formation of radioactive isotopes and the release of neutrons and radiation. Separation of the radioactive fission products and the interaction of some of the neutrons with target materials to form new isotopes with special properties, are two of the main sources of radiochemicals which have many applications and great potential in medicine, non-destructive examination in manufacture, research, agriculture, space devices and food preservation. This is a vast field and the section of this chapter dealing with radiochemicals will of necessity be selective.

The bulk of the energy from nuclear reactions is released in the form of heat. The emphasis so far in using the nuclear based heat has been in the large scale generation of electricity, but nuclear heat can be used instead of fossil fuels for many other industrial and commercial requirements. The temperatures that can be achieved for application depend on reactor design, the form of the fuel and the coolant.

The temperatures achievable range from those required for district heating, at about 100 °C, up to those necessary for coal gasification or steel making processes, about 1000 °C. The use of by-product steam from combined heat and power systems and the various end uses categorized by temperature are considered separately in sections 5.5 to 5.9.

In the military area there has been considerable use of nuclear energy for ship propulsion and this is described in section 5.10. For submarines the long endurance of the fuel and the fact that no air is required by the nuclear engine has enabled these ships to remain submerged for long periods. Outside the role for military shipping the only real success has been in ice-breakers where the problems with land-locked harbours have been the incentive for the Soviet Union to build three ships and a fourth is under construction.

The non-electrical uses of nuclear energy are therefore very diverse and except for nuclear shipping they have not been the subject of public controversy.

5.2 Peaceful Nuclear Explosives

Military fission explosives based on plutonium and uranium–235 were developed first by the United States (1945), then USSR (1949), United Kingdom (1952), France (1960) and China (1964). In these countries the building and testing of fusion explosives based on isotopes of hydrogen followed some years after the first explosion of a fission device. Several other countries claim that they have nuclear explosive ability.

The Indian nuclear explosive development (1974) has demonstrated that devices can be produced without recourse to large reactor systems and costly development programmes.

Nuclear explosives developments can have peaceful as well as military applications. Some of these are similar to the existing uses of chemical explosives but new possibilities arise because of the enormously greater scale of energy release in compact nuclear devices.

Programmes of development of civil nuclear explosives were undertaken by the USA, USSR, Britain and France and these have been the subject of public conferences in 1964 and 1970. The Non-Proliferation Treaty of 1963 places an obligation on the nuclear weapon states to make available the benefits of the peaceful applications of nuclear explosives to non-weapon states.

However, the developments in the United States which had started in 1957 under the global title 'Plowshare' ended in 1973 and effectively there has been no practical work in the 'West' since then, although the Soviet Union has continued the development to the stage where it appears to have fully established procedures for many applications.

Nuclear explosions are rated in terms of the weight of the chemical explosive, TNT, having the same release of energy. Peaceful nuclear explosive devices have typically been in the range 1–40 kilotonnes of TNT. In their experimental programmes the Americans reported about ten experiments, the French thirteen in the Sahara and to date there appear to have been over 80 in the Soviet Union.

Possible applications described in the literature include:

1. Major civil engineering projects such as a new Panama Canal or diversion of north flowing Soviet rivers towards the south. This type of development was investigated by Schooner and Buggy in the 1960s.

   However, following the introduction of restrictions on surface or airborne nuclear explosions in 1963, there has been no further work on schemes of this kind.
2. Prospecting for oil, gas and mineral deposits by analysis of signals from nuclear explosions several kilometres beneath the surface. This is a more powerful development of a technique already used extensively by oil companies. Nuclear prospecting is employed in USSR and is playing an important role in building up information on oil and gas fields there.
3. Increasing the output and over-all recovery from existing oil and gas wells. Increases in output from US oil wells from 30 per cent to 300 per cent have been reported while figures nearer the bottom of this range appear to be typical for oil, with gas the increases in flow are higher. In addition the over-all recovery can be icreased from the 30 per cent which is normal in oil and gas wells to 60 per cent of the fuel in the field.

   The level of radiation induced in the fuel appears to be acceptably low.
4. Large underground storage caverns can be produced using nuclear explosives. Self-sealed caverns are formed in salt domes (unlike rock formations) and in salt beds have been used for natural gas storage in several locations in the Soviet Union. The high level of corrosive hydrogen sulphide makes underground storage more satisfactory and more economic than storage in surface tanks.
5. Nuclear explosives appear to have been used to extinguish fires in oil and gas wells in USSR.

5.3 Radio Chemicals

5.3.1 *Introduction*

Radiochemicals are unstable isotopes of

elements which change into stable forms by the emission of charged particles or alternatively by absorbing an electron into the nucleus. These changes may also be accompanied by the emission of electromagnetic radiation. Sensitive methods of detection and accurate measurement of radioisotopes lead to practical applications in engineering, non-destructive testing, process control, medicine, agriculture and research.

An enormous range of radiochemicals is available commercially. Currently the sales from Amersham International are of the order of £100 million per annum increasing at above 20 per cent per annum.

Basically there are three methods of making radiochemicals-neutron—bombardment of specimens in a nuclear reactor (the major source); separation of radioactive species from the fission products in spent fuel, and the bombardment of target materials with heavy charged particles in specially built accelerators.

The applications of radiochemicals are extensive and for the present purpose a short selection of topics has been made to indicate the scope.

5.3.2 *Medical Applications*

Radioisotopes are used widely in research, diagnosis and treatment. For example suitable radio-nucleides emitting γ-rays or positrons can be administered to a patient to locate a tumour, in which they are selectively absorbed, or to provide information on blood distribution. Over the past few years the use of radioisotopes with new computer analysis techniques has enabled the functioning of the heart to be analyzed and the preparation of three-dimensional maps of the distribution of the radioactive material in the body.

Radio-immunoassay is used to study hormones and other essential biochemicals in the body in minute quantities. The high sensitivity of detection methods for radiochemicals and the fact that both the radioactive and the non-active isotopes behave in the same way chemically is the basis of the technique. The development has two features, the use of materials labelled with a radioisotope which react specifically with the biochemical to be measured and the measurement of the small amounts involved outside the body. It plays a part in diagnosis of 10–20 per cent of the patients in hospitals in developed countries. The technique can cover diverse substances, hormones, drugs, vitamins and materials released by diseases and is now being extended to non-radiochemical tracers.

Radiation from external radioisotope sources can give a lethal concentration of radiation at a cancerous growth in the body with little damage to surrounding tissues. Alternatively tiny radiation sources can be inserted in the body or selectively absorbed radio isotopes may be used to give localized radiation fields for treatment (e.g. radioiodine for thyroid treatment).

5.3.3 *Industrial Applications*

Application of radioactive sources to replace heavy and expensive X-ray equipment in inspection of heavy components is widespread. Nucleonic gauges are used in product control e.g. paper and plastic thickness, leak detection, or fluid level measurement, or identification of the moving boundary between two fluids pumped along a pipeline in order to actuate a valve. Neutron techniques have been used to monitor in line, the calorific value and ash content of coal and many other applications have been developed. Industrial use of radiation techniques includes sterilization of needles used for injections, and of food packaging material, vulcanization of rubber, production of heat resistant wires and electrical insulation. West Germany and the USA have reported the use of Cobalt-60 or Caesium-137 for sterilization of sewage sludge so that it can be used as a soil conditioner.

There has been wide application of nuclear isotope technology for the remote measurement of the level of grout poured around piles set in the seabed for oil production platforms. Major development was needed to produce instrumentation robust enough for the hostile environment.

Radioisotope tracers have been used to measure the flow of rivers, to study the movement and deposition of sediments, and to detect and trace water seepage. Nuclear techniques are used in tracking rainfall and estimating subterranean water reserves. They have also been used to follow the direction of travel and extent of dispersion of the flue gases from the high stacks of Midland's power stations as part of a study of acid rain.

5.3.4 *Food and Agriculture*

Nuclear techniques have a major role to play in agriculture and food production and preservation:

1. By inducing mutants to improve the characteristics of cultivars. These developments have led to over 300 improved plant species developed since 1950 including grains, vegetables, fruit as well as many industrially useful and decorative plants. Apart from higher outputs, mutants have been produced which show high resistance to disease, ready adaptation to new locations and acceptance of high fertilizer levels.
2. Development of more efficient practices for biological nitrogen fixing by measuring uptake of isotopically labelled fertilizers.
3. Improvement in animal productivity and health, in many cases using radio immunoassay techniques to diagnose infections. Tracers contribute to studies on management and feeding practices and radiation sterilization in the production of vaccines for immunization against parasite diseases.
4. Pest control is important because insects cause losses of over 10 per cent to crops and livestock. A technique used for control of pests such as tsetse fly and medfly, which mate only once, in Africa and Mexico is the release of radiation sterilized males in the field, a method which could completely eliminate these pests from some areas.

   Radioisotopes can be used to investigate the feeding habits of insect pests and the factors that can reduce their attacks on the crops. Radioactive markers are useful in investigating the distribution of pesticides and the advantages of various spraying techniques.

   Radioisotopes can also be employed to kill insects and other pests in crops in storage.
5. Food storage can be improved by irradiation treatment which can inhibit sprouting (e.g. potatoes, onions and garlic), delay ripening or reduce the number of superficial micro-organisms which can cause food to deteriorate (fruit, fish and meat). The treatment can also control pathogenic organisms and parasites in food, reducing microbe contamination of spices and dry foods.

   These techniques are approved for potatoes, onions and spices in many countries while more limited clearance in a few countries covers fish (Netherlands), wheat and flour (Canada) and rice. Bangladesh, Chile, South Africa and the Netherlands have the widest range of food cleared for irradiation preservation. The technique is being assessed in the United Kingdom.

5.4 The Market for Nuclear Heat

Heat Production

In nuclear reactor power systems, in common with other thermal power plants generating electricity, less than half of the heat released in the fuel appears as electricity.

The system employed in most power stations is to use the source of heat to generate steam which is then passed to a steam turbine linked to an electrical generator. In passing through the steam turbine the steam pressure and temperature fall.

The low pressure steam at the turbine outlet is condensed by passing it across tubes through which cold water is pumped. The cooling water becomes tepid and its disposal often presents environmental problems. This 'condenser' heat loss is the major energy loss from the system. The condensed steam is then pumped back as feed water to the evaporator. For thermodynamic reasons it is advantageous to use a small proportion of the steam flowing through the turbine to heat the water pumped back to the boiler. This is '*regenerative feed water heating*'.

The amount of the electrical energy that can be generated is governed by the laws of thermodynamics and the economics and efficiency of the plant components but is between 30 and 40 per cent of the energy from the fuel. If the energy is used for heating purposes the efficiency of use will be much higher than this.

There are basically three ways of employing heat produced in nuclear reactors:

1. Use of the heat that is normally thrown away in the condenser cooling water. The main problems here are the low temperatures (15–25 °C) and the enormous quantity of low-grade energy available.
2. Parallel generation of electricity and process heat i.e. co-generation or combined heat and power. This is accomplished in one of two ways, either generating electricity and then using all of the medium pressure steam from a 'pass out' turbine for heating processes, or drawing off some of the steam from one or more tappings in the low pressure turbine in addition to the steam used for regenerative feed water heating.
3. Making direct use of all of the nuclear heat and employing hot water, steam or hot gas to transfer it to where it is to be used.

Table 5.1 gives an indication of the amounts of energy used in USA for various purposes. The industrial heat use as a function of temperature level is given in Table 5.2 with examples in Table 5.3. Nuclear reactor temperature levels are given in Table 5.4.

Table 5.1 Energy Use in USA

	x 10^{18} Joules 1983	Per cent
Electricity	26	35
Other industrial	18	24
Transport	20	27
Commercial and residential	10	14
Total	74	100

There has been an increasing interest, particularly with fossil fuels, in greater use of the waste energy from power stations (e.g. greenhouse heating at the coal power station at Drax in UK using some of the circulating water from the condensers) and in co-generation such as in the plant built for the East Midlands Electricity Board at Hereford. There are similar cases to these in the Nuclear field.

Table 5.2 Industrial heat use

Temperature level °C	Steam per cent of total	Direct heat per cent of total	Total per cent
175	28	1	29
175–260	23	1	24
260–400	4	5	9
400–540	1	5	6
540–925	–	12	12
925	—	20	20

Table 5.3 Typical Heat Use Temperatures

	°C
Fish farming	10–20
Greenhouse heating	20–40
District heating	
Commercial and domestic	
Low temperature	60–90
High temperature	120–150
Desalination	40–300
Oil recovery from tar sands	250–400
Catalytic reforming	400–600
Vinyl chlorine	500–650
Town gas	500–700
Ethylene	750–850
Olefins	600–900
Roasting iron ore	850–950
Ammonia synthesis gas	350–950
Hydrogen	600–1000
Bauxite roasting and lignite gasification	900–1150
Bituminous coal gasification	1050–1400
Cement Glass	800–1800

Table 5.4 Reactor Coolant Temperatures

Reactor type	Reactor coolant temperature range °C	Coolant
Swimming pool	60–90	Water
Organic cooled reactors	60–400	Organic
Low pressure 'safe' water reactor	120–230	Water
Candu	200–250	Heavy water
PWR/BWR	280–330	Water/steam
Magnox	150–400	Carbon dioxide
AGR	350–650	Carbon dioxide
HTGCR	350–750–950	Helium
Fast breeder reactor	350–600	Sodium

5.4.2 *Energy Parks*

Nuclear plants are generally most economic in large sizes and plants of up to 1500 MWe are in use with a total heat release from the fuel of about 4800MWt. There are no individual industrial processes that require this quantity of energy and one solution to the problem of using the large quantity of thermal energy would be to associate a nuclear plant with an industrial park with many medium size industrial users. This has the additional advantage that there is likely to be a wide spread of uses and a more even distribution of load over the day than with a single user. Furthermore since the economic and practical life of a nuclear plant is at least 20 years, and power utilities in the United States are even considering a life of 50 years for future nuclear water cooled reactors, most industrial plants would require replacement, major modifications or complete changes several times during the reactor lifetime. Thus effective and satisfactory use of the energy from a nuclear source would best be achieved with a multiplicity of users.

With most forms of direct use of heat the area of use has to be fairly close to the nuclear heat source. The greater the distance the hot water, steam or gas has to travel the greater the heat and the pumping (or pressure) energy losses. In most cases nuclear source heating to sites more than 30 km from a reactor is unlikely to be economic. However, a study for the Loviesa station in Finland indicated that an 80 km transmission line would provide heat energy competitive with a locally sited station.

5.4.3. *Technical Factors*

The reactor coolant cannot be used directly for distributing heat to the load because it becomes radioactive. There is therefore a need for an intermediate heat exchanger and a secondary circuit as is customary in most electricity generating reactor types. This inevitably leads to losses in the temperature levels available for distribution which is important with atmospheric pressure reactors or with high temperature applications where the application is sensitive to the temperature levels. The temperature difference between the reactor coolant and the secondary circuit can be reduced by increasing the size of the intermediate heat exchanger but there are often space and cost limitations. The secondary coolant pressure is normally higher than the reactor coolant so that in the case of a leak between the circuits in the heat exchanger the distributed coolant cannot become contaminated with radioactive primary coolant.

A domestic and commercial heat load will vary between day and night and also between the seasons. Industrial loads may also vary and this may make the economics of nuclear heat unsatisfactory and control difficult for individual applications. Many processes cannot accept a break in the energy supply and these will require several reactors or standby fossil plant which can be brought in rapidly when a reactor trips. Spare or standby plant can be costly.

Most nuclear fuels and components have not been designed to accept frequent and rapid changes in output. There can also be problems in attaining full output when required if the neutron absorbing fission product xenon build-up affects the core reactivity and there is not adequate reactivity in the fuel. This may be the case with small reactors when the fuel is near the end of its active life and the normal procedure is to refuel by replacing the complete core. In some cases the load could be maintained by burn up of burnable poisons, adjustment of soluble poison levels or reducing the temperature in the case of water reactors.

5.4.4 *Electrical Heating*

One of the ways in which nuclear energy can meet some of the low and high temperature energy demands is by electricity, as nuclear generation becomes a large proportion of the total installed capacity. Whatever the temperature levels of the reactor coolant, the electricity produced can generate high temperatures for special applications e.g. electric arc furnaces for making special steels. These uses can be remote from the nuclear generating stations.

5.4.5 *Other Factors*

In Britain the function of the electricity boards fixed by Parliament is to generate electricity economically and the interest in co-generation or supply of steam or heat for industry or commercial heating from either fossil or nuclear plant has been low. It is interesting to note that, in Canada, a special Act was passed in 1981 to extend the mandate for Ontario Hydro, the largest electrical utility in Canada, to the supply of 'Heat Energy' in addition to electricity. This has led to the active marketing of nuclear heat for commercial and industrial purposes there.

5.5 Combined Heat and Power (CHP)

Using a central steam supply from a nuclear source, particularly if it is linked to a bled steam system with a number of tappings from a turbine driving an electricity generator, allows a wide degree of flexibility of steam pressure level and the possibility of increasing electricity generation when the factory or district heat load drops.

Ontario Hydro in Canada is actively marketing steam heat uses from its nuclear plants at Bruce, Darlington and Pickering as well as from fossil fuelled stations. The advantages quoted for nuclear heat sources for industrial customers are:

1. Lower steam or heating costs than with oil or gas, but site dependent when compared with coal.
2. Saving in capital expenditure for steam raising equipment necessary in individual plants.
3. Firm prices for long-term contracts—predictable energy prices.
4. Reliability of supply as there are several reactors at the power station and there would also be stand-by fossil generation.
5. Because the fuel and operating costs are only a small proportion of the energy cost, nuclear heat (and electricity) is likely to increase in cost with time less than fossil fuels. This is important during periods of high inflation.
6. In Switzerland, where atmospheric pollution is important, particularly in the valleys where most of the potential heat loads are situated, nuclear heat sources are often more acceptable than fossil fuelled alternatives.

At Bruce, Canada, the nuclear station was designed from the outset to produce steam for the production of heavy water as well as to generate electricity. However, the anticipated expansion in the need for heavy water for future nuclear plants is no longer required because the projected rise in electricity demand has not occurred. There is therefore an excess of steam production and a considerable marketing effort for steam sales. Heat for a greenhouse and for fish breeding is already being supplied on a small scale.

In the United Kingdom the Calder Hall Magnox power station in addition to generating electricity also supplies process steam to the neighbouring nuclear fuel processing plant at Sellafield.

In Switzerland there are two operating combined heat and power nuclear plants. These demonstrate the ease of slow and steady extension of steam usage with bled steam. At Gosgen-Daniken the KWU 920 MWe PWR completed in 1979 supplies 1200 t day^{-1} of process steam at 220 °C (50 MWt) to a board and cotton mill with an electricity generation loss of 10 MWe. There are plans to extend steam use to district heating.

At Beznau, built on an island site, there are two Westinghouse PWR units of 350 MWe completed in 1969 and 1972. District heating is being installed in stages to an area within a 13 km radius of the power station. Although all of the users will not be connected to the system for some time the first stage was completed during 1985. The bled steam from the turbine is at 2.6 bar and 128 °C condensing and delivering heat to the water district heating circuit at 16 bar at a temperature in winter of 110 °C and 75 °C in summer. The full extension to the system is not intended to be finished until 1990 when the reactors will be about 20 years old.

Since 1963 the Stade nuclear station in West Germany has been supplying 60 t of steam per hour at 270 °C and 8 bar to a factory 1.5 km away.

In the USSR all nuclear turbines have tappings for supplying district heating water circuits operating at design temperatures of 150 °C and a return temperature between 35 °C and 70 °C.

An interesting case of co-generation is the Soviet 120 MWe sodium cooled fast reactor at Shevtshenko. The steam from one of the three circuits is used to supply heat for the production of 50 000 t of fresh water per day in a desalination plant.

The 60 MWt organic cooled heavy water moderated pressure tube reactor plant at Whiteshell in Canada produces no electricity but some of the steam generated (15 MWt) is used to heat the station buildings in the winter. The rest of the heat is rejected to a river. This plant has been operating since 1966. Coolant outlet temperatures of 400 °C can be achieved with the organic coolants and these are significantly higher than those available from water cooled reactors.

5.6 Reactors for Low Temperature Heat

Power reactors produce inconveniently large quantities of low temperature heat and the co-generation types of application use only a proportion of the heat energy available. Nuclear reactors in general only become fully economic in large sizes and the most numerous water reactors are only acceptable to the public when they are sited remotely, which is not an ideal situation for their use as a heat source.

In spite of this, there has been considerable interest, particularly in countries which have a long cold winter and have remote isolated communities, in reactors which operate at atmospheric or relatively low pressures, have a modest heat rating, and have very low radioactive releases to the atmosphere. These reactors would be much lower in heat output than power reactors and have an output from a few megawatts to a few hundreds of kilowatts of heat. Their proponents consider that they are likely to be so safe that they could be publicly acceptable in almost any location.

During the 1960s about 150 swimming pool reactors were ordered and built worldwide, principally in universities. Many of these have since been shut down as they are costly for universities to run, needing constant attention and manning, and the range of experiments for which they are suited is limited. The Canadian swimming pool was called Slowpoke and this type of reactor was designed to be self limiting in output and able to be left in operation without constant attendance. Its output was small but was relatively inexpensive and very reliable. The Canadians have now designed a small heating reactor based on their successful experience with 6 research swimming pool reactors. (Table 5.5).

Much of the technology of the research reactors has been retained, the main changes being uprating from a few kilowatts to 2 MW, allowing local boiling in the core, using low enriched uranium rather than 93 per cent ^{235}U

Table 5.5 Small Reactor Parameters

		SLOWPOKE	TRIGA
Output	MWt	2	10–53
Circulation		Natural	Forced
Pressure		Atmospheric	0.7 bar(gauge)
Fuel		UO_2 pellets	UZrH rods
Enrichment %		5	19.9
Control		Reflector	Absorber and reflector
Size diameter m.		4	2.4–2.7
Coolant temperature °C			
Reactor outlet		90	142
Return		—	102

as oxide in rods rather than plates, control by neutron reflector rather than by absorbers.

Small and portable heating reactors built from sections which can be transported by air or lorry were developed early in the Soviet Union. A transportable 1.5 MWt system (TES3) was available in 1961 and this could be moved by caterpillar vehicles from one site to another. SEVER–2 replaced this and further experimental reactors employing boiling water or organic cooling (ARBUS) have led to the concept a 'nuclear boiler house' (ATU15). Organic coolants have the advantage that temperatures in excess of 100 °C can be achieved at atmospheric pressure, with extremely low activity levels in the coolant and the possibility of using carbon steel in the reactor circuit. There are many small communities in remote regions of the Soviet Union where a reactor heating system with its infrequent fuel changes must be attractive.

Elsewhere about 50 TRIGA atmospheric research reactors have been built and the largest of these is a 15 MWt reactor operating in Romania. Studies are in hand at General Atomics in the United States on designs of outputs up to 53 MWt. Many of the features of the research reactors are being maintained for the larger units such as the use of fuel in the form of uranium-zirconium hydride (UZrH) which contributes a large prompt negative temperature coefficient to the core, and natural circulation cooling. Control is by both beryllium reflector and cruciform boron carbide plates.

The main change from the research reactor is the pressurization of the primary circuit to 0.7 bar gauge. There is a double tank system with the inner tank containing the hot water and the core surrounded by a second tank of cooled water which also acts as a source of coolant if the primary circuit fails. Details are given in Table 5.5.

There are three water circuits, the primary which conveys the heat from the core to an intermediate circuit which then transfers the heat to the district or industrial heating circuit. A water temperature to the district heating system of about 130 °C appears to be achievable. Preliminary estimates indicate a cost of heat lower than can be achieved using oil or gas in the United States.

5.7 Pressurized Water Reactor Systems

There are a number of water reactor systems which can supply a few hundreds of MWt at temperature levels suitable for district heating or low temperature and pressure steam. In fact any of the standard PWR systems have versions in sizes of 500–1700 MWt which could be used for heating purposes but these are too large for most applications.

In France the THERMOS reactor has been designed in 100 MWt and 200 MWt versions (Table 5.6). The reactor is a pressure vessel system installed at the bottom of a pond filled with ambient water and surrounded by concrete shielding. The primary heat exchanger is above the core while the control rods are inserted from below.

Table 5.6 Low Temperature Heat Reactors

		Thermos	Secure LH	AST
Output	MWt	100–200	200–400	300 and 500
Circulation		forced 3 or 4 loop	forced	natural
Reactor pressure MP(a)		1.2	0.7	2
Fuel		UO_2 plates	UO_2 plates	UO_2 pellets
clad		zircaloy	zircaloy	zircaloy
Vessel		steel	prestressed concrete	steel
diameter m		4.7	13.4	—
height m		10.8–11.3	37	—
Coolant temperature				
core outlet °C		144	120	200
return °C		125–131	90	120

The normal temperatures employed for district heating in France has the supply at 160 °C with the return at 80 °C. The THERMOS system would give temperatures lower than this. However, the geothermal district heat systems in the Paris region operate at 50–75 °C so the THERMOS level of 130 °C is about midway between the high and low figures. The number of potential sites for THERMOS in France is however, somewhat limited and so far it has been unsuccessfully considered for Grenoble, where a refuse heating plant was preferred, and the nuclear research establishment at Saclay where total costs were too high.

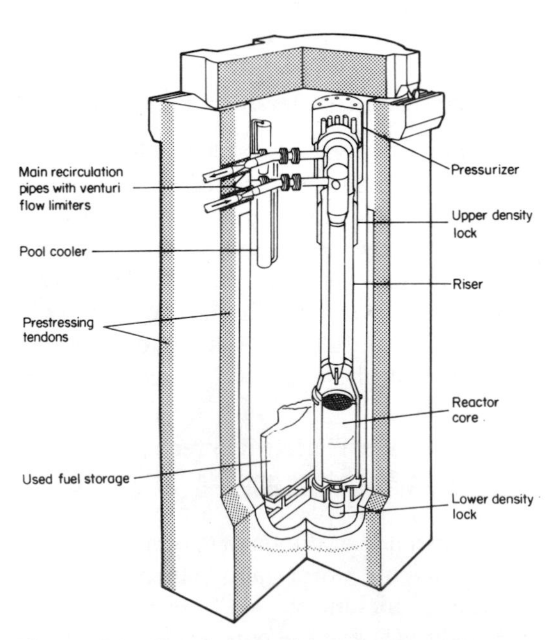

Figure 5.1 Heating version of secure reactor. Source: ASEA Atom

The larger CAS system which has an output of 300–1000 MWt has been operating as a combined heat and power plant at the sodium cooled fast reactor research station at Cadarache but is unlikely to find application for heating only.

The SECURE series of reactors developed in Sweden (Figure 5.1) have three projected modes of operation. The power version (called PIUS in the United States) is intended to be an ultra safe, environmentally acceptable plant of a total output of 500–600 MWe with several modules. The same principles are employed in the low heat (LH) versions suitable for district heating with a water temperature of about 100 °C and a medium temperature heat (H) version with a temperature of about 190 °C. Both of these are suitable for district heating and the higher temperature version could also supply low pressure steam heating. The main parameters of two sizes of the low temperature scheme are given in Table 5.6.

Considerable attention has been paid to the inherent safety of the system with the primary circuit and core inserted in a tank containing borated water which is drawn into the core region in the event of a circulating pump failure. The pressure vessel is of pre-stressed concrete. The wide ranging safety features and the low environmental impact have been specially developed in an attempt to make the system acceptable in Sweden where there is well organized and powerful opposition to nuclear power employing the traditional water reactor systems. At this stage there is no indication that SECURE units are yet acceptable in Sweden or elsewhere.

The Soviet Union is building systems based on water reactor technology in the cities of Gorky and Voronezh. (See Figure 5.2).

In Gorky the two 500 MWt units will provide about half of the district heating load for the city replacing oil and an inefficient boiler plant. Large open hot water storage tanks are used to level out the peaks and troughs of the daily heat load. The nuclear heat sources are relatively low pressure and the radiation levels and safety are considered to be suitable for siting as close as 2 km from the limits of the future built up areas of cities and 10 km from the area of use. Table 5.6 gives the main parameters of these reactors. The fuel and core design are low output versions of PWR Soviet technology. Similar systems are being considered for other Eastern European countries.

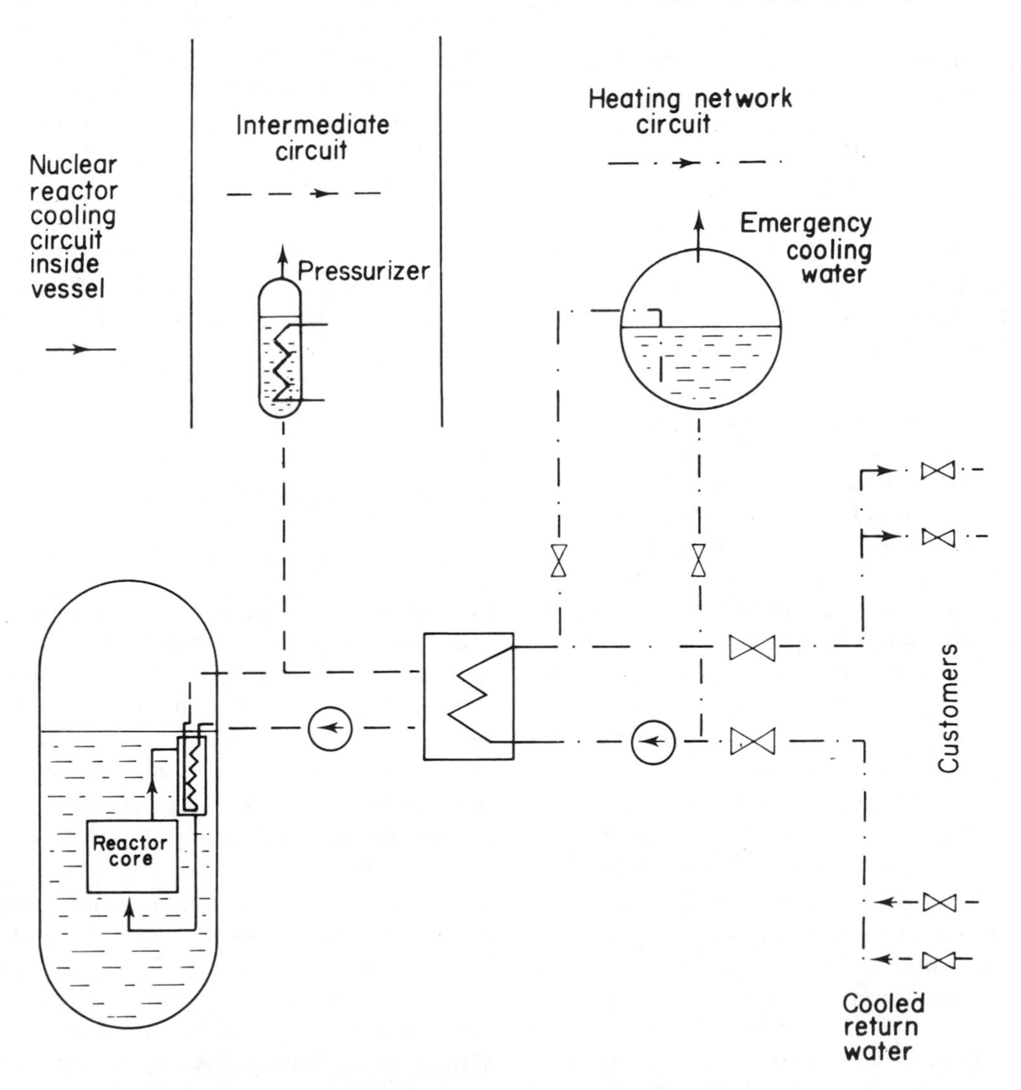

Figure 5.2 Nuclear district heating (e.g. Gorky and Voronezh, USSR)

5.8 Medium Temperature Reactor Applications

There are large deposits of high viscosity heavy oil and bitumen in the world which can be recovered if a copious supply of medium temperature high pressure steam can be made available. The steam would be required to emulsify and heat the oil and push it to the surface where it can be treated and upgraded. The oil sands of Alberta, Canada and deposits in Venezuela and Colombia are examples of heavy oils of this type.

The Canadians have studied the relative merits of coal or gas-fired steam raising boilers and several types of nuclear reactor for the recovery of oil from the sands in Alberta. As might be expected, the capital cost of fossil plants is much lower than nuclear but there is a potential steam cost saving using nuclear steam source.

The use of nuclear energy is potentially viable because the heat available in the oil recovered is about three times the heat supplied by the reactor. The nuclear reactor can also supply steam for electricity generation.

CANDU reactors operate at relatively low temperatures and pressures (280 °C and 40 bar) and consequently are only suitable where the oil is relatively light and does not require high pressures and temperatures for recovery. For heavier, more viscous oils, steam compressors are required and the system efficiency is then poor (about 50 per cent) and the economics and the problems normally associated with large scale steam compression make such a scheme technically and economically unattractive.

The PWR would be somewhat better as it can supply steam at 300 °C and 65 bar but even these levels are not always adequate.

A developed form of the organic cooled heavy water reactor which has been operating at Whiteshell in Manitoba can supply steam at the appropriate temperature and pressure 350–400 °C and 150 bar. An alternative would be a gas-cooled reactor, AGR or perhaps Magnox, although the latter would only supply a proportion of the steam at the required pressure. The National Nuclear Corporation in the United Kingdom investigated in 1982 the potential for using Magnox reactors, based on the plant installed at Oldbury on Severn, for recovering heavy oil in Venezuela.

The availability of the high temperature and pressure steam makes the system practical for recovery of oil from tar sands.

The HTGCR conditions would enable very long (100 km) pipelines to be practical and if 950 °C heat is available, could supply all the needs of a refinery too.

Steam at 400 °C and 60 bar can be used extensively in an oil refinery and a large refinery requires a reliable source of heat and electricity equivalent to 2000 MWt. A group of modular gas reactors would appear to be able to meet these requirements. A study on these lines was carried out in 1984 for the Port Arthur refinery on the Gulf Coast of USA.

The interest in such schemes as these has declined with the copious supply for the market of more easily recovered oil supplies. Steam is employed to increase the level of recovery from oil wells but there has been no suggestion of using nuclear heat for this.

5.9 High Temperature Heat

5.9.1 *Gas Reactors*

In the United Kingdom the Magnox reactors had temperature limits associated with the fuel and the cladding (Chapter 3) and a maximum coolant temperature of less than 400 °C. Retaining the carbon dioxide coolant but changing over to oxide fuel and austenitic steel cladding allows an increase in the coolant temperature outlet to about 650 °C in the AGR.

It has long been felt that cycles using a medium like steam–water, particularly where the pressure of the water was several times that of the reactor coolant, are basically unsatisfactory. However the steam–water system uses a Rankine cycle with the medium compression occurring in the liquid phase requiring relatively little power. The alternative gas turbine has to recompress the medium in the gaseous phase and this absorbs a lot of the energy generated. For this reason a gas-turbine cycle will only have a higher over-all cycle efficiency than a steam cycle with inlet gas temperatures in excess of 850 °C—well above the capability of the AGR.

The development of the high temperature gas-cooled reactor (HTGCR) had therefore as its prime target, after proving the system with conventional steam cycles, the achievement of a direct cycle gas turbine system.

The main features of the HTGCR system, which was first demonstrated in the Dragon plant at Winfrith (UK), is the use of the inert coolant gas helium and the development of a fuel in which the fissile material is integrated with the graphite moderator either in hexagonal fuel elements or as balls in the pebble bed system. (See Chapter 3).

As a power system generating steam for steam turbines the small 15 MWe AVR has been running at Julich in West Germany since the end of 1967. Fort St Vrain in the USA has been operating since 1976 and the 300 MWe THTR (using Thorium as a breeding material in the fuel) is being commissioned at Hamm-Uentrop in North-Rhine Westphalia. The main parameters of the systems are given in Table 5.7 together with details of the projected plant and schemes for small unit and/or modular concepts.

The 500 MWe scheme being designed in Germany is intended to be a joint electricity process steam plant. The 100 MWe scheme is intended either as a small unit suitable for countries with small networks or, if used for process heat, as a multiple reactor installation giving the degree of reliability of heat essential for many high temperature processes. There are schemes of this type using the spherical fuel elements under development in USA and Germany.

As can be seen in Table 5.7, although the AVR has operated for many years at a core exit temperature of 950 °C (suitable for gas turbine power generation or for use in reforming, steel making and solid fuel gasification) all of the immediate developments are designed for steam generation with a core gas exit temperature of 700 °C or slightly higher. This is only 50 °C in excess of current AGR designs in the United Kingdom. The first heat schemes for the HTR are therefore likely to be for combined heat and power using subcritical steam and reactor gas outlet temperatures up to about 725 °C.

The systems which can be considered are:

1. Electricity generation with process or heating steam tapped from turbine.
2. Electricity generation in parallel with topping turbines passing steam to process.

Table 5.7 High temperature gas-cooled reactors

	AVR Julich	Fort St Vrain (design)	THTR Hamm	500 MWe	100 MWe
Core power MWt	46	842	750	1265	256
Gross electrical Power MWe	15	342	300	500	100
Gas pressure bar	10.8	50	39	50	70
Gas temperature:					
Core outlet °C	950	785	750	723	700
Core inlet °C	275	405	250	266	253
Steam:					
Pressure bar	72	170	177.5	180	190
Temperature °C	500	540	530	525	530
Reheat:					
Pressure		42	46		
Temperature		540	530		
Fuel	spherical	hexagonal	spherical	spherical	spherical

3. Heat and steam for recovery of oil from deposits in oil shale, tar sands, etc., with use of steam and 'syngas' in refining.
4. Steam, heat and 'syngas' in coal liquefaction or gasification and subsequent refining.
5. Steel making processes
6. Other metal extraction e.g. zinc or aluminium.

The first two of these schemes requires little comment. It is worth noting that with the power station levels of steam temperature the effect of tapping off steam for process use has a considerably smaller penalty on electricity generation than with the lower pressure water reactors.

It is not, of course, possible to describe all of these types of process and some of them have extremely complex heat and material flow charts. They have been the subject of many papers presented at conferences and meetings. The principal references are given at the end of the chapter.

There are, however, a number of general points that apply to all schemes.

For safety reasons and because the nuclear reactor system has to be kept clean to avoid corrosion and the build-up of radioactive material in the gases, an intermediate circuit would normally be used with helium usually at a pressure higher than the nuclear core coolant. The steam generator as well as the process vessel should be in the intermediate rather than the primary circuit because if the boiler is in the reactor circuit the inevitable steam and water leaks are likely to be amongst the most frequent causes of nuclear plant shut-down with these high temperatures.

With processes requiring very high temperatures, the gas from the process will still be at a temperature which is too high for the nuclear core inlet or the gas circulator. In all of the processes steam generators will be installed in series with the process itself. Reactor core inlet temperatures will be about 250 °C (see Table 5.7). The steam would be used to generate electricity and supply heat or steam to the plant. These systems are therefore combined heat and power plants in which the electricity and steam are produced from what is in effect a 'waste-heat' boiler. The proportion of the total heat which goes to the process and steam generation will depend on the process but Figure 5.3 gives an indication of the proportions for coal gasification with a gas outlet temperature of 800 °C.

Because of its complexity and the probable use of several nuclear reactors and process plants in parallel the control of the whole system is likely to be complex.

5.9.2 *Potential applications*

Four potential applications of high temperature heat are reviewed below:

5.9.2.1 Methane Reforming ('syngas') The high temperature reaction of methane with steam gives a gas containing hydrogen, and the two oxides of carbon. The reaction is carried out at high temperatures in the presence of a catalyst. To obtain a sufficiently fast reaction, temperatures in excess of 700 °C are necessary. An input of heat is required because the principal reaction absorbs energy. Because the volume of the product is greater than that of the reactants the process

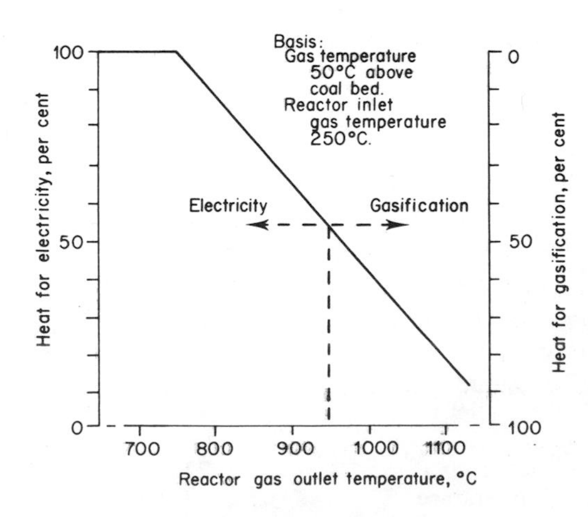

Figure 5.3 Relationship between heat for gasification and electricity production as a function of reactor outlet temperature

operates best at relatively low pressures, typically 30 bar.
The main reactions are:

$$CH_4 + H_2O \rightarrow CO + 3H_2 \quad (5.1)$$
$$CO + H_2O \rightarrow CO_2 + H_2 \quad (5.2)$$

Methane is shown in the gas reaction (5.1) but in practice mixtures of volatile saturated paraffin hydrocarbons can be used as feed.

With a high proportion of steam in the feed a high yield of hydrogen gas can be produced by scrubbing out the carbon dioxide and condensing the steam.

The gaseous product is useful in a wide range of processes; some of which are also suitable users of nuclear sources of heat. A selection of the follow-on processes are:

(a) *Hydrogasification* This yields high calorific value gaseous fuels from coal or lignite:

$$C + 2H_2 \rightarrow CH_4 \quad (5.3)$$

This process requires 800 °C and operates best at high pressures, typically 50–100 bar with a reformer gas product with a high concentration of hydrogen. To supply this the feed materials to the reformer should be in the proportion of between three and four parts of steam to one of methane.

(b) *Direct Reduction of Iron Ore* With a high hydrogen gas the reaction:

$$Fe_2O_3 + 3H_2 \rightarrow 2Fe + 3H_2O \quad (5.4)$$

requires a temperature of 500 °C. The process operates at the same pressure as the reformer and requires a high steam to methane ratio.

The other process requires a temperature of 800 °C but involves both the hydrogen reduction of operation (5.4) and carbon monoxide reduction:

$$Fe_2O_3 + 3CO \rightarrow 2Fe + 3CO_2 \quad (5.5)$$

This would be a low pressure process at about 3 bar.

(c) *Methanol Synthesis* This is a low temperature process at about 300 °C with a high pressure of about 100 bar and a gas with a somewhat higher level of hydrogen than carbon monoxide:

$$CO + 2H_2 \rightarrow CH_3OH \quad (5.6)$$

(d) *Hydrocracking* This is an important process in building up high octane fuel. The feed to the reformer can be any of the volatile hydrocarbons. The process generates unsaturated olefine chains by splitting heavy oils into radicals at moderate temperatures and high pressures. The process requires a high hydrogen gas from the reformer.

(e) *Ammonia Synthesis* A very high hydrogen gas reacts with nitrogen gas in the Haber process at 400 °C and very high pressures (500 bar) to give ammonia gas, a source of nitrogenous fertilizers:

$$N_2 + 3H_2 \rightarrow 2NH_3 \quad (5.7)$$

(f) *Coal Hydrogenation* Here a high hydrogen product gas reacts with coal to give a product which can be used as a basis for transport fuel. The process takes place at a low temperature (500 °C) and a much higher pressure than hydrogasification. There are in fact a wide variety of coal gasification processes.

A novel proposed use of the 'syngas' is to distribute it to energy users at ambient temperature where the reaction can be reversed to release heat which can be used locally before returning the steam and methane to the reformers.

This 'closed cycle' is an alternative to distribution of the hot gas from the nuclear reactor. Heat recovery from the gas from the reformer would be by generating steam some of which would be used in the reformer and some for electricity generation.

Steam reforming of gaseous hydrocarbons for several of these processes is a fully established technology. The fuel used is normally natural gas and the outline of the process is shown in Figure 5.4. The temperatures and

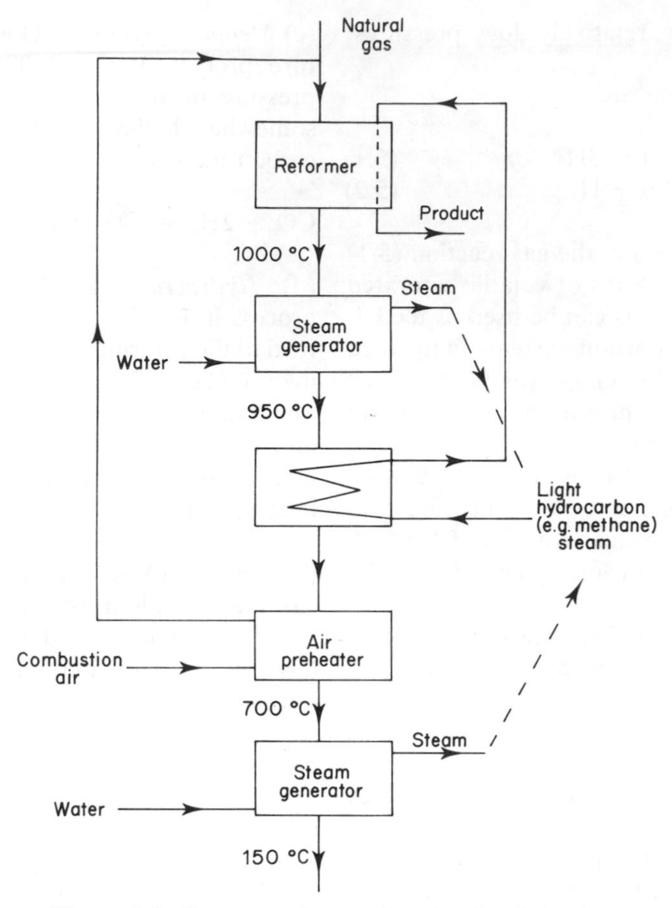

Figure 5.4 Steam–methane reforming–fossil heating

flow arrangements will vary from plant to plant.

In a nuclear heated plant the process could be as shown in Figure 5.5.

The advantage of using nuclear heat for the process is that it avoids employing a high grade fuel. Figure 5.5 shows the use of reactor heat directly in the reformer. Problems which could arise in such an installation have been investigated at Julich. On hydrogen and tritium diffusion through the tube walls, it was found that the steam used in the reforming gave a protective layer which reduced the diffusion of tritium into the 'syngas' and hydrogen to the nuclear coolant to acceptable levels. However any practical arrangement with an intermediate helium circuit would eliminate these potential problems.

The nuclear reformer design operates at lower and more uniform temperatures than with conventional combustion and provides comparable heat fluxes and total heating surface because the high pressure helium gives a higher heat transfer coefficient than the atmospheric pressure combustion gases.

An oil refinery with reforming and 'cat-cracking' uses large quantities of heat, steam and hydrogen. The amount of hydrogen required increases as heavier oils are used. A refinery processing 350 000 barrels of crude oil per day would require, allowing a 10 per cent margin, about 2400 MWt of nuclear reactor heat, probably supplied by several units for security of supply, with 85 MWe electricity and 1450 MWt of steam.

5.9.3 *Gas Production from Solid Fuels*

With shortages of liquid and gaseous fuels

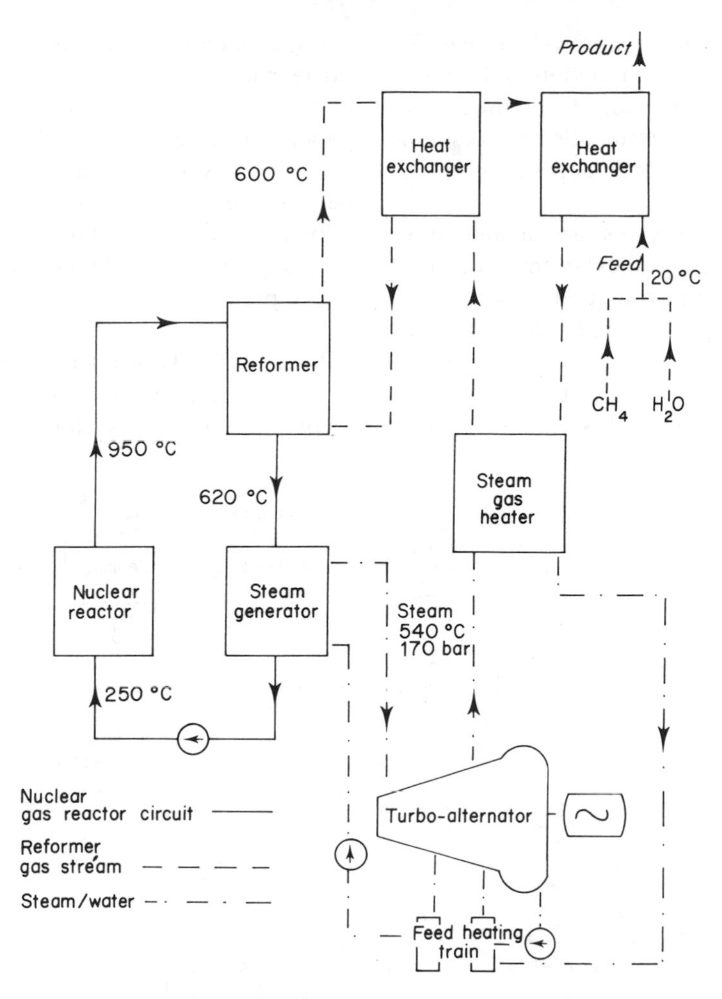

Figure 5.5 Nuclear steam–methane reforming

forecast for the long term and apparently enormous resources of solid fuels—coal and lignite—the use of nuclear heat for coal gasification and liquefaction has been studied. The aim of the investigations is to save the fossil fuel that would normally be burned to keep the process going. So, as in many nuclear applications, the considerable extra capital cost of the nuclear source, including a secondary circuit, has to be balanced against the saving in fossil-fuel costs. Most processes would have some electricity generation by a boiler and steam cycle. The most attractive coals from the processing point of view i.e. low-grade lignites are those that can be gasified at the lowest temperatures. However because these are relatively cheap, the fuel-cost savings to be balanced against the high cost of the HTGCR would also be low. In general lignites have to be used close to the mines because their calorific value is low and they are costly to transport.

In gasification the use of a limited amount of air yields a low calorific value 'producer gas' which contains, principally, carbon monoxide and dioxide and nitrogen, with small amounts of hydrocarbons. External heat is not required as the main reactions are exothermic.

If steam or another hydrogen-compound is added to the air the reaction produces hydrogen as well as carbon monoxide, by reaction directly with the solid material (as 'char') and also in the gas phase. This gas has a higher calorific value than producer gas. To get a reasonable rate of reaction and, hence, an economic size of plant, the temperature of reaction for these gasification processes is typically 800 °C or above, depending on the type of coal.

Several basic processes are available to deal with the main problems: ensuring good contact between the coal and the gasifying media in a way that provides a continuous feed of coal, and removal of the incombustible residue. The details of these processes are outside the scope of this review.

Experimental work connected with use of nuclear heat for gasification has been mainly in West Germany, although there have been paper studies elsewhere. The favoured type of gasifier using nuclear heat is the fluidized bed. There are two schemes, and for each a pilot plant has been built and operated for several years. Both of these processes operate under pressure.

5.9.3.1 Coal Gasification Experiment The HTGCR nuclear heat is transferred to a secondary helium system with an outlet

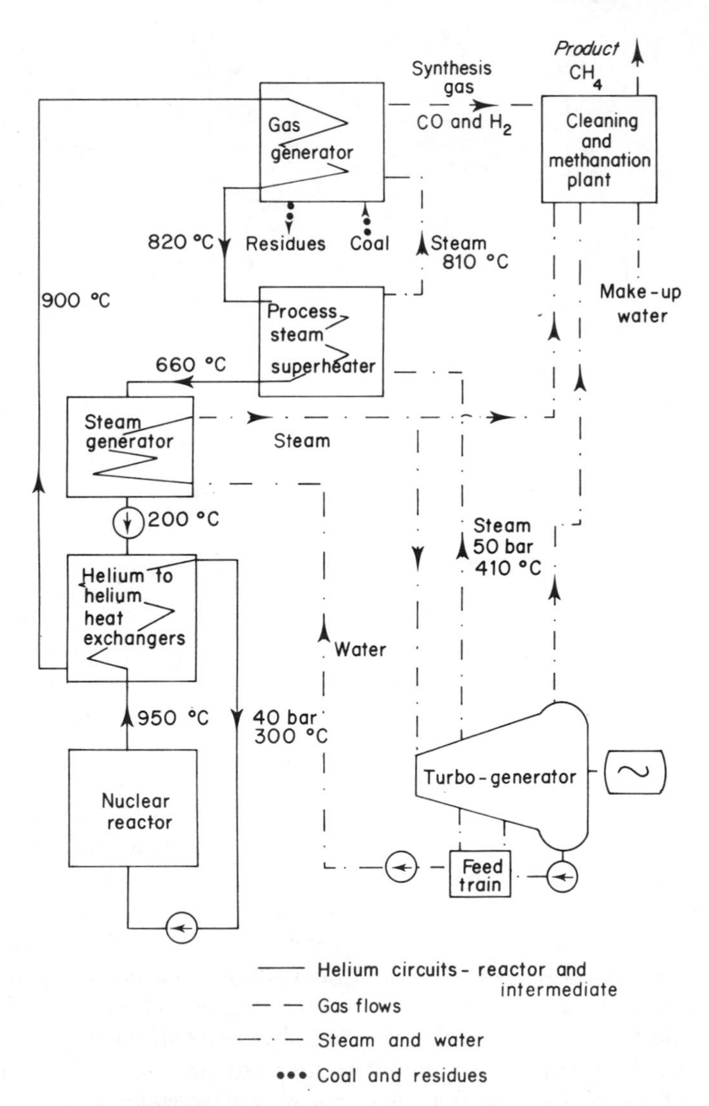

Figure 5.6 Steam gasification of coal using nuclear heat. — Helium circuit—reactor and intermediate --- Gas flows. -·-·- Steam and water. . . . . . . Coal and residues

temperature of 900 °C. After passing through the fluidized-bed heating tubes, the gases produce superheated steam in a boiler. Some of the steam from the steam generator is used both to fluidize the coal and react with it, using some of the heat from the secondary helium circuit. The rest of the steam drives a turbine to generate electricity.

A small experimental pilot plant using electricity to heat the helium has been operating since 1976. Programmes of development are associated with heat-exchanger tube material both for the intermediate heat exchanger and also for the arduous conditions in the fluidized bed. Data on reaction rates are also being obtained.

The strategy is to build a larger pilot plant (30 th^{-1} coal throughput) before tackling the more difficult task of designing a nuclear reactor system.

The process is shown in diagrammatic form in Figure 5.6.

Results suggest that, to obtain satisfactory reaction rates with some coals, the addition of a catalytic material such as potassium carbonate is necessary.

Experience with the operation of the pilot plant has given an indication of the probable emissions. Sulphur appears in the gases in the form of the easily removable hydrogen sulphide, all tars are destroyed in the bed and dust emissions are restricted to those in the coal handling plant.

The economics of the process have also been studied—the use of nuclear heat reduces the gasification cost by 20–30 per cent as compared to the use of coal combustion. However the current cost of natural gas from the European grid is lower than the calculated gasification figure.

The programme is being carried out by Bergbau-Forshung GmbH in Essen.

5.9.3.2 Hydrogasification The target product of this process is a gas with a high methane content. The basis of the development has been the fossil-heated pilot plants in Australia and the USA. The active agents are hydrogen (produced in the plant) and steam. This process is intended to use brown coal (lignite) and to generate a 'substitute natural gas' (SNG) to augment or replace natural gas supplies. A pilot plant has been operating and the timescale and future plans are similar to those of the gasification scheme.

The main pilot plant study covers the investigation of the gasification process using steam and hydrogen. The pilot plant is operated by Rheinische Braunkohlenwerke, Cologne. The two forms of the over-all process proposed are shown in Figures 5.7 and 5.8, one uses a methane-steam reformer using some of the methane product as feed while the other has carbon monoxide mixed with hydrogen and steam in the gasifier.

The work on these schemes for gasification and possible larger scale operation continue. Both the reactor and the gasification processes will be established by the early 1990s. Whether they will find practical application depends, amongst other issues, on long term availability of natural gas and its price level.

5.9.4 *Steel Making*

In the 1970s with the concern over energy resources and the cost of energy, processes like steelmaking, where enormous quantities of energy are required, were the subject of studies to see where nuclear energy could be employed and whether it could reduce costs as well as conserve fossil fuels. Steel works, like oil refineries, use heat, steam and electricity and so there is scope for nuclear plant.

The nuclear system can either be integrated with the steel works or remote from it. The Japanese have preferred an integrated system while European studies preferred having the nuclear plant away from the steel works and to supply heat, electricity and 'syngas' to other plants as well as the steel works. The development of smaller reactors may favour adoption of an integrated scheme as a number of reactors could be then associated with each steel works to ensure reliability of supplies to the steelmaking plant.

The normal coke fired blast furnace process followed by basic oxygen conversion to steel

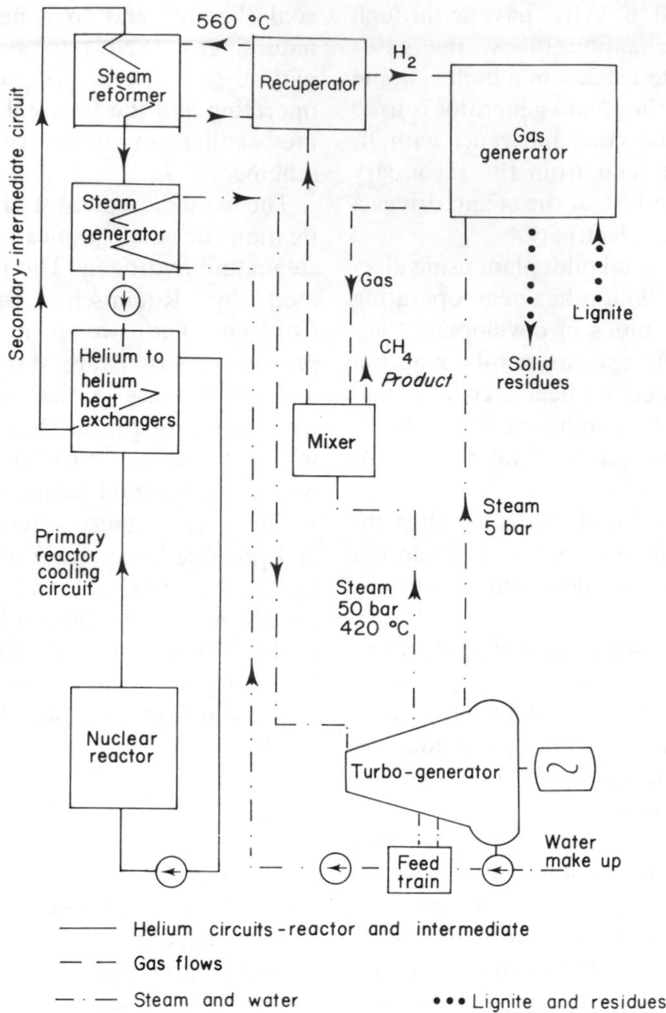

Figure 5.7 Lignite hydrogasifcation using nuclear heat
— Helium circuit—reactor and intermediate.
--- Gas flows.
-·-·- Steam and water.
. . . . . Lignite and residues

requires temperatures of 1400 °C and would allow only 13 per cent of the energy requirements to be met by nuclear heat. If hot hydrogen or 'Syngas' supplements the coke reduction, the nuclear source can supply over 50 per cent of the total energy requirement.

The electric arc process starts from ferrous scrap, augmented by iron pellets produced from direct reduction of ore in a fluidized bed or from passing reducing gas over pellets of ore in a shaft furnace. Clearly the electricity can be from a nuclear source but in addition the ore reduction requires a temperature of only 850 °C, well within the capability of a nuclear heat source. This direct reduction-electric arc furnace route allows 100 per cent of the energy to be supplied from an HTGCR system. The reducing gas is from a methane-steam reformer. Figure 5.9 illustrates the process.

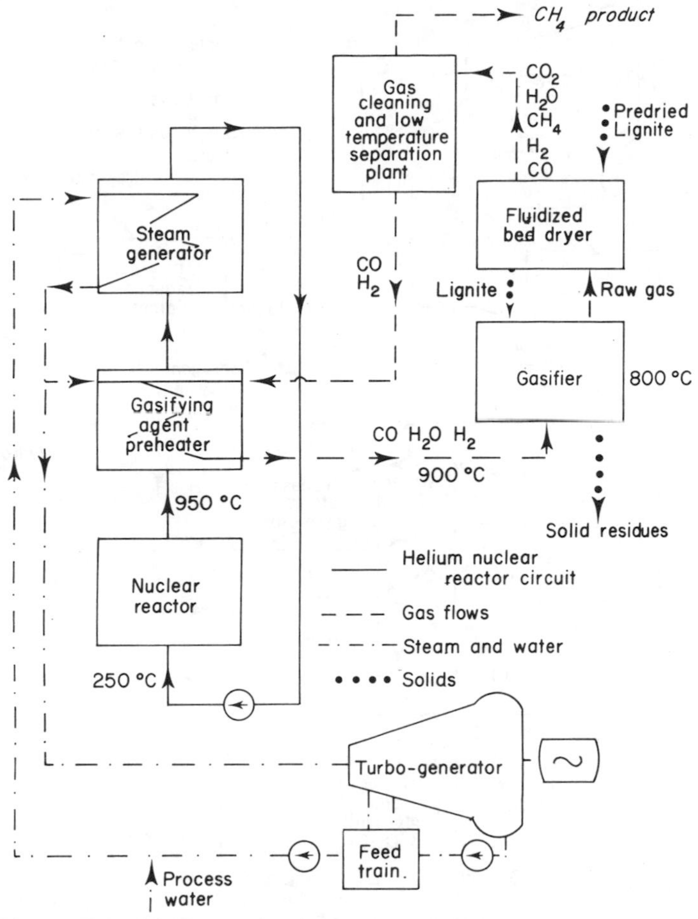

Figure 5.8 Lignite hydrogasification–gasifying agent preheated using nuclear heat
— Helium nuclear reactor circuit
--- Gas flows
-·-·- Steam flows
. . . . . Solids

5.9.5 Hydrogen

Hydrogen can be the main product of the natural gas reforming process, and nuclear heat from an HTGCR is a suitable source of energy for this. There has also been some interest in the possible use of hydrogen as a substitute for natural gas when supplies of this become too expensive to use. The use of nuclear energy for making a substitute gas from coals is an alternative approach.

In addition there was considerable interest in the 1970s in hydrogen as the ultimate fuel for a wide range of uses from transport, where it certainly has some advantages for aircraft, to an alternative to electricity or natural gas in an energy distribution grid. Hydrogen is a clean fuel with no emission problems but it diffuses through metal containers, and can embrittle metallic materials.

If it is heated to high enough temperatures water dissociates:

$$2H_2O \rightarrow 2H_2 + O_2 \quad (5.8)$$

but the temperatures required are high and there are major problems in separating the

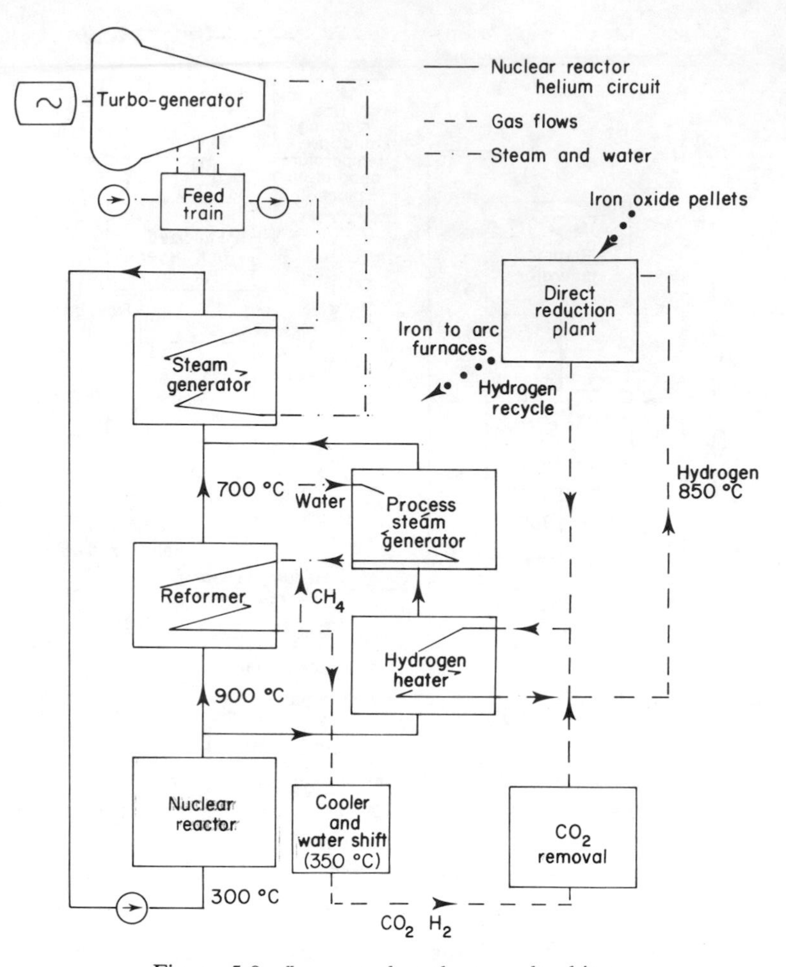

Figure 5.9 Integrated nuclear steelmaking

hydrogen and oxygen so that they do not recombine.

The simplest way of splitting water into its constituent elements is by electrolysis which was the main method of making hydrogen in the years before cheap natural gas and steam reforming became available. Off-peak nuclear electricity could well provide an attractive means of providing inexpensive energy for electrolysis. However, for this to be useful on any scale there is a requirement for large, reliable, relatively inexpensive electrolysis cells able to operate for a life comparable to power plant. Using thermally generated electricity, however, the electrolysis method is likely to have an over-all efficiency of only about 30 per cent.

Chemical splitting of water to give hydrogen with oxygen as a by-product has been the subject of investigations in France, United States and Italy.

A chemical route for splitting water is, theoretically, attractive. The general scheme follows the pattern:

$$X + H_2O \rightarrow XO + H_2 \quad (5.9a)$$
$$2XO \rightarrow O_2 + 2X \quad (5.9b)$$

The chemical X would be recycled and the only material to be supplied would be water or steam.

A large range of chemical processes are possible, most of them involving more than the ideal two stages in reactions (5.9a) and (5.9b). Table 5.8 gives five examples of systems that have been investigated experimentally.

Table 5.8 Chemical routes for hydrogen production from water

	Reactions	Temperature °C	Thermal chemical efficiency %
(1) Iron–sulphur dioxide			57
	$Fe_3O_4 + 2H_2O + 3SO_2 \rightarrow 3FeSO_4 + 2H_2$	125	
	$3FeSO_4 \rightarrow 3/2\ Fe_2O_3 + 3/2\ SO_2 + 3/2\ SO_3$	725	
	$3/2\ Fe_2O_3 + \frac{1}{2}\ SO_2 \rightarrow 2Fe_3O_4 + \frac{1}{2}\ SO_3$	925	
	$2\ SO_3 \rightarrow 2SO_3 + O_2$	925	
(2) Chlorine–steam	$2CrCl_2 + 2HCl \rightarrow 2CrCl_3 + H_2$	325	46
	$2CrCl_3 \rightarrow 2CrCl_2 + Cl_2$	875	
	$H_2O + Cl_2 \rightarrow 2HCl + \frac{1}{2}O_2$	850	
(3) Calcium bromide			39
	$Hg + 2HBr \rightarrow HgBr_2 + H_2$	250	
	$HgBr_2 + Ca(OH)_2 \rightarrow CaBr_2 + HgO + H_2O$	200	
	$CaBr_2 + 2H_2O \rightarrow Ca(OH)_2 + 2HBr$	725	
	$HgO \rightarrow Hg + \frac{1}{2}O_2$	600	
(4) Iron chlorine	$3FeCl_3 + 4H_2O \rightarrow Fe_3O_4 + 6HCl + H_2$	650	41
	$Fe_3O_4 + 3/2\ Cl_2 + 6HCl \rightarrow 3FeCl_3 + 3H_2O + \frac{1}{2}\ O_2$	120	
	$3FeCl_3 \rightarrow 3FeCl_2 + 3/2\ Cl_2$	420	
(5) Copper chlorine	$2Cu + 2HCl \rightarrow 2CuCl + H_3$	100	27
	$4CuCl \rightarrow 2CuCl_2 + 2Cu$	100	
	$2CuCl_2 \rightarrow 2CuCl + Cl_2$	600	
	$Cl_2 + Mg(OH)_2 \rightarrow MgCl_2 + H_2O + \frac{1}{2}O_2$	80	
	$MgCl_2 + 2H_2O \rightarrow Mg(OH)_2 + 2HCl$	350	

The complexity of the process can be judged from the flow chart of the iron-chlorine cycle in Figure 5.10.

Although it appeared to be possible to achieve a workable process the abundant supplies of relatively cheap natural gas and other gasification processes removed the main requirement of an early need for copious supplies of hydrogen. It is difficult to see how chemical processes could maintain steady and economical production on a tonnage scale over many years.

5.10 Nuclear Marine Propulsion

For naval use, nuclear power has many advantages for large ships and submarines. Nuclear ships do not require frequent refuelling, and can undertake long voyages without fresh fuel supplies. They are instantly available for despatch to any station. For submarines there is the additional advantage that the propulsion unit requires no air so the vessels can remain submerged for long periods without their presence being known. Their range is virtually unlimited.

The application of nuclear power to naval shipping clearly has advantages that could be applied to passenger and cargo shipping. There have been nuclear merchant ships built in USA (*Savannah*), Germany (*Otto Hahn*) and Japan (*Mutsu*) and extensive paper studies carried out in UK in the 1960s.

In the USSR the role for nuclear propulsion was for an ice-breaker, the *Lenin*. This ship has three PWRs with turbo-electric machinery and can operate in 2 m of ice. In many respects an ice-breaker seems to be an ideal role for nuclear ship propulsion.

The Soviet Union with many ice-bound harbours has a need for a powerful ice-breaker with considerable endurance. It is clear that nuclear propulsion is reliable enough to meet such a need and that the experience with the first ship the *Lenin* was sufficiently encouraging to justify building three further nuclear ice-breakers of greater power.

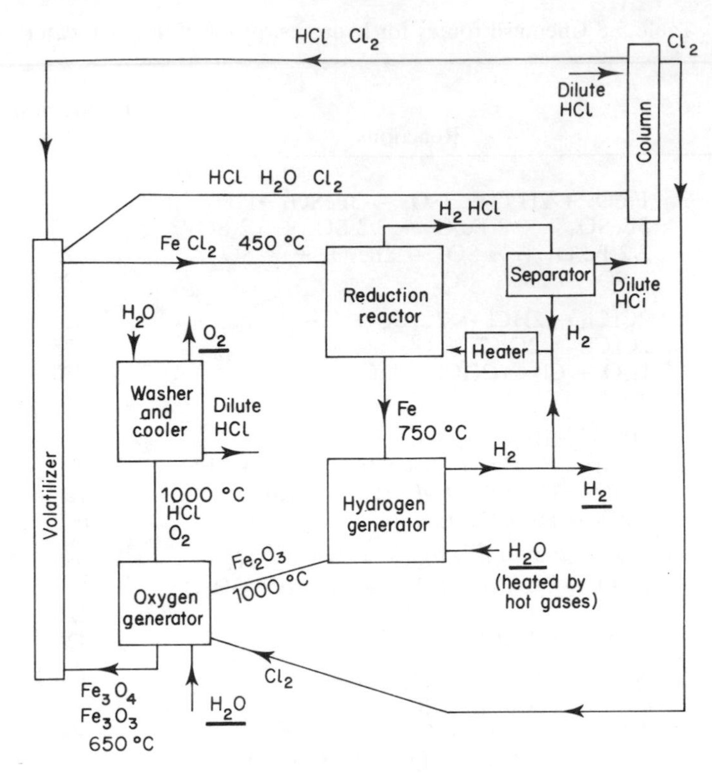

Figure 5.10 One scheme for iron–chlorine cycle for splitting water—simplified flow diagram

There may well be a role for nuclear ice-breakers and perhaps nuclear tankers and ore-carriers if the mineral potential of the Antarctic continent is to be fully realized.

Nuclear merchant ships only require refuelling every two or three years, permitting a high proportion of time at sea. Because the nuclear reactor does not require a large volume of fuel space, a nuclear ship can carry more cargo than a fossil-fired vessel. A considerable degree of automation is practicable. However, the nuclear propulsion unit is heavy and requires specially strengthened structural ship sections.

There are commercial and political problems. Few ports allow nuclear ships to berth, and the limited routes available to them were among the factors that hampered the commercial activities of both the *Savannah* and the *Otto Hahn*.

A nuclear propulsion plant is extremely expensive, and it takes many more years to build the propulsion plant than the ship. This means that the shipowner has to take a decision on the ship and its role many years earlier than for a fossil-fired ship, requiring money on which interest has to be paid. The return on this investment only appears many years later in the form of fuel oil cost savings, and, to justify a decision to build, a merchant ship requires a guaranteed high cargo loading with most of the time spent at sea.

Although considerable ingenuity has been shown in producing ideas for successful nuclear ships, the continuing economic problems of the shipping industries, the heavy over production of oil, the interest in the alternatives such as coal firing and wind assistance, and the acceptance of low speeds reducing fuel consumption has halted any serious interest in nuclear merchant shipping.

5.11 Conclusions

The review has revealed widespread present uses of nuclear energy for processes for non-electrical applications. Radiochemicals have an assured position in modern society and their use could spread to new areas where their low toxicity make them preferable to chemicals.

Nuclear steam and low temperature heating is already widespread. There are plans for reactor systems to supply only steam or hot water. Their success in the Soviet Union appears to be assured but there are no firm plans elsewhere.

Nuclear explosives appear to have a number of uses in the Soviet Union—in particular to ensure that fossil fuels are found and exploited. The only successful non-military marine propulsion use again appears to be in the Soviet Union where the ability of nuclear ships to operate in frozen seas is being exploited.

With high temperature reactors only steam reforming and refinery plants appear to be applications in the foreseeable future. Cheap fossil-fuels will ensure that nuclear gas making either from coal or the decomposition of water will not be required for many years.

These endeavours surely indicate that whatever the apprehensions about the ultimate exhaustion of fossil fuel supplies, there are nuclear alternatives that can supply many of the world energy and fuel needs and probably at costs not greatly in excess of those at present.

ACKNOWLEDGEMENTS

This material is based on the author's paper and article on 'The Potential use of Nuclear Energy for Hydrogen Production' in *Gas Engineering and Management*, October 1977 and later amplified in the Watt Committee on Energy book on *Nuclear Energy*.

The other general sources of information are the publications of the British Nuclear Energy Society (*Nuclear Energy*), European Nuclear Society (*Nuclear Europe*), the *International Atomic Energy Agency Bulletin*, ASME, Institute of Energy and Modern Power Systems.

Other information has been received by private communications from, Atomic Energy of Canada (Whiteshell), Ontario Hydro, ASEA (Sweden), Kernforshungsanlage Julich, Bergbau-Forschung GmbH, Essen, British Steel, Hochtemperatur-reaktorbau GmbH, Mannheim, and Beznau, Switzerland.

The author acknowledges the help of staff at Babcock Power Ltd and in particular Mrs E. Spallin for assistance in obtaining information and organizing its transformation into its final form.

BIBLIOGRAPHY

Section 5.2

'Engineering with nuclear explosives' *American Nuclear Society*, 14–16 January 1970, 4 volumes.

Summary in *British Nuclear Energy Society Journal* Volume 9, Number 3, July 1970, p. 150

'Borg, I., Nuclear explosives, the peaceful side', *New Scientist*, 8 March 1984, p. 10.

Section 5.3

Radiochemical Manual (now out of print), Radiochemical Centre, Amersham, 1966.

Annual reports of Amersham International.

Publications of the *International Atomic Energy Agency* such as their Bulletins; Conferences organized by IAEA and UNESCO and the Food and Agriculture organization of the United Nations (FAO), e.g. 'Isotope hydrology', Vienna, September 1983, 'Food irradiation in Asia and the Pacific', Tokyo, November 1981; 'Improving crops in Latin America', Lima, October 1982; 'Radiation and industry', Grenoble, September 1981.

Section 5.4

Kupitz, J., and Podest M. (1982) 'Nuclear Heat Applications—World Overview', *IAEA Bulletin*.

Panesco, A., Sycher, V.G., and Mensel, K. (1984). 'A Promising one for collaboration', *IAEA Bulletin*, **December 1984**.

'Symposium on Nuclear Energy Applications other than Electricity Production'. Jülich Conference, December 1976.

Gambill, W. R., and Kasten, P. R. 'National need for utilizing nuclear energy for process heat generation'. *ASME*, 84 JPGC NE 15.

Kolb, J. O. 'Technical feasibility and economics

of retrofitting an existing nuclear power plant to cogeneration operation for hot water district heating', *ASME* 84 JPGC NE5.

Section 5.5

Mayanan, S. A. (1983). 'WRI Reactor—a Western Canadian success story', *WNRE*, 530, Sept. 1983.

Leffer, W. R., and Barnstable, A. G. 'Use of low grade heat from Electrical Generating Stations', *OHS* 1979–16.

Section 5.7

Secure LH Nuclear Power for district heating, ASEA–ATOM brochure.

Hulst, J. (1983). 'Nuclear District Heating', *Nuclear Energy*, (*Journal of BNES*), **22, 4**,

Duffy, J., 'The Little reactor that can' AECL reprint from *Ascent 3*, **November 1981**.

Hanners, K. (1984). 'SECURE reactors goals and principles'. *Nuclear Europe*, **October 1984**.

Section 5.8

Bancroft, A. R. 'Nuclear Energy for oil sands—a technical and economic feasibility study', *AECL*, 7677

Perrett, R. J. (1981). 'High temperature fossil fuel processing Appendix 1', *Nuclear Energy*, **20, 6**.

'Recovering oil with Magnox steam', *Nuclear Energy International*, **September 1982**.

Section 5.9

Three symposia have a number of papers covering topics in this section:

1. 'The high temperature reactor and process applications' —Conference of the *British Nuclear Energy Society*, November 1974.
2. 'Nuclear energy applications other than electricity production', Julich, April 1976. (*Jul. Conf*. Dec. 1976).
3. 'Nuclear Heat for High Temperature Fossil Fuel Processing', *Institute of Energy*, London and South Eastern Section, April 1984.

Other sources

McDonald, C. F., Goodjohn, A. J., and Silady, F. A. 'Small modular gas-cooled nuclear unit with pebble bed reactor', *Modern Power Systems*, **December–January 1984/85**.

Kubeak, H., Schroter, H. J., Sulimma, A., and van Huk, K.H. (1983). 'Application of K_2CO_3 catalysts in the coal gasification process using nuclear heat', *Fuel*, **1983**.

Schroter, H. J., Schendler, W., and Weber, H. (1984). 'High temperature corrosion of metallic heat exchanger materials in a fluidized bed gasifier', in *High temperature materials corrosion in coal gasification atmospheres*, Applied Science Publishers Ltd.

Kubiak, H., van Huk K. H., and Juntgen, H. (1984). 'Mathematical modelling of a fluidized bed gasifier for steam gasification of coal using high temperature reactor heat'. *XVI ICHMT Symposium, Dubrovnik*, September 1984.

Johnston T. A., and Quode, R. N. 'Co-generation applications of high temperature gas-cooled reactors'.

Goodjohn, A. J., McMam A. T., Jr., Johnston, T. A., and Quode, R. N. *ASME* paper 84 JPGC–NF–1

Shimokawa, K. (1977). 'Present status of research and development of nuclear steel making in Japan', *Bulletin Japan Institute of Metals*.

Barnes, R. S. (1977). 'Nuclear Energy and the steel industry', *Steel Times Annual Review*, **1977**.

Sealy, T. (1982). 'Challenge to the blast furnace', *Financial Times, London*, **September 8, 1982**.

'Recent development of HTR in Germany', *BNES and Institution of Nuclear Engineers*, meeting 14 February 1985.

SECTION II
AREAS OF PUBLIC INTEREST

Nuclear Power: Policy and Prospects
Edited by P. M. S. Jones

6

Radiation

P. A. H. SAUNDERS
UKAEA, Harwell

6.1 Introduction

The impact of a major technology on the environment has rightly become an important determinant in the decision making process. Nuclear power is no exception; indeed, environmental issues probably loom larger in the nuclear debate than in the debate about most other advanced technologies. There are two features of nuclear power which, from the environmental viewpoint, distinguish it from fossil fuel power, both features being inherent in the fission reaction itself. The first is the great concentration of this energy source and the second the associated radiation and production of radioactive materials. One ton of uranium, in a thermal reactor, produces as much energy as over 20 000 tons of coal. The environmental effects of this are almost entirely beneficial: mining, transport, land use, fuel storage and volumes of waste produced are all very much less for nuclear than fossil fuel stations. It is the association of nuclear power with radiation that has resulted in the extensive study and public concern about its environmental impact. This chapter is concerned with radiation and its effects, and the measures that are taken to control exposure of radiation workers and members of the public to radiation.

6.2 Radiation

There is nothing new about the radiation associated with the nuclear industry. Man has evolved in a naturally radioactive environment, exposed to radiation from outer space and from the earth, food, water and air. Since the beginning of the twentieth century he has added to this natural background, through medical and industrial uses of radiation, mainly in the form of X-rays, through the testing of nuclear weapons, and through the increase in air travel since the intensity of cosmic radiation is greater at higher altitudes.

In fact, natural background radiation is responsible for nearly nine-tenths of the total, with the great majority of the balance resulting from medical X-rays. Other sources of 'man-made' radiation, including nuclear power, account for less than 2 per cent of the total (Hughes and Roberts, 1984).

The term 'radiation' now embraces electromagnetic waves, such as light, radio waves and X-rays, and the particles emitted by radioactive materials as they disintegrate or decay. These particles and the more energetic electromagnetic waves produce electrically charged particles called 'ions' in the materials they strike. This ionization frequently results in chemical changes which, in living tissue, can lead to injury in the organism. The non-ionizing radiations, such as those produced by ultra-violet lamps, lasers, radio transmitters and even sunshine, are only hazardous in special circumstances. Only the ionizing radiations are considered here.

The key discoveries were made at the very end of the nineteenth century (Henry, 1969).

X-rays were discovered by Röentgen in 1895. Becquerel, in 1896, reported the first observation of radioactivity from measurements of the ionization caused by pitchblende (an ore of uranium) and later that year Pierre and Marie Curie announced the concentration of the small fraction of radium in this ore. Three years later, Rutherford demonstrated that the radiation from pitchblende had two components of different penetrating powers which he labelled α and β. Becquerel showed that the β-rays were electrons and Rutherford later showed that the α-particles were the nuclei of the element helium. During this period Villard identified a third component, γ-rays, which were later shown to be energetic electromagnetic radiation. These radiations, together with neutrons, discovered by Chadwick in 1932, comprise the main radiations of importance in the nuclear industry and in medical and industrial applications of radiation and radioactive materials.

The unit used to measure doses of radiation to individuals is the sievert (Sv). Doses of tens of sieverts are used in radiotherapy to destroy cancerous growths. The average dose from the natural background in the UK is about 0.002 Sv, or 2 mSv (10^{-3} Sv). A typical chest X-ray gives about 0.02 mSv. The average dose received by members of the UK public from the activities of the nuclear power industry is about 2 μSv (10^{-6} Sv) a year, about the same as the extra dose resulting from cosmic radiation during a return flight from London to Geneva. About 4 μSv are received each year in the UK from emissions from the combustion of coal, since all coal contains traces of radioactive minerals (Hughes and Roberts, 1984).

6.3 Biological Effects

There was very little delay between the discovery of ionizing radiations and observations of their effects on human tissue. In 1897 a US court awarded damages to a patient affected by over-exposure to X-rays and in 1903 Rutherford visited the Curies and recorded: 'We could not help but observe that the hands of Professor Curie were in a very inflamed and painful state due to exposure to radium rays' (Eve, 1939). Marie Curie later died of leukaemia, and enhanced cancer rates were found among early workers with X-rays and radium.

The chemical changes that follow the absorption of radiation are highly complex; in living matter the DNA macromolecules are probably the most critical targets. These structures, present in every cell of the body, carry the information required for the development and division of cells and for the growth, proper function and reproduction of the organism. Radiation may alter a small part of the DNA molecule, such as a single gene, or it may break one or both strands of the DNA in one or many places, destroying or altering some of the information carried. Damaged DNA can to a considerable extent be repaired by enzymes in the cell. However, in some cases the cell survives in an unrepaired state, which can then be transmitted to large numbers of daughter cells by the normal processes of cell replication. It is for this reason that damage to DNA is probably more important than damage to other parts of living matter. If the damage results in the death of the cell, it will be absorbed or rejected by the organism. This is significant only if many cells are killed, since most organs contain far more cells than are needed to maintain normal function.

Cells that have been transformed or 'mutated' by radiation and survive do not necessarily lead to any deleterious effects. Indeed many such cellular changes occur normally during the lifetime of any organism. However, in some cases damaged cells can multiply in such a way that a cancer results or, if the damage occurs in a germinal cell that is itself later involved in the reproductive process, effects can be seen in later generations. Thus there are three kinds of biological effects of radiation to be considered: cell killing, cancer, and genetic.

6.3.1. *Cell killing*

Cell killing is only important if sufficiently large doses of radiation are received in a sufficiently brief period. A dose of 10 Sv or more delivered to the whole or a substantial part of the body within a few minutes is almost invariably fatal. A single dose of about 4 Sv will result in a one in two chance of death in the absence of medical treatment. The same dose delivered gradually over a year, however, would probably be tolerated because of the action of the body's natural repair processes. The ability of radiation to kill cells is, of course, the basis of radiotherapy where localized doses of tens of sieverts are used to treat cancers and other growths. Cell killing effects are characterized by a threshold below which no significant effects occur; there are in general no observable effects below about 1 Sv. The only important exception to this is in the developing embryo, where at certain stages only a few cells may be involved in important stages of growth and doses as low as a few hundredths of a sievert may result in abortion or serious malformation. Even in this case, however, there is probably a threshold below which no harm results (Pochin, 1983).

6.3.2. *Radiation-induced Cancer*

Unlike the cell killing effects, which in general appear within a relatively short time after exposure and exhibit a threshold, cancer and genetic effects are delayed and it must be assumed that there is no threshold below which one can be certain that no harm will result. However, as discussed later, there is no evidence of effects at low doses (below a few tens of millisieverts), and the universal and inescapable natural radiation background and the 'natural' prevalence of cancer and genetic defects are such that it will probably never be possible to prove the existence or absence of a threshold.

There are vast amounts of data on the effects of radiation on cells and animals and there are a rather limited number of cases where a small excess cancer incidence has been found in groups of people exposed to sufficiently high doses of radiation (UNSCEAR, 1982). However, the chain of events leading from radiation-induced damage in a cell to a developed cancer is far from well understood. Even when a link has been established between a radiation exposure and a subsequent cancer the radiation exposure itself may be only a necessary and not a sufficient cause of the cancer. Subsequent radiation exposures, exposures to chemicals or metabolic changes may be required before a tumour results. Such processes take many years; latent periods are typically five to ten years for leukaemias and 15–30 years for solid cancers.

While most but not all cancers can be induced by radiation, it is not possible to distinguish between a radiation-induced cancer and any other cancer. The only way in which the number of cancers that may be caused by an exposure to radiation can be predicted is on the basis of epidemiological studies of large groups that have been exposed to radiation in the past; the numerical estimates that result from such studies are discussed later.

6.3.3 *Genetic Effects*

The basic process that can ultimately lead to a cancer or a genetic defect is similar—damage to a DNA molecule. As with cancer, however, the mechanism by which a mutation in a cell leads to a genetic defect is complex and not fully understood. If the mutated cell dies or is not actually involved in the fertilization process there will be no genetic effect. Also if the new individual created at conception is not viable and dies at an early stage of embryonic development, it will probably not be detected. Most species seem to have evolved a protective mechanism by which faulty embryos are rejected; chromosome alterations are frequently found in spontaneous abortions.

A mutation does not necessarily show up in the first generation. Indeed a mutation can be transmitted through many generations without apparent effect before being expressed in a distant descendant. The reason for this is that chromosomes, which are made up of DNA, occur in pairs, one of each pair coming from each parent. Both chromosomes usually act together in controlling the characteristics of the individual through the information carried by the genes. However, if the corresponding genes on the two chromosomes are different the influence of the gene on one chromosome may dominate over the influence of the corresponding gene on the other chromosome. Thus genes can be 'dominant' or 'recessive'; a recessive gene is not normally expressed except when carried by both chromosomes of a pair. A dominant gene is expressed in every generation while a recessive gene is only expressed when inherited from both parents. Thus a lethal or disabling recessive gene can be transmitted from generation to generation without effect on the carrier. Some mutations are not fully recessive, and their occurrence may be advantageous or deleterious to the carrier.

It can be seen that the link between a mutation and its expression as a defect in an individual can be difficult to trace. As with cancers, it is not possible to distinguish between a radiation-induced defect and any other. Unlike cancers, however, there is no reliable human evidence of genetic effects following radiation exposure at any dose level; even among the survivors of the Hiroshima and Nagasaki bombs, no excess of genetic defects has been found in the children subsequently conceived by exposed parents. All estimates of genetic effects of radiation in humans, therefore, have to be based on evidence from animal studies, and extrapolation from animals to humans is associated with major uncertainties. However, knowledge of the fundamental mechanisms of radiation injury at the genetic level is perhaps more complete than knowledge of the mechanism of radiation-induced cancer, and there is reasonable agreement on the estimates that can be made from such extrapolation.

6.4 Risk Estimates

At high dose levels where cell-killing is the most important process, the effect depends on the size of the dose. The risk falls to zero below the threshold dose. In contrast, the delayed effects, cancer and genetic defects, are probabilistic in nature. They only occur in a small proportion of those irradiated, in an apparently random way. The size of the dose governs the probability of the effect occurring, not its severity. And with no evidence of a threshold, it must be assumed that any dose, however small, carries some risk.

6.4.1 *Risk of Radiation-induced Cancer*

Estimates of the risk of radiation-induced cancer are based on observations of a rather small number of groups that have received sufficiently high doses of radiation for the effects to be detectable against the high 'natural' incidence of cancer (about one in five deaths in the UK result from cancer). These are the survivors of the Hiroshima and Nagasaki bombs, patients who have received large doses of radiation for therapeutic or diagnostic purposes, and workers who have been exposed to high levels of radiation, such as radiographers, watch-dial painters using radium, and uranium and other hard-rock miners. The distribution of doses and the number of people exposed as a result of the Chernobyl accident in the USSR in April 1986 are not yet known in sufficient detail to enable any reliable estimates to be made of the possible long-term consequences. Preliminary indications, however, suggest that a sufficient number of people may have received high enough doses to result in detectable increases in cancer incidence. It will clearly be of great importance to monitor the health of these people carefully over a long period. The existing evidence is, however, sufficient to enable an upper limit to be put on the risk of radiation-induced cancer with a high degree of confidence.

Studies of the groups in the past show that the most significant cancers are leukaemia and solid cancers of the thyroid, female breast, red bone marrow and lung. The evidence has been exhaustively reviewed and assessed by national and international scientific bodies (UNSCEAR, 1982; ICRP, 1977; BEIR, 1980). The consensus view of these reports is that there is about a one in 100 risk of a fatal cancer developing for each sievert of radiation dose received (over and above the doses received from the natural background), with an uncertainty in this risk of about a factor of two; for safety purposes this uncertainty is of no significance. The one in 100 figure is an average over all ages and both sexes. Females are more at risk than males because of breast cancer; this risk is higher for exposure during adolescence and up to the age of 30. For certain types of cancer, particularly leukaemia and thyroid cancer, children may be more at risk than adults. The risk factor for a foetus is about twice that for the average adult.

Most of the evidence relates to doses of 1 Sv or more. There are a few cases where excess cancers have been found following doses of about 0.1 Sv, but there are not enough data to provide quantitative risk estimates. There is no evidence of any effects associated with the doses received from the natural background, which results in lifetime doses of about 130 mSv in the UK. There are many parts of the world where background radiation levels are higher than the UK average, as a result of local geology or high altitudes. Attempts have been made to find correlations between natural cancer incidence and background radiation levels in such areas and no statistically significant effects have been found. There have also been detailed studies of cancer incidence among workers in the nuclear industry, where typical radiation exposures are 2 mSv a year, or less than 0.1 Sv in a working life, which is less than the natural background dose. No consistent pattern emerges from these studies, indicating that any effect from radiation is too small to be detectable. Several studies purporting to find statistically significant links between occupational exposure and enhanced cancer rates have been refuted by subsequent analysis (BEIR, 1980).

These negative results confirm that the risk estimates derived from observations following doses of 1 Sv or more are unlikely to seriously underestimate the risks at much lower levels. The question remains, what is the true risk at low levels? Does a dose of a few millisieverts, a typical annual occupational dose, result in a risk of one-thousandth of that at a few sieverts, typical of the doses at which cancers have actually been observed? And does a dose of a few microsieverts, the average dose received by members of the UK public from the activities of the UK nuclear industry, result in any risk at all? Such questions may never be answered by observation, because of the size of the population that would have to be studied to give statistically significant results. They can only be answered indirectly through an understanding of the basic processes involved in radiation damage and its repair.

Radiation damage has been clearly demonstrated in laboratory studies. However, the body appears to be remarkably resistant to radiation. The natural background will in just one week result in at least one ionization event occurring in every cell in the body. Yet background radiation cannot be the cause of more than a very small fraction of all cancers, or different cancer rates would be found to correspond with different background levels. Thus either the damage must almost all be innocuous, or there must be a very effective repair mechanism. Our understanding of DNA damage and repair processes (Pochin, 1983) suggests that, for X-rays, γ-rays and electrons, much of the damage caused by low doses of radiation is repaired, typically within minutes, and the risk associated with such doses is less than would be inferred from observations at high doses. α-particles and neutrons appear to produce damage that is more difficult to repair, and the risk is more nearly proportional to the radiation dose. The relationship between dose and resultant damage, the so-called dose–response relationship, is illustrated for these two types of radiation in Figure 6.1. This picture is confirmed by studies of the way cancer incidence varies with dose, at the higher dose levels where excess cancers are actually observed. The dose–response relationship for such cancers is a straight line for α-particles and neutrons, indicating a strict proportionality between dose and effect. For X-rays, γ-rays and electrons the relationship is curved, like the lower line in Figure 6.1, suggesting that low doses are less damaging per unit dose than higher doses. For the purposes of

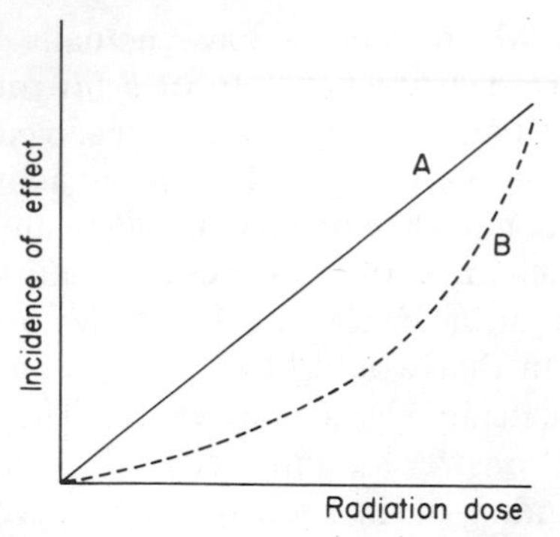

Figure 6.1 Dose–response relationship for cancer induction by α-particles and neutrons (A) and X-rays, γ-rays and electrons (B)

radiological protection where it is clearly right to err on the side of caution, it is generally assumed that strict proportionality applies to all types of radiation. However, the International Commission on Radiological Protection stresses that in choosing between a practice involving radiation exposure and an alternative practice that does not, the risk estimates should be used only with great caution and with explicit recognition of the possibility that the actual risk at low doses may be lower than that implied by the deliberately cautious assumption of strict proportionality (ICRP, 1977).

If it is accepted that the risk from radiation is strictly proportional to dose, with no threshold below which doses can be neglected, it follows that the total detriment, in terms, say, of number of fatal cancers, resulting from a given amount of radiation spread over a population is independent of the size of that population. So the number of cancers resulting in a population of 1000 people each receiving 1 Sv (or a 'collective dose' of 1000 man-sieverts) would be the same as that resulting in a population of one million people each receiving 1 mSv (also 1000 man-sieverts). The existence of repair processes and the departure from strict proportionality already noted make this unlikely.

6.4.2 *Risk of Radiation-induced Genetic Defects*

While there is no doubt of the ability of radiation to produce genetic defects, the absence of evidence of such effects in humans means that risk estimates have to be based on results obtained with other species, notably mice. There are two ways in which such estimates can be made, the 'direct' method and the 'doubling dose' method.

In the direct method, data on the rates of induction of specific kinds of mutation in mice are used, with appropriate corrections, to estimate the rate at which the same kinds of mutation are induced in man. It is difficult to obtain estimates of over-all genetic risk by this method because information is only available on a few categories of genetic damage. On the basis of the limited amount of information available, however, the frequency of defects in the first generation is estimated to be between 0.5 and 2.1 per 1000 children per sievert received by the parent (Pochin, 1983).

In the 'doubling' dose method the experimental data are used to calculate the dose required to double the rate at which mutations arise spontaneously in mice. It is then assumed that the same doubling dose applies to man, and knowledge of the rate at which defects occur spontaneously in man allow estimates to be made of the rate at which additional defects will occur as a result of a given radiation dose. In addition to the assumption that mice and men have similar doubling doses it is also assumed that all spontaneous defects can in principle be induced by radiation and that the sensitivity of a gene to a spontaneous mutation and to a radiation-induced mutation is the same. There is some experimental evidence to justify the last of these assumptions, but only in fruit-flies (Shukla, 1979).

The doubling doses derived from studies of different types of radiation-induced defects in mice range from 0.4 Sv to 2.6 Sv (Searle, 1979). The absence of any evidence of excess defects in the offspring of the Hiroshima and Nagasaki bomb survivors suggests that the doubling dose is unlikely to be below about 1.5 Sv, which indicates that the use of a doubling dose of that order for estimating effects in men is probably erring on the side of caution. In order to provide an extra degree of caution the UNSCEAR report assumes a rather lower doubling dose of 1 Sv (UNSCEAR, 1982).

The doubling dose method expresses radiation-induced genetic defects in terms of naturally occurring defects; the latter represent an equilibrium between new mutations and mutations that are lost from the population by death or failure to reproduce. A doubling dose has to be maintained for five to ten generations before equilibrium is reached and the number of radiation-induced disorders equals the number of naturally occurring disorders (BEIR, 1980). The first generation incidence is between one-fifth and one-tenth of the equilibrium incidence.

The spontaneous incidence of genetic defects that are likely to be increased by radiation is about 1.5 per 100 live born children (UNSCEAR, 1982). At equilibrium, therefore, a doubling dose of 1 Sv would result in 1.5 defects per 100 live born children, with one-fifth to one-tenth of that incidence in the first generation.

The results of the direct and doubling dose methods are thus in reasonable agreement, in spite of the uncertainties involved. As with radiation-induced cancers, the evidence on which the estimates are based results from observations at doses of around 1 Sv and above. In order to assess the risks at the far lower levels of radiation received by workers or members of the public as a result of the activities of the nuclear industry it is again necessary to make assumptions about the shape of the dose–response relationship. Although the existence of repair processes may result in a departure from strict proportionality at low doses, at least for X-rays, γ-rays and electrons, the uncertainties involved are such that it is clearly sensible to err on the side of caution and assume a linear relationship.

These estimates refer to the risk of a genetic defect occurring following exposure of a hypothetical population of fertile individuals irradiated prior to conception of their offspring. In practice, some of the dose to which a real population is exposed will have no genetic significance because it will be received when no more children are likely to be conceived. Taking this factor into account reduces the risk, expressed in terms of a given total lifetime radiation dose to a given real population, to 0.4 times the figures given above.

6.4.3 *Summary of Risk Estimates*

In summary, a dose of 1 Sv is associated with a one in 100 risk of fatal cancer in the exposed individual and a somewhat lower risk of a serious genetic defect appearing, spread over all subsequent generations. The risk of the defect appearing in the first generation is between one-fifth and one-tenth of this. These figures are unlikely to be in error by more than a factor of two.

On the basis of these estimates, and on the deliberately cautious assumption of strict proportionality between dose and effect, in the UK about 1300 cancer deaths per year and about 300 genetic defects per year might be ascribed to natural background radiation, although there is of course no evidence that such effects occur. It is worth noting that cancer kills about 140 000 people in the UK each year. On the same basis, just over one cancer death per year and, if the exposure continued for enough generations for an equilibrium to be established, about one genetic defect a year might be ascribed to the exposure of the population to the activities of the UK nuclear industry. The existence of repair processes within the body probably means that the true risks are even lower, and at such low levels of exposure a zero risk is not incompatible with the evidence.

6.5 Control of Radiation Exposure

The extremely low average public exposures associated with the nuclear industry results from the application of extensive measures to control radiation, which originated long before nuclear fission was discovered. The early evidence of damage to tissues and cancers caused by X-rays led to the creation of national committees and, later, to the setting up in 1928 of an International Committee on X-ray and Radium Protection, subsequently renamed the International Commission on Radiological Protection (ICRP). This body formulates regulatory principles and makes specific recommendations on the basis of the best available knowledge of the effects of radiation.

Protection recommendations were originally based on the concept of a threshold; limits on exposure were set to ensure that the threshold was not exceeded. However, it is interesting to note that even in the early days the recommendations included such words as '. . . should place himself as remote as practicable from the X-ray tube . . .' (Taylor, 1979). In the late 1930s and early 1940s it became clear that a safe threshold could not be proved for genetic effects and the term 'tolerance dose' that had been used to express limits was replaced by the term 'permissible dose'. When the ICRP was re-formed after the Second World War it was able to review the large amount of information that had by then accumulated. The importance of protection against both cancers and genetic effects was realized. Since no safe thresholds could be assumed, control could not be based simply on the setting of dose limits. ICRP recommendations evolved during the following decades as new data emerged from studies of the bomb survivors and other exposed groups and from a massive amount of research on animals and cells in the laboratory. Their most recent recommendations are based on three central principles (ICRP, 1977):

1. All practices involving radiation exposure should show a net positive benefit (*justification*).
2. Radiation doses should be as low as reasonably achievable, economic and social factors being taken into account (*optimization*).
3. All radiation exposures should be within the recommended dose limits (*dose limits*).

This approach relegates the dose limits to the role of a protection backstop, focusing most attention on the concept of optimization.

The first principle, justification, recognizes the possibility of harm resulting from any dose of radiation, however small. It has resulted in the banning of trivial uses of radioactive material, such as in toys and jewellery, and in discontinuing the use of X-ray machines in shoe shops, where the minimal benefits did not justify even the low risk.

The second principle, optimization, requires doses to be reduced as far as is reasonable, and proper interpretation of the word 'reasonable' is one of the current challenges in radiological protection. One approach to the problem that is now gaining acceptance is the use of cost-benefit analysis. Where the cost of reducing radiation exposure is less than the resultant harm, then the exposure should be reduced. The application of the technique involves placing a monetary value on a unit of collective dose and thus on any human harm—disease, death, or injury to subsequent generations—that might result. While this process may appear repugnant, such judgements frequently have to be made, either explicitly or implicitly. The question is that of the best way to apply society's resources (Roberts, 1984). Estimates of the amount of money spent in saving one life range from several million pounds in the chemical industry, through a few thousand or tens of thousands of pounds for many medical treatments, to much lower figures for the prevention of starvation in third world famine areas. A figure of £120 million per life saved is associated with the recent decision of British Nuclear Fuels plc to reduce further the discharges from the Sellafield plant to the Irish Sea (Avery, 1984), a figure that is difficult to justify on any rational grounds.

In the UK, the National Radiological Protection Board, the body responsible for advising the government on radiological protection, has proposed monetary values of collective dose that depend on the level of radiation dose received (NRPB, 1981). They suggest that more resources should be devoted to the protection of individuals at the highest risk, on the reasonable basis that while it must be assumed that risk is directly proportional to dose, it does not follow that the expenditure which should be associated with averting risk should also be directly proportional to dose. The values proposed by NRPB correspond to £200 000 per life saved at the lowest doses to £5 million per life saved at doses approaching the dose limit. It should be remembered that if the true risks at low doses are substantially below those calculated on the basis of strict proportionality between dose and effect, as seems likely, the corresponding monetary values would be substantially higher.

The third principle, dose limits, applied in conjunction with the second principle, ensures that individuals are not exposed to unacceptable risks. The limits are currently set at 50 mSv per year for radiation workers and 1 mSv per year for members of the public, although 5 mSv is permissible in any particular year provided that the average annual exposure over a lifetime does not exceed 1 mSv. These limits do not constitute a sharp boundary between 'safety' and 'danger'; this is clear from the earlier discussion on the effects of radiation and indeed from the fact that there are two limits. Acceptability is based on the generally accepted view that some industries are considered 'safe' and that some level of public risk is considered to be 'acceptable'. Application of the principles to radiation workers in the UK nuclear industry has resulted in average doses of about 2 mSv in a year. The associated estimated risk of total cancer is about one in 40 000 per year, a level that is typical of 'safe' industries such as textiles, foods or timber, and well below that associated with industries such as construction or coal mining.

The average annual dose to members of the public in the UK from the nuclear industry (2 μSv) corresponds to an estimated risk of less than one in 10 million per year. The ICRP recommendations, however, are designed to protect individuals as well as the population as a whole. In order to ensure the protection of each and every individual member of the public it is necessary to understand in detail the routes or 'pathways' which can lead to the irradiation of people when radioactive materials are allowed to enter the environment. The subject is a complex one involving the measurement of the concentration of radionuclides in the environment, foodstuffs, drinking water, etc and the modelling of the movement of radionuclides along all possible pathways from sources to people (Johns, 1983). Generally it is found that one or two pathways will contribute most of the exposure and these are referred to as 'critical pathways'. The population affected may be a small group of people or even one person—the 'critical group'. A discharge of radioactive material is only authorized (in the UK by the Environment Departments and the Ministry of Agriculture, Fisheries and Food) if it can be shown that the ICRP dose limit is not exceeded even for the critical group. Regular assessments carried out by the regulatory authorities have shown that critical group doses have never significantly exceeded ICRP limits.

6.6 Accidental Releases of Radioactivity

The risks associated with routine emissions from the nuclear industry are thus clearly small. There remains, however, the possibility of a release of radioactive material following an accident. Depending on quantities of material released, wind, weather and population distribution, certain accidents could lead to radiation doses to some groups exceeding the ICRP limit or to large collective doses. Techniques for estimating and reducing the risk of such accidents are discussed in Chapter 7.

There have been two nuclear reactor accidents that resulted in members of the public receiving small doses of radiation: Windscale and Three Mile Island, and one accident in which large quantities of radioactive material were released: Chernobyl.

At Windscale, in October 1957, a fire occurred in an early reactor built for the sole purpose of producing plutonium for the British nuclear weapons programme. The reactor had little similarity with electricity-generating reactors; in particular it was cooled by forced air which was emitted through a filter tower to the atmosphere. The fire resulted in the release of 10 000–20 000 curies of iodine–131 and about 230 curies of polonium–210, together with other radionuclides of no radiological significance. While no individual received a dose larger than that received in a year from the natural background, a very large number of people received small doses. On the basis of the ICRP assumption of strict proportionality between dose and effect, down to the lowest dose levels, the release could have resulted in about 33 serious health effects (fatal cancers plus serious genetic defects) (Crick, 1983). However, on the basis of the arguments given in section 6.4.1, the figure may be very much lower.

The Three Mile Island accident, in March 1979, resulted in the release of only 16 curies of iodine–131. The average radiation dose to a person living within three miles of the site was less than one-tenth of the annual background dose and on the basis of the ICRP assumption there is about a one in three chance that one cancer death will occur as a result of the release.

The Chernobyl accident in April 1986 resulted in the release of a significant fraction of the volatile materials in the core. The latest available figure of 31 deaths, mainly among those fighting the fire, suggests that these people received doses of many sieverts. Radiation levels within a few tens of kilometres of the site were judged by the Soviet authorities to be sufficiently high to require people to be evacuated, and food restrictions were applied over a large area. At the time of writing, no figures are available for the actual radiation levels that resulted in these measures being taken. The passage of the cloud of radioactive material was detected in many countries (Fry *et al.*, 1986; Webb *et al.*, 1986). Estimates of the radiation doses received suggest that doses at distances of over a few hundred kilometres from the site are unlikely to exceed a small fraction of the recommended annual dose limit for members of the public of 1 mSv per year.

REFERENCES

Avery, D. G. (1984). *Liquid Discharges from the Sellafield Site, Evidence to the Sizewell Inquiry*, BNFL/P/1 (ADD 12).

BEIR (1980). *The Effects on Populations of Exposure to low levels of Ionizing Radiation: 1980 (The 'BEIR III report')*, National Academy Press, Washington DC.

Crick, N. J., and Linsley, G. S. (1983). *An assessment of the Radiological Impact of the Windscale Reactor Fire, October 1957*, NRPB-R135 and Addendum.

Eve, A. S. (1983). *Rutherford*, Cambridge University Press, Cambridge.

Fry, F. A., Clarke, R. H., and O'Riordan, M. C., (1986). *Nature*, **321**, 193.

Henry, H. F. (1969). *Fundamentals of Radiation Protection*, Wiley–Interscience, Chichester.

Hughes, J. S., and Roberts, G. C. (1984). *The Radiation Exposure of the UK Population—1984 Review*, NRPB-R173.

ICRP (1977). Recommendations of the International Commission on Radiological Protection, ICRP Publication 26, *Annals of the ICRP* **1(3)**, 1–53; and other ICRP publications.

Johns, T. F. (1983). *Nuclear Power Technology, Vol 3: Nuclear Radiation*, Oxford University Press, Oxford.

NRPB (1981). *Cost-benefit Analysis in Optimizing the Radiological Protection of the Public; a Provisional Framework*, NRPB-ASP 4

Pochin, E. (1983). *Nuclear Radiation: Risks and Benefits*. Oxford University Press, Oxford.

Roberts, L. E. J., (1984). *Nuclear Power and Public Responsibility*, Cambridge University Press, Cambridge.

Searle, A. G. (1979), 'Hereditary damage', *Radiation and Environ. Biophysics*, **7**, 41.

Shukla, P. T. (1979). 'Is there a proportionality between the spontaneous and X-ray induction rates of mutations?' *Mutation Research*, **61,** 299.

Taylor, L. S. (1979). *Organization for Radiation Protection*, Office of Health and Environmental Research and US Department of Energy, Washington DC.

UNSCEAR (1982). *Ionizing Radiation: Sources and Biological Effects*, United Nations Scientific Committee on the Effects of Atomic Radiation, New York.

Webb, G. A. M., Simmonds, J. R., and Wilkins, B. T. (1986). *Nature*, **221,** 821.

Nuclear Power: Policy and Prospects
Edited by P. M. S. Jones

7

Safety

P. M. S. Jones
Department of Economics, Surrey University

7.1 Introduction

The radioisotopes in the core of a nuclear reactor, in a nuclear fuel reprocessing plant or in radioactive spent fuel or waste stores would represent a considerable health hazard to the workforce and to members of the public if they were not safely contained.

Since low levels of radioactivity have always been a part of man's environment with no demonstrable adverse effects, minor additions, comparable to the variations in natural background, might be regarded as unimportant (see Chapter 6). However, the release of a major part of the radioactive contents of a reactor would be totally unacceptable. The safety of nuclear installations is therefore a matter on which great stress has been placed since the earliest days of nuclear power.

This chapter is concerned with the general principles employed in the design of nuclear plant to ensure that the risks of major releases are reduced to negligible proportions, and with the principles of safety assessment, which seeks to confirm the safety of the design or to indicate ways in which improvements could be made. In practice the two processes are iterative and although probabilistic risk analysis is comparatively new it only represents a systematic and more exhaustive approach to thought processes that accompanied the earliest designs.

7.2 The Principles of Nuclear Safety

A major objective of nuclear plant design is to ensure that all radioisotopes and radiation are contained, with releases to the environment restricted to carefully controlled and monitored quantities that are considered by the national authorities to pose no threat to the workforce or public.

Radiation release is controlled by surrounding any radioactive source with sufficiently thick barriers (screens) to absorb the radiation. The barrier could be tissue paper for low energy α-particles or metres of concrete or lead for high energy γ-rays. The design of screened buildings or equipment has to ensure that any access routes for ducting, entrances, etc., do not compromise the screening and that access by personnel to a screened area is carefully controlled.

The first line of containment for radioisotopes may be the walls of pipes or vessels housing the radioisotopes or in the case of nuclear fuel it may be the crystal lattice of the fuel itself. If this containment is not perfect or if it could be breached by accident or in the course of normal operation, then secondary or even tertiary containment would be employed. Thus vessels could be double walled, plant could be placed in isolating ventilated compartments, nuclear fuel can be contained in fuel cans. For many purposes this will be sufficient when coupled with the use of suitable monitoring procedures which could quickly detect any failure of containment so that it could be remedied safely.

Simplicity is rapidly lost, however, when one moves to reactors or chemical plant. In a power reactor, for example, the fuel is used to produce heat which must be removed using coolant. At high temperatures and high coolant flows in the presence of high radiation fluxes the materials used for containment may corrode or erode or fail due to thermal or mechanical stress. If coolant is lost the temperature will rise rapidly and this could lead to the fuel can, or even the fuel itself, melting: it may also lead in some systems to chemical reactions producing explosive gas mixtures. Research and development are used to define the material specifications and the safe range of operating conditions that need to be maintained. The designer then seeks to ensure that the reactor can not, under any conceivable set of circumstances, deviate from the safe range.

In the case of a reactor the nuclear fission reaction has to be brought to a rapid end if anything fails (such as the cooling system) that could jeopardize the safety of the plant. This is done using control rods that absorb the neutrons and, in liquid cooled reactors, this can be supplemented by injection of neutron absorbers such as boron into the coolant.

In the earliest experimental reactors used in the Manhattan project such systems were crude but effective. The world's first reactor, the pile at Chicago University, had a shutdown neutron absorbing rod suspended on a rope and a glass vessel containing boric acid solution on top. In the event of any unforeseen contingency a man with an axe was to cut the rope and another with a hammer was to smash the bottle. The early reactors used in the weapons programme were also located in sparsely populated areas so that in the unlikely event of a release of radiation few, if any, members of the public would be at risk.

In a power reactor the quantities of heat being released by fission products in the fuel after reactor shutdown is still substantial and means of removing this heat have to be provided if large temperature rises are to be avoided.

In process plant or stores containing fissile materials (including fuel stores) a principle design concern is to avoid circumstances in which a critical mass of material could be formed (see section 2.3.3). This is done by monitoring their concentrations in process fluids, by limiting the volume of pipework and storage tanks, by physical separation of packages or vessels or by incorporation of neutron absorbing materials or poisons. Operating rules and fissile material accountancy provide additional safeguards. These procedures, if properly practiced, ensure that there can be no sudden bursts of radiation which could result in excessive exposure of plant operators and no heat effects which could damage the plant. Plant design has also to guard against normal chemical and physical hazards which could lead to leakage or dispersal of radioiosotopes into the environment.

7.3 Reactors

7.3.1 *Control of Reactors*

The separate reactor sections in Chapter 3 describe some of their inherent safety features and the means adopted to guard against major releases of radioactivity. The first general point to be made is that no reactor can explode like an atomic bomb to release large amounts of fission energy instantaneously. In the very worst conceivable case with breach of containment and loss of coolant a major part of the fission products contained in the core could be released to the environment together with some of the more volatile actinides. The only case where this has happened is the 1986 Chernobyl disaster where fuel overheated, and the containment was breached by a chemical explosion (see 7.5.6).

In normal operation a reactor's power is controlled and stabilized by negative feedback loops, a process which is facilitated by the few percent of fission neutrons that are released after a delay ranging up to tens of seconds. This allows time for the control rods to be moved and the other self compensating physical changes to take effect.

Any rise in reactor power will lead to a rise in fuel temperature which increases the neutron capture cross section of uranium–238 in the fuel (the so called Doppler effect). This then reduces the number of neutrons available to produce fissions and the reactor power is reduced.

A rise in reactor power can also increase the coolant and moderator temperatures reducing their density and their moderating effect. For water coolant or moderator this leads to a reduction in the number of fissions and reduces the reactor power, although it can have the opposite effect if significant quantities of plutonium–239 are present in the fuel, due to fission and capture resonances at around 0.3 eV. In the case of reactors with two phase coolants (e.g. the BWR) a rise in temperature can reduce the volume of liquid coolant in contact with the fuel and this too reduces the moderating effect. With sodium coolant in fast reactors any voidage in the coolant increases reactivity but also increases neutron leakage from the core. The latter together with the effects associated with temperature rises in the fuel, give an over-all strong negative power coefficient.

In thermal reactors in particular the build up of fission products with high parasitic neutron capture, particularly xenon–135, can exert a flattening effect on reactor power.

The precise magnitude of these effects depends on the fuel, the reactor and fuel geometry and the nature of the coolant, moderator and other reactor materials. Negative feedback is supplemented by the control rods which can be inserted or withdrawn to maintain the reactor power at any desired level up to the full design power. The rods themselves are controlled using measurements from instruments within the reactor. These internal and external feedback mechanisms constitute the first line of defence against accidents. (For a more detailed technical discussion see Thomson and Beckerley, 1973.)

The second line of defence in the power reactor is the so called scram system which is designed to shut down the reactor automatically in the event of any power excursion above the reactor's design level. Detectors in the system immediately release the neutron absorbing shutdown rods in the event of such an excursion. The detectors, shutdown rod control and their feed mechanisms have to be designed to be fail safe so that in the event of power failure or damage to the circuits the reactor is immediately shut down even if a proportion of the rods jam. To avoid unnecessary costly shutdowns the power supplies and monitors can be replicated.

7.3.2 *Heat Removal*

The core of a reactor continues to release heat even after shutdown due to the decay of fission products within the fuel. Immediately on shutdown this heat corresponds to about 8 per cent of normal operating power, declining to 2 per cent after 15 minutes, 1 per cent after a few hours and about 0.5 per cent after 24 hours. If severe damage to the core and fuel are to be avoided this heat must be removed. In normal circumstances this is easily done by the reactor's cooling system.

In the PWR, which normally operates at 160 bar pressure there is a special residual heat removal system which comes into operation after the reactor has been shut down for maintenance or refuelling. This operates at 28 bar and 177 °C.

Both the normal cooling and residual heat removal systems (PWR) are operated on two or more separate circuits with their own circulating pumps, piping, heat exchangers and electrical supplies so that in the event of any failure only part of the cooling capability would be lost. AGRs for example have eight gas circulators and modern PWRs three or four separate coolant circuits.

If coolant flow were to be interrupted for any reason, temperatures within the reactor could rise, eventually leading to some or all of the consequences listed in section 7.2. Some reactor systems, however, such as the liquid metal cooled pool type fast reactor, have large thermal inertia and can get sufficient cooling from natural convection and passive heat exchange to the atmosphere so that circulating pump failures are not serious. Other reactors such as the AGR or Magnox have a large graphite heat sink so that the temperature rises only slowly and there is ample time to restore services at the levels required to cater for decay heat. The pressure tube reactors like CANDU also have a large thermal inertia in the bulk moderator.

If the coolant containment is breached for any reason there is little leakage from a low pressure system like the sodium cooled fast reactor, so that residual heat removal is not compromised. In the gas cooled systems pressures are relatively low (40 bar) and at worst can fall to atmospheric pressure (1 bar). Natural circulation is adequate to remove the residual heat produced in the core provided the boilers contain water. The main concern is to prevent excessive air ingress since air could react with the hot graphite moderator. This would be achieved, where necessary, by injecting additional carbon dioxide from the multiple storage systems so as to maintain pressure in the reactor above 1 bar.

The AGRs are designed so that they can be intentionally depressurized and carbon dioxide coolant vented to the atmosphere. The coolant inevitably contains some gaseous radioactivity (argon–41 and sulphur–35 plus possible traces of activation products and fission products) so discharges, whether deliberate or as a result of leakage, are passed through filters and released from high stacks in quantities that pose no threat to the public or workforce.

The water cooled reactors operate at higher pressures (160 bar in the PWR or 72 bar in the BWR, see Chapter 3). If the coolant circuit is breached the pressure drops and the coolant boils, so that in principle all the coolant could be lost quite rapidly given a large breach. Such a situation could not be tolerated so that the water reactors are provided with a number of separate emergency core cooling systems.

The first of these for the PWR is a passive injection system based on a number of large tanks of cool borated water under pressure. These vent under gravity into the reactor core through non-return valves immediately the pressure in the main reactor vessel falls below a critical value.

A second system (LPI—low pressure injection) operates at low pressure following major failure and feeds water from the refillable refuelling tank storage system. Water from the sump in the containment building, where the leaking water would collect, can also be recirculated to the core through heat exchangers to provide cooling over a long period.

A third system (HPI—high pressure injection) operates at high pressure to deal with small leaks. It injects concentrated boric acid solution into the system and then uses high pressure pumps to feed water in from the refuelling tank store. Both of these active systems have replicated pumps and piping systems to allow for any possible failures.

Typically the passive accumulator system can remove all the decay heat produced in the first hour after shutdown, the HPI could operate on three circuits each capable of removing 70 MW and the LPI on two circuits each capable of removing 470 MW. This compares with the 50 MW of heat being produced a few hours after shutdown in a 1 GWe PWR.

In the event of a major loss of coolant accident (LOCA) in PWR the drop in pressure would lead to steam production in the core and a rise in fuel cladding temperature. After about 2 s the HPI and LPI pumps would be switched in and the HPI would begin to operate after about 14 s. The accumulator system would function when the pressure dropped to about 41 bar and the LPIS after 30 s as containment and coolant circuit pressures equalized. At this stage the fuel is still poorly cooled and the fuel and cladding temperatures will have risen considerably with some possible rupture of the cladding and fission product release into the containment. However, the water in the system will by then be rising and will gradually quench the fuel until complete immersion is accomplished over a period of some 5 min. The fuel and cladding temperatures will then have dropped to below 200 °C and the situation will be under control, with subsequent decay heat being removed via the LSI and the heat exchangers.

The BWR would lose its coolant more slowly than a PWR. It too is equipped with a range of emergency core cooling systems with built in redundancy.

7.3.3 *Containment*

The bulk of the radioactivity in an operating reactor arises from the fission products produced within the fuel. For uranium oxide or mixed plutonium–uranium oxide fuelled reactors the fission products are largely (>98 per cent) retained in the ceramic fuel matrix which is chemically stable and does not melt until its temperature is well above 2500 °C. Only the more volatile fission products like krypton, xenon and iodine diffuse out of the pellets into the voidage in the fuel can in normal operation.

The fuel pellets are contained in a sealed can (Chapter 4) made of zirconium alloy for water cooled reactors or stainless steel for AGRs, with sufficient voidage to retain the volatile fission products. Fuel and fuel cans, coolant and moderator are chosen to be chemically compatible under normal conditions (see Chapter 4). Failure of a very small proportion of fuel cans due to metallurgical flaws or mechanical defects can be tolerated and reactor systems are designed to cope with this. The fuel would normally be removed and replaced.

The fuel and coolant have to be contained within a pressurized circuit or pressure vessel for the reactor to function and any small quantities of fission products released into the coolant are retained. In the case of the Magnox reactors, AGR, PWR and BWR the moderator is also within the pressure containment. For the gas cooled reactors this is a prestressed concrete vessel with a multiplicity of steel tendons that provide considerable redundancy and which can be individually checked and replaced if need be. The compression in the vessel is considerably greater than the normal internal working pressure and it is inconceivable that such a vessel could fail catastrophically. The pressure vessel surrounds a gas tight mild steel liner which is water cooled on the outside and insulated on the inside to keep coolant in and to protect the concrete pressure vessel from thermal stresses. The 5 m thick pressure vessel provides the necessary radiation shielding.

In the case of the PWR the pressure vessel is a massive thick walled (20 cm) steel vessel clad internally with corrosion resistant austenitic steel which is designed to withstand pressures well above the normal working pressure. This pressure vessel, together with the remainder of the primary coolant circuit, including the steam generators, is enclosed in a large steel lined containment building of reinforced concrete capable of withstanding pressures of 3.5 bar. This building houses the crane for refuelling and would contain any water or steam and fission products lost from the primary circuit if it failed. The building is provided with alkaline cooling sprays to help trap fission products (particularly iodine), if any were released, and to keep the building cool in the event of an accident.

A great deal of discussion has centred on the safety of the PWR pressure vessel which has to be designed and manufactured to specifications which ensure that it can not fail catastrophically. Such failure might arise from an internal explosion from fuel--cladding–coolant interactions at abnormally high temperatures, from failure of the vessel support system, or from a mechanical weakness exposed by operational stresses. The first contingency is guarded against by the design of the control and emergency cooling arrangements and the second by mechanical design.

The mechanical integrity depends on both design and the metallurgical and chemical characteristics of the steel, including its resistance to radiation, thermal cycling and shock, fatigue and chemical attack. These topics have been studied in great detail, particularly the question of fatigue crack growth where minor flaws in the initial vessel could act as centres of stress and grow to the point where they might lead to sudden major failure. Sufficient is now believed to be known to permit estimation of the critical crack size below which a crack will remain stable under the radiation, thermal cycling and stresses to which the pressure vessel is exposed over its lifetime, including any major failure of the primary cooling system.

To ensure that defects above the critical size are absent the materials have to be produced and fabricated to extremely tight specifications. They then have to be checked both before and whilst in service by visual and ultrasonic methods to demonstrate that no critical defects exist. Provided the design intentions are met using well-characterized materials using tested procedures backed by quality assurance checks and rigorous inspection, the vessels are believed to have a failure rate during their operating lives of less than one in a million reactor operating years.

The pressure tube reactors like CANDU (see Chapter 3) have their pressurized primary coolant and fuel contained in a large number of separate tubes passing through the moderator. These tubes are separated from the moderator by gas filled annular tubes which can be monitored to detect any leakage from the pressure circuit. It was believed that the design of the pressure tubes was such that any failure would be gradual and detectable so that a defective channel could be shut down and the tube replaced.

Sudden tube failures due to hydrogen embrittlement have, however, occurred (see Chapter 13) due to a materials mismatch and this has led to a programme of retubing those CANDU reactors affected. The separation of the fuel channels and the large heavy water heat sink mean that tube failures whilst undesirable are not catastrophic in the way that a major failure of a PWR pressure vessel would be.

7.3.4 *Other Safety Features*

A main feature of nuclear plant safety design is defence in depth. The reactors are designed to be benign and easily controlled but at the same time should any failure or series of failures occur the reactor would shut down, cooling would be maintained or restored and fission products would be retained. Essential measuring instruments, power supplies, circulatory pumps, coolant circuits are replicated and care is taken to ensure that cable and pipe runs and safety devices are such that common mode failures are not possible whereby nominally independent systems could be put at risk by a single event such as a mains power failure, a localized fire or localized mechanical damage.

7.4 Nuclear Plant

The essential features of safe plant design have been outlined in section 7.2. In general plant or other facilities containing radioactive material for storage or transport have to be designed so that the radioactivity is safely contained taking account of heat generation, criticality, chemical reactivity and the range of accidental damage or other failures that might occur in the particular circumstances.

Thus transport containers for spent fuel have to be capable of withstanding the impact of a traffic accident and a possible fire without losing their integrity, whilst at the same time dissipating passively any heat produced from the fuel.

Spent fuel or fissile material stores are designed to ensure physical separation of materials so that even in the event of ingress of moderating materials (e.g. water) a critical mass could not be formed. Forced or convection cooling with appropriate back up systems are incorporated where necessary.

In nuclear processing plant the vessel geometries and protective measures, including operational rules and monitoring systems (with due redundancy), are designed to ensure that critical concentrations can not be produced. Plant also has to be designed so that in the event of any physical damage resulting in a leakage of radioisotopes, the radioactivity is contained and can be dealt with safely.

The quantities of radioactivity that could be released in even a major incident at a chemical plant are small compared with those from a reactor and the main or only impact would be very localized. In addition to the normal physical protection through containment, in which can be included the building housing the plant, the radioactivity alarm systems enable speedy location of leaks and rooms or buildings can be evacuated if need be, with subsequent operations being conducted employing suitable protective equipment.

Some concern surrounds the routine dispersal of radioactive materials to the environment as part of nuclear plant operations. The levels of release are carefully monitored and subject to strict limits imposed by government to ensure that the most exposed individuals are not subjected to significant additional risks from these practices (see Chapter 6). Since effluents can be fed to delay tanks and monitored before release there is no reason why these arrangments can not be scrupulously observed.

7.5 Historical Incidents

7.5.1 *General*

There have been a number of serious incidents in experimental reactors which have resulted in damage to the reactor core. The majority of these have been in the USA and have not involved any significant release of radioactivity to the environment. The list includes the Fermi–1 reactor in Detroit, the experimental breeder in Idaho, the sodium reactor experimental facility in California and the Westinghouse test reactor. Further incidents have occurred at the Canadian NRX reactor at Chalk River and the Argentinian RA-II reactor near Buenos Aires. In the latter incident a technician changing the fuel configuration in the zero energy experimental facility was killed as a result of breaches in procedures. These resulted in a prompt criticality incident which produced a burst of radiation but did no physical damage.

More serious accidents involving major physical damage to plant have occurred in the plant used for defence purposes at Idaho (USA) and Windscale (UK) and in the civil power reactors at Browns Ferry (Alabama), Three Mile Island (TMI) and Chernobyl. Brief descriptions follow below.

7.5.2 *The SL–1 Reactor (1961)*

This experimental 3 MW t BWR, fuelled with highly enriched uranium–aluminium alloy, had its central control rod withdrawn whilst it was shut down with its head on. The sudden power excursion melted the fuel and led to an explosion which lifted the reactor vessel several feet and blew steam, water and core debris into the reactor building killing three operators.

Despite the fact that there were few engineered safety features and the reactor building offered no special containment, subsequent analysis showed that only about 0.3 per cent of the core inventory of iodine–131 and 0.03 per cent of strontium–90 and caesium–137 was released to the external environment. In total 0.01per cent of the total fission product inventory was released compared with some 5–10 per cent which had escaped into the reactor building itself, illustrating the considerable margins of safety built into accident studies which assume that all fission products released from the fuel find their way into the environment.

7.5.3 *Windscale (1957)*

The No. 1 plutonium producing air cooled graphite moderated reactor at Windscale normally discharged cooling air through filters and 400 ft stacks. A fire starting in the fuel channels set the graphite alight and had to be extinguished using water. The fire took two days to get under control, during which time 20 000 curies of iodine–131 together with other radioisotopes were released to the atmosphere. The release was only a small part of the core content and it was greatly diluted by dispersion in the atmosphere. The levels of iodine–131 were such that sales of locally produced milk were stopped for a period and radiation levels in all foodstuffs were monitored until the small hazard had passed. The radiological implications for the public are discussed further at the end of Chapter 6.

7.5.4 *Browns Ferry (1975)*

A lighted candle being used to check air leaks

in the containment building of the three BWR plants set fire to electrical cables and spread, taking several hours to extinguish. The two 1100 MWe reactors at full power at the time were shut down safely but the fire immobilized the residual heat cooling system and emergency core-cooling system for unit–1 and alternative means had to be found rapidly for effecting cooling. This required depressurization of the unit and came close to exposing the top of the core.

In the event there was no damage to the fuel and no release of radioactivity, but the electrical damage took over a year to repair at a cost of $10 million.

This fire drew attention to the dangers of common mode failures and led to physical separation of back up cable and pipe runs in subsequent designs.

7.5.5 *Three Mile Island (1979)*

One of the 960 MWe PWRs on the site near Harrisburg, Pennsylvania, was operating at near full power when a combination of valve problems in the feed water system caused the main pumps to trip which in turn caused the main turbine to trip automatically. The reactor perfomed properly with a rise in the primary coolant circuit pressure leading to automatic shutdown.

With the rise in reactor temperature and pressure due to the decreased cooling a relief valve opened to release excess pressure and stuck in that state for 150 min before the fault was realized. Auxiliary feed pumps started up but delivered no water because the isolating valves had been wrongly left closed after earlier maintenance. Steam and water were discharged to the drain tank and containment building and the pressure fell to 110 bar bringing in, correctly, the HPI system. This was throttled back by operators, wrongly, because of a high water level signal.

Two-thirds of the total primary circuit water inventory was discharged to the containment building which was not properly isolated and some was pumped to an auxiliary building. Nevertheless the fuel was immersed, although coolant was boiling, and decay heat was being removed. At this stage operators realized the pressure relief valve was stuck and took action to stop the loss of water. Inadequate flow was provided and the core was exposed temporarily with fuel reaching 1800 °C and reactions between zircalloy cladding and steam produced hydrogen.

At about 200 mins. the HPI system was turned fully on and reflooded the core although hydrogen and steam impaired coolant flow. Some 10 hrs were required to get the reactor into a stable condition and a further 5 days to remove the hydrogen which it was feared could have caused an explosion.

A large part of the fuel cladding was destroyed and fuel at the top disintegrated and collapsed partly blocking coolant flow paths. The majority of the noble gas fission products were released from the fuel into the environment but little of the iodine-131 (1 part in 10 million). Consideration was given to local evacuation and pregnant women and children were advised to leave as a precautionary measure which, in the event, need not have been done. The radiation effects on the local populace were negligible (see end of Chapter 6).

The owners of the plant are now removing the fuel from the damaged reactor under carefully controlled conditions. The total financial costs in terms of lost output and clean up will be in the region of $1 Bn.

The lessons of TMI were mainly concerned with better operational practices, better core instrumentation, and the avoidance of arbitrary interference with the built in safety systems.

7.5.6 *Chernobyl (1986)*

The accident at the Soviet 1000 MWe RBMK boiling water cooled pressure tube graphite moderated reactor was due to the deliberate overriding of safety and control systems and disregard of safety instructions by operators conducting tests on turbogenerators. The tests had not been cleared by the safety authorities. The reactor, which has no direct civil parallels in other countries, has design features which, under the circumstances that prevailed, led to a runaway reaction and the destruction of the reactor.

The experiments were intended to check whether or not the reactor could continue to power emergency control systems during run-down in the absence of other sources of mains supply and until such time as stand-by diesel generators could be started up.

Power run-down for a scheduled shut-down was begun 12 hours before the accident and the emergency core cooling system was switched off contrary to safety rules. A further reduction was made 10 hours later but reactor power was erroneously allowed to fall to 30 MWt, well below the minimum target of 700 MWt at which operation is not permitted due to known stability and control problems in this reactor type. The reactor was brought back to 200 MWt by manual control some 20 minutes before the accident. Due to the build-up of neutron absorbing fission product xenon in the fuel following power reduction, the control rods had to be withdrawn further than regulations permitted, and the reactor was brought to a dangerously unstable condition.

Failing to realize the risks, the operators began the test and increased the main cooling flow which decreased reactivity and led to further withdrawal of the control rods and a drop in water level in the steam drums which led the operator to switch off their reactor trips.

The operator, believing that he had stabilized the reactor, turned off the final reactor trip so that he could repeat the test if necessary—the trip that would have closed the reactor down safely when he tried the fatal test.

When he initiated the test the coolant flow pumps slowed, steam formed in the pressure tubes, the voidage created increased reactivity and the temperature of the fuel rose, producing more steam in a positive feed-back loop. The rising reactivity alerted the operator who activated the manual shut-down system, but the control rods were too far out to have immediate effect so the power rose within seconds to some 100-times normal full power in a prompt criticality excursion (Chapter 2) before the control rods were able to stop the chain reaction.

The overheated fuel burst the cans and turned remaining coolant to steam which in turn burst a great many of the pressure tubes and lifted the reactor shield letting air into the hot (700 °C) graphite. A further chemical explosion, possibly caused by hydrogen from high temperature steam–zirconium reactions, wrecked the containment building and this, allied with the fact that the graphite had by then caught fire, led to the release of large quantities of fission products.

The fire took several days to extinguish using sand, boron and lead dropped from the air. The 'blanket' and the damage to the cooling systems made removal of decay heat difficult and a complete core meltdown was feared. After about a week some cooling was effected and fears eased. The reactor has now been underpinned by a reinforced concrete raft to prevent any leakage of radioactivity to the ground waters.

The fire was accompanied by a release of high levels of fission products and some actinides which raised atmospheric radiation levels and levels in food in most European countries. A major evacuation of people within a 20 mile radius of the site was necessary and some of the area may be unsuitable for residential use for a long time. Some 30 people were killed fighting the fire or died from acute radiation effects within the first few weeks after the accident. Some hundreds were seriously affected by radiation and many thousands have had doses from fallout that could affect their life expectation.

The incident has led to calls for more effective alerting procedures for events with trans-national implications and a call for further international safety studies. It offers the prospect of a major improvement in the basic statistics of low level radiation effects on humans and the environment if proper monitoring is instituted.

The Chernobyl disaster arose from deliberate breaches of safety rules in an experiment which had not been adequately planned or scrutinized. Several of the reactor's own defences, which would have prevented the power excursion, were overridden. The reactor design had shortcomings in terms of its ability to get into a positive reactivity feed-back loop and in terms of the opportunity afforded to shut-off safety systems. The Soviet Union has taken steps to remedy the design deficiencies on existing and future plant and to strengthen its operating procedures. It is confident that given these changes the remaining RBMK reactors, including those at the Chernobyl site, can be run safely. (Gittus, 1987).

7.6 Inherently Safe Reactor Concepts

The increasing complexity and cost of engineered safety in the water cooled reactor systems which dominate world markets has led in recent years to consideration of alternative approaches. For water reactors concern has centred on the question of fission product heat generation and ways in which this can be dissipated from a shut down reactor. If the reactor is small and the design allows for sufficient heat loss through convection or thermal radiation, the need for complex safety cooling might be greatly reduced. This condition is, of course, one that is approached more closely by the gas cooled reactors and the pool type fast reactor than the PWR and BWR.

Various designs have been produced that go some way to meeting the criterion for passive safety. The Swedish SECURE in its power version PIUS is a PWR immersed in a large tank of cold borated water inside the reactor pressure vessel. Any temperature disturbance in the system would link the pool into the primary coolant circuit and shut the reactor down (see Chapter 5 and Klueh, 1986).

The HTGR (Chapter 3) can also be designed so that its negative temperature coefficient of reactivity would stabilize its temperature in the event of coolant loss and, in small sizes, the fission product decay heat could be dissipated without coolant flow.

One characteristic of these systems is their small size which loses benefits of scale but there might be compensating gains from replication of the larger numbers required and reduced safety costs. The smaller size also has attractions to smaller utilities for logistic and financial reasons. (See Chapter 21.)

The liquid metal cooled pool type fast reactor also has considerable attractions, even in large plant. As mentioned earlier, the reactor has a strong negative power coefficient which ensures stability in normal operation and, in the event of loss of pumped coolant flow with the reactor remaining on power, it would ensure that the reactor temperature stabilized, on the basis of natural thermal convection, at temperatures below the boiling point of the coolant. The fuel would therefore suffer no damage.

The submergence of the whole core and primary coolant circuit under liquid sodium in a double walled vessel, with no penetrations below the coolant surface, combined with operation at near atmospheric pressure, means that loss of primary coolant is incredible and that the reactor can survive loss of all decay heat removal for some 10 h after it has been shut down. The provision of multiple shut down and decay heat removal systems provides added assurance, particularly since one of the latter operates on natural circulation and heat removal to atmosphere via a sodium–potassium alloy circuit where the coolant is liquid at room temperature.

The sodium coolant in the fast reactor also has a high affinity for fission products such as iodine–131 and this provides a chemical safety barrier which is additional to the physical containment.

The pool type fast reactor therefore inherently possesses many of the safety features being sought in the development of ultrasafe thermal reactor systems.

7.7 Safety Assessment

7.7.1 *General*

There is no such thing as absolute safety. Everyone is perpetually at some risk of injury whether he realizes it or not and every system has some potential for failure due to inherent weakness or the incidence of external events. In favourable circumstances the risk may be infinitesimally small, in others it may be large and obvious.

This section addresses two matters: the assessment of the level of risk and the apparent acceptability of different risk levels. The question of risk in nuclear as opposed to other energy systems is dealt with in Chapter 8.

7.7.2

7.7.2.1 Preliminary Hazard Analysis In any engineering or plant design one of the first things done is to identify the obvious hazards and their consequences and then to structure the design process to eliminate any that are considered unacceptable. This procedure leads to a set of general design principles for the case under consideration.

The approach can be systematized to look in more detail at complex plant or equipment by identifying possible accidents involving the outline design, their consequences, their effects on other plant items, and, where the consequences are unacceptable, to identify changes to the design or process which will overcome the problem. The approach is essentially qualitative although a quantitative examination could be undertaken of consequences and the costs of remedial action, were it of particular significance.

7.7.2.2 Failure Mode and Effects Analysis Consideration of individual components or sub-systems can indicate the ways in which they might fail and can assess the implications of such failure. Thus failure of a thermostat could lead to overheating or a shut down of heating systems depending on its mode of operation. The penalty for such failure might be economic, might be of no real consequence, or might give rise to serious safety problems depending on the heating system and its use.

Clearly if a failure mode leads to a dangerous situation then steps would be needed to provide back up control devices or to change the system design. Similarly if the penalty is economic, for example through needless shut down of a production line, then alternative devices of greater reliability might be incorporated.

The over-all objective is to ensure that the likelihood of failure of components or systems is kept to a level at which the consequences are acceptable.

7.7.2.3 Hazard and Operability Studies These studies use a more systematic approach to examine both the design and operation of a plant in an attempt to embrace all possible malfunctions and maloperations.

For a plant, the normal operating conditions are defined and the operating instructions are taken into account. Possible deviations from the norm are then systematically considered using a system of checkwords such as 'more', 'less', 'no', 'part of', 'as well as', 'other than', 'reverse'. These are applied to each process variable for each section of the plant.

Similar checkwords can be applied to operations: viz. 'starting', 'stopping', 'early', 'late'; either for operations in isolation or in relation to responses to plant parameter deviations.

The implications of deviations singly or in combination can thus be thought through, ideally by a multidisciplinary group. Any that are undesirable indicate the need for appropriate remedial action in the plant design, the plant control systems or in the operating instructions. The technique is claimed to save considerable time in plant commissioning but to have weaknesses in the identification and analysis of low probability events caused by unlikely fault sequences.

7.7.2.4 Event Trees The event tree is a more comprehensive methodology which also starts by identifying a possible primary failure or event called the initiating event and then explores the chain of consequences that might follow, culminating in the worst case with some undesirable hazard.

The event tree, as indicated by its name, starts with the initiating event and branches sequentially depending on the occurrence or non-occurrence of further consequential events. The latter may involve inbuilt automatic responses or operator intervention, or the simultaneous existence or occurrence of other conditions in the plant or its environment. Figure 7.1 illustrates the process for a gas leak. The numbers on the branches represent illustrative assessments of the probability of each branch being followed. The over-all probability of any outcome of a given chain of events can be derived by multiplying the sequential probabilities backwards from the outcome to the initiating event. The sum of all sequences leading to a dangerous outcome is the over-all risk which, in the case of Figure 7.1 is a probability of 4.75 x 10^{-2} that the leak will lead to an explosion.

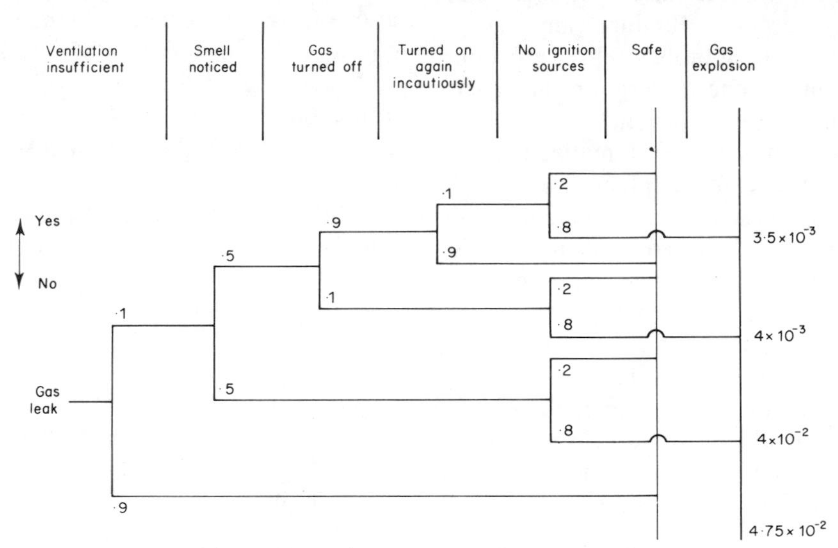

Figure 7.1 The event tree for a gas leak

The assigned probabilities may be subjective judgements or they may reflect statistical experience with similar systems. If the over-all risks appear high steps can be taken to reduce them. For example, a gas detector and alarm might be installed to enhance the likelihood of detection. This might also be used to trigger automatically some additional safety measure, like a shut down in gas supply or enhanced ventilation, in order to eliminate the need for a human presence and intervention in the short term.

In the case of nuclear plant the technique can be applied to the events following failure of individual components or their combination, to trace the consequences and set numerical risk factors to them.

7.7.2.5 Fault Trees These are the reverse of event trees. They begin by identifying a 'top event', which might be the gas leak referred to in the previous section, and then identify faults that might give rise to it, singly or in combination. These so called first level faults may have arisen in turn from other faults and these are traced back in a branching tree until a point is reached where the fault is a component failure for which statistical data are available or can be obtained. This failure is then termed a 'basic event'.

For large systems such as nuclear reactors fault trees can be large and very complex so that a great investment of effort is required to ensure that any study is comprehensive. They have to look at faults that can arise by alternative routes (in which case the over-all probability is the sum of the alternatives) or they can require the coincidence of earlier occurrences (in which case the over-all probability is the product of the precursors). In a comprehensive study the probability of the top event occurring can be assessed and the relative contribution of different routes to the event can be seen.

From the designer's standpoint various

ways in which undesirable top events can have their probability of occurrence reduced are exposed by the fault tree. He can then assess the feasibility of reducing or eliminating dominant fault paths.

7.7.2.6 Common Mode Failure The need to avoid common mode failures was discussed earlier in this chapter. In probabilistic risk assessment an overall low probability of risk is often the result of multiplying a sequence of probabilities for necessary precursor events together, on the assumption that they are independent of each other. If, however, a single accident (like the Brown's Ferry fire) can simultaneously affect a number of ostensibly independent systems, then the simple analysis is invalidated and risk levels can be increased significantly.

To overcome this problem attention must be paid to the possibility of external events, such as fire, flood or impact, or operator intervention or error, or other ways through which common mode failures could be induced. In critical safety systems good practice requires that all items identified in fault trees are identified with their location and the tree searched to see whether an event in any location can affect interacting systems.

7.7.2.7 Probabalistic Risk Assessment The use of fault trees and event trees, singly or in combination, can give an indication of the probability of particular modes of failure or undesirable events arising for specific designs of plant. The techniques have become well established in the past quarter century and computer techniques assist their application to large complex systems.

In complex systems, however, the techniques have some limitations. Fault trees can be hard to understand and do not follow the designer's flowsheet or plant layout. Event trees are not well suited to dealing with parallel sequences of events. Both are restricted to binary logic, black or white, on or off, and can not accommodate continuously variable events or faults easily. Mistakes and omissions can be hard to spot and common mode failures are an added complication.

In treating quantified risk the probabilities assigned to events or faults are not themselves certain but have a probability distribution. In some cases, like standard component failure, both the mean probability and its distribution can be determined statistically, but for subjective probabilities assigned to rare or unique events the range itself has to be judgemental.

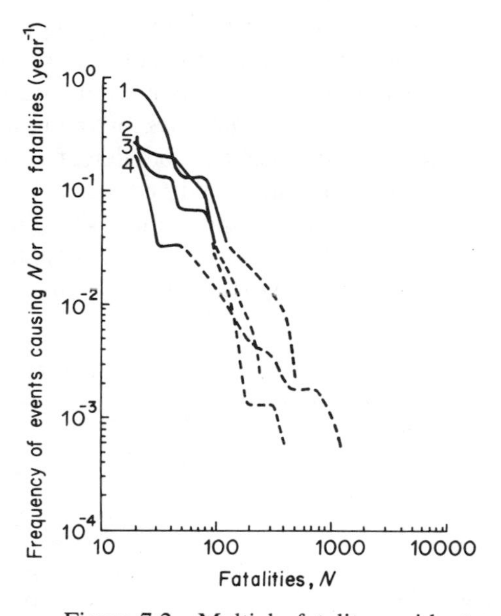

Figure 7.2 Multiple fatality accidents.
1. Aircraft accidents.
2. Mining accidents.
3. Railway accidents.
4. Fires and explosions.
— Data on accidents occurring in the UK
--- Extrapolation based on world data

The distribution functions for probabilities can be incorporated into analyses to yield distributions for specific top events or outcomes. In such cases computer sampling techniques such as Monte Carlo or latin hypercube methods come into their own.

The end result of full probabilistic risk assessments will be a series of risk distributions linked to specific groups of fault sequences and specific consequences. The overall risk index for a system will be the bounding envelope on a frequency–consequence diagram (Figure 7.2) provided sufficient fault-event sequences have been followed in depth.

7.7.2.8 Human Intervention Human intervention can affect the risk associated with plant and equipment in many ways. Operators and maintenance staff can make errors of commission or omission which contribute to plant failure or inactivate built in safety systems. Where such events can be anticipated the plant designer can attempt to build in additional automation or warning systems to eliminate the risk or to alert operators to their occurrences. The inadvertent use of wrong materials or techniques of manufacture and maintenance can be detected by careful inspection and quality control.

For nuclear plant and other situations where life may be at risk (e.g. aircraft) staff training and supervision with the use of computer based simulators to familiarize staff with hazard conditions and the remedial actions are essential.

In general human intervention can be beneficial, even in accident conditions, provided staff are properly trained and have time to react. In both the Windscale and Chernobyl reactor fires the effects following the initial incident were greatly reduced by human intervention despite the very considerable difficulties. On the other hand, where time is short or where actions are based on an incorrect assessment of a situation, human intervention can lead to or exacerbate problems as at TMI (although here the situation was also eventually rectified by human intervention). It is clearly difficult to build such aspects into probabilistic assessments yet they should not be overlooked in event tree or failure mode analyses.

Table 7.1 Incidence rates of fatal accidents (deaths per million per year)

	No. of industries reported	Low rate	Median rate	High rate
Canada, 1975–76	11	10, finance and insurance	150	1240, fishing, hunting and trapping
France, 1979	15	12, clothing manufacturing	75	405, transport and maintenance
Federal Republic of Germany, 1978	35	19, health and welfare 22, textiles and clothing	100	850, inland waterways
India, 1976–78	10	44, cotton textiles	155	440, mining
Japan, 1977–78	33	10, communications	40	1040, coal and lignite mining
Philippines, 1976	12	120, textiles	670	830, mines and quarries 900, saw mills
Sweden, 1979–80	9	12, education and hospitals	90	255, mining
Switzerland, 1968–72	49	15, textiles 25, precision instruments	250	1140, forestry 1340, foundries
South Africa, 1968	23	28, textile manufacturing 34, leather manufacturing	190	980, transport 2000, fishing
UK, 1980	25	3, clothing and footwear	30	365, quarrying
USA, 1982	8 major groups, 40 subgroups	(8 subgroups less than 10)	113 (mean)	550, mining and quarrying

Reproduced by permission from ICRP (1985). 'Quantitative basis for developing an index of harm', ICRP 45, *Annals of the ICRP*, Copyright (1985) Pergamon Press, Oxford.

7.7.3 Acceptability of Risk The preceding paragraphs have indicated that despite its inevitable limitations, probabilistic risk assessment is capable of giving some broad quantitative indication of the level of risk associated with particular plants. It also provides a basis on which the designer or operator can identify and modify specific risks through redesign, replication of sub systems, or specified operational procedures or regulations.

Table 7.2 Incidence Rates of Fatal Accidents in the UK (deaths per million at risk per year 1970–1980)

Industry	Mean
Clothing and footwear	3
Instrument engineering	9
Electrical engineering	9
Textiles	19
Vehicles	19
Paper, printing and publishing	24
Food, drink and tobacco	32
Metal goods (other)	29
Leather, leather goods and fur	29
Mechanical engineering	33
Timber, furniture, etc.	37
Chemical and allied industries	67
Bricks, pottery, glass, cement, etc.	70
Shipbuilding and marine engineering	113
Metal manufacture	118
Coal and petroleum products	148

Reproduced by permission from Annual reports of Chief Inspector of Factories and Health and Safety Executive.

However the question remains as to whether the residual estimates of risk are such that the plant should be acceptable to the public. There are as yet, and may never be, universally accepted standards, although it would not be unreasonable to feel that there should be groups of clearly acceptable and clearly unacceptable risks separated by a grey area where opinions might differ.

Tables 7.1 to 7.4 indicate the risks of premature death attached to different occupational groups in different countries and within the UK; and the risks of accidental death from a range of causes in the UK and USA. The figures appear to show significant variations between countries but the statistical bases may not be identical. The relativities within a country are not subject to this error.

Table 7.3 Incidence Rates of Fatal Accidents in UK and USA (deaths per million per year)

Cause	Incidence of fatalities	
	UK	USA
All accidents	340	630
Road accidents	140	250
Falls	110	100
Fire	18	40
Drowning	11	33
Electrocutions	2.4	6.3
Lightning	0.2	0.5

Table 7.4 Incidence of Fatal Cancers from Occupational Exposure to Chemicals (Deaths per million at risk)

Occupation	Form of cancer	Estimated annual mortality (per 10^6 at risk)
Shoe manufacturing (press and finishing rooms)	Nasal	130
Printing trade	Lung and bronchus	about 200
Work with cutting oils (Arve district)	Scrotum	400
Wood machinists	Nasal	700
Coal carbonizers	Lung (and bronchitis)	2800
Rubber mill workers	Bladder	6500
Mustard gas manufacturing (1929–45)	Bronchus	10 400
Cadmium workers	Prostate (incidence)	14 000
Nickel workers (pre-1925)	Nasal sinuses	6600
	Lung	15 500
β-naphthylamine workers	Bladder	24 000

Reproduced by permission from ICRP (1985). 'Quantitative basis for developing an index of harm', ICRP45, *Annals of the ICRP*, Copyright (1985) Pergamon Press, Oxford.

In the UK risk levels of about 100 fatalities per million at risk per year appear to be accepted without great concern. This likelihood corresponds to at worst a few incidents from any specified cause in an individual's sphere of contacts during his working life. In other countries or in earlier times in the UK, much higher levels of risk have not been regarded as untoward.

Public attitudes are influenced, however, not only by the frequency of events but by the number of people involved in any single event and by its nature. Thus a single though infrequent aircraft crash killing 500 attracts far greater attention than the daily carnage on the roads, and a nuclear incident involving no deaths gets far greater media coverage than a house fire involving several deaths, despite the fact that the latter are far from uncommon occurrences. The frequencies of events involving multiple deaths are illustrated in Figure 7.2.

A reasonable goal for the nuclear industry would seem to be to keep the level of risks associated with its activities within the bounds of what the public and workers accept from other causes. Thus the average risk of premature death for individual nuclear workers might be set at 10^{-4} per year. A value of 10^{-6} for members of the public would place nuclear risk in a category similar to death from lightning, and well below 1 per cent of the total risk of death from all other potential accidents to which he is exposed. Other suggestions appear elsewhere (Chicken, 1986). A further aim would be to keep the probability of accidents involving larger numbers of casualties at worst comparable to other accidents. Such a criterion was laid down in 1967 in the so called Farmer curve. A detailed analysis of comparative risks on an international basis has been published by the International Commission on Radiological Protection (1985) as a contribution to discussion on this issue. The analysis includes both accidents and risks from radiation and occupational exposure to chemicals which are cancer inducing (Table 7.4).

The US Nuclear Regulatory Commission Commissioners issued a policy statement on safety goals for the operation of nuclear power plants in June 1986. Their qualitative safety goals were: 'Individual members of the public should be provided a level of protection from the consequences of nuclear power plant operation such that individuals bear no significant additional risk to life and health. Societal risks to life and health from nuclear power plant operation should be comparable to or less than the risks of generating electricity by viable competing technologies and should not be a significant addition to other societal risk.'

Quantitative objectives were to be used in determining achievement of the above safety goals. These were that 'The risk to an average individual in the vicinity of a nuclear power plant of prompt fatalities that might result from reactor accidents should not exceed one-tenth of one percent of the sum of prompt fatality risks resulting from other accidents to which members of the US population are generally exposed. The risk to the population in the area near a nuclear power plant of cancer fatalities that might result from nuclear power plant operation should not exceed one-tenth of 1 per cent of the sum of cancer fatality risks resulting from all other causes.'

A number of major probabilistic risk assessment studies have been undertaken on reactors and plant. The first major study to use realistic risk assessments was that by Rasmussen (USNRC, 1975). This demonstrated that the risks from even 100 US LWRs was small compared with other individual or societal risks (Figure 7.3). The risks of a major core melt were estimated to be 5×10^{-5} per reactor year but the consequences of such a melt down were seen as being minor in terms of acute or latent illness or property damage. The Rasmussen study was reviewed by Lewis some years later, who concluded that the methodology was sound but that it failed to attach sufficient weight to the uncertainty in the probability estimates.

A subsequent study for the Federal Ministry of Research and Technology of the Federal Republic of Germany (1979) confirmed the general conclusion of

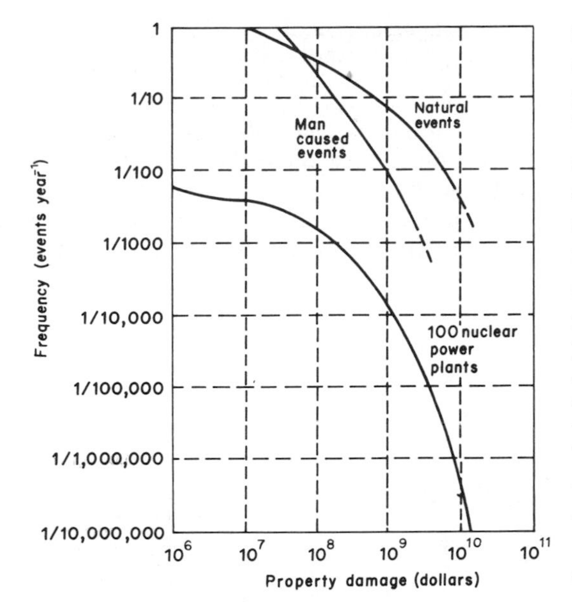

Figure 7.3 Frequency of property damage due to natural and human caused events. Source: WASH-1400. Risks covered include fires, explosions, dam failures, air travel, toxic chemicals, hurricanes and earthquakes

Rasmussen although differing on points of detail.

Similar analyses have been done for reactors such as the UK Sizewell B PWR. The generating board's target for serious accidents is that there should be less than a one in 10 million chance per reactor year of operation of any particular combination of events leading to a core melt and an over-all chance of core melt from all causes of less than one in a million per annum. This target is below the assessed performance of the LWRs considered by Rasmussen and well below the figure of once in 10 000 years that was mentioned for Soviet reactors in the aftermath of the Chernobyl incident.

The methods of probabilistic risk assessment offer the prospect of exposing the assumptions to public scrutiny (although few other than specialists are capable of following the complex analyses), and of drawing comparisons between risks and setting rational standards. This aspect is considered further in Chapter 8.

7.7.4 Other Aspects The importance attached to nuclear safety is such that all see it as an area where the maximum exchange of experience and data is desirable. The international agencies (the IAEA and NEA) have been involved in setting up exchanges, studies and international programmes of co-operative R and D so that all can benefit from the collective experience. All major incidents in the West (like TMI) have been fully studied and the lessons learned passed on so that mistakes should not be repeated. The USSR has indicated that it would in future like to participate more fully in such exchanges and has promised to make its findings on the Chernobyl disaster available. Such co-operation can only serve to make what is already a very safe industry even safer.

There is a need, however, to ensure that expenditure on nuclear safety is not totally out of proportion to the risk and consequences of possible releases. For this reason the International Commission on Radiological Protection has laid down guidelines for what it considers to be an appropriate and cost-effective approach (see Chapter 6). The use of cost-benefit analysis in the implementation of their ALARA criterion seeks to ensure that resources of the economy are efficiently deployed. Up to this point in time the approach is finding application in plant design and operation, but there are problems with its application to accident avoidance (Brown *et al.*, 1987).

REFERENCES

Brown, M., Blackman, T.E., Jones, P.M.S., and McKeague, R. (1987). 'The Application of ALARA', in *Proc. Conf. Inst. Nucl. Engineers, Sept. 1986*, Adam Hilger, London.

Chicken, J. (1986). *Risk assessment for hazardous installations*, Pergamon Press, London.

Federal Ministry of Research and Technology of the Federal Republic of Germany (1979). *The German Risk Study*, Bundesministerium für Forschung und Technologie, Bonn.

Gittus, J. H. (1987). *The Chernobyl Accident and its consequences*, HMSO, London.

International Commission on Radiological Protection (1985). 'Quantitative basis for developing a unified index of harm', *Annals of the ICRP*, Pergamon Press, Oxford.

Klueh, R. (1986). 'Future Nuclear Reactors—safety first', *New Scientist*, **1986 No. 1502**, 41–45.

Lewis, E. E. (1979) *Nuclear Power Reactor Safety*, John Wiley and Sons Ltd, New York.

Lindley, R. A., and Stader, J. E. (1986). 'American designers achieve reduced costs', *Nucl. Eng. International*, **31** 31–33.

Thomson, and Beckerley, (1973).

USNRC (1975). *Reactor Safety Study: an assessment of accident risks in US commercial nuclear power plants*. WASH–1400 (NUREG 75/014), United States Nuclear Regulatory Commission.

Nuclear Power: Policy and Prospects
Edited by P. M. S. Jones

8

Is Nuclear Riskier than other Energy Forms?

HERBERT INHABER
Risk Concepts Inc.

8.1 Introduction

C. Northcote Parkinson, the author of Parkinson's Law and other spoofs, never commented directly on nuclear risks, as far as I know. However, in his writing we can find at least one indirect reference:

> 'It might be termed the Law of Triviality. Briefly stated, it means that the time spent on any item of the agenda will be in inverse proportion to the sum involved.'

Although Parkinson's jests may seem far removed from comparative risk analysis, in this particular case he was accurate. Much of risk analysis concentrates on relatively small hazards, while large ones either go unnoticed or are dismissed.

By way of partial proof, consider merely the total amount of research effort and funding put into study of risk by energy source. While no full accounting has ever been made, most observers would be surprised if the risks of nuclear energy had not been studied more than that of all other energy sources combined. If it were universally agreed that nuclear had by far the greatest risks of known sources, this would have been completely understandable. Yet, as will be shown below, most observers agree that nuclear risk is one of the lowest, if not the lowest, of the major sources of the future. Parkinson may have intended his remark to apply to items in a committee agenda, but the inverse relation between research effort and level of risk fits his statement as well.

Before proceeding further, a brief definition is in order. By 'risk' we mean danger to human health, by death, accidents or disease. The last two categories can, of course, produce the first. While environmental risk, or impact on other living things, is associated with all forms of energy, the subject is too vast to be handled in a short space.

A key to what follows is the notion of comparison. While in principle energy risk analysis is done so that one aspect can be compared easily to another, giving decision-makers the luxury of options, in practice this is often not carried out. Results applicable to one energy system, or even one phase of that system, are often segregated from those of other systems. As a result, both decision-makers and the public frequently find it difficult to use the data of risk analysis, gathered at great cost and scientific effort. The rest of this Chapter will emphasize comparison, because it is by this means that we learn most of what we know.

For example, if we regard someone on the street as either tall or short, it is because we are making a subconscious comparison with everyone else we know. We do this comparison so quickly that we are unaware of it. Because risk analysis is considerably more complex than estimating heights, we have to be more analytical in our approach.

8.2 Some Major Assumptions and Areas Covered

Comparative risk analysis is, if one takes the comprehensive view, an enormous subject. It touches on health physics, human perception, statistics, occupational safety, probability, energy analysis, physics, many branches of engineering and so on. An effort must be made to simplify and clarify by appropriate assumptions. A few are as follows:

1. In this chapter, emphasis will be placed on considering only the major conclusions of risk analysis. This is not then basically a review article, but primarily a summary of major issues with specific comparisons added. Inhaber (1982), Paskievici (1982) and Etnier and Travis (1983) will be the major quantitative sources, although others will be mentioned as well.

2. Wolf Häfele, the director of a major study on energy for the International Institute for Applied Systems Analysis, called the differences of opinion between those who favour nuclear power and others who demand non-conventional or 'soft' energy, a 'holy war'. If there is such a conflagration going on, part of the battleground is the risk, either to the public or workers, produced by the competing energy systems.

   It would be inadequate to dismiss these 'soft' energy sources as those of the distant future, although they may well be. For many, especially in the United States, they are sources which will be used when society gets through the troublesome teething phase of nuclear power and fossil fuels. Again for many, the sooner this happens the better. For them the teething ring should be discarded quickly.

   Most comparative risk analyses have concentrated on what could be termed conventional sources, i.e. those which are now in widespread use. In practice, this has come to mean fossil-fuels, nuclear and hydroelectricity. Non-conventional commercial energy sources produce a very small proportion of total energy in almost all countries, with the possible exception of Israel. Fuelwood and similar sources make a major contribution to heating and cooking in many third world countries but are a small part of commercial energy. Because of the intense interest, almost mediaeval faith, in them on the part of substantial segments of society, it would be unreasonable to eliminate them from a risk comparison.

3. It seems logical, though not provable, to consider the entire fuel or energy cycle for each of the energy systems we consider. This avoids unanswerable questions which arise when we compare one part of system X to another of system Y. For example, we may find that the occupational risk of building and installing a given system may be greater for X than the public risk of Y. But suppose the public risk of operating it were greater than that of X. We can always subdivide the total risk into sections which are greater or smaller than the corresponding parts of another system. As a result, advocates of one or another energy system have often held up a favourable risk result as if it were part of the Holy Grail, claiming that the data showed that their system was the best. Risk results to date show that no energy system has the lowest risk in all phases of activity, from occupational to public. The only way to avoid these conflicting claims is to sum all the risks produced, from mining the raw materials to disposing of the final wastes, if any.

4. The time frame of a risk analysis carries a curious dichotomy. On one hand, we wish to use the most current data available. Very few risk analyses, if any, deliberately use outdated or superseded information. Yet the nature of the data is such that it is spread out over years, if not decades. For example, occupational risk data can often be some years out of date. If we wish to use an average,

instead of a less representative current value, older data may have to be employed.

As another example, for a number of years, the only analytic calculations of public risk from nuclear accidents was the Rasmussen report (1975), issued in 1974–75. Estimates of these risks would have had to have been based, at least in part, on this report until later probabilistic risk analyses were developed in the late 1970s.

A second point on the time frame of risk analyses concerns what can be called incremental risks. The choice confronting decision-makers is not whether they are going to build 30–year old coal-fired plants or reactors, but new ones. As a result, data based on a part of any energy system which is not completely up-to-date should not be incorporated into the calculations.

At first sight, this proposition seems attractive and reasonable. However, it quickly runs into data problems. For most industrial installations, occupational safety is not subdivided by the age of the factory or mine. For example, in the United States coal mining accidents are shown by the Bureau of Labor Statistics for all pits, and not broken down by when they opened.

There is anecdotal evidence that newer installations generally have a better safety record than old ones, but this information is not yet available from official bureaux. While no complete investigation has yet been undertaken, it is highly likely that this condition prevails in Europe as well. In other words, newer is better.

On the other hand, one can easily suggest instances where newer is not necessarily better. For example, the Three Mile Island accident of 1979, the most serious nuclear reactor accident affecting the public of which we had direct knowledge up to Chernobyl, took place not in an ageing and decrepit installation, but in an almost new one. The Vajont dam disaster of 1963, as far as I know the largest single energy-related accident in world history, again did not occur in a cracked and broken-down site, but in a modern one. If there is one rule we can make about the relative risk of new or old installations or entire energy systems, it is that there are no rules.

John Updike once said: 'Americans have been conditioned to respect newness whatever it costs them'. The fragmentary evidence presented above suggests that it can cost them and others in risk to life and limb.

5. In terms of the time frame of risk, we can carry the argument one step forward to estimate what future risk from energy systems will be. For some, the future is what Ambrose Bierce once defined, "when our affairs prosper, our friends are true and our happiness is assured." In terms of risk, it will be when danger to health from energy drops monotonically to zero.

There are some indications that part of this prediction may come true. For example, coal mining death rates have been declining all over the world. The latest probabilistic risk analyses for US nuclear reactors show even lower public risk than the early ones, which had already calculated extremely small values. And the web of safety regulations in the last decade or two at least holds out the possibility or reducing energy risk substantially.

However, there are a number of leaves turned into the wind, rather than blown in its direction. A few suggest themselves immediately, and others probably would turn up on more detailed investigation. For example, non-fatal coal mining accidents have decreased much less rapidly than the fatal variety in many nations. In the United States, coal productivity, as measured in tonnes per hour of work, has declined in the last few years. As a result, the risk of non-fatal accidents per unit energy output has actually gone up in some recent years.

Consider offshore oil production. Until the discovery of oil under Lake Maracaibo in Venezuela in the 1930s, all oil production was done on land. Now an increasing proportion of petroleum exploration and production is done offshore. The North Sea has probably generated the most publicity of these fields.

While data are still not completely definitive, all indications are that offshore production is riskier, per unit output, than onshore. Certainly the record of onshore production does not have the black marks against it that were produced by the capsizing of oil rigs in the North Sea and off the east coast of Canada. As onshore resources become depleted, there will be increasing pressure to produce riskier offshore petroleum.

A third example is related to the previous one. Books like *Limits to Growth*, popular in the 1970s, spread the idea that mineral resources were running out. Many of these concepts proved to be defective, but there is agreement that no mineral will last forever. Already for some like copper the grades of ore being mined are much lower than in the past. All other factors being equal, this will generate greater risk per ton of metal produced. Since metals and minerals are used for all energy systems, in turn this factor will increase the risk per unit energy for all, or almost all, energy systems.

Given the present state of knowledge, it is not possible to say which way—lower or higher—the scales of risk will tilt in the future. Results are likely to be highly localized, with some regions or nations experiencing increases and others decreases. If anything, this discussion should dispel the myth that in terms of risk, 'every day in every way we are getting better and better'.

6. Catastrophic versus non-catastrophic risk forms one of the great divides in the field. For some, the former is the only type worth considering. The 1984 gas explosion at San Juan, near Mexico City, followed within a few weeks by a chemical leak in Bhopal, India, reinforced the belief for many. Certainly these events got far more newspaper headlines than so-called 'routine' accidents which occur in energy facilities.

   In terms of research effort, it seems likely, though not provable, that more work has gone into estimating and reducing catastrophic risks of energy than the ordinary variety. Certainly the massive and expensive probabilistic risk analyses for nuclear reactors, designed to estimate the rate of catastrophes, have entailed much more effort than reducing smaller sources of risk.

   However, Inhaber (1980) has shown that when the entire fuel or energy cycle is considered, the proportion attributable to catastrophes—ascribed by one source (US Bureau of the Census, 1977) to cases where five or more are killed—is low. For those systems, such as hydro or nuclear power, where catastrophes either could take or have taken place, the proportion of catastrophic to non-catastrophic risk is still small, of the order of a few per cent or less.

   So we have here another example of the misplaced attention of society. If we are only concerned about reducing risk, we would concentrate on reducing non-catastrophic risk.

7. Allied to the point above on conventional and non-conventional energy sources is the question of centralization and decentralization. In a sense, the former can be viewed as a subset of the latter. Non-conventional energy systems can be regarded as inherently decentralized or dispersed in nature, although systems like solar thermal electric—the so-called 'tower of power'—with massive capital requirements, would likely be used exclusively by large public or private utilities. Nonetheless, if by

non-conventional one means the solar rooftop collector popular in some parts of the United States, one is clearly referring to a decentralized energy system.

The question then is, how does the degree of centralization affect the risk calculated? By posing this question, we are perhaps tripping over a variety of definitions. First of all, we have no real measure of centralization, or at least one which is reproducible. Clearly, a large coal-fired system or nuclear reactor, looming over the landscape, is about as centralized as one can get, although in the palmy days of nuclear enthusiasm in the 1940s there were claims that one day people would drive fission-powered cars.

Conversely, rooftop collectors scattered on the attics of a city seem inherently decentralized. Yet most of the materials from which they are constructed—glass, steel, copper—come from centralized mills, mines and factories. If these centralized facilities did not exist, the economic costs of supposedly decentralized energy sources would be far higher than at present.

While data are by no means clear, such that do exist suggest that smaller industrial establishments have a higher occupational risk per unit output than larger ones. One possible reason for this lies in the extensive occupational health and safety programmes in most large industrial facilities. These are often scanty or non-existent in smaller establishments.

As a result, all other factors being equal, decentralized energy sources will usually have higher occupational risk than centralized sources. Note that this statement applies solely to risk suffered by workers. Risk to the general public may or may not be higher, depending on the nature of the system.

8. How much risk is voluntary, and how much is involuntary? The question has perplexed risk analysts and policy-makers for decades. If the subject could be easily settled, it would have been resolved long ago.

The reason why the debate continues is, in part, because the definition of voluntariness remains loose. Take, for example, a case of presumably clearcut involuntary risk: the irradiation of the public around the Three Mile Island reactor in Pennsylvania. If ever there was a case of involuntary risk, this is it.

Or is it? Most, if not all, of the surrounding population were aware of the fact that a reactor was in their midst. Those who felt endangered by it could have moved, and some in fact did so when the reactor was being built. The United States has a more mobile population than most other countries, with some estimates showing the average stay in one location at about five years.

So while the risk suffered by the population around TMI was in one sense involuntary, in another it was at least partly voluntary. It is true that it may not be fair to expect people to move from an area when they see an allegedly hazardous facility being constructed, but thousands do this around the world.

Take a counter example of voluntary risk: underground mining. Since the Second World War in Britain, when the 'Bevin boys' were sent down into the pits by law, nobody anywhere has been required to engage in this hazardous occupation. It is then completely voluntary.

Or is it? Most miners are the sons of miners, or at least from the mining district. They often have no alternative to mining in terms of jobs. They are often somewhat aware of the riskiness of the work, thus making its acceptance fairly voluntary, but risk perception studies have shown the prevalance of the 'it can't happen to me' syndrome. In that sense, risks are ignored. This then makes the risk more involuntary.

The two preceding examples show that drawing the dividing line between voluntary and involuntary danger is difficult. In principle, we should concentrate solely on reducing involuntary risk, allowing those who accept voluntary risk—with its presumed side-benefits—to fend for themselves. In practice, the two are so intertwined as to make separating them almost impossible.

9. The power produced by a nuclear reactor is about as reliable as that of the fossil fuel plants it replaces. There are inevitable refuellings and servicing of equipment, but the over-all reliability of the two types of power systems is similar.

Not so with systems dependent on mother nature. When the sun does not shine and the wind does not blow, the solar collector or windmill is of no use to anyone, unless some type of back-up (of conventional energy sources) and/or storage has been provided in advance. These cessations of 'fuel' supply are arbitrary in time, in contrast to the planned outputs of thermal stations. Of course, no power source is perfectly reliable, and this applies to those dependent on mother nature as well, even with their attachments of back-up and storage.

What does this have to do with risk calculations? If we are to assume all energy systems must have about the same reliability, then for some non-conventional energy systems we need storage, back-up or both. This is in addition to the requirements for collectors, windmills and the like.

The precise ratio of storage to back-up is still a matter of some debate, and will depend to some extent on weather conditions—sun and wind—at the site considered. Inhaber (1982) used a value of six hours of storage for solar thermal electric, solar photovoltaic and windpower. Other studies have used higher or lower values.

If we are comparing system A with a given level of reliability with system B at a substantially lower level, we are contrasting the proverbial apples and oranges. Risk enters into the discussion because building and operating additional storage and back-up facilities, to bring system B up to the standard of A, produces both occupational and public risk.

10. Occupational and public risk are sometimes viewed as identical to voluntary and involuntary, respectively. However, for the reason that voluntariness comes in degrees, it is more appropriate to use the occupational–public dichotomy. In fact, this is used in many of the discussions of quantitative risk. Kates and Kasperson (1983) have listed almost all risk surveys published to that date.

In this sense, occupational risk is incurred by everyone who builds and operates an energy system. Public risk is incurred by everyone else. While this definition, like all others, is imperfect, it has a certain logic. It will be used here.

11. For some people, the risks evaluated here are irrelevant to the key questions of energy hazards. They state, for example, that the real risks of civilian nuclear energy lie in the possible proliferation of nuclear weapons due to the spread of uranium and plutonium. Whatever the merits of these claims, it is clear that the deaths from nuclear war would dwarf any of those considered in this chapter, or any other study of energy risk.

Conversely, conjectures have been made that war, either in the Middle East or elsewhere, could break out over dwindling supplies of petroleum. Nobody can predict the chances of this happening, but the Middle East would not be, as the headlines often put it, the powderkeg of the world, if it were barren of oil. In this sense, petroleum—or its lack—could also cause world war.

Predictions of war directly or indirectly attributable to solar energy have been rare. However, the French did use what was then the world's largest solar furnace to develop materials for their nuclear weapons programmes. Some observers have said that if solar energy were widely adopted, its heavy materials requirements would spark conflicts similar to those we see over fossil fuels today.

What do all these contentions have in common? They cannot be quantified in an ordinary sense. Heising, at the Massachusetts Institute of Technology, has made a valiant beginning in assessing the likelihood of nuclear proliferation, but these efforts have a long way to go before they are universally accepted. This poses a dilemma for risk analysts, one that is not likely to dissipate soon. It has been claimed that we evaluate the infinitesimal, while ignoring the infinite.

To put it in another perhaps more colourful way, some have said that risk analysts are in the position of the tipsy gentleman who was noticed searching for his car keys under a street lamp. When asked where he had lost them, he said that it was down the street. Confronted by this contradiction, he said that the light was much better where he was looking. Risk analysts hope that they are not in the same position as that misguided searcher.

12. The units used in risk calculations can determine what conclusions are drawn. For example, compare a small backyard windmill to an imposing fossil fuel plant or nuclear reactor. Without knowing any of the risk factors involved, it would seem intuitively obvious to most observers that the larger facilities would have greater occupational risk, and probably public risk as well.

Arriving at this conclusion would ignore the fact that the larger facilities would also produce much more energy. The reasonable way to handle this is to divide the risk produced by the energy generated. In this way, a small facility is put on the same basis as a large one.

When this is done, our perspective changes. Because a nuclear reactor produces so much energy, the denominator for the risk formula is enormous. The risk per unit energy is then usually small. Conversely, the energy produced by a small solar collector or windmill is tiny. The denominator is then small, and the risk per unit energy will likely be large, other factors remaining constant.

This reversal of what at first glance seemed obvious is no mere mathematical sleight of hand. The common currency of the subject is surely total energy produced in the course of a given time period. Just as we measure the economic cost of energy by having the latter quantity in the denominator, we do the same in terms of risk.

13. Finally, perception. There is little doubt that there is often a divergence between what risk analysts conclude and what the public—or at least some parts of it—think or believe. As will be shown below, with almost no exceptions risk analysts have concluded that nuclear energy has among the lowest risks of any energy system. Many members of the public would dispute this.

Exactly why this is so remains controversial. Slovic and his colleagues in Oregon have attributed this to a 'dread factor', but how this came about and how much of it is based on fact as opposed to conjecture remains unknown.

Some of this dread is undoubtedly attributable to the mistaken belief that reactors can explode like nuclear weapons. Surveys in the United States have shown that about half the population have this conviction, based on a condition which is physically impossible. The proportion holding this belief has remained about constant over the decades, in spite of endless statements from the nuclear industry—and even some from anti-nuclear activists—that the hypothetical explosion could not happen.

Risk analysts are then faced with a situation in which they are told that, regardless of the accuracy of their work, substantial portions of the public are skeptical of their conclusions. The implication is then that much of their work is vitiated. However, one cannot build a scientific enterprise by letting it be swept about by the rising and falling tides of public opinion.

Bill Clark of Oak Ridge Associated Universities and the International Institute for Applied Systems Analysis has devised an analogy to the way in which nuclear risks are calculated and how they are perceived. Witches in the Middle Ages and later were murdered by the thousands. Yet no objective evidence for their existence was ever developed. Strange events were attributed to them, and public perception of the day demanded that those who caused these events be punished. Today anyone believed to cause bizarre events would likely be the subject of an approving story on the evening news. Yuri Geller's apparent ability to bend spoons from a distance turned him into an international celebrity in the 1970s. Thus has public perception changed, while the facts remain the same. Does the debate over nuclear power hold any similarities?

8.3 How Does Nuclear Compare in Risk?

The above discussion has, of necessity, been general. Unfortunately, many papers on relative risk end at about this point, noting that all energy systems have at least some risk, but refraining from ranking them. A typical example is Rom and Lee (1983), but there are many others.

Even worse, occasionally specific conclusions are drawn from data which do not warrant it. An example of this type of *non sequitur* is again Rom and Lee (1983). After discussing many energy systems, most in a non-quantitative way, they conclude:

> 'We encourage government research and development funding for solar and nuclear fusion projects, which appear to have less potential for adverse environmental and health impacts, rather than for nuclear fission or synthetic fuels where the private sector is already active.'

Leaving aside the question of whether research funding should go to high-risk or low-risk energy sources—a logical case can be made either way—the truth is that this paper did not show that solar or fusion do have 'less potential' for risk than other energy sources. Exactly what 'potential' for risk means, as opposed to actual risk, is unclear. Finally, the question of what proportion of an energy system is publicly or privately owned or financed is fascinating, but irrelevant to the question of risk analysis. In terms of fission alone, accidents have taken place at public (Windscale) and private (TMI) facilities.

Paskievici (1982) is probably the most complete compilation to date of the risks of conventional energy systems. He considers a total of nineteen studies, ranging in time from 1974 to 1980 (these studies are listed in the Bibliography). Many of these studies draw on previous ones, so it is unlikely that any of them is completely independent. Some were done in the United States, Canada and France, but others have an international flavour, such as those performed by the Organization for Economic Co-operation and Development and the World Health Organization. It would take far too much space to describe all of the assumptions and databases. Paskievici himself does not attempt this in the space of 25 pages.

Paskievici confines himself only to the conventional sources, and only electricity derived from them: coal, oil, gas and nuclear. Although hydroelectricity is in common use in many parts of the world, its risks have only rarely been calculated. Inhaber (1982, Appendix K) finds that the public risk of hydro is between three and thirteen times that of nuclear. Its total risk, both occupational and public, is between the relatively low values of nuclear and the high values of coal.

Paskievici divides the data into occupational and public risks, and then subdivides it even further. For example, in terms of public deaths from diseases in the coal cycle, he divides the data into two sections: that from conversion, i.e. burning the fuel, and waste management, i.e. risk due to ash-heaps or other wastes. The former case is primarily due to air pollution.

Only some of these purportedly comprehensive risk analyses fill in all the boxes in all the tables. That is, there are still considerable gaps in our knowledge. Yet when all nineteen studies are compared, we get a synthesis impossible with just a handful. The blanks in some of the risk tables can be viewed as inadvertent omissions on the part of some investigators, and those who supplied values taken as having approximately correct data. For example, of the nineteen studies, only seven had values for the number of occupational disease deaths in the uranium fuel cycle for transportation.

This would indicate a major gap in our knowledge if the seven recorded values were both large and highly variable. They are not. In this particular instance, the values range from zero to 0.004 deaths per gigawatt-year. We can compare these values to the range for all occupational disease deaths for the uranium cycle, about 0.1 to 0.3. The point made here is that we do not need every study to estimate every aspect of every energy system. There is enough agreement among the various studies to allow broad conclusions to be drawn.

Before discussing these values in some detail, some note should be made of the limitations of Paskievici's analysis. First, his work is only as good as the studies which he evaluates. However, many of the studies do not explicitly list their assumptions, so the problem of comparing different studies is severe. To take one example at random, some of the studies include the risk attributable to building the mines and mills from which the raw materials—uranium, coal, etc.—flow. Others do not. Even if all aspects of the studies were similar, they would diverge in this respect.

Paskievici has not included values based on the probabilistic risk studies of reactors which have filled the literature. This may seem like an odd omission, given the fact that considerable effort has gone into these PRAs. He states that he: '. . . deals only with real accidents, (so) risks from hypothetical accidents are not further discussed'.

Certainly the risk of reactor accidents forms only a small part of total uranium cycle risk. Inhaber (1982, Table D–2) found that reactor accident risk formed from 5 per cent to 10 per cent of total uranium risk. This was after multiplying reactor risk (derived from the Rasmussen report, 1975) by a factor of about eight. Using a more realistic value would have put reactor accident risk at about 1 per cent of total uranium risk. Then the omission of Paskievici seems reasonable.

In Table 8.1, the trans-scientific risks such as weapons proliferation and resource wars are excluded. The reasons for this were noted earlier. The two categories—those noted in Table 8.1 and those discussed earlier—seem to encompass all that is known about risk. As Harold Pinter wrote in *The Homecoming:* 'Apart from the known and the unknown, what else is there?'

What conclusions can we draw from Table 8.1? First, coal and oil have by far the highest risks compared to natural gas and uranium. Of the four, gas appears to have the lowest values. However, its total number of non-fatal events is about the same as those for uranium. As will be discussed below, the uncertainties in all the values are large enough to produce an overlap between gas and uranium in terms of fatalities.

Second, public risk dominates for coal and oil, but occupational risk for gas and

Table 8.1 Accident and Disease Hazards per gigawatt-year of electrical energy (adapted from Paskievici, 1982)

	Accidents				Diseases					
	Workers		Public		Workers		Public		Total	
	fatal	non-fatal	fatal	non-fatal	fatal	non-fatal	fatal	non-fatal	fatal	non-fatal
Coal	1.4	60	1.0	1.8	1	3	10	2000	13	2100
Oil	0.35	30	?	?	?	?	10	2000	10	2000
Natural gas	0.20	15	0.009	0.005	~0	~0	~0	~0	0.2	15
Uranium	0.20	15	0.012	0.11	0.1	1	0.1	0.1	0.4	16

uranium. In turn, most of the public risk of coal and oil is due to air pollution effects.

Third, the ratio of non-fatal effects to fatalities varies strongly from one category to the next. This can pose a computational problem if one wishes to draw specific conclusions. In the final two columns, the number of non-fatal cases for coal and oil are much higher than those of uranium and gas, as are comparable values for fatalities. However, it is not difficult to envisage cases where the non-fatal cases for one energy form were higher than those of another, and the fatal cases were opposite in proportion. In that event, it would be difficult to draw general conclusions.

Inhaber (1980) tried to solve this problem by reducing all effects to man-days (or person-days) lost. In this system, all non-fatal accidents of diseases would be assigned values corresponding to those used by insurance companies in actuarial tables. Fatalities are obviously in a different category from non-fatal events, yet occupational safety specialists often use a value of about 6000 man-days lost for these events. The International Commission on Radiological Protection (1977) also used this value. The number is based on the approximate number of days lost from work for a typical industrial accident.

Inhaber was able to develop, on this basis, over-all risk values for different energy systems. He was also able to show that while the total risk of each system varied with the number of man-days lost assigned to each fatality, the relative ranking of each system did not. That is, the sensitivity of the rankings to changes in the importance assigned to death was small.

Let us now turn to the question of uncertainty. As W. R. Inge wrote in *Assessments and Anticipations:*

> 'What is worth knowing is mostly uncertain. Events in the past may be roughly divided into those which probably never happened and those which do not matter.'

It would clarify matters greatly if we could assign a specific and statistically valid degree of uncertainty to each of the values in Table 8.1. At present, we cannot. Paskievici made an attempt to do so, based primarily on his judgement as opposed to statistical tests. He assigned uncertainty factors to most of the accident values (the first four columns) of this table.

To conserve space, these factors are not repeated here. All were either 1.5 or 2. This would mean, for example, that the number of non-fatal accidents suffered by workers in natural gas production varied from 30 (= 15 x 2) to 7.5 (= 15 ÷ 2). The assignment of these factors is useful in drawing attention to the fuzziness of the data, but until they are put on a firmer statistical basis they will be of limited value.

In spite of this limitation, if the uncertainty factors are only about two, the conclusion that coal and oil are riskier than gas and uranium still holds. There is also overlap between gas and uranium, so there is some chance that the former is riskier than the latter. Clearly the two are about equal in occupational risk; the difference lies in public risk.

The evidence so far then states that fossil fuels, with the exception of natural gas, are considerably riskier than nuclear power. This applies to both occupational and public risk. The evidence is so strong, being based on many independent studies, that it overcomes the inevitable uncertainties.

8.4 Nuclear and Non-Conventional Energy Risk

The next question to ask is, how does nuclear risk compare to that of non-conventional energy sources, such as solar and windpower? Answering this question is more difficult than comparing nuclear to fossil fuels, because comparatively little is known about non-conventional energy sources. The Mitre report (1977, see Bibliography) was one of the first to attempt even a semi-quantitative approach to this question.

Clearly, leaving aside such bizarre events as someone falling off a rooftop collector on to a passer-by or a windmill blade flying to strike an innocent victim, there are no obvious risks from non-conventional sources. However, the preceding discussion on conventional sources showed that non-obvious sources, such as mining and milling, can outweigh the obvious ones, such as reactor accidents. In order to evaluate non-conventional energy risk, we must then consider that which does not immediately meet the eye. In any case, as P. J. Kavanaugh wrote in *The Perfect Stranger*, 'the deafeningly obvious . . . is always news to somebody'.

The concept that non-conventional energy sources can have substantial risks per unit energy output is an old one. Petr Beckmann of Colorado and Czechoslovakia was one of the first to state the case qualitatively. The Jet Propulsion Laboratory in the United States (Caputo, 1977) was one of the first to evaluate some non-conventional systems quantitatively, although its approach suffered from a number of defects. Apparently Inhaber (1978; see Bibliography) was one of the first to analyse quantitatively a wide variety of non-conventional systems, evaluating a total of six.

While it may be imagined by producers of probabilistic risk assessments that their reports are often controversial, if anything these heights are often surmounted by the fierce debates on the risks of non-conventional systems. It would take far too much space to enumerate all the issues involved, let alone clarify them. An indication of the heat, if not light, which the subject has generated is shown in Inhaber (1982). His Appendix Q, devoted to some of the controversy, fills more than a third of the entire book!

As noted above, part of the differences arise from the fact that comparatively little is known about how non-conventional systems operate, at least in relation to conventional ones. On the other hand, at least one non-conventional source, geothermal energy, was producing electricity in Lardarello, Italy, decades before the first nuclear plant was even conceived. There obviously is a mix of experience in non-conventional systems, just as there is among conventional ones.

To put the preceding on a more concrete basis, consider the example of load factors of windmills. The load factor can be defined as the ratio of total energy produced in a fixed period to the maximum theoretically possible in that period. Some theoretical studies have assumed a load factor of 33 per cent; Inhaber (1982) assumed 20 per cent; studies of working windmills often have 5–10 per cent. In general, the higher the load factor, the lower the risk per unit energy, since the denominator of the risk formula will increase. As we gain more experience with non-conventional systems, the range of uncertainty will probably decrease.

This said, what have been the results? To keep the discussion relatively brief, we will only compare non-conventional risks to those of nuclear. It seems clear from the discussion above that coal and oil are substantially riskier than nuclear. Whether or not non-conventional sources are more or less risky than these two fossil sources is probably not of great consequence for the present book.

Etnier and Travis (1983), using the results of a seminar held in 1981, list some cross-comparisons between nuclear and non-conventional sources. The problem of comparison is made complicated by the fact of considerable overlap between values. For example, if energy system A has a range of risk of 0.2–0.5 (in arbitrary units), and B has a range of 0.3–0.6, what can we say about their ratio? To answer this question, we would have to know each of their statistical distributions within that range, a knowledge which is rarely if ever granted to us.

Cox *et al.* (1983) tried to solve this problem by assuming a linear distribution within each range. That is, there is as much chance of finding the true value at one point as another inside the range. This simplifies the mathematics considerably, although its validity is unknown.

If one makes this assumption, we can go on to say that, on the basis of Etnier's compilation, that it is 100 per cent certain that solar thermal electricity is riskier occupationally than nuclear; 100 per cent certain that solar photovoltaic is; 94 per cent certain that windpower is; 100 per cent certain that solar rooftop collectors are; 100 per cent certain that geothermal is; 100 per cent certain that biomass from forestry is; 88 per cent certain that ocean thermal is; 84 per cent certain that biomass wastes are; and so on. The conclusion one can draw from this is that, from the viewpoint of occupational risk alone, non-conventional energy systems almost certainly have a higher risk (as measured in total person-days lost) than nuclear. The evidence varies from one technology to the next, but the conclusions seem clear.

Cox *et al.* (1983), in an independent study, considered a smaller number of systems. They evaluated only electricity producers, thus omitting solar rooftop collectors, which are probably the most widely used non-conventional systems throughout the world. In any case, they found, again in terms only of occupational risk, that it was 88 per cent certain that solar thermal electric was riskier than nuclear, and 86 per cent certain that wood biomass was. They also showed a value of 75 per cent certainty for wind energy *vis-à-vis* nuclear. However, this last value, the lowest in this and the preceding paper by Etnier, is of some doubtfulness because Cox does not show the experimental assumptions on which it is based. The value is substantially lower than the value of 94 per cent derived from Etnier. Until the relation of assumptions to field data is clarified, the appropriate value cannot be specified.

It is curious that the two preceding papers deal only with occupational risk. Controversies about the relative risk of nuclear power generally have centered on its public, not its occupational risk. If we are to make a fair assessment of the over-all risk of non-conventional systems, we have to weigh their public risks in addition to their danger to workers.

It might be thought that we are engaging in over-kill here. The ratio of non-conventional occupational risk to that of nuclear was seen in the two preceding papers discussed to be high, as borne out by the percentages close to 100. However, if nuclear has a high public risk and non-conventional energy systems a low one, the percentage may creep closer to zero.

We saw in Table 8.1 that nuclear public risk is comparatively low, at least compared to coal and oil. At first thought, it might appear that non-conventional systems have a vanishing public risk.

However, Inhaber (1982), drawing on the work of the Jet Propulsion Laboratory (Caputo, 1977) has shown this is definitely not the case. There are three sources of public risk from these systems:

1. *Back-up made up of conventional sources*. As shown in Table 8.1, these conventional sources all have a non-zero public risk. Of the six non-conventional systems that Inhaber considered, he assumed back-up to be required for only three: solar thermal electric, solar photovoltaic, and windpower.
2. *Transportation risk*. Materials for all energy systems have to be transported,

producing both occupational and public risk. In general, non-conventional systems require much more materials per unit energy output than conventional ones (see Inhaber, 1982, Figure 6), so their public transportation risk will usually be higher than that of nuclear. Etnier and Travis (1983) confirm this. As an example of their results, and using the methodology immediately preceding, it is 85 per cent certain that public transportation risk of solar rooftop collectors is higher than nuclear, 98 per cent certain for solar photovoltaic, and so on.

3. *Emissions from materials use*. All energy systems require metals like steel and copper. In their production, air pollution is generated. In turn, this yields a risk to the surrounding population. These emissions effects should be counted against all energy systems, not just non-conventional ones. However, as noted above, non-conventional systems have greater materials requirements than conventional ones, so their public risk from emissions will be correspondingly greater.

There is some irony in this aspect of non-conventional public risk, since these systems are often promoted on the basis of their 'non-polluting' nature. There is pollution all right, but earlier in the process and less apparent as energy from sun or wind is generated.

All this said, what do we find? For all six non-conventional energy systems that Inhaber considers—solar rooftop collectors, solar thermal electric, solar photovoltaic, wind, ocean thermal and methanol—he finds no overlap between the highest total values for nuclear and the lowest for these systems. That is, when we combine occupational and public risk, on the basis of the overlap principle noted above, Inhaber implies that it is 100 per cent certain that non-conventional energy systems have total risk greater than nuclear.

We can make the situation more favourable for those systems which require a conventional back-up. We can assume that the back-up will not be used, even though this would make the resulting energy output highly erratic. Doing so then reduces the risk as well. However, this omission does not affect the results: the three non-conventional systems which require back-up are still 100 per cent certainly riskier than nuclear when their back-up is removed.

8.5 Conclusions

What can we learn from all of this? In spite of the vast publicity surrounding nuclear risk, there is strong—one is tempted to say overwhelming—evidence that its total risk is less than that of the fossil fuels against which it will be competing in coming decades. A possible exception is natural gas, which apparently has a slight edge. However, in most parts of the world natural gas in future years will be used more as a feedstock for petrochemicals rather than to produce electricity.

It is possible that eventually the world will turn more to non-conventional sources such as sun and wind. Based on the imperfect knowledge we have now, nuclear appears a winner in this risk race as well. We do not have complete certainty of this contention, but the percentages presented in the preceding section, when they are not actually at 100 per cent, are remarkably close.

All of this may seem surprising to those with only a casual interest in risk, whose attention is caught all too easily by newspaper headlines. Yet, as Sir Arthur Conan Doyle wrote in *The Beryl Coronet*, 'It is an old maxim of mine that when you have excluded the impossible, whatever remains, however improbable, must be the truth'.

REFERENCES

Caputo, R. (1977). *An initial comparative assessment of orbital and terrestrial central power systems*. Jet Propulsion Laboratory, Pasadena, Cal., report 900–780.

Cox, L. A., Fiksel, J., Kalelkar, A. S., and Ricci, P. F. (1983). 'Occupational risks of energy production', *Nuclear Safety*, **24**, 459–470.

Etnier, E. L., and Travis, C. C. (1983). 'Risk of energy technologies', *Nuclear Safety*, **24**, 671–677.

Inhaber, H. (1980). 'Risk and consequences in energy production', *Interdisciplinary Science Reviews*, **5**, 304–311.

Inhaber, H. (1982). *Energy risk assessment*, Gordon & Breach, New York.

International Commission on Radiological Protection (1977). *Problems involved in developing an index of harm*, Pergamon Press, London, publication no. 27.

Kates, R. W., and Kasperson, J. X. (1983). 'Comparative risk analysis of technological hazards (a review)', *Proceedings of National Academy of Sciences*, **80**, 7027–7038.

Paskievici, W. (1982). 'Health hazards associated with electric-power production: a comparative study', in B. N. Kursonoglu, A. C. Millunzi, A. Perlmutter, and L. Scott (eds.) *A Global View of Energy*, Lexington Books, Lexington, Mass., pp. 249–274.

Rasmussen, N. (1975). *Reactor safety study: an assessment of accident risks in US commercial nuclear power plants*. US Nuclear Regulatory Commission, Washington, DC, report WASH–1400.

Rom, W. N., and Lee, J. (1983). 'Energy alternatives: what are their possible health effects', *Environmental Science & Technology*, **17**, 132A–143A.

US Bureau of the Census (1977). *Statistical Abstract of the United States, 1977*. Government Printing Office, Washington DC, p. 73.

BIBLIOGRAPHY

American Medical Association Council on Scientific Affairs (1978). 'Health evaluation of energy-generating sources', *Journal of American Medical Association*, **240**, 2193–2195.

Belhoste, J. F., Durant, B., and Maccia, C. (1979). *Risques sanitaires et ecologiques d'energie electrique. Cycle nucleaire (PWR), fuel, charbon*. Centre d'etudes sur l'evaluation de la protection dans le domaine nucleaire, Fontenay-aux-roses, France.

Comar, C. L., and Sagan, L. A. (1976). 'Health effects of energy production and conversion'. *Annual Review of Energy*, **1**, 581–600.

Committee on Nuclear and Alternative Energy Systems (CONAES) (1980). *Alternative energy demand futures*. US National Academy of Sciences, Washington, DC.

Hamilton, L. D. (1974). *The health and environmental effects of electricity generation—a preliminary report*. Brookhaven National Laboratory, Upton, NY, report BNL–20582.

Health and Safety Commission (1978). *The hazards of conventional sources of energy*. Her Majesty's Stationery Office, London.

Hittman Associates Inc. (1974). *Environmental impacts, efficiency and cost of energy supply and end use*. Columbia, Maryland, report HIT–593.

Inhaber, H. (1978). *Risk of energy production*. Atomic Energy Control Board, Ottawa, Canada, 1st edition, report AECB 1119.

Inhaber, H. (1980). *Risk of energy production*. Atomic Energy Control Board, Ottawa, Canada. 3rd edition, report AECB 1119/Rev–3.

Institute for Energy Analysis, Oak Ridge Associated Universities . (1979). *Economic and environmental impacts of a US nuclear moratorium*. MIT Press, Cambridge, Mass.

MITRE Corp. (1977). *Accidents and unscheduled events associated with non-nuclear energy resources and technology*. Environmental Protection Agency, Washington, DC, report EPA–600/7–77–016.

Morris, S. C. (1977). *Comparative effects of coal and nuclear fuel on mortality*. Brookhaven National Laboratory, Upton, NY, report BNL–23 579.

Morris, S. C., Novak, K. M., and Hamilton, L. D. (1979). *Databook for the quantitation of health effects from coal systems* (draft). Brookhaven National Laboratory, Upton, NY.

Pochin, E. E. (1976). *Estimated population exposure from nuclear power and other radiation sources*. Nuclear Energy Agency, Organization for Economic Co-operation and Development, Paris.

Ramsay, W. (1978). *Unpaid costs of electrical energy: health and environmental impacts from coal and nuclear power*. Johns Hopkins University Press, Baltimore, Maryland.

Schurr, S. H. *et al.* (1979). *Energy in America's future: the choices before us*. Johns Hopkins University Press, Baltimore, Maryland.

Smith, K. R. *et al.* (1975). *Evaluation of conventional power systems*. University of California, Berkeley, report ERG 75–7.

US Atomic Energy Commission (1974). *Comparative risk-cost-benefit study of alternative sources of electrical energy: a compilation of normalized cost and impact data for current types of power plants and their supporting fuel cycles*. Government Printing Office, Washington, D.C., report WASH–1224.

World Health Organization (1978). *Health implications of nuclear power production*. WHO Regional Office for Europe, Copenhagen, Denmark.

Nuclear Power: Policy and Prospects
Edited by P. M. S. Jones

9

Waste Management

M. MÅRTENSSON
Studsvik Energiteknik AB

9.1 Introduction

Radioactive wastes are generated as undesired by-products at most stages of the nuclear fuel cycle. In mining and milling of uranium, radioactive materials of natural origin are removed from the earth in such a way that about 70 per cent of the radioactivity originally present in the ore is left in the tailings. Small quantities of radioactive waste also arise during enrichment and fuel fabrication. The largest quantities of radioactive waste—and those potentially of greatest concern—are however produced by the nuclear processes occurring in reactor operation. Almost all of this radioactivity remains in the irradiated fuel which is periodically discharged from the reactor. This category of waste includes the only high-level radioactive wastes resulting from nuclear electricity generation. Although small in volume, it accounts for about 99 per cent of the radioactivity produced in the nuclear fuel cycle.

The radioactivity contained in spent fuel derives from a great variety of radioactive elements which can be divided into a number of classes depending on their nature and properties. With respect to their physical formation, they can be divided into two broad categories, fission products and activation products. The fission process gives rise to over a 100 different isotopes of about 30 different chemical elements some of which are gaseous. The chemical composition of the fission products varies to a certain degree with the type of reactor, the type of fuel and the burn-up of the fuel. However, these differences do not affect crucially the subsequent management of the waste (see Chapter 4).

The fission products decay with the emission of β-particles and γ-rays. They have half-lives varying from a few days up to about a million of years. However, the great majority of the potentially hazardous radionuclides have half-lives in the range up to 30 years, and their radioactivity will accordingly decline to innocuous levels within a 1000 years.

Activation products are formed when neutrons are absorbed (captured) by the structural material of the reactor, the coolant and the nuclear fuel cladding. They decay with the emission of β-particles and γ-radiation and most of them are rather short-lived. The total radioactivity due to this type of activation product is very much less than that of the fission products.

An important, special group of activation products is made up of the actinide elements which are produced as the result of neutron capture by uranium. This family of radioactive elements includes plutonium, neptunium, americium and curium. They decay mainly with the emission of α-particles and their half-lives are commonly long. Materials contaminated with significant quantities of actinides must therefore be isolated from the biological environment for many thousands of years.

Using quite another principle of classification, radioactive waste products can be divided into three different categories: reprocessing waste, spent fuel (when treated as waste) and reactor waste. This particular method of categorization reflects the origin of the wastes in the nuclear fuel cycle rather than the process by which they have been formed. The distinction between reprocessing waste and spent fuel, 'when treated as waste', is particularly important because these two types of waste arise from different kinds of fuel cycles: the so-called reprocessing and once-through cycles (see Chapter 4).

Reprocessing involves chopping up the fuel, dissolving the spent fuel core with nitric acid and subjecting the fuel solution to a series of chemical extractions designed to recover and purify uranium and plutonium. The main liquid waste stream resulting from these operations is the high-level liquid waste containing over 99 per cent of the non-gaseous fission products.

Reprocessing also results in the creation of new categories of waste contaminated with radioactivity from the original spent fuel: fission product gases which are released during the chopping and dissolving operations, cladding and fuel assembly materials some of which are contaminated with long-lived actinides, other forms of solid waste arising from discarded process materials and failed equipment, and, finally, large volumes of intermediate- low-level liquid waste containing comparatively small concentrations of fission products and actinides.

In the other of the two basic fuel cycle options, the 'once-through cycle', the spent fuel is not reprocessed but is disposed of after a period of storage and appropriate conditioning.

The fundamental choice between the reprocessing and the once-through cycle has to be made on the basis of a complicated mixture of economic and national policy considerations. From a pure waste management point of view, neither of the two options can be said to have a definite advantage over the other. An obvious drawback of the reprocessing alternative is that it results in the formation of several categories of waste each of which requires a special treatment. The potential hazard and risk of failure during the management of the waste before its final disposal are therefore greater in the reprocessing than in the once-through alternative. On the other hand, reprocessing and recycle of plutonium provide a way to reduce long-term risks by reducing the amounts of long-lived actinides in the wastes.

As has already been emphasized, most of the fission products and actinides formed during reactor operation remain within the fuel and are transferred with the spent fuel when it is removed from the reactor. However, small amounts of fission products may enter the cooling system because of leaks in the fuel cladding. In addition, the cooling water will be contaminated by radioactive materials produced outside the fuel elements by neutron activation of structural material, corrosion products and the coolant itself. A small part of this contaminated water will spread to different parts of the plant by leakage and, as a consequence, radioactivity will appear in various liquid streams outside the primary cooling system. This type of waste is usually called 'reactor waste'. It is characterized by a short-lived, low- and medium-level radioactivity deriving, for the most part, from β—and γ—radiation.

Apart from the principles of waste categorization used above, there exist a number of classifications which are based on the techniques required for the management of the wastes. A common classification, which has in fact already been presupposed above, is to divide the wastes according to the degree of radioactivity into high-, medium-, and low-level wastes. This categorization is primarily of interest from the point of view of handling since it determines the radiation shielding requirement. (Thus, high-level and medium-level wastes require shielding during normal transport and handling while low-level waste does not.) Other basic distinctions are between solid, liquid, and gaseous wastes and between waste that needs to be cooled and that which does not.

Table 9.1 Classification of wastes

Type of waste	Shielding required	Cooling required	Deep underground disposal generally required
HIGH-LEVEL WASTE			
High-level liquid reprocessing waste	Yes	Yes	Yes
Spent fuel when treated as waste	Yes	Yes	Yes
NON-HIGH-LEVEL WASTE			
Medium- and low-level, liquid and solid reprocessing wastes with significant levels of long-lived α-emitting elements	May be	No	Yes
Medium- and low-level, liquid and solid reprocessing wastes with insignificant levels of long-lived α-emitting elements	May be	No	No
Reactor wastes	May be	No	No
Refining and fuel fabrication wastes	No	No	No

Yet another division of wastes can be made on the basis of the methods used for their final disposal. For some types of wastes, a final disposal is required that ensures a safe isolation from human environment for a very long period of time. A deep geological repository is the kind of final disposal that is usually considered necessary for the fulfilment of this condition. High-level reprocessing waste and reprocessed spent fuel, when treated as waste, are the two main types of waste that require to be finally disposed of in this way. However, low- and medium-level wastes containing significant amounts of long-lived α-emitting elements must also be regarded a potential hazard comparable to high-level waste and should therefore be disposed of in a similar way.

On the other hand, there are several types of wastes which do not need the long-term isolation provided by deep geological disposal (or equivalent methods of disposal). In these cases, economic considerations lead to the selection of less expensive disposal concepts, such as shallow-ground repositories, emplacement in natural or man-made cavities and dumping in the sea.

In the following review of waste management methods, the division of radioactive waste according to final disposal requirements is used as the main criterion of classification. It simply means that the wastes are divided into two broad categories as is shown in Table 9.1.

9.2 High-Level Waste Management

9.2.1 *General Considerations*

The management of high-level waste includes a number of consecutive operation stages, the most important of which are interim storage, conditioning and final disposal. The appropriate timing of these operations—as well as the individual nature of each of them—is determined in the first place by the radioactivity and the heat emission of the waste and the way in which these properties vary with time. These fundamental characteristics are illustrated by means of Figures 9.1, 9.2 and 9.3.

Figure 9.1 shows, for a typical example, how the radioactivity of the most important radionuclides in spent fuel varies with time after discharge from reactor. Figure 9.2 shows, in a similar way, radioactivity as a function of time for high-level waste from reprocessing, and Figure 9.3 shows the variation with time of heat emission from both

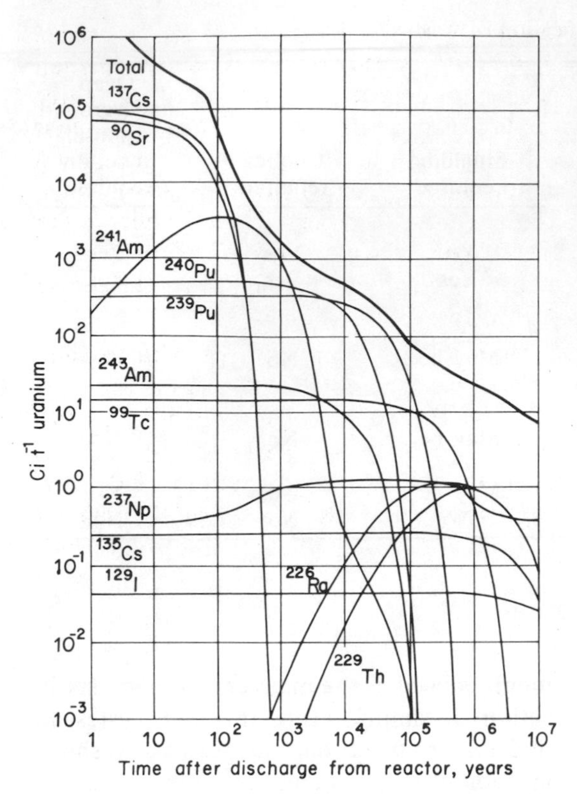

Figure 9.1 The activity of the radioactive elements in spent pressurized water reactor fuel as a function of time. *Reproduced by permission of the Swedish Nuclear Fuel and Waste Management Co.*

spent fuel and high-level waste from reprocessing.

Some of the problems encountered in storage and disposal of high-level waste can be elucidated starting from Figures 9.1 to 9.3. The very high radioactivity and associated heat output of irradiated reactor fuel is of course the primary reason why special techniques are required in high-level waste management. But the figures show that a steady decline of radioactivity and heat emission takes place continuously in course of time. This pattern of decline explains, among other things, why storage is used as an interim stage in high-level waste management.

In fact, two separate storage periods are included in the handling sequence of high-level waste. In both the once-through and the reprocessing cycle, spent fuel is first stored for a few years after discharge in a storage facility adjacent to reactor. The purpose of this initial storage is to facilitate the subsequent handling of the fuel by making use of the very rapid decrease of radioactivity and heat emission occurring immediately after discharge. In addition, a second storage period with a length of several decades is used for spent fuel or reprocessing waste prior to final disposal. The purpose of this longer-term storage—which usually takes place in a centralized facility—is to reduce the complexity and cost of the final repository by taking advantage of the continuous decrease of the heat emission of the wastes.

From Figures 9.1 and 9.2 it can be seen that the unceasing process of diminution of radioactivity can be divided into two periods, an initial period of short length and rapid decrease of radioactivity followed by a much longer period with slower decrease. During

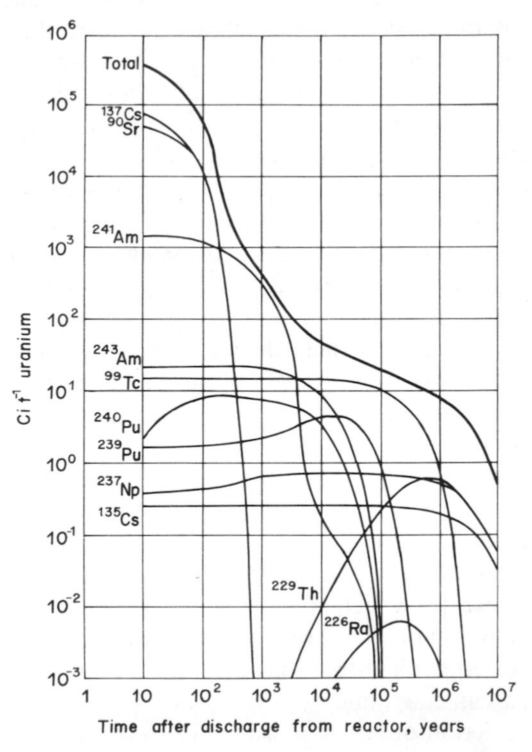

Figure 9.2 The activity of the radioactive elements in high-level waste arising from 1 t of pressurized water reactor fuel as a function of time. It is assumed that reprocessing takes place ten years after discharge of the spent fuel from reactor. *Reproduced by permission of the Swedish Nuclear Fuel and Waste Management Co.*

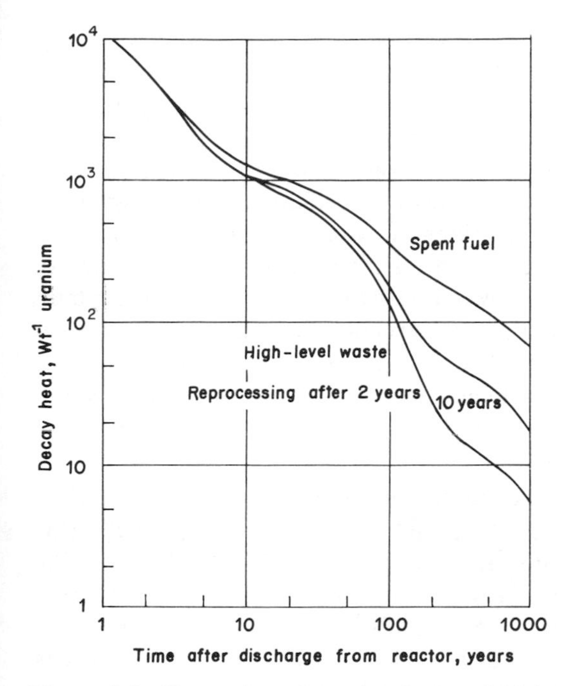

Figure 9.3 Decay heat in spent fuel and high-level waste from pressurized water reactor. *Reproduced by permission of the Swedish Nuclear Fuel and Waste Management Co.*

the first period, which has a length of about 1000 years, the greatest contribution to the radioactivity comes from strontium–90 and cesium–137, i.e. from fission product elements emitting β-particles and γ-radiation. During the second period, the length of which from a practical point of view may be considered almost infinite, the greatest part of radiation comes from the very long-lived α-emitting actinides and their daughter products. However, even though the total radiation during the second period is a factor of about 1000 lower than during the first one, it is still great enough to have harmful effects. This extreme longevity of the hazard arising from nuclear waste explains why, in the management of this waste, a time horizon must be adapted that is very much longer than is otherwise used in the planning of technical activities.

9.2.2 *Conditioning of High-level Waste*

One of the most important interim steps included in waste management of the reprocessing cycle is solidification of the high-level liquid waste arising from reprocessing. When leaving the reprocessing section of the reprocessing plant, this liquid stream is first concentrated by evaporation and then temporarily stored in stainless steel tanks. The high concentration of radioactivity in the solution leads to the emission of substantial amounts of heat and the tanks must therefore be cooled.

This technique has proved quite feasible and has been practiced on a routine basis for more than 30 years in the United States and the United Kingdom. However, in order to facilitate the further handling and final disposal of the waste it is generally planned to transform the liquid waste into solid form.

Several approaches to solidification have been developed, e.g. calcination, vitrification and incorporation of waste into crystalline ceramics and synthetic minerals. The most straightforward of these processes is calcination which means that the liquid waste is sprayed through an atomizer at high temperature and converted to a dry, granular solid. This process has been used since 1963 at Idaho Falls in the United States for the solidification of defence waste. It is quite simple but has to be supplemented with some kind of further treatment in order to prepare the waste for final disposal.

The most thoroughly investigated of the solidification processes is vitrification, i.e. the conversion of the liquid waste into a glass block which is cast into a stainless steel container. The kind of glass usually considered for this purpose is borosilicate glass which has a number of advantages as solidification medium (e.g. great accommodation ability for all kinds of radionuclides, low leachability, high mechanical and irradiation stability).

Several different variants of the vitrification process have been developed. In all of them, the vitrification proper is preceded by an evaporation–calcination stage the purpose of which is to remove water and convert the radioactive elements present in the waste into oxides. The calcined waste is then mixed with a borosilicate glass frit and the mixture is melted at temperatures in the range 1100–1200 °C.

These operations may be performed either in one or two main process stages and, since in each of these two alternatives, the operations may be carried out either discontinuously or continuosuly, there exist four essentially different variants of the process (Sombret, 1984).

For example, the British HARVEST process and the Indian WIP process, which use only one main process vessel both for evaporation, calcination and vitrification, can be characterized as discontinuous, one-stage processes. On the other hand, the PAMELA process, developed by German industry and research institutes in collaboration with Eurochemic, is a typical example of a continuous, one-stage process. In this process, the main equipment consists of a ceramic-lined, electric melter to which the waste solution, premixed with glass frit powder, is fed continuously. The heating of the melter is accomplished by means of metallic electrodes which are immersed directly in the molten glass ('joule heating').

For the two-stage processes, several choices of apparatus are possible both for the calcination and for the vitrification stage. Thus, for the calcination stage fluidized beds, rotary kilns or spray towers can be used while a number of possibilities exist for the vitrification stage, e.g. pot melters provided with joule, induction, or microwave heating. The FIPS process being developed at Jülich in the Federal Republic of Germany and the French AVM (Atelier de Vitrification de Marcoule) process are examples of two-stage, discontinuous and continuous processes, respectively. In both these processes, a rotary kiln is used for the calcination stage and an induction-heated melter for the vitrification stage.

The French AVM process is the most advanced vitrification process available for the time being. The method has been under development since 1957 and has been tested in successive steps in the Gulliver and Piver pilot plants. In 1978 an industrial-scale facility based on the AVM process was put into operation at Marcoule. Between June 1978 and July 1982 this facility vitrified 490 m^3 of high-level fission product solution corresponding to more than 9000 t of spent metal fuel. Based on the experience acquired at Marcoule, two new vitrification facilities will be constructed at La Hague in France. Each of these facilities, which are closely connected with the two light-water reactor fuel reprocessing plants under construction at the same site, is designed to handle a quantity of fission products corresponding to the annual reprocessing of 800 t of fuels (Bastien Thiry *et al.*, 1984).

A few other countries are also well on the way to industrial realisation of vitrification processes. In the United Kingdom, work is in progress on the construction of a vitrification plant at Sellafield. This facility—which will use a modification of the French AVM process—is scheduled to be in operation in 1989 (Smith, 1985). In India, the Waste Immobilization Plant (WIP) at Tarapur, which commenced active operation in 1983, will be followed by an industrial-scale plant at Trombay designed to treat 400 m^3 per year of high-level waste and scheduled to be operational in 1990 (Sunder Rajan *et al.*, 1984). In Belgium a demonstration plant based on the PAMELA process is being built at the Mol site by the German 'Gesellschaft für Wiederaufarbeitung von Kernbrennstoffen'. This plant will be operational in 1986 (Höhlein *et al.*, 1985).

As an alternative to glass, several kinds of ceramics have been proposed as media for incorporation of high-level waste. One of the ceramic candidates is SYNROC, an Australian developed material obtained by heating calcined waste and a mixture of oxides of titanium, zirconium, calcium, barium and aluminium to temperatures in the range 1000–1200 °C. SYNROC is more complex and more expensive to manufacture than glass produced by conventional vitrification processes. Its main advantage is a very low leachability, even at temperatures above 300 °C. It can therefore be buried underground immediately after manufacture and does not need any additional encapsulation or overpacks of the kind that are usually considered in the case of vitrified waste (Reeve and Ringwood, 1984).

9.2.3 *Interim Storage*

As has already been pointed out, interim storage is a necessary step both in the once-through and the reprocessing cycle. Although the main purpose is still to let the waste cool down before further handling, the function of interim storage has undergone some important changes in recent years. In the early days of nuclear power it was generally believed that storage of spent fuel would be limited to one or two years before its transport to a reprocessing facility. However, a number of factors, e.g. steeply increasing costs of reprocessing, unforeseen delays in the start and construction of reprocessing plants, and increasing public opposition to the reprocessing concept, have in many countries resulted in a reassessment of earlier plans. In the United States, in particular, the ban placed in 1976 on reprocessing completely upset existing plans to move spent fuel from nuclear power reactors to reprocessing plants.

As a result of these unexpected occurrences, two new policies have developed. In some countries, e.g. the United States, Sweden and Finland, earlier decisions to reprocess the spent fuel have been reversed and replaced by a decision to prepare the spent fuel for final disposal without any preceding reprocessing. Obviously, this policy means a switchover from the reprocessing cycle to the once-through cycle.

The other course of action adopted to meet the new situation has been to postpone reprocessing until sufficient reprocessing capacity will be available. This policy is followed in countries where reprocessing still is looked upon as an important part of an integrated waste management program, e.g. the United Kingdom, the Federal Republic of Germany and Japan.

Irrespective of which of these two policies is pursued, it has been necessary to construct additional storage capacities to meet the build-up of spent fuel. Two different routes have been followed for this purpose. One is to increase the capacity of existing on-site or 'At Reactor' (AR) storage facilities which, although originally designed for a storage period of a few years only, are part of the normal auxiliary equipment of all nuclear power plants. In most cases, such AR spent fuel stores are constructed according to the 'wet storage' principle which simply means that the fuel assemblies are arranged in special racks which are placed in water-filled pools, the water serving at the same time as coolant and radiation shielding. An extension of AR cooling pools can be achieved by several means, e.g. by a closer spacing of the spent fuel assemblies, by adding a second level of storage racks above the existing racks, or by expansion of the volume of existing pools (IAEA, 1984a).

The other way of providing additional interim storage capacity is to build central facilities—so-called Away From Reactor (AFR) storage facilities—exclusively dedicated to storage of spent fuel. Such AFR storage facilities can either be constructed according to the 'wet storage' concept—i.e. by means of the method already used at most of the existing AR storage facilities—or by applying one of several 'dry storage' techniques, the common feature of which is that the decay heat is removed by means of natural or forced air circulation.

The CLAB facility in Sweden is an example of an AFR storage facility based on the wet storage principle (see Figure 9.4). In this facility, spent fuel will be stored for a period of about 40 years prior to disposal. The storage section of the facility consists of underground water pools located in an excavated cavern, the roof of which is about 30 m below ground level. The present storage capacity of CLAB is about 3000 t, but provisions have been made for an extension of the capacity to 6600 t by blasting additional rock caverns parallel to the first one (PLAN, 1982).

Another example of a centralized spent fuel storage facility based on the wet storage principle is the special facility for reception

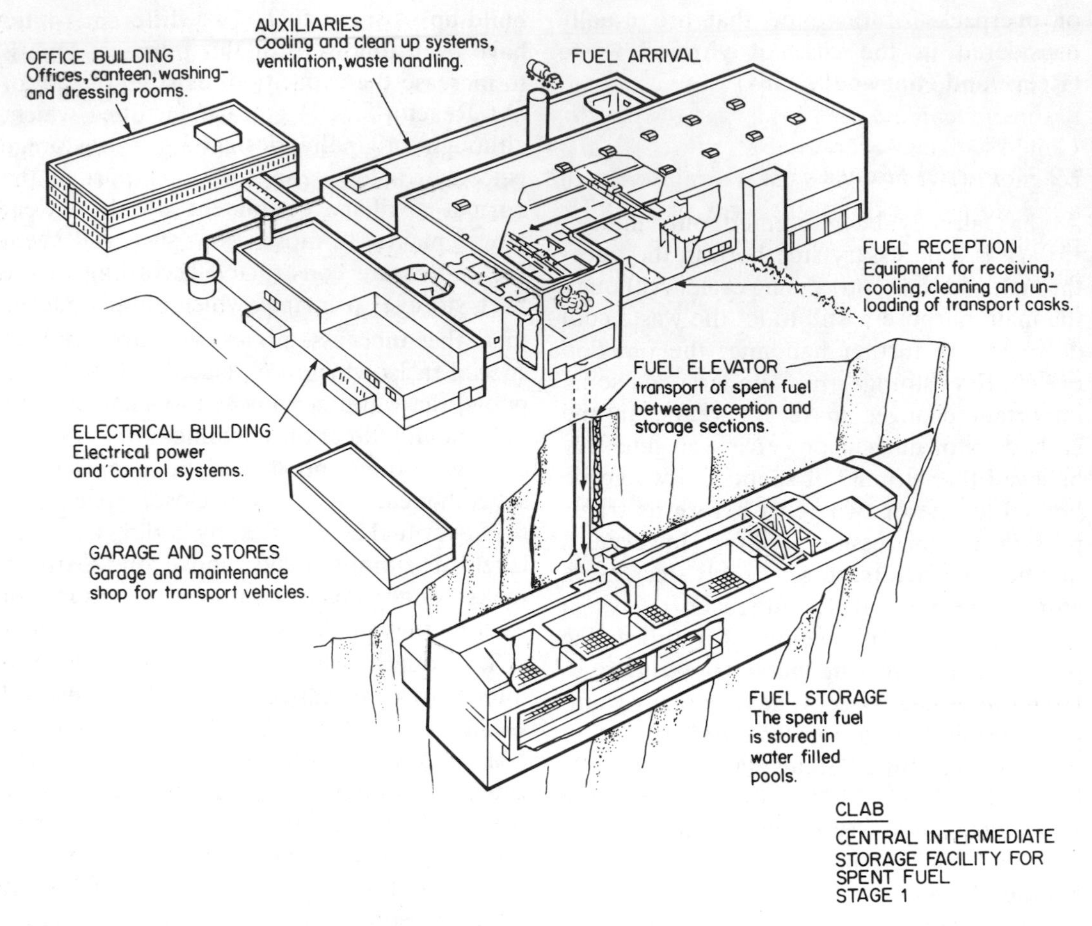

Figure 9.4 The design of the Swedish interim storage facility for spent fuel, CLAB. The four underground pools, each of which can hold 1000 t of spent fuel, are located in a vault 120 m long, 21 m wide and 27 m high, and with the roof 30 m below ground level. *Reproduced by permission of the Swedish Nuclear Fuel and Waste Management Co.*

and short-term storage of spent fuel which is currently under construction on the site of the La Hague reprocessing plant in France. In the storage part of this installation, spent fuel assemblies will be stored in water-filled ponds for two or three years while awaiting reprocessing. Two ponds, each with a storage capacity of 2000 t, are already in operation and two more ponds with the same capacity are under construction (Chometon and Cantin, 1983).

Up to now, dry storage has not been applied to the same extent as wet storage. However, the dry storage method has several potential advantages from a technical and economic point of view and can be used both as an alternative to AR water-cooled ponds and for short-term or long-term AFR storage. A great number of different dry storage projects have already been tested and systems for large-scale dry storage of spent fuel are presently being marketed in Switzerland, the Federal Republic of Germany, the United Kingdom and the United States.

The dry storage technique affords a number of different possibilities: storage in casks, vaults, drywells or concrete silos. The cooling may be achieved by natural convection or by forced air or inert gas circulation. Depending on the method used, dry storage facilities may be located on the surface or underground. A particularly attractive form of dry storage consists of a system where the same container is used both for short-term on-site storage, transport, and extended interim storage pending final disposal.

The feasibility of the dry storage method has been demonstrated for a long time in the United Kingdom where, already in 1963, the Magnox station at Wylfa was equipped with a dry storage facility of the vault type. Other possibilities of using dry store technique for storage of spent fuel have been tested or developed in Canada (concrete canisters for storage of spent CANDU fuel), in the United States (ground surface drywells and dry storage cask systems) and Switzerland (vault storage and combined storage–transport casks). In the Federal Republic of Germany, the development has been concentrated on combined transport–storage cask systems, which is now a licensed technology with a 1500 t capacity AFR spent fuel storage facility commissioned in 1984 and a second one of the same capacity under construction (Baatz *et al.*, 1983).

In this survey of options and methods of interim storage, stress has been put on the storage of spent fuel. This treatment of the subject is determined in first place by the natural sequence of the different stages of the fuel cycle but, accidentally, it also reflects present trends because, in recent years, storage of spent fuel has become a much more urgent problem than storage of vitrified high-level waste.

On the other hand, there is no particular need for a special treatment of interim storage of vitrified waste because the methods applied are essentially the same in this case as for interim storage of spent fuel. It should be observed, however, that in the case of spent light water reactor fuel, storage in water-cooled pools can hardly be avoided for an initial period of a few years due to the high heat emission, while such a restriction does not apply to vitrified waste. In the latter case, emphasis has therefore been put on dry storage because this method affords greater flexibility, easier technology, greater reliability and, probably, lower cost than the wet storage method.

For the time being only two industrial facilities for interim storage of vitrified waste are in operation, viz., a facility associated with the AVM vitrification plant at Marcoule, and a facility associated with the Waste Immobilization Plant (WIP) at Tarapur. Both are based on the vault storage concept but differ in so far as forced air cooling is used at Marcoule while a natural draught air cooling system is used at Tarapur. However, three other interim storage facilities—associated with each of the vitrification plants at Mol, La Hague, and Sellafield—are currently being constructed or designed. They are all based on the air-cooled vault concept (Heafield, 1984).

9.2.4 *Transportation*

Transportation is a necessary component of all nuclear fuel cycle activities, and is especially important as a link between different process stages in the handling sequence used for the management of high-level waste. Over the last 30 years, the transport of radioactive material has become a routine activity and there are now several millions movements each year. Although no accidents resulting in serious consequences to the public have occurred, there is still some concern in certain countries over the safety of the transport of radioactive materials. In particular, this concern is focused on the transport of spent fuel which is the most frequent type of movement of high-level waste occurring for the time being.

The very good record in transportation has not occurred by chance. All transport of radioactive wastes on public roads must be carried out in accordance with rigorous national and international regulations. The purpose of these regulations is to protect the public, transport workers and property from any damage resulting from the carriage of radioactive material. The guiding principle of the regulations is that, rather than relying on human intervention with its associated potential for error, the transport equipment itself should be designed to a maximum degree of safety so as to reduce to a minimum any possibility of accident.

For the transport of spent fuel, this objective is achieved by means of a transport system, the central part of which is a massively constructed, continuously welded container—the so-called 'flask'—weighing 50–100 t or more. The lid of the flask is usually held in position by several large diameter steel bolts, and leakage is prevented by a system of compression seals. The flask is also normally provided with fins for external cooling. The transport of the flask—e.g. from the reactor to an AFR storage facility or from the reactor to a harbour for further transport by sea—is carried out by specially designed vehicles, which, however, do not differ extraordinarily from those used for moving other heavy loads.

The high quality of the technology used for the development of high-level waste transportation equipment is demonstrated by the experience acquired up to now of transport of spent fuel by road, rail and sea. It is estimated that more than 15 000 shipments of spent fuel have been made in Europe and North America since the beginning of commercial nuclear power. Very few accidents or incidents have occurred during these shipments, and none has resulted in a release of contents or significant exposure to workers or members of the public (ENS, 1984).

9.2.5 *Final Disposal*

Many methods have been proposed for the permanent disposal of high-level nuclear waste. These proposals fall into four different categories: deep geologic disposal, seabed disposal, ice sheet disposal and extra-terrestrial disposal. Up to now, almost all research and development work has been concentrated on the deep geological disposal method by which the conditioned waste would be placed in carefully selected geologic formations several hundred metres below the surface of the ground.

Geologic disposal has often been described as a 'multi-barrier' concept. Provided that the geologic repositories are located in stable geologic formations, the only way in which radionuclides can migrate to the biosphere is by dissolution of the waste and transport in ground water. The selection of a suitable host medium for a geologic repository and the design of the repository itself should therefore be performed in such a way that two broad objectives are fulfilled: the prevention of any contact between waste and circulating ground water, and the minimization of the rate of migration of any radionuclide that may nevertheless be released. In the multi-barrier model, the realization of these objectives is assumed to be dependent on the joint action of two kinds of 'barriers', usually called the engineered, or man-made, barriers and the geologic barriers, respectively.

The primary task of preventing waste materials from leaving the repository zone lies with the engineered barriers which consist of three different components: the waste form itself, i.e. the ceramic pellets of spent fuel or the vitrified high-level waste from reprocessing, the canister surrounding the primary waste form, and the buffer and backfill materials used to fill up the space between the canister and the surrounding rock.

Each of these components contributes independently to the function of waste isolation but operates at the same time as interrelated parts of a whole. The very low solubility of spent oxide fuel and of vitrified waste—the first of the components of the engineered barriers—is a guarantee that the fuel matrix, or the solid glass structure, will remain intact and that the waste products will be kept effectively immobilized for a very long time.

A further barrier against leakage is provided by the canister. The choice of canister material has to be adapted to the type of waste and also to the type of hosting geologic medium. Thus, according to Swedish investigations, copper canisters are considered the most suitable choice for the disposal of spent fuel in granite (KBS–2, 1978) whilst for the disposal in granite of reprocessing waste solidified in glass, a multiple encapsulation system consisting of the stainless steel container used as mould in the vitrification process and an overpack consisting of lead surrounded by a canister of titanium has been considered (KBS–1, 1977).

The function of the third component of the man-made barrier, the buffer material, is not only to fill up remaining cavities in the repository, but also to reduce the movement of ground water and to delay the migration of any radionuclide released from the waste package. For the accomplishment of this two-fold objective, bentonite is generally considered to be the best choice. Bentonite is a clay formed by the decomposition of volcanic ash having the ability to absorb large quantities of water and to expand to several times of its normal volume. If ground water were to seep into the deposition holes, the bentonite would therefore swell and fill all cracks and cleavages with a plastic clay having a very low water conductivity. In addition, the bentonite will provide a large ion exchange capacity and by that means cause a substantial delay of movement of most radionuclides through the buffer zone.

A very large number of independent investigations have been carried out into the stability and durability of the range of different components that can provide the engineered barriers, and different countries have different preferences. On the basis of the great volume of knowledge acquired by means of this research, it seems reasonable to conclude that the engineered barriers will provide effective containment more than long enough for the highly-radioactive but short-lived fission products to decay to a harmless level (see Figures 9.1 and 9.2). However, in the very long run, i.e. over periods of time of the length of several tens of thousands of years, the behaviour of the man-made barriers is difficult to predict because human experience of these materials under different conditions is too short to make possible extrapolations of this magnitude. Thus, whereas the man-made barriers combine to a kind of first-line defence against the release of the short-lived fission products, it is the geologic barriers that provide the ultimate means of protection against possible radiotoxic effects caused by the more far-off dispersion of the lower-level, but longer-lived transuranium elements.

The geologic barriers are made up, in the first place, of the host rock in which the repository is located but include, in a broader sense, also the geologic formations of the whole area chosen as site for the repository. As already mentioned, the host medium should be able to satisfy the double requirement of acting at the same time as a *physical* barrier against water movement and as a *chemical* barrier capable of assimilating any radionuclide that happens to be released and dissolved in the ground water. The efficiency of the host rock as a physical barrier depends on a number of factors, including its porosity, the occurrence of fractures and fissures and their characteristics, and the rate and pattern of ground water flow in the vicinity of the repository. The assimilation by the host rock of radionuclides, already dissolved in the ground water, can be effected by means of a variety of processes, e.g. precipitation, surface adsorption, ion exchange and the formation of solid solutions, and the capability of the host rock in these respects is therefore a complicated function of its physicochemical properties.

Plasticity, high heat conductivity and good mechanical strength are other desirable properties of the host rock. A high degree of plasticity may ensure that fractures, which develop as a result of mechanical and thermal disturbances, will be self-sealing and that initially excavated openings to the repository will gradually close. On the other hand, a high plasticity of the host rock will make the design and construction of the repository more difficult.

A high heat conductivity ensures a rapid dissipation of the thermal energy emitted by the radioactive waste and prevents, in the neighbourhood of the waste package, the formation of steep temperature gradients which can produce stresses and deformation and give rise to fracturing and creep. A good mechanical strength is an obvious advantage from the points of view of construction and long-term integrity of the repository.

The types of geological formations that have been considered for disposal of high-level waste are: salt formations, either in the form of bedded salt or salt domes, crystalline rocks such as granite and gneiss, sedimentary formations such as clay and shale, and volcanic rocks such as basalt and tuff. Each of these types of rock has characteristics which makes them potentially suitable as host media.

The main advantages of salt—which is the most thoroughly studied of the different options—are plasticity, high thermal conductivity and low permeability (to non-circulating water). The crystalline rock that attracts most interest is granite. Granites have some outstanding attributes for waste repositories such as high structural and chemical stability, low porosity and permeability and good heat tolerance. The main advantages of clays and shales are low porosity and high sorptive capacity. Clays, in addition, possess a high degree of plasticity which makes them self-healing for cracks and fractures.

Basalt and tuff are two geologic formations which have been produced by volcanic processes. Basalt has low permeability and moisture content and is very hard and strong and is therefore favourable for maintaining unsupported mined openings. Tuff exists in two forms, a high-density rock with low porosity and moisture content ('welded' tuff) and a low-density rock, the main advantage of which is a high sorptive capacity for the most important radionuclides ('silicic' tuff).

In selecting a suitable site for a geologic repository, the properties of the surrounding geologic formations must also be taken into account. A prime requirement is that the repository should be located in a geologic environment with very low seismicity and, correspondingly, a high tectonic stability. The possibility that the function of the repository will be jeopardized by long-term erosion, including glaciation, and other events occurring on the surface of the ground must also be considered. By locating the repository at a sufficient depth, the influences of such phenomena can of course be avoided. On the other hand, an upper limit of the depth of the repository is set by the geothermal gradient and by the growth of the rock stresses which may either lead to unacceptably high operating temperatures or to a breakdown of the rock vaults during the construction and deposition phases. The results of investigations carried out for repositories in Swedish bedrock indicate, for example, that all detrimental effects of surface phenomena will be avoided if the repository is located at a level deeper than 400–500 m, and that the geothermal gradient and the rock stresses will pose no problem as long as the repository is located above the 1000 m level (KBS–3, 1983).

For the time being, two fundamentally different concepts exist for the design of high-level waste repositories: the deep-hole concept and the mined-tunnel concept. A deep-hole repository will involve the emplacement of waste canisters in the lower part of boreholes drilled to depths of several thousand metres. In the mined-tunnel repository, waste will be placed either directly into tunnels, or in boreholes drilled into the floors of tunnels, which are mined and serviced from one or several shafts.

The main advantage of the deep-borehole concept is that the radioactive waste will be emplaced at such an extreme depth that all possible disturbances by climatic or surface changes, ground water transport of waste to the surface, or human intrusion will be reduced to a minimum. The concept is applicable in the first place to disposal in sedimentary formations or salt deposits where standard drilling techniques can be used to reach depths of at least 6000 m.

Of the two alternatives for repository design, the mined-tunnel concept has attracted the greatest interest. Conceptual designs of mined-tunnel repositories have been studied in several countries, e.g. the United States, the United Kingdom, Canada, the Federal Republic of Germany, the Netherlands, Belgium, Switzerland and Sweden, and have been applied to all the host media which are considered at present.

Disposal in salt formations has been studied in the United States, the Federal Republic of Germany, the Netherlands and Denmark. In the United States, feasibility studies and pilot research projects have been carried out since 1965 in abandoned salt mines situated in Kansas (Project Salt Vault) and Louisiana (Avery Island Mine). The most advanced of the US waste disposal pilot projects is the Waste Isolation Pilot Plant (WIPP) facility which is located in a bedded salt deposit in southeastern New Mexico and designed to demonstrate the safe disposal of nuclear waste from the US defence programme (Stein and Collyer, 1984).

In the Federal Republic of Germany, work into the possibility of using salt as a disposal medium has been concentrated to the Asse salt mine which has been used as a test and evaluation facility for the underground disposal of all categories of radioactive waste. An extensive research program has also been carried out in the Gorleben salt dome, which is also intended for the disposal of all categories of radioactive wastes.

Work on crystalline rocks as an option for the disposal of high-level waste is being done in the United States, Canada, Switzerland, Finland, Sweden and elsewhere. The most advanced underground pilot research facility which has been built to date in the United States is located on the Nevada Test site. The purpose of this facility, known as the Climax Spent Fuel Test Facility, is to evaluate the behaviour of granite under repository conditions and to demonstrate the safe and reliable packaging, handling and storage of spent fuel (Stein and Collyer, 1984). In Canada, research and development on nuclear fuel waste disposal is concentrated on disposal in plutonic rock of the Canadian

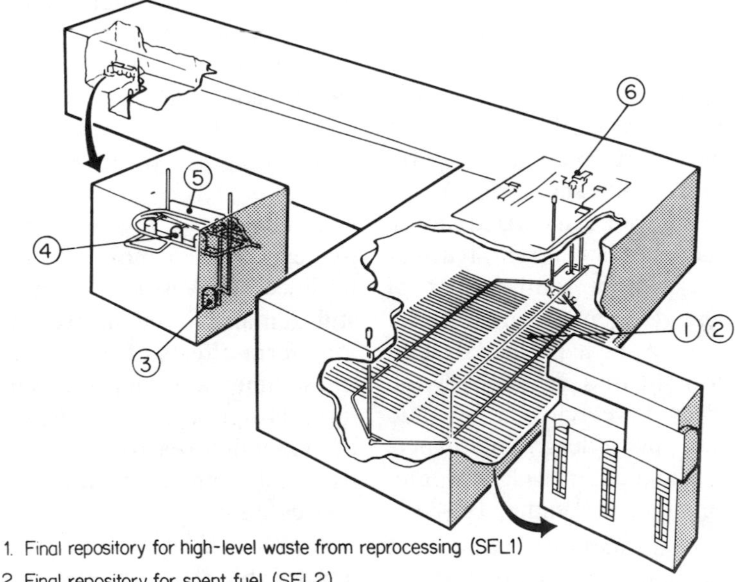

Figure 9.5 Conceptual design of an underground repository intended for location into Swedish granitic bedrock. The main part of the facility (1 and 2) is constructed according to the mined-tunnel repository concept with the waste canisters emplaced in boreholes drilled in the floors of a system of tunnels. This part of the repository is designed for the disposal of 6400 t of spent fuel together with high-level waste arising from reprocessing of about 900 t of spent fuel. *Reproduced by permission of the Swedish Nuclear Fuel and Waste Management Co.*

Shield and is concerned with disposal of both intact irradiated fuel bundles and high-level reprocessing wastes. In Sweden, the planning of the waste management system is based on the assumption that most of the spent fuel will be disposed of in unreprocessed form (see Figure 9.5). The construction of an underground repository is planned to start around the year 2010 and, up to now, field investigations have been carried out at a number of different potential sites located in various parts of the country. The choice of crystalline rocks as host media is to a great extent conditioned by the regional geological situation: like Finland, Sweden lies in an area of crystalline bedrock—the Fennoscandian Shield—dominated by granitic and gneissic formations with an age of over one billion years and a very high stability (Bjurström and Svenke, 1984).

Although sedimentary formations also have been considered in the United States as one of several candidate media for underground repositories, most work on this group of minerals has been performed in two European countries, viz. Italy and Belgium. In Italy, investigations have been carried out on a thick clay formation in the Trisaia Nuclear Research Centre area in the southern part of the country. The Belgian final disposal programme is focused on a particular sedimentary formation situated underneath the Mol-Dessel nuclear research station—the Boom clay—where an underground experimental facility at about 225 m below ground level has been constructed (Bonne, 1984).

The two volcanic formations, basalt and tuff, are studied mainly in the United States where deposits of these rocks can be found in significant amounts. The work on basalt is related to a particular locality close to the Hanford site in the state of Washington. The work on tuff is associated with the tuffaceous rock at Yucca Mountain on the Nevada Test site where field operations have been performed in a previously-built tunnel system at a depth of 455 m below ground (Stein and Collyer, 1984).

Thus, while a considerable volume of research has been devoted to the development of different options for disposal in geologic formations under the land, the work up to now on other disposal concepts has been comparatively limited. Next to disposal in geological formations with regard to feasibility and attractiveness comes seabed disposal. This concept includes several different possibilities, e.g. emplacement of waste into deep ocean sediments, on the ocean floor or into sub-sediment bedrock. It has also been proposed that waste could be placed in deep ocean trenches where the containers might actually be drawn down into the earth because of the phenomenon called 'subduction'.

However, the only seabed disposal concept that is being seriously considered at present for high level wastes is emplacement into deep ocean sediments which is supposed to be accomplished either by means of so-called free-fall penetrators or by lowering the canisters in initially drilled bore holes (NEA, 1984). The thick sub-ocean clay materials considered for this purpose offer a number of advantages as disposal medium, e.g. availability in large areas free of seismic or tectonic activity, low porosity and high sorptive capacity for most of the relevant radionuclides. However, a number of questions still remain to be answered regarding the long-term effects of heat and radiation upon the sediments. Other uncertainties surrounding this option are the risks associated with the extended sea transport of the waste and its emplacement in water of several kilometres depth.

Another disposal concept, the feasibility of which still has to be demonstrated, is emplacement of waste in glaciers or ice sheets. Theoretically, ice disposal would offer geographical and long-term environmental isolation, and could be implemented by means of existing technology. However, uncertainties regarding the long-term behaviour of ice sheets give rise to serious doubts about the practicability of this option.

The most speculative of existing proposal for final disposal is projection into space. This concept would involve putting waste into a solar orbit or sending it into outer space.

However, such an extraterrestrial disposal would be very expensive and possible only for the small volume of long-lived radionuclides (i.e., the actinides). The feasibility of space disposal is therefore dependent on reprocessing of spent fuel and extensive partitioning to remove short-lived fission products which have to be disposed of separately. Another essential condition is the development of a completely reliable space flight system as well as a high-integrity capsule to assure no break of containment even for the worst accident.

Another concept which, like space disposal, would afford only a partial solution of the high-level waste management problem, is the partitioning/transmutation concept. This option means that the long-lived actinides are separated from the short-lived fission products at the time of reprocessing and, subsequently, eliminated by transmutation, i.e. conversion into fission products by neutron bombardment. This could be accomplished by the neutron field of a nuclear power reactor in which case the energy released would add to the normal energy output of the reactor. Since the transmutation would not be complete, several successive reprocessing and transmutation stages would be required to eliminate all the remaining actinides.

A number of studies carried out in several countries indicate that, although the partitioning–transmutation concept appears technically feasible, it would need a large and lengthy development effort for its implementation and would give only limited long-term radiological benefits which might well be outweighed by the short-term detriments arising from the partitioning and separation processes (WHO, 1982).

9.2.6 *Safety Assessment for Underground Disposal*

Safety assessments of proposed final disposal systems are necessary in order to estimate the expected performance of the system and to compare it with existing acceptability criteria. There are two general types of safety evaluations, one dealing with the operational safety during the period of loading, and the other with the long-term safety after disposal operations have been completed.

In the operational type of safety assessment, such problems as protection of workers, exposure of members of the general public to direct irradiation and the risk of equipment failures during the phase of operation have to be considered. It is obvious that safety evaluations of these kinds do not differ fundamentally from those made for other types of nuclear installations.

The objective of safety assessments for the post-loading phase is to make predictions, over a very long space of time, of radiation doses to the general public caused by natural, human, and waste- or repository-induced processes. As it is evidently impossible to demonstrate by experimental means the behaviour of a final disposal system on such long time scales, the necessary confidence in the system must be provided by means of analytical techniques based upon mathematical models which are designed to quantify the ways in which significant phenomena are expected to occur.

The general approach used in such long-term safety analyses is to estimate the probability that a release of radioactivity will occur, and to combine this estimate with an estimate of the consequences of the release in order to obtain a measure of the resulting deleterious effect. When used for the assessment of the long-term safety of a deep underground repository, the first step of the analysis is to identify all types of occurrences that could lead to the release of radionuclides. In doing so, appropriate consideration must be given both to discrete events, caused either by human activities or by sudden natural changes, and to continuous physical processes, such as erosion and ground water flow, which might lead to a slow but incessant release of radioactivity.

By this means, a list of 'release scenarios' is produced. For each scenario, the probability of occurrence and the corresponding radiation doses to man are calculated. These probabilities and consequences are then multiplied together and the resulting *risks* are summed and integrated over time to yield a value for the total detriment.

Numerous studies of this kind have been made in different countries for the evaluation of the potential risks to future generations from radioactive wastes placed in geological repositories. These studies have been based on both generalized and specific cases of repository siting, and differ considerably with respect to the analytical technique, the waste characterization, the proposed repository design and the properties of the host medium.

One of the conclusions that can be drawn from this accumulated body of research is that disruptive events, such as earthquakes, giant meteorite impacts or volcanic explosions, which could result in direct and sudden release of waste into environment, are extremely unlikely for a repository which is located at an adequate depth in a suitably selected area. Although the short-term consequences of such events would be significant, the *risk* entailed can therefore be regarded as negligible.

Another group of risks which has often caused much concern in the public mind is related to accidental or even purposeful intrusion of disposal sites by humans. For example, an accidental intrusion may be caused by inadvertent drilling into the burial ground by a future civilization looking for minerals. However, even in the very unlikely situation that a future society would be capable of drilling to such depths without being able at the same time to recognize a waste repository and take appropriate action, the consequences of such a coincidence would hardly be catastrophic as it must be assumed that the intrusion takes place at some future date when the activity of the waste has decreased to a harmless level.

The deliberate intrusion by saboteurs or terrorists is another scenario which has often been contemplated. In this case, it must be assumed that the intrusion takes place at a time when the site of the repository is still well known and therefore can be located exactly. Under such circumstances, it is hard to imagine that an illegal penetration of the bedrock to a depth of several hundred metres—and a retrieval of the waste no matter for what purpose—would be possible without any action taken by the society. It seems therefore that human intrusion, whether unintentional or deliberate, does not represent any problem.

Thus, eliminating the scenarios that are either very unlikely or associated with small consequences, the only remaining process by which significant amounts of radionuclides might find their way back to biosphere is slow transport by groundwater. As has been outlined above, the likelihood of such an occurrence depends on the performance of a series of barriers which are specially designed to stop the movement of radionuclides. Numerous studies based on complex computer models, where all relevant barrier and process parameters are taken into account, have been performed in order to evaluate the rate and time at which radionuclides from buried waste would penetrate the system of barriers and reach the biosphere. The results of these studies show unequivocally that for a properly designed repository and a reasonable site location, there are no plausible mechanisms whereby release can occur earlier than 1000 years after disposal. By this time, the fission products that dominate early risk will have decayed to harmless levels.

This result means in fact nothing but a definite confirmation that the system of man-made barriers will really fulfil the purpose they are designed for: the effective containment of the waste for at least 1000 years. The whole problem of long-term safety of deep underground repositories can therefore be reduced to one single question: What will happen in a situation when the man-made barriers are no longer intact while some of the waste constituents (i.e. the long-lived actinides) are still harmful? This particular question has been the subject of a great number of assessment studies which cover a wide spectrum of disposal conditions including both 'pessimistic' assumptions, such as early failure of all the engineered barriers in a poorly selected site, and 'less pessimistic' ones, e.g. adequate performance of at least one engineered barrier in an average site.

A common feature of all long-term safety assessment studies is that the uncertainties in the calculations of the risks increase with time after disposal. On the other hand, the risk never actually vanishes because, although radioactivity decays, it never becomes zero. A view that is commonly taken in order to cope with these difficulties is that the risks should be seen in the context of the natural risks from exposures to radiation already present, such as from naturally deposited uranium ore bodies, and that the predicted risk associated with disposal of high-level radioactive waste could be considered acceptable when it is comparable with that of naturally occurring uranium.

However, depending on their location, the risks currently experienced from natural uranium deposits vary in a wide range, from several times natural background in some areas to many orders of magnitude below natural background in others. On the other hand, the potential risks to future generations quantified by assessment studies also vary considerably depending on the underlying assumptions. Now, a broad comparison shows that the ranges of risk are quite similar in the two cases. It means in fact, that the 'pessimistic' risk estimates for high-level waste disposal are roughly equal to the highest risks associated with natural uranium deposits while the 'less pessimistic' risk estimates derived from assessments studies are comparable with the lowest risk currently experienced from natural uranium ore bodies. (IAEA, 1982).

9.2.7 *Different National Approaches to High-level Waste Management*

It has been shown in previous sections that the technical approaches to high-level waste management made by different countries are essentially similar. For example, there is a widespread conviction that deep underground repositories afford the best solution of final disposal of high-level wastes. It is also widely agreed that spent fuel is an acceptable form for safe, ultimate disposal, and that glass—preferably borosilicate glass—is the best medium for the immobilization of liquid reprocessing waste.

However, in spite of the considerable degree of international agreement that exists on engineering issues, there remain wide policy differences between countries. Apart from the basic choice between the once-through and the reprocessing cycle, these differences are related mainly to the schedule to be followed in repository development, i.e. to the timing to be applied in interim storage operations. On this issue, two different national policies can be observed. Some countries feel that an early demonstration of safe and effective waste disposal is desirable for reasons of public opinion and as a necessary prerequisite for the further successful development of nuclear power. According to this view, interim storage should not be used to delay actions on waste disposal. Another group of countries take the view that, since there is technically no urgency to proceed to final disposal, the definite decision on this matter can be postponed until better understanding has been obtained of the technical problems connected with ultimate disposal. The proponents of this 'go slow' approach also claim, completely in disagreement with the advocates of early disposal, that the public will appreciate the technical advantage of delayed disposal, taking into account that interim storage is an already demonstrated technology with very low risk of accidents and concomitant environmental impact.

The group of countries that have established firm timetables for repository development includes the United States, the Federal Republic of Germany, Switzerland, Sweden and India. In the United States, a long period of indecision and uncertainty on waste management policy was brought to an end by the passage of the Nuclear Waste Policy Act in 1982. This act sets forth general guidelines, policy and timetables for management and disposal of commercial spent fuel and high-level waste. In particular, the new law aims to pave the way for operation of a permanent geological repository by the mid–1990s, and firm deadlines are therefore established for federal agency decisions and for the selection of repository sites. However, the act also provides for the development of proposals for the construction of facilities for the long-term 'monitored retrievable storage' (MRS) of spent nuclear fuel and high-level waste. Such facilities should be developed on a parallel track with geologic repositories, and a firm timetable is also established for the selection of the first site of an MRS installation.

The MRS concept is an interesting feature of present US waste management policy. While not regarded as an alternative to a geologic repository in a strict sense, an MRS facility differs on the other hand from an 'away from reactor' (AFR) storage facility with respect to the length of time it is designed to operate. By definition, an AFR storage facility is explicitly temporary in nature. MRS facilities, on the other hand, must be capable of storing waste safely for extended periods into the future. In fact, the MRS system is supposed to be able to function long into the future in the event that no alternative storage or disposal programme will be implemented (Cooper, 1984). The introduction of the MRS concept as a component of the US waste management system appears therefore to indicate a certain weakening of the traditional boundaries between interim storage and ultimate disposal.

In the Federal Republic of Germany, the policy for the management of nuclear wastes is based on the so-called 'integrated waste management concept', the key elements of which are reprocessing of spent fuel, recycling of plutonium (either in light water reactors or in fast breeders), and disposal of low-level, medium-level and high-level waste in deep geologic formations. In the time schedule for the realization of this policy, it is envisaged that all plants pertinent to the integrated concept will be in operation on an industrial scale by the year 2000. It should be noted, however, that West German authorities in 1979 called for the investigation of other waste management technologies, especially the direct disposal of spent fuel. This resolution requires that by the mid–1980s a decision has to be made as to whether the direct disposal of spent fuel will involve some decisive advantages in terms of safety compared with the integrated waste management concept (Reuse, 1984).

In Sweden, the waste management policy is governed by the principle that the owner of a reactor, under Swedish law, is obliged to demonstrate that used fuel wastes can be *safely* managed and disposed of before the reactor is allowed to be charged with nuclear fuel. In conformity with this legislative requirement, an extensive waste management programme is under development, including an interim storage facility for spent fuel, which will be in operation in 1985 (CLAB), and a deep geologic repository for spent fuel which is planned to start operation in 2025. The programme also provides for interim storage during 40 years of a limited amount of vitrified waste and the final disposal of this waste beginning in 2020 (Bjurström and Svenke, 1984).

In a similar way as in Sweden, the Swiss waste management policy is based on a 'project guarantee' which requires the utilities to show by the end of 1985 that safe disposal of nuclear waste can be guaranteed. The main components of present Swiss waste management plans are interim storage of spent fuel in a central storage facility, reprocessing in a foreign reprocessing plant, and final disposal of the vitrified waste in domestic repositories which probably will be developed in deep granite formations (Niederer, 1984).

Although less definitely than Sweden and Switzerland, India should also be included in the group of countries where firm timetables for final disposal have been established. The Indian nuclear power programme is based upon an entirely self-sustained fuel cycle, using indigenous resources, and includes therefore reprocessing as an important component. According to present plans, the high-level reprocessing waste will be stored for about 20 years after immobilization and then ultimately disposed of, probably in domestic gneiss or granite formations. The Indian programme also envisages the setting up of an engineering-scale pilot repository towards the late 1990s in order to develop optimum techniques and methods for the transport and disposal of solidified radioactive waste in deep underground geological media (Sethna *et al.*, 1984).

The group of countries that base their waste management policy on the philosophy of delayed disposal comprises the United Kingdom, Canada, France and Japan and other countries. In the United Kingdom, reprocessing and recycling of plutonium, probably in future fast reactors, are regarded as indispensable stages of the nuclear fuel cycle. According to present plans, vitrified high-level waste, arising mainly from Magnox and AGR fuel, will be stored for at least 50 years before ultimate disposal. The long-term storage of used fuel elements is also considered as a means of providing flexibility in the timing of the reprocessing of used oxide fuel. Although no firm plans exist for final disposal, a research and development programme is maintained to provide future generations with information of the options for safe disposal whenever the decision is made. Geological disposal is regarded as the main option but an alternative option of disposal beneath the deep ocean is also considered (The Uranium Institute, 1984).

The French waste management policy is similar to that of the United Kingdom, i.e. based upon reprocessing and interim storage of vitrified waste during a period of indefinite length. Deep geological repositories are likely to be chosen as the final disposal method and, with this option in view, studies have been made on a number of different management scenarios, with the duration of intermediate storage before final disposal varying between 30 years and over 150 years. A demonstration disposal centre, probably located in a granite formation in the Massif Central, is planned to be operational around 1993 (Lefevre, 1983).

Also in Japan, the reutilization of uranium and plutonium contained in spent fuel is considered essential for the promotion of nuclear energy development, and reprocessing is therefore regarded as a necessary stage of the fuel cycle. Up to now, the high-level radioactive waste produced in the existing reprocessing plant (at Tokai Mura) has been stored in liquid form in stainless steel tanks. However, solidification processes are being developed, and a pilot plant for vitrification will start operation in the latter half of the 1980s. In addition, research and development activities for geologic disposal are being performed under a long-term plan with a view to establishing the technology for disposal as soon as possible after the year 2000 (Okui, 1984).

The Canadian concept for waste disposal is focused on interim storage of used fuel for an indefinite period (up to 50 years) followed by immobilization and subsequent disposal in deep geologic repositories. Unlike most other countries, the Canadian nuclear power programme is based on heavy water, natural uranium reactors and the incentive for reprocessing is therefore less than in the case of light water reactors. Technologies are nevertheless being developed for the immobilization of both unreprocessed spent fuel and recycle waste, so that options are maintained for the disposal of either form of nuclear fuel waste (Rummery *et al.*, 1984).

An interesting conclusion that can be drawn from this survey of national waste management policies is that the practical implications of the differences of opinion that exist about the urgency of repository development do not seem to be very far-reaching. In those countries aiming at an early demonstration of final disposal, the implementation of this policy will nevertheless be delayed for several decades because interim storage is anyhow required for an extended period. On the other hand, in those countries favouring the delayed disposal concept, extensive development programmes are pursued in order to prepare for final disposal and to identify candidate sites for repository locations (Raudenbush, 1983).

9.2.8 *Economics of High-level Waste Management*

The economics of the back-end stages of the nuclear fuel cycle differ in several respects from those of the front-end stages. In the latter case, a commercial market exists for all needed products and services, such as natural uranium, enrichment, conversion and fuel element fabrication, and the calculation of the corresponding unit costs, expressed for example in dollars per kilogram fuel, can therefore be based on prevailing market prices and estimated future trends. For most of the components of the back-end of the fuel cycle, the situation is different. Unlike the front-end components, where a mature technology already exists, the technology of most of the back-end operations, in particular those involved in the long-term management of used fuel and high-level waste, is still under development and will not be definitively demonstrated for many decades. In these cases, therefore, the unit costs must be calculated on the basis of estimates of expenditures some of which will be incurred in the distant future. In addition, because of their *a priori* nature, these cost estimates are very often limited to an evaluation of the undiscounted costs of the goods and services involved. In such cost estimates, no present worth calculations of future expenses are performed, nor are any conclusions drawn about the prices that might possibly develop in a future commercial market for waste management services.

These difficulties were observed in a study of the economics of the nuclear fuel cycle performed in 1984 by a working group of experts under the auspices of the OECD Nuclear Energy Agency (NEA, 1985). This study was based primarily on fuel cycle data gathered by means of a questionnaire submitted for reply to seventeen OECD countries. In analyzing the data collection obtained in this way, the working group had to consider that most of the back-end cost data were given as undiscounted costs while the front-end cost data were derived from existing market prices. In order to make the former group of data consistent with those of the latter group, the NEA study made use of the so-called 'levelized discounted cost method' (see Chapter 21 for a fuller description).

Using this method when appropriate, the NEA working group calculated the fuel cycle component costs for different types of reactors, including the Canadian CANDU reactor and the heavy water moderated, light water cooled advanced thermal reactor (ATR) under development in Japan. For conventional light water reactors (of the pressurized water type), the working group made separate calculations for the once-through and the reprocessing cycle alternatives. The principal results of these calculations (for the back-end components of the fuel cycle) are given in Table 9.2 which also shows the fuel cycle costs in mills per kilowatt-hour for the different back-end components of the fuel cycle.

A large number of assumptions underlie the results of calculations shown in Table 9.2. One of the most important assumptions is that of the discount rate. For the reference case calculations, the results of which are shown in Table 9.2, the NEA study adopted a discount rate of 5 per cent but employed also variants of 0 and 10 per cent in order to explore the sensitivity of the fuel cycle costs to the discount rate assumptions. Another important assumption is the lagtime for the different back-end operations. By definition, lagtime specifies the date at which the payment for a back-end service occurs after the date of fuel discharge from reactor. The reference fuel cycle costs given in Table 9.2 were obtained on the assumptions of a lagtime of five years for reprocessing and 40 years for final disposal of spent fuel and reprocessing waste.

The NEA study confirms the prevailing opinion that the costs of the back-end stage of the fuel cycle are relatively small in comparison with those of the front-end stage. For the total fuel cycle costs the NEA study gives the reference values 7.78 mills kWh^{-1} for the once-through cycle and 8.56 mills kWh^{-1} for the reprocessing cycle (see Table 9.2). This means that the back-end contributes about 12 per cent of over-all fuel cycle

Table 9.2 Reference unit costs in $ per kilogram heavy metal and corresponding fuel cycle costs in mills per kwh (at January 1984 price level) for the back-end components of the reprocessing and the once-through fuel cycles (NEA 1985)

Component	Reprocessing cycle Reference unit cost $\$\ \mathrm{kg^{-1}HM}$	Reprocessing cycle Fuel cycle cost mills $\mathrm{kwh^{-1}}$	Once-through cycle Reference unit cost $\$\ \mathrm{kg^{-1}HM}$	Once-through cycle Fuel cycle cost mills $\mathrm{kwh^{-1}}$
Transportation of spent fuel	40[1]	0.14	40[1]	0.14
Storage of spent fuel	40+4y[2]	0.17	40+4y[2]	0.65
Reprocessing	550 }	2.18 (Reprocessing + Vitrification)		
Vitrification	200[3] }			
Conditioning			200 }	0.18 (Conditioning + Disposal)
Disposal	150	0.08	150 }	
Total of back-end		2.57		0.97
Uranium credit		−0.54		
Plutonium credit		−0.28		
Total fuel cycle cost (incl front-end)		8.56		7.78

[1]within the European area
[2]y = storage period in years
[3]including conditioning and predisposal storage

costs in the once-through cycle and about 20 per cent in the reprocessing cycle (with the uranium and plutonium credits taken into account). The impact of the back-end on the total economics of nuclear power is of course still smaller. Thus, taking into account that the total fuel costs contribute about one-third of the total production cost of nuclear power, it follows that the back-end part of this contribution amounts to only about 5 per cent on an average for the once-through and the reprocessing fuel cycles.

The NEA study also shows that the back-end fuel cycle costs are relatively insensitive to variations of the basic assumptions with respect to the unit costs and lagtimes of the back-end components. For example, a variation in the range from 150 to 550 $\$\ \mathrm{kg^{-1}\ HM}$ of the sum of the unit costs of conditioning and disposal of spent fuel (see Table 9.2) will only result in a 3 per cent variation of the total fuel cycle cost. For another main component of the back-end of the once-through cycle, interim storage of spent fuel, a variation of the cost (for a 40–year storage period) in the range from 100 to 300 $\$\ \mathrm{kg^{-1}\ HM}$ will result in an 8 per cent variation of the total fuel cost.

It should be noted, however, that the effects on the back-end costs of variations of the discount rate are considerably greater than those of variations of unit costs and lagtimes. Thus, due to the long discounting periods involved, an increase in the discount rate brings about a rapid decrease in the costs of long-term storage and final disposal. However, these effects are outweighed by the increase of the front-end costs that takes place at the same time. The combined effect of a discount rate variation in the range 0 to 10 per cent is that the total fuel cycle cost first decreases as discount rate increases up to 3 to 4 per cent and then increases for the further increase of the discount rate up to 10 per cent. Compared to the reference 5 per cent discount rate, the resultant maximum increase of the total fuel cycle cost is about 5 per cent for the reprocessing cycle and about 8 per cent for the once-through cycle.

9.3 Non-High-Level Waste Management

The term non-high-level waste covers a large range of wastes most of which are produced in liquid form as the result of different kinds of wet processes used in fuel cycle operations (see Table 9.1). The primary aim of treatment of these wastes is volume reduction. The processes available for this purpose generally fall into three main categories:

evaporation, ion exchange and chemical precipitation.

Evaporation is a process whereby the volume of a solution or a slurry is reduced by means of vaporization of the solvent, which usually is water. Although it is a fairly simple operation which has been successfully applied in the conventional chemical industry for many years, the application of evaporation in the treatment of radioactive waste involves a number of problems, such as corrosion, scaling and foaming.

To withstand corrosion, evaporators are usually constructed of stainless steel and operated at lowest possible temperature. The main disadvantage of scaling—i.e. the formation of deposits on the metal surfaces inside the evaporator—is that it reduces the heat transfer efficiency and so increases the energy consumption of the evaporation process. Since this means an increase of costs, scale has to be removed periodically, either mechanically or chemically. Foaming also reduces the efficiency of the evaporation process. It is a major problem in the evaporation of most radioactive liquid wastes but can be avoided or reduced by means of foam-breaking coils inside the evaporator or by the addition of antifoaming agents.

The use of ion-exchange methods for the treatment of liquid waste means that the radionuclides are removed from the solution and bound to a solid adsorbent. When the ion-exchanger has become fully loaded with radioactive ions, it must either be removed from service and treated as radioactive waste, or be regenerated by means of strong acids or bases in which case a radioactive liquid waste with a high salt content will be obtained. A wide variety of systems particularly designed for the use of ion-exchange methods in radioactive waste treatment have been developed, and a great number of ion-exchange plants are currently being operated at nuclear power plants, reprocessing plants and central installations for managing liquid wastes from research centres and hospitals.

The aim of chemical precipitation is to produce, within the liquid medium, a separable solid phase which entrains the radioelements by precipitation, co-precipitation or sorption. This can be achieved by adding specific chemical reactants or by adjusting the acidity (pH) value of the solution. The floc which is formed by any of these processes has to be collected by settling or filtering and is then handled as the radioactive waste concentrate. The separation achieved in chemical precipitation is never complete and the decontamination factor is relatively low. The method is therefore suitable only for low-level wastes. On the other hand, it is by far the cheapest procedure available for the concentration of liquid non-high-level wastes (IAEA, 1984b).

As already mentioned, a great variety of solid wastes also arise in the course of fuel cycle operations (see Table 9.1). For the treatment of these kinds of wastes also, the primary objective is volume reduction. This can be achieved by means of three processes: incineration, compaction, and surface decontamination. Incineration is applied to solid, combustible wastes with a wide range of activity levels. It results in a significant volume reduction and leaves an ash which must be handled as radioactive concentrate. Compaction is used widely for low-level solid wastes and is often preceded by other mechanical treatment techniques such as cutting, crushing and grinding. Decontamination is applied to surface-contaminated reactor components and process equipment in order to reduce their actual radioactive waste volume prior to conditioning and disposal. It can be achieved either chemically by means of acids or other suitable solvents or mechanically by sand, metal or slag blasting.

Volume reduction by means of the methods described above usually results in products that either still contain appreciable amounts of water or may be quite easily leached or dissolved by water. A further conditioning of these products is therefore necessary in order to transform them into leach-resistant and compact forms suitable for transport, storage and disposal. For this purpose, three different methods are available: cementation, bituminization and polymer processes.

Cementation consists of mixing a cement

material—usually ordinary Portland cement —with the waste, be it as a solution, slurry or solid, within a container. The cement mix is then allowed to set. This process has developed as a standard technique in waste management and has been commonly used on an industrial scale for several years in most nuclear power countries.

Bituminization consists of mixing waste solutions or sludges with bitumen at elevated temperatures. The water contained in the waste is evaporated and the radioactive residues are thoroughly mixed with the bitumen. The bitumen mixture is then drawn off in suitable containers where it cools down and solidifies. The bituminization process, which can be performed either batch-wise or continuously, has been under development since the early 1960s and has proved to be a very useful immobilization process for a wide variety of radioactive wastes. Large-scale installations for bituminization now exist at a great number of nuclear power plants and research centres.

Methods using polymers as matrix material for the incorporation of radioactive waste are relatively new and still in the pilot-plant stage or in a stage of early industrial application. Unlike the cementation and bituminization processes, polymeric processes do not really solidify the wastes but operate in such a way that the long-chained molecules of the organic polymer are linked together to form a porous sponge that 'traps' the waste (IAEA, 1983).

When the waste has been immobilized, it will be packaged in a container, the purpose of which is to facilitate further handling and to prevent the spread and release of radionuclides during transport and storage and, in particular, after disposal. A large number of types of container with different shapes and volumes ranging from some litres to several cubic metres are currently in use. The materials employed are usually mild carbon steel, stainless steel and concrete.

Unlike high-level waste, where extended storage periods are used in order to take advantage of the activity decrease with time, non-high-level wastes need only be stored when final disposal facilities are not immediately available.

The techniques for storage of non-high-level wastes can broadly be divided into two categories: area storage and engineered storage. Area storage simply means that the waste is stored on the ground surface in the open air or possibly with a simple cover as a protection against rain and sunshine. By contrast, an engineered storage facility consists of specially prepared structures like covered concrete trenches, buildings or engineered rock cavities. Depending on national regulations and on the physical and chemical properties and the activity content of the waste package, these engineered stores can be built to different degrees of complexity with respect to ventilation, temperature and humidity control, radiation protection, installation for remote handling and activity monitoring etc.

For the final disposal of non-high-level waste two fundamentally different concepts are available: underground disposal and sea dumping. As far as underground disposal is concerned, two different types of system are generally considered suitable: shallow-ground repositories, and repositories in abandoned mines or man-made, or natural, rock cavities. Deep geological repositories of the same kind as those used for final disposal of high-level waste are of course also applicable for non-high-level wastes but are much more costly than shallow-ground repositories and provide more isolation than necessary. They are therefore normally not considered for non-high-level waste disposal.

Shallow-ground burial is the simplest and most economic method of disposal of the large volumes of low-level wastes resulting from nuclear activities. The method involves emplacement of the wastes in repositories above or below the ground surface, the final protective covering being a few metres thick. When shallow-ground disposal is used, site selection is of considerable importance. An acceptable site should have climatological, geological and hydrological characteristics which ensure that radionuclides will not leach from the disposal site in unacceptable quantities during the period the wastes remain toxic. Shallow-ground disposal is used in a large number of countries, e.g. France, India, the United States, the United Kingdom, Hungary and Czechoslovakia.

Rock-cavity disposal is important for countries where favourable conditions for shallow-ground disposal do not exist. A well-known example of a rock cavity used as a repository for low- and intermediate-level wastes is the Asse salt mine located in the Federal Republic of Germany. Abandoned mines are also being used as repositories in the German Democratic Republic, Czechoslovakia and Spain. Rock-cavity repositories in hard rocks are under construction, planned or studied in Sweden, Switzerland, the United Kingdom and other countries.

The other mode for disposal of conditioned low-level wastes is to dump them into the deep sea. This disposal process is controlled by the London Convention on the Prevention of Marine Pollution by Dumping of Wastes and Other Matter, which, after having been adopted by an international conference in 1972, came into force in 1975. This convention places an absolute ban on ocean dumping of high-level waste and stipulates that an appropriate national authority in each country that wants to dump radioactive waste (other than high-level waste) into the sea has to issue a special dumping permit for each dumping operation. It also gives the International Atomic Energy Agency the responsibility to define 'high-level radioactive wastes or other high-level radioactive matter unsuitable for dumping at sea' and to make recommendations regarding the dumping of radioactive matter not included in this definition.

From 1967 through 1976, the OECD Nuclear Energy Agency organized and supervized sea-dumping operations in which, at one time or another, eight OECD Member Countries participated. Since 1977 dumping operations by OECD Member Countries have been conducted under an OECD Multilateral and Consulative Surveillance Mechanism, being administered by NEA, which is designed to further the objective of the London Dumping Convention, taking into account the IAEA definitions and recommendations (IAEA, 1983).

The different steps included in non-high-level management—conditioning, interim storage and final disposal—fulfil essentially the same purposes as the corresponding steps in high-level waste management. However, because of the lower activity levels and the less stringent requirements for shielding and cooling, the technology of management of non-high-level waste is much simpler and less costly than that of high-level waste management. In addition, several of the operations needed for non-high-level waste management are generally not regarded as independent processes in the same sense as the corresponding operations on high-level waste management. For example, the solution of the problems related to reactor waste management is generally considered to be entirely the responsibility of the reactor owner, and the facilities required for the accomplishment of these operations are therefore usually regarded as part of the normal auxiliary equipment of the reactor installation.

As a consequence, the costs of management of low- and medium-level liquid and solid wastes produced during reactor operations are not accounted for as separate cost items but included in the total reactor capital, operating and maintenance costs (NEA, 1985). In fact, no detailed analyses of the costs of processing and disposal of radioactive wastes from nuclear power plants have been published up to now. There are, however, some data available which specify the share of the total capital and running costs of nuclear power that has to be assigned to reactor waste management (Dlouhý, 1982). These data indicate that the total costs of processing and disposal of reactor operation wastes are within the range 2–5 per cent of the total cost of production of electricity from nuclear power.

A similar accounting practice as for reactor waste is usually applied also to non-high-level reprocessing wastes. The facilities required for conditioning of these wastes form an integrated part of the reprocessing plant, and their capital and operating costs are therefore included in the total reprocessing cost. Such an approach is applied, for example, in the above-mentioned NEA study (see Table 9.2), which also concludes that the costs of disposing of low- and medium-level reprocessing wastes are a relatively small part of the total back-end cost (NEA, 1985).

REFERENCES

Baatz, H., Einfeld, K., and Malmström, H. (1983). 'Dry cask AFR storage in the Federal Republic of Germany', *Nuclear Europe*, **4:2**, 15–16.

Bastien Thiry, H., Laurent, J. P., and Ricaud, J. L. (1984). 'French experience and projects for the treatment and packaging of radioactive wastes from reprocessing facilities', in *Radioactive Waste Management*, International Atomic Energy Agency, Vienna, Vol. 2, pp. 219–237.

Bjurström, S., and Svenke, E. (1984). 'The Swedish program for spent fuel and waste', *Nuclear Europe*, **4:10**, 26–28.

Bonne, R. A. (1984). 'Clay: Evaluation of geological disposal of radwaste in Belgium', *Nuclear Europe*, **4:2**, 29–30.

Chometon, P. L., and Cantin, P. (1983). 'Centralized disposal of spent fuel elements in France', *Nuclear Europe*, **3:2**, 12–15.

Cooper, B. S. (1984). 'Monitored, retrievable storage: Priority needed', *Nuclear News*, **27:14**, 118–121.

Dlouhý, Z. (1982). *Disposal of radioactive wastes*, Elsevier Scientific Publishing Company, Amsterdam, Oxford, New York.

ENS (1984). 'Safe transport of spent nuclear fuel. A public policy statement of the European Nuclear Safety', *Nuclear Europe*, **4:5**, 10–11.

Heafield, W. (1984). 'Handling and storage of conditioned high-level wastes', in *Radioactive Waste Management*, International Atomic Energy Agency, Vienna. Vol. 3, pp. 69–92.

Höhlein, G., Tittman, E., and Wiese, H. (1985). 'PAMELA: Advanced technology for waste solidification', *Nuclear Europe*, **5:2**, 16–18.

IAEA (1982). *Nuclear power, the environment and man*, International Atomic Energy Agency, Vienna.

IAEA (1983). *Conditioning of low- and intermediate-level radioactive wastes*, International Atomic Energy Agency, Vienna.

IAEA (1984a). *Guidebook on spent fuel storage*, International Energy Agency, Vienna.

IAEA (1984b). *Treatment of low- and intermediate-level liquid radioactive wastes*, International Energy Agency, Vienna.

KBS–1 (1977). *Handling of spent nuclear fuel and final storage of vitrified high level reprocessing waste*, Swedish Nuclear Fuel and Waste Management Company, Stockholm.

KBS–2 (1978). *Handling and final storage of unreprocessed spent nuclear fuel*, Swedish Nuclear Fuel and Waste Management Company, Stockholm.

KBS–3 (1983). *Final storage of spent nuclear fuel*, Swedish Nuclear Fuel and Waste Management Company, Stockholm.

Lefevre, J. (1983). 'Nuclear waste management policy in France', *Nuclear Technology*, **61**, 455–459.

NEA (1984). *Seabed disposal of high-level radioactive waste*, Nuclear Energy Agency, Organization for Economic Co-operation and Development, Paris.

NEA (1985). *The Economics of the Nuclear Fuel Cycle*, Nuclear Energy Agency, Organization for Economic Co-operation and Development, Paris.

Niederer, U. (1984). 'Waste management policy and its implementation in Switzerland', in *Radioactive Waste Management*, International Atomic Energy Agency, Vienna, Vol. 1, pp. 203–211.

Okui, Y. (1984). 'Radioactive waste management policy in Japan', in *Radioactive Waste Management*, International Atomic Energy Agency, Vienna, Vol. 1, pp. 159–166.

PLAN 82 (1982). *Radioactive waste management plan PLAN 82*, Swedish Nuclear Fuel and Waste Management Company, Stockholm.

Randenbush, M. H. (1983). 'Looking at Waste Management Worldwide', *Nucl. Eng. International 28*; 8, pp. 30–33.

Reeve, K. D., and Ringwood, A. E. (1984). 'The SYNROC process for immobilizing high-level nuclear wastes', in *Radioactive Waste Management*, International Atomic Energy Agency, Vienna, Vol. 2, pp. 307–324.

Reuse, B. (1984). 'Waste management policy and its implementation in the Federal Republic of Germany, in *Radioactive Waste Management*, International Atomic Energy Agency, Vienna, Vol 1, pp. 149–158.

Rummery, T. E., Lisle, D., Howieson, J., and Charlesworth, D. H. (1984). Radioactive waste management policy and its implementation in Canada', in *Radioactive Waste Management*, International Atomic Energy Agency, Vienna, Vol. 1, pp. 175–190.

Sethna, H. N., Ramanna, R., Meckoni, V. N., and Sunder Rajan, N. S. (1984). 'Waste management policy and its implementation in India', in *Radioactive Waste Management*,

International Atomic Energy Agency, Vienna, Vol.1, pp. 89–99.
Smith, W. (1985). 'Vitrification of Sellafield wastes', *Atom*, **341**, 3–7.
Sombret, C. G. (1984). 'Etat d'avancement des différents procédés de vitrification des solutions concentrées de produits de fission', in *Radioactive Waste Management*, International Atomic Energy Agency, Vienna, Vol. 2, pp. 67–75.
Stein, R., and Collyer, P. L. (1984). 'Pilot research projects for underground disposal of radioactive wastes in the United States of America', in *Radioactive Waste Management*, International Atomic Energy Agency, Vienna, Vol. 3, pp. 311–328.
Sunder Rajan, N. S. (1984). 'Treatment and conditioning of radioactive wastes arising from reprocessing plants in India: A review of the status of projects', in *Radioactive Waste Management*, International Atomic Energy Agency, Vienna, Vol. 2, pp. 259–278.
The Uranium Institute (1984). *Management and disposal of used nuclear fuel and reprocessing wastes*, The Uranium Institute, London.
WHO (1982). *Nuclear power: management of high-level radioactive waste*, World Health Organization, Regional Office for Europe, Copenhagen.

Nuclear Power: Policy and Prospects
Edited by P. M. S. Jones

10

Decommissioning

A. N. Knowles
United Kingdom Atomic Energy Authority, Risley

10.1 Introduction

The term 'decommissioning', of nineteenth century naval origin, has acquired a special meaning in the nuclear power industry. It now implies not only the long term shut-down of an installation, a shut-down which might or might not be followed by scrapping, but also the safe demolition, the conditioning of the resulting wastes and any other measures needed to return the site to unrestricted re-use.

Radioactive plants have been cleaned up, modified or re-equipped as a routine ever since the advent of the industrial use of radioactivity. Progressively, as experience accumulated, a pattern of safe procedures has been established and these techniques are the foundation upon which the present concepts for the decommissioning of very large plants such as a commercial sized nuclear power station are based.

It almost goes without saying that standard industrial demolition practices will not in general be suitable for those parts of a radioactive installation where there is a radiation hazard. They may not even be acceptable for those non-radioactive parts of the plant which are close to active areas. However, with plants like nuclear power stations there will be other very large areas where the demolition is non-active and the demolition techniques which would have been used for, say, a coal-fired power station will be fully acceptable. As with other demolitions, subject to the overriding requirement that the work be carried out in a safe fashion, the need is to carry out the operation as cheaply as possible, taking full advantage of recycling the scrap materials, wherever this is economic.

10.2 Experience

While the decommissioning of small processing plants has occurred from time to time, the decommissioning of very large installations has not hitherto been attempted. The group of early nuclear power stations will probably be the subject of the first application of decommissioning techniques to very large active plants. In the UK, for example, many of these power stations are now past the 20 years' life originally used for accounting purposes, and will, in a few more years, reach 30 years, the life for which most were designed. Decommissioning plans are therefore being reviewed and pilot scale nuclear power plant demolitions are being put in hand in several countries. Accordingly, the main part of this chapter will deal with nuclear power station decommissioning, the much more varied situation with active process plants being less amenable to a general treatment.

Table 10.1 lists the more notable experimental and prototype reactors already decommissioned; some have been only partially decommissioned and on several work is still in progress. Of outstanding

Table 10.1 List of the Principal Experimental and Prototype Reactors which have been Decommissioned

Reactor	Operation	Present state
BEPO (Harwell)	1948–1960	Stage 2
Windscale piles	1950–1957	Virtually Stage 1
DMTR (Dounreay)	1958–1969	Stage 1
DFR (Dounreay)	1959–1977	Approaching Stage 1
Dragon (Winfrith)	1964–1976	Stage 1
WAGR (Windscale)	1963–1981	Progressing to Stage 3 (1993)
Hallam (Neb)	1963–1966	'Entombed' (approx Stage 2)
Elk River (Minn)	1964–1968	Stage 3 1974
EBRI (Idaho)	1951–1955	'Deactivated' (approx. Stage 2)
Peach Bottom I (Penn)	1967–1974	'Mothballed' (approx. Stage 1)
SRE (California)	1967	'Mothballed', now to Stage 3
Shippingport	1957–1982	Progressing to Stage 3 (1988)

importance are the Elk River, Windscale AGR and Shippingport reactors. The first, an indirect cycle BWR of 50 MWt, was completely demolished and the significant parts of the site cleared in the 1970s. This represented an exceptionally valuable pioneering demonstration which was carried out without meeting any major difficulty (United Power Association, 1974). The second important project is the Windscale AGR. This 30 MWe prototype advanced gas cooled reactor is being decommissioned as an R and D project aimed at exploring techniques and providing data on effort and costs. The reactor operated successfully for 17 years, much of the time as a test facility for experimental fuel, and the programme for its complete demolition and site clearance is on a generous time-scale so that opportunities for research can be exploited. Shippingport (70 MWe, PWR) is being decommissioned on a rather different basis. Here there is a legal obligation to clear the site and the work will be done as quickly and as cheaply as possible. However, this operation will also produce valuable information, especially as the plan is to dispose of the main reactor vessel in one piece by burial in a specially constructed trench.

Not included in Table 10.1, because firm plans have not been announced, are comparable pilot scale decommissioning projects which are under consideration in France, Germany, Italy, Japan and Canada. All these projects will, through the medium of existing international collaboration agreements, serve to create the basic data and the background of experience upon which sound technical and economic judgements will depend. However, even at the present time the broad outlines of the technology have emerged and can be described.

10.3 Some Definitions

Preceding this description, it is necessary to define a few of the terms in general use. The first two are not peculiar to decommissioning:

1. *Contamination* in this context means the presence of surface deposits of radioactive material. The deposit could be merely a layer of dust loosely adhering to the surface or the deposited material might have penetrated the surface or chemically combined with a surface layer. Decontamination therefore involves processes ranging from brushing or washing to the removal of the surface layer by high pressure jetting, shot blasting, flame scarphing, or even more strenuous treatments.
2. *Activation* is distinguished from contamination because it is caused not by a surface deposit of activity but is the result of the material being exposed to penetrative radioactive fluxes for a period of time.

These produce active isotopes within the material in accordance with the degree of penetration and the time of exposure.

The next terms are really only descriptive and are somewhat arbitrary in definition. However, they were produced by international agreement (in the IAEA committee, 1975) and the fact that they are in general use is a measure of their convenience. They have already appeared in Table 10.1.

Decommissioning is conceived as being applied in three stages:

1. *Stage 1.* The installation is permanently shut down, and all portable sources of radiation removed. Openings in the existing barriers are sealed, atmospheres are controlled and monitored and surveillance and maintenance schemes devised and instituted. In the case of a power station, these actions involve the removal of fuel and control rods, the permanent sealing of openings in the biological shield, the monitoring of all internal atmospheres, the regulation of access and the creation of a regular schedule of inspection and maintenance.
2. *Stage 2.* The installation is reduced to minimum size, all easily dismantled parts being removed, possibly involving some decontamination. The remaining radiation barriers are reinforced and the need for atmospheric monitoring and surveillance reduced as far as possible. In the case of a power station, virtually all plant outside the main biological shield is dismantled or converted to other use, the residual plant being permanently protected by passive barriers, with the need for maintenance reduced to occasional checking of external surfaces.
3. *Stage 3.* All materials, equipment and parts of the plant, the activity of which remains significant, are removed and the site released to unrestricted re-use. In the case of a power station this involves the demolition of the reactor and its biological shield and the packaging and transport of the resulting radioactive debris.

The completion of each stage is conceived as representing a logical pause in decommissioning. Typically, the first stage is relatively inexpensive and the waste produced is of the same type as that which arose in normal operation of the plant. Stage 2, which can involve substantial inactive demolition, would usually release equipment, buildings or valuable space to offset the cost. Since Stage 2 can well involve decontamination, some small amount of radioactive solids or liquids at low levels may require conditioning and disposal. Stage 3, involving the dismantling of the most active parts of the plant, probably requiring remotely controlled demolition and waste packaging, usually represents the major cost as well as the major source of active waste. The Stages may be carried out with long or short breaks between them or blended into a continuous operation. Each plant will have its own economic and environmental features which will control the strategy adopted for that case.

10.4 Timing

One of the major considerations governing the timescale of decommissioning is the balance of benefit in delaying long enough for the activity to decay. A review of all the known isotopes (Felstead and Woollam, 1984) has produced a list of those which might be important for decommissioning. Out of more than 2600 known isotopes, the large majority can be eliminated either on the grounds of having a short half-life, or of not arising in significant concentration in the materials used in plant construction. To some extent the short list of isotopes worthy of consideration is specific to the plant concerned, but a list of only 20 or 30 may be expected in the case of a power reactor.

Of the isotopes on a typical short list, ^{60}Co, ^{94}Nb and ^{108}Ag are often seen as the key γ-producers in activated material; ^{137}Cs is of importance as a γ-active contaminant. Several more are important as β-emitters, but β-emission is mainly of concern from a disposal inventory point of view. The γ-emitters govern the potential dose to operatives and therefore determine the shielding and hence the cost of dismantling, packaging and transport. As the following curve (Figure 10.1) shows, in the typical case the ^{60}Co (5.27 years half-life) tends to dominate the picture.

This domination by the shorter-lived isotopes naturally suggests that there would be an advantage in delay, especially from delaying Stage 3 when the bulk of the active material has to be handled. To illustrate this by an actual example, in the case of the Windscale AGR the total active inventory two years after shut-down is estimated to have been about 200 000 Ci (7700 TBq), at seven years it will be about 70 000 Ci (2700 TBq), at ten years 46 000 Ci (1770 TBq), while at 100 years it will be only about 3000 Ci (115 TBq).

Against this is the cost of maintenance, which experience shows becomes considerable after a few years. In general, this is because industrial buildings are commonly constructed of low cost, non-durable materials and would become hazardous ruins if totally neglected for more than a few years. This is particularly true of buildings on coastal sites where exposure to strong winds and air-borne salt particles produces rapid deterioration. In the case of active plant, surveillance and the maintenance of monitoring equipment are also major components of the annual cost.

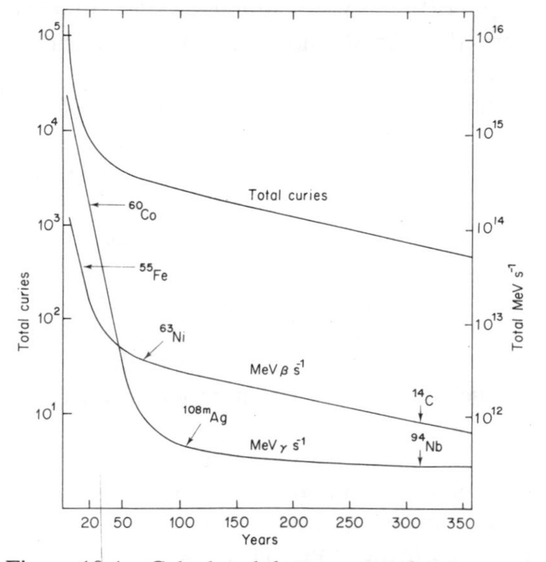

Figure 10.1 Calculated decay curves for the total radioactivity in WAGR following shutdown

Another factor against delay is the proven deficiencies of archived records of the original construction. After one or two decades the knowledge of the plant residing in the operating team (and taken for granted in the operational phase) will be dispersed by redeployment and death. Ostensibly, the written records kept were complete; in fact, for plants constructed up to the present time, the maintenance of a sufficiently detailed record, especially the inclusion of modifications made during the later phases of construction or during service, has normally been pushed aside by more immediate concerns. More attention is being given to the completeness and the durability of records for future plants, so it can be expected that this penalty for delaying decommissioning will diminish by the time plants currently under construction reach decommissioning.

It will be apparent that the formation of a judgement of the break-even point between early demolition and various periods of delay for a particular plant will often have to be made without much assistance from formal numerical assessments of benefits and disbenefits. A few paper studies have been carried out for nuclear power stations and these are of interest in giving a numerical perspective to the situation, although they are probably too specific to be applied generally except as a guide. Figure 10.2 presents a synthesis of such studies. It illustrates one range of strategies for decommissioning a large nuclear power station in which the main variable is a delay between Stage 2 and Stage 3. Since it merely shows a simple addition of unit costs without allowance for inflation or any discounting, it gives a measure of the typical technical benefits of delaying Stage 3 for periods up to 95 years. The reduction in the cost of the demolition and clearance of the most active parts of the station, which results from the decay of the radioactivity, is somewhat offset by the accumulating expenditure on maintenance and surveillance. The resulting envelope of technical benefit with delay of Stage 3 is consequently rather flat.

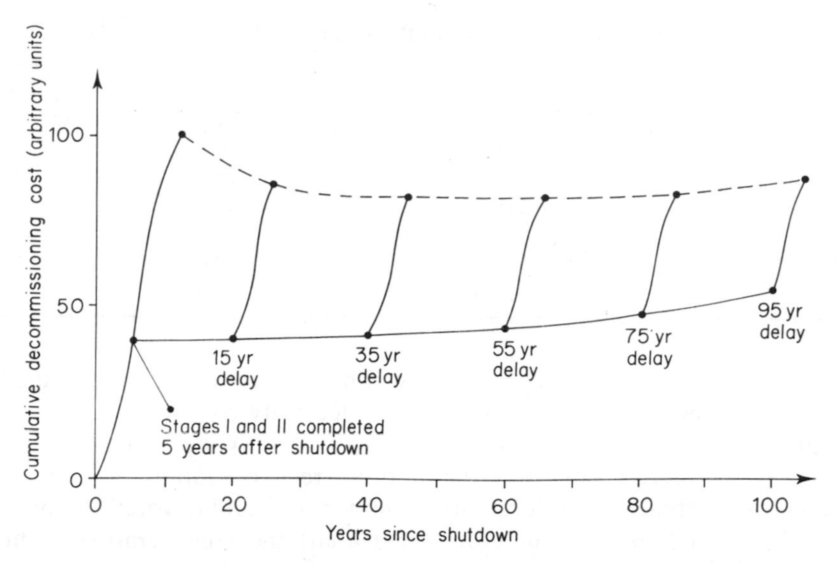

Figure 10.2 Costs with Stage III commenced after various delays

Such estimates of cost, which merely record a running total of the undiscounted cost of each operation, imply that the money has to be found at the moment it is needed. With any project involving a spend spread over a long period, more sophisticated management of the funding would be applied, taking advantage of long term and short term investments as well as providing for inflation. Unfortunately, when an operation is spread out over a very long period (such as the 50 or more years sometimes considered for the final stage in power station decommissioning) the selection of valid discount and inflation rates for prediction purposes becomes difficult. However, their importance is such that they cannot be neglected, as is readily seen from Figure 10.3, in which a constant net discount rate of only 2 per cent has been applied to the previous figures.

In this illustration it was assumed that a fund was accumulated out of the income produced by the power station and the costs quoted in Figure 10.3 are the necessary magnitude of that fund at the date of shut down. Only a very small utility would manage its finances in this simple way so the signifi-

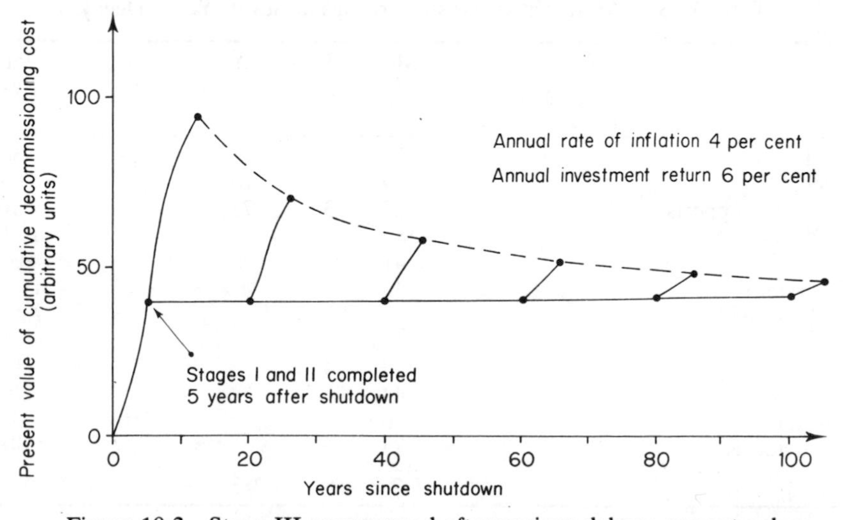

Figure 10.3 Stage III commenced after various delays: present values

Table 10.2 Decommissioning Wastes from the Complete Demolition of WAGR (t)

	Non-radioactive solids	Low active solids	Intermediate level active waste
Concrete	~8000	750	
Graphite	—	150	130
Steel MS	~4000	260	470
SS		15	74
Miscellaneous	~2000	50	
Total	~14 000 t	~1200 t	~700 t

cance of Figure 10.3 is strictly qualitative. It is enough to show, however, that consideration of the method of finance is an important factor, potentially completely swamping the effects of the various technical considerations illustrated in Figure 10.2. It must therefore be a major concern in deciding upon a decommissioning strategy. In this connection, it should be recognized that in any commercial undertaking of a size appropriate to centralized power generation which is operated for a profit (i.e., is not subsidized from an external source) discount financing is a fact of life. Even if a fund of the type envisaged above is not set up in a formal way, the normal operations of the undertaking constitute a profitable investment. Discount financing is therefore an inherent feature of successful commercial operation and in connection with long term options, such as those arising in the case of decommissioning, will have a real effect in reducing the cost to the electricity consumer. On the other hand, if the plant concerned is not part of a commercial operation (it might be government owned and devoted to non-commercial research) the consideration of financing its eventual demolition will be on a different basis.

10.5 Waste Arisings

For decommissioning, the conditioning, packaging and disposal of the radioactive material is likely to be a determining factor in the strategy and a major part (15–25 per cent) of the cost. In the context of power station decommissioning, enough detailed studies have been made to permit a generalized treatment of the topic.

Taking the Windscale AGR as an example, the active and inactive wastes arising from a complete demolition of the reactor and its

Table 10.3 WAGR Active Waste Arisings at Seven Years Decay

Item	MS t	SS t	Graphite t	Other t	Total Curies	Approx. Peak Ci t^{-1}
Pressure vessel	210			19	480	<1
Core and reflector	9	8	210		24 300	1300
Neutron shield and supports	27	13	73		6310	1100
Thermal shield	186				6270	50
Hot box	41	15			<1	<1
Core support system	48				4140	220
Core support bearings		2			7000	3500
Control rods	2.5				6500	2500
Loop tubes		4			9000	4000
Other items in RPV	44	47		35	1200	470
Concrete bioshield	90			750	325	3
Other items outside RPV	75				48	3
Total	732.5	89	283	804	65 574	

Note: Ms Mild Steel; SS Stainless Steel; RPV reactor pressure vessel.

containment (but excluding the turbine hall, heat exchanger and ancillary buildings) will be as shown in Table 10.2.

Several points can be made which will be generally applicable. First, about 90 per cent of the wastes will be inactive. Second, most of the active wastes will be low level and, third, there will be no high level wastes nor will there be any significant quantity of α-emitting waste. This last point is one of the main differences between power station decommissioning and the decommissioning of most process plants.

The next Table (Table 10.3) gives a breakdown of the Windscale AGR wastes at the intermediate level of activity. It can be seen that a few items having rather high specific activities tend to dominate the list. Some of these, representing something less than 10 t in mass, may require special packaging or a decade or so of storage, to meet the transport regulations.

The waste arisings from power station decommissioning are strongly dependent upon the type of power station. This stems from the major differences in relative size. The power density of the core is the main source of the size difference. Magnox cores are typically 1 MWt m^{-3}, AGRs are 3 MWt m^{-3} and PWRs are 100 MWt m^{-3}.

Acting counter to the core power density differences are various factors affecting the quality of the wastes. The most important arises from the intensity of activation which clearly increases with the power density. Thus a PWR would produce solid wastes having much higher spectrum of specific activity than a Magnox station. Also, since the LWRs generally suffer significantly more contamination within the main biological shield (and in some cases, outside it) than the gas cooled reactors, higher arisings of low level liquid wastes (from decontamination procedures) will also occur.

Giving all these factors due weight, the typical relative situation if 1200 MWe stations could be compared is shown in Figure 10.4.

Again this is generalized and comparative. Of course no Magnox station of 1200 MWe exists and there is reason to believe that the

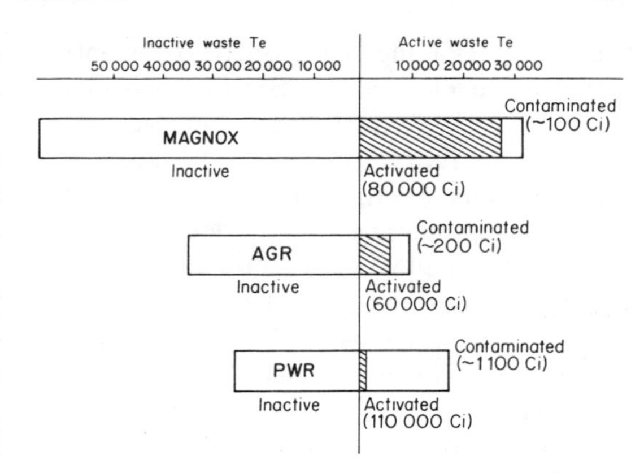

Figure 10.4 Typical arisings of decommissioning waste (1200 MW stations)

variation of decommissioning waste arisings with station size is not linear with electrical power; intuitively, the variation would be expected to follow about the same law as the variation of capital cost.

10.6 Cost of Decommissioning

These comparisons do, however, lead to the expectation that decommissioning a PWR station will be comparatively less costly than decommissioning a Magnox or an AGR station. This expectation is supported by estimates made by CEGB of the possible cost of decommissioning the proposed Sizewell PWR (1100 MWe) which is compared with the cost for a 450 MWe Magnox station (Gregory, 1983). These figures suggest a figure of about £150 million for the PWR and £290 million for the Magnox. In either case, these are large sums, justifying the attention now being given to decommissioning. However, if these costs were related to the value of the electricity produced, the addition to the cost of each unit of electricity sold is small, especially if a positive net discount rate and a few decades of delay are invoked (e.g. for 4 per cent and 50 years, the increment required per KW is 0.03p or 0.12p respectively).

10.7 Packaging

The packaging of active solid waste from decommissioning is in principle no different from the packaging of any other low or intermediate level waste. Existing practice, however, has established standard steel drums in the range 200– 500 l as the preferred package units. Such small packages are obviously rather unattractive for decommissioning, where smaller size implies more cutting for most classes of waste. The cutting of large activated components, which usually has to be done remotely and is followed by remotely controlled handling and packing, is time consuming, requires costly equipment and each cutting operation produces secondary active waste which also has to be collected and packed. There are very stong economic incentives for the reduction of such operations to the minimum, consistent with the practical limits. These limits turn out to be limits on size for transport and disposal and limits on total mass produced by cranage, etc.

Accordingly, for the Windscale AGR the UKAEA have produced a large package of approximately cubical form with a side of 2.2 m. In its present embodiment, which is under consideration by the regulatory authorities, it comprises a reinforced concrete box of 23 cm wall thickness sheathed externally with 12 mm of rust resistant mild steel. The contents are consolidated by a free flowing liquid concrete after the payload is placed (Figure 10.5). This package is suitable both for disposal in the deep ocean and at land disposal sites. Its consolidation and the steel envelope impart an outstanding resistance to impact damage. It is also suitable for prolonged storage. When used for the highest activity material from decommissioning, high density concrete is required for shielding. In this form the maximum weight of each unit is about 50 t.

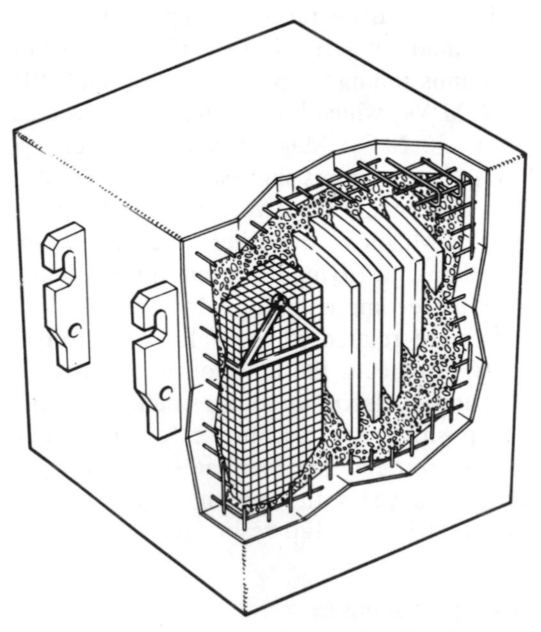

Figure 10.5 The large container for WAGR ILW decommissioning waste (2.21 m × 2.21 m × 2.34 m)

These packages are for the intermediate level waste. Low level solids suitable for trench burial do not require additional shielding and can be packed loose in skips or in purpose-made non-returnable containers of similar form.

Studies by NIREX include the development of a range of reinforced concrete packages for ILW, of which that described above is a part. Some proposed packages in this range have returnable shielding, unlike the one illustrated. Economic assessments are being made which should identify the most cost effective package format for specific circumstances. A key parameter in such assessments is the amount of waste (the 'payload') which can be fitted into a given package. With the more active wastes the payload may be limited by surface activity limits from the transport regulations, although advantage would be taken of self-shielding by placing less active material to the outside, but for the general run of material activity limits are not approached. For this material the limit on the payload arises from the practical problems of placement under conditions of remote operation, bearing in mind the need to properly consolidate the completed package with liquid concrete.

The limits on payload revealed by work for the Windscale AGR package (gross volume 11.23 m^3 and internal capacity of 5.75 m^3) show considerable variation, dependent upon the nature and density of the waste. These wastes range from segments of thick steel plate (over-all packaged density about 1.5 t m^{-3}) and graphite core blocks (0.7 t m^{-3}) to bulky items such as thermal insulation and contorted steel reinforcement from activated concrete (0.4 t m^{-3}). For the intermediate level wastes from WAGR, the over-all mean is expected to be only 0.6 t of waste per m^3 of package, but naturally attempts will always be made to improve on this.

The low level solid waste, since it requires no shielding, is easier to package, attaining around 1.25 t m^{-3}.

Initial cost estimates based upon a strategy for disposing activated core graphite to sea using either 200 l drums or the large package described above seem to show a substantial advantage in cost effectiveness for the large package. Specific studies of this type will be required to confirm that generalized results, such as that mentioned, are sustained under the particular conditions for a given waste and its complete processing and disposal strategy. Over-all, waste processing and disposal is likely to represent between 15 per cent and 25 per cent of the total cost of decommissioning.

10.8 Whole Item Disposal

The use of large waste packages helps to reduce the creation of secondary wastes from cutting and helps to keep down the increase in total volume which packaging always entails. To take this trend to its logical conclusion, some large components of plant, such as the heat exchanger units from the early commercial power stations, could be disposed of without any dismemberment. This strategy was investigated some years ago with respect to the four heat exchangers from the Windscale AGR (each 21 m long, 5.3 m diameter, 170 t, about 30 Ci) and has already been mentioned as the declared method for the Shippingport PWR.

The Shippingport PWR (70 MWe) has a thick-walled steel vessel of similar dimensions to that of WAGR, but of greater mass and since the package includes the core structure, of a much higher activation (280 t, 14 000 Ci, cf. 230 t, 500 Ci). The plan is to consolidate the interior steel work, provide the necessary external shielding and to ship the whole package (some 770 t) by barge from Shippingport to Hanford, where a special disposal pit is planned. This entails only short overland movements, the main movements being by sea, rivers and the Panama Canal. It is claimed that this procedure is not only cheaper but environmentally more acceptable since the activated components are retained in their original compact array which affords a high degree of self-shielding. However, it seems unlikely that this technique could be applied to items much more massive than the Shippingport RPV.

10.9 Packaged Volumes

Reverting to the more general route of packaging decommissioning wastes in containers having a gross volume of about 15 m^3, the active waste arisings, given previously in tonnes for typical Magnox, AGR and PWR power stations, suggest the following approximate packaged volumes for stations of about 1 GWe size:

Magnox 20 000 m^{-3}
AGR 7000 m^3
PWR 1000 m^3

This covers both ILW and LLW waste categories. The subdivision in the case of Magnox would be about 40 per cent LLW and 60 per cent ILW by volume.

These very approximate figures give a guide to the size of the surge in waste arisings which will occur when a nuclear power station is decommissioned.

10.10 Active Process Plants

The great variety of these plants makes their decommissioning rather difficult to describe in general terms. Some generalizations are however valid.

The first point is that activation is normally restricted to plants devoted to the reprocessing of spent fuel elements. Even here, the consequences of activation are normally relatively less important than contamination from actinides and fission products. Decontamination is therefore the prime technique. It is aimed at reducing activity levels to an extent which permits 'hands-on' demolition under suitable controls. Occasionally decontamination may be uneconomic and remotely controlled deconstruction has to be followed by packaging appropriate to the level of activity.

A study of decontamination techniques shows that many of them produce low level liquid wastes. The convenience and cost of disposing of this is another general factor deciding the strategy for process plant decommissioning. In the event that existing facilities for the treatment and disposal of low level liquids are inadequate, dry methods or methods which produce very small amounts of active liquid must be considered.

Considerable research into decontamination is taking place; it is a key technique in the industrial use of nuclear energy, but its importance to decommissioning has influenced the context of this research. Decommissioning requirements not only involve the decontamination of small items of equipment and localized areas, but they also require the consideration of treating very large areas.

USDOE (1980), Halter and Sullivan (1980) and Rogers (1980) provide a general introductory review of decontamination techniques, although in view of the developments which can be expected as a result of the general increase in interest in decontamination, it will always be necessary to check the latest bibliographies before planning an actual operation.

One of the problems encountered in decommissioning process plants is that the original designs were usually conceived against specifications which, if they included the need for eventual decommissioning at all, gave it the lowest priority. In the future, certain obvious and inexpensive features will be incorporated which will make decommissioning easier and cheaper. Before listing these, it is worth remarking that the section on the economics of decommissioning commercial nuclear power stations earlier in this chapter makes it clear that if deferred decommissioning is in mind, only low cost changes to ease future decommissioning are likely to be justifiable.

10.11 General Features to Simplify Plant Decommissioning

As a general list these comprise:

1. Design for easy decontamination:
   a. Use smooth non-absorbant and chemically inert surface finishes.
   b. Avoid crevices, projecting fasteners and corners where access for brushing, washing or swabbing is restricted.
   c. Compartmentalize the plant so that decontamination of one part can be effected without the risk of carry-over from other parts which are still active.
   d. Where possible, keep complex equipment, such as motors, actuators and the metering and recording parts of instruments outside the areas subject to contamination.
   e. Provide effective porting arrangements for replacement of stirrers, choppers, monitoring heads and anything else which is essential within an active compartment.
2. Design for low cost demolition:
   a. Use fastenings which will be simple to undo or sever when the time comes to demount part of the plant. Avoid continuous welds unless, as in the case of sheathing, they are essential for sealing.
   b. Avoid the use of monolithic reinforced concrete walls and framing if this will become contaminated or activated and will need demolition before local dose levels are reduced to 'hands-on' limits. Consideration should be given to making concrete shield walls demountable, e.g. by using a keyed block construction with weak mortar jointing provided that seismic resistance can be made adequate. In general, design the

main load bearing structure to be outside, and protected from, the radioactive contents of the plant.

3. Select Materials:
   a In activation conditions, avoid constituents which give rise to γ-active isotopes (Felstead and Woollam, 1984). Especially avoid as far as possible materials containing cobalt and niobium.

It will be observed that many of these features are as important for plant maintenance as for decommissioning. Modern active process plants may be designed for low cost decommissioning without incurring more than marginal increases in their initial cost, provided a decommissioning plan is conceived at the outset, along with the plans for operation and maintenance.

The procedures required for efficient and low cost decontamination have, of course, been widely studied. In the United Kingdom, the interested parties have jointly produced Codes of Practice which are endorsed by consensus opinion and published by the United Kingdom Atomic Energy Authority. Two of the most relevant, which contain a critical survey of commercially available methods, as well as a review of the fundamental chemical and radiochemical phenomena involved, are AECP 1002 (Coatings) and AECP 1057 (Decontamination).

10.12 Experience in Process Plant Decommissioning

This is very varied; however, there is a relative paucity of published information. Perhaps of outstanding interest among the operations actually carried out and for which detailed descriptions are available in the open literature, was the modification of the small fast reactor fuel reprocessing plant at the UKAEA's establishment at Dounreay, Caithness, Scotland.

This plant, usually known by its building number, D1206, had been designed for the reprocessing of the metallic uranium fuel irradiated in the Dounreay Fast Reactor. This was conceived as a small power producing plant (15–20 MWe) and started operation in 1959 with the main objective of studying the commercial embodiment of the fast breeder fuel cycle. The process plant was therefore an essential part of the enterprise, treating the niobium clad fuel bars and extracting the bred plutonium, processing about 10 t of fuel in a period of 15 years. In the early 1970s a second stage of this breeder programme started with the commissioning of the 300 MWe Prototype Fast Reactor. As this used oxide fuel in stainless steel cans and as the original Dounreay Fast Reactor was to be closed down, it was decided to modify D1206.

This modification effectively involved a Stage II decommissioning of D1206 and is described by Barrett and Thom (1978). This plant although of modest size (6 t of PFR fuel per year designed throughput) is complete in that every feature essential for fuel reprocessing is contained within it. Thus it includes active fuel reception, breakdown caves, dissolvers, separation of uranium, plutonium and fission products and storage of active raffinates prior to further processing.

10.12 Prospects for Decommissioning

From the previous pages it will be seen that decommissioning, even the decommissioning of old and highly active installations such as the large nuclear power stations now approaching the ends of their service lives, presents no particularly novel problems. Established techniques in decontamination, remotely controlled operation and waste conditioning and packaging will be combined in new ways and may require extending. The R and D required for this is chiefly to enable well understood practices to be operated on a larger scale.

This matter of scale, whether in connection with power stations or active process plant, does, however, lead to problems of organization and logistics. Larger teams than used hitherto will be required; it will no longer be possible to make temporary use of the valuable core of skilled and experienced operatives, well accustomed to working with radioactivity, who ran the plant during its service. Large teams of new specialists will require training and these might be recruited either from the industrial demolition field (for a power station 90 per cent of the work will, in any case, be inactive industrial demolition) or from the nuclear construction contractors. The matter of scale will also affect the waste disposal requirements. Decommissioning a large nuclear power station to Stage 3 will produce an appreciable surge in low level and intermediate level active waste arisings which must be integrated with the national and perhaps international waste strategies. However, since in principle the timing of a Stage 3 deconstruction is constrained by economic considerations and hardly at all by immediate issues of safety, there is likely to be freedom to plan each operation so that it takes place at a time which suits the over-all strategy.

In the early days of nuclear power, decommissioning of the power stations and the associated process plants was seen as feasible, but only at heavy cost. Now the measure of the cost has been taken and found to be far less than feared; indeed it represents only a small addition to the over-all cost of each unit of power produced. Nevertheless, when the time comes for the money set aside to be spent, large sums, of the order of a tenth of the original construction cost, will be involved, making active plant demolition an attractive new commercial venture, involving specialization, skill and high grade management.

REFERENCES

AECP 1002 (1984). 'The Coating of Surfaces Requiring Decontamination', Atomic Energy Code of Practice 1002.

AECP 1057 (1983) 'Radioactive Decontamination', Atomic Energy Code of Practice 1057.

Barrett, T. R., and Thom, D., (1978). 'The Decommissioning and Reconstruction of the Fast Reactor Fuel Reprocessing Plant, Dounreay', IAEA SM 234/9, *International Symposium on the Decommissioning of Nuclear Facilities*, Vienna.

Felstead, L. D., and Woollam, P. B. (1984). *An assessment of all known isotopes to determine which might be important in the decommissioning of thermal nuclear reactors*, TPRD/B/0386/N84, Berkeley Nuclear Laboratories, Gloucester.

Gregory, A. R. (1983). *Evidence to Sizewell B Public Inquiry*, Proof CEGB/P/24.

Halter, J. M., and Sullivan, R. G. (1980) *Equipment for the removal of contaminated concrete surfaces*, PNL-SA–8855.

IAEA (1975). *Decommissioning of nuclear facilities*, IEAE 179.

Rogers, L. N. (1980). *The cleaning of radioactively contaminated surfaces by means of particle impact processes*, RD/B/N4784, Berkeley Nuclear Laboratories, Gloucester.

United Power Association (1974). *Final Elk River Reactor Program Report*, COO–651–93, Revised, United Power Association, Elk River, MT.

USDOE (1980). *Decommissioning Handbook*, USDOE/EV/10128–1.

SECTION III
WORLD EXPERIENCE

Nuclear Power: Policy and Prospects
Edited by P. M. S. Jones

11

The United States of America

J. J. Taylor
Electric Power Research Institute, Palo Alto, California

11.1 Stagg Field to Shippingport: The Initial Development of Nuclear Power in the USA

11.1.1 *The Second World War Phase (1940–1947)*

The primary purpose of the initial efforts of the United States in nuclear technology was to develop a nuclear bomb in the world war against the Hitler tyranny. The fact that nuclear power was born from that military effort has had profound implications on the course of that development, particularly in the United States. That the technology was fostered because of its military application is not new or unique to nuclear power. The airplane, radar, computers and many modern technologies either originated or were strongly enhanced by military research and development funding. But the awesome nature of this military application and the fact that the bomb was first developed and used by the United States has resulted in a far more pervasive government involvement in nuclear power and greater difficulty in gaining public acceptance.

The story of the development during the Second World War is well known. Recognizing that the fission process discovered by Hahn and Strassmann in 1937 had the potential to be made into an enormously powerful weapon, Einstein appealed to President Roosevelt to initiate a major project to develop the bomb. A team of the best scientific and industrial talent the country could provide was assembled and the 'Manhattan Project' was formed to win the race with the Germans to develop the bomb. Among the project's first successes was the first experimental demonstration in 1942 that a fission chain reaction could be sustained and controlled in the Fermi 'pile' of uranium and graphite built under the stands of Stagg Field, the football stadium of the University of Chicago. On that day nuclear power was born.

The opportunities for the peaceful use of atomic energy was clear in the minds of these scientists, including the potential to 'breed' an almost unlimited supply of nuclear fuel for power application. But they had to put all of their ingenuity and energy into the race for the bomb. Thus, they developed the diffusion process to enrich the isotape ^{235}U in uranium, reactors for plutonium production, and a chemical process to separate uranium and plutonium from the spent fuel of the production reactors. When the development of the bomb was completed in 1945, the major basic processes essential to the development of nuclear power technology had also been successfully completed.

11.1.2 *Post-war Redirection (1947–1954)*

After the Second World War, the United States took vigorous action to pursue the development of the peaceful atom, on legal, administrative, and technical fronts. The Atomic Energy Act was passed in 1946, putting continued military development firmly under civilian control and providing funding for the development of peaceful applications. The Atomic Energy Commission (AEC) was established to carry out these functions. Power reactor development was initiated at the laboratories that had been formed during the Manhattan Project: the Argonne, Hanford, and Oak Ridge National Laboratories. A seminal policy decision was made during that period to separate the development of commercial power reactors from the continued development of reactors whose purpose was to produce plutonium for weapons. We will see later in this chapter the salutary effect that policy decision has had in the light of both the Three Mile Island and the Chernobyl accidents.

Many design studies of reactor types that would be capable of producing electricity were carried out in this period but the lack of technical knowledge and serious questions as to the economic potential kept those developments for the most part in the 'paper' stage. What emerged as the first priority goal of power reactor development in the United States was submarine nuclear propulsion. The General Advisory Committee to the AEC established this priority on the basis that the successful achievement of nuclear propulsion would result in a revolutionary new submarine, the capability of which would justify the cost of propulsion and the strategic value of which would justify the massive development funding which would be required. The Naval Reactor Programme was thus set up under the leadership of Admiral Rickover, an obscure naval captain who had participated in a special training programme in nuclear power at the Oak Ridge National Laboratory and had become a zealot for the development of the nuclear submarine.

The story of the development of this second major phase of nuclear power for naval propulsion is well recorded. Although the existing national laboratories were a vital part of the early development of submarine nuclear propulsion, two new laboratories were set up in 1947 and 1948: the Knolls Atomic Power Laboratory, operated by General Electric, with the goal to develop a sodium cooled submarine reactor and the Bettis Atomic Power Laboratory, operated by Westinghouse Electric, to develop a pressurized water-cooled submarine reactor. In an amazingly short time the full-scale prototype of the first nuclear propulsion plant for a submarine was built and operated successfully by 1953 at the National Reactor Testing Station in Idaho. This was followed by the launching in January, 1955, of the first nuclear submarine, the USS *Nautilus*. The *Nautilus* was the forerunner of a vast fleet of nuclear-propelled attack and missile submarines as well as surface ships for the US Navy. These were all powered with a pressurized water-cooled reactor (PWR). The sodium-cooled system was successfully run as a prototype and in the second nuclear submarine, *Sea Wolf*. From then on, the naval programme concentrated on the PWR-type.

In parallel with this development, the prospect for a 'breeder' reactor which would provide abundant nuclear fuel for power generation for many centuries also justified and attracted development funding. Led by the Argonne National Laboratory, a small experimental sodium-cooled breeder system to determine feasibility was built at the Idaho test station within a few miles of the *Nautilus* prototype. The experimental breeder reactor (EBR-I) was brought to power in December, 1951, and produced a token amount of electricity from fission for the first time in history.

These impressive accomplishments stirred major enthusiasm for nuclear power and prompted President Eisenhower's 'Atoms For Peace' speech in December, 1953. He proposed an international programme for sharing with other nations the technology for the peaceful use of atomic energy in return for their pledge not to develop nuclear weapons and with a reciprocal pledge by the nuclear weapons nations to halt the nuclear arms race. The International Atomic Energy Agency (IAEA) was also proposed to institute international controls over nuclear weapons proliferation. The Atomic Energy Act was amended in 1954 to support this international co-operation, to provide both the stimulus and the control for the development of civilian nuclear power, and to permit industrial participation in that development, including private ownership of nuclear power plants.

This enthusiasm, combined with the observation that the United Kingdom was moving rapidly into the large scale production of electricity from nuclear power with their Calder Hall station, prompted the US Congress to authorize construction of its first nuclear power plant for the production of electricity, a pressurized water reactor (PWR) patterned after the PWR type of reactor used on the *Nautilus*. Authorized at the same time to develop alternate types of nuclear electric generators were a small experimental boiling water reactor plant (EBWR), an experimental sodium-cooled, graphite-moderated reactor, a second sodium-cooled experimental breeder reactor (EBR-II), and an experimental aqueous homogenous fuelled reactor.

The Congress judged that the team that had so successfully developed the *Nautilus* would have the experience and skill to develop the nation's first large scale electric generating station and so Admiral Rickover and the Bettis Atomic Power Laboratory were given the job. It was decided to build the plant at Shippingport, Pennsylvania, near Pittsburgh, on the Duquesne Electric Company network. Duquesne Light would provide the site and the turbine generating facilities and the government (AEC) would provide the nuclear steam supply system. The Shippingport plant began producing electricity for commercial use in 1957, again in an impressively short time frame. It became the forerunner of all but one of the 100 nuclear power plants operating in the United States today for the generation of electricity, many of them over 1000 MWe in power output. The 60 MWe Shippingport plant also set the stage for the next step in the US nuclear power development story.

11.2 Commercial Origins: The Nuclear Electric Power Demonstration Programme (1954–1961)

Following the authorization of the Shippingport plant and to further stimulate nuclear power development, Congress authorized the AEC in 1955 to enter into a 'Power Demonstration Reactor Programme.' In response to the solicitation of the Atomic Energy Commission, three proposals were submitted and accepted. Yankee Atomic, a consortium of thirteen New England Utilities, joining with Westinghouse as the reactor supplier, proposed a 100 MWe PWR of the Shippingport type. Detroit Edison proposed a 100 MWe fast breeder reactor and formed a broad-based industrial consortium, Atomic Power Development Associates, to design and build it. The Consumers Public Power District in Nebraska, with Atomics International as reactor supplier, proposed a 75 MWe sodium-cooled, graphite moderated reactor. Two projects were also undertaken by utility companies with private financing only: the Nuclear Power Group headed by Commonwealth Edison in Chicago, with General Electric as the reactor supplier, proposed a 180 MWe boiling water reactor (BWR) based on the pioneering development work with the EBWR carried out by the Argonne National Laboratory. The Consolidated Edison Company of New York, with Babcock and Wilcox as the reactor supplier, proposed a 236 MWe pressurized water reactor but based on thorium fuel rather than uranium fuel. Commonwealth Edison's plant (the Dresden–1 plant) near Chicago began operating in 1959 and remained in operation for about 20 years. The Yankee Atomic plant (the Yankee Rowe plant) near Grove, Massachusetts, went on line in 1960 and still continues to operate. The other three systems were also completed and brought to power but did not achieve sufficient technical and economic success to warrant continued long term operation.

The power demonstration programme, in subsequent rounds, sponsored a wide variety of other reactor types and carried them through varying stages of development: the high temperature gas-cooled reactor (HTGR), the organic cooled reactor, the heavy water moderated, light water cooled reactor, and the molten salt reactor. A small demonstration plant or experimental facility was built for each of these types. In the case of the HTGR, the development subsequently progressed from the demonstration phase to the construction of a 400 MWe power plant, the Fort St Vrain plant of Public Service of Colorado. In this same period other applications of nuclear power were being pursued with vigour. A military aircraft nuclear propulsion system, a nuclear rocket for space exploration, a mobile nuclear power plant for remote military bases, and a nuclear propulsion system for commercial shipping. But the rapid success of the light water reactor (LWR) systems resulted in a focusing of industrial interest on the LWR plants, both the PWR and BWR types. The remaining story of US experience will therefore concentrate on LWR nuclear plants.

11.3 Early Commercial Nuclear Electric Power (1961–1968)

11.3.1 *LWR Plants in the US: Successful Initial Experience*

The successful operation of the PWR in the Yankee Rowe plant and the BWR in the Dresden–1 plant led to authorization of plants with higher unit power output to improve economic competitiveness against coal. The economy of scale achievable in the 500 MWe range was needed to reduce the capital cost of the nuclear plant per kw produced. Thus, in a continuation of the power demonstration program, the San Onofre 1 plant was sponsored by Southern California Edison with Westinghouse as reactor supplier; the Connecticut Yankee plant was sponsored by the Yankee group, again with Westinghouse as the reactor supplier; and the Dresden 2 plant was sponsored by Commonwealth Edison with General Electric as supplier. Each of these plants operated in the 400–500 MWe unit power output range. All have had highly successful operating experience, and are continuing in operation today.

11.3.2 *US Projects Overseas: Successful Initial Experience*

In this period, the European community had also developed substantial interest in nuclear power, the story of which is provided elsewhere in this book, but the US participation in this European development was in itself an important part of US power development. Westinghouse took on contracts, working with the industrial organizations in the respective countries, to build pressurized water reactors in the 200–300 MWe power output range in Belgium, Spain, Italy, and Japan: the Ardennes plant (320 MWe) in France, a joint Belgian–French enterprise, the Trino plant (270 MWe) in Italy, the Jose Cabrera plant (150 MWe) in Spain, and the Mihama–1 plant (320 MWe) in Japan. Similarly, General Electric contracted for the construction of boiling water reactor plants in Germany, (the 240 MWe KRB–1 plant), in Spain (the 440 MWe Burgos plant), in Italy (the 150 MWe Senn plant), in India (the 200 MWe Tarapur 1 and 2 plants), and in Japan (the 340 MWe Tsuruga–1 plant). Again, these projects were completed successfully. These plants have all had successful operating experiences over many years.

It should be noted that all of these projects, both those built in the United States and those built in Europe through contracting arrangements with US firms, were built in short times and at low costs by comparison with today's record. Four years and less was typical for the time between authorization and completion of the plant. Capital costs were in the range of $150 dollars per kilowatt electric. Although the governments in the United States and Europe provided substantial development support for these projects, the utilities and private industrial groups carried the major costs of building and operating the plants. As an example, the contribution which the US government gave to the construction of Yankee Rowe, was $8 million, in return for which the government was given the rights to the data obtained from operation of the nuclear fuel.

11.3.3 *Introduction of the 'Turnkey' Nuclear Plant*

In spite of these impressive results, nuclear power had not won the economic race it was engaged in during this period. It still was not projected to generate electricity at a total cost equal to or less than coal, the widely used alternative for the generation of electricity. Because the costs were too high and because the technology was new and unfamiliar, the US utilities had little interest in taking the next step after the demonstration program, that is, to authorize a nuclear power plant on a straight commercial basis.

Two significant events caused a breakthrough in this stalemate. Cost competitiveness was achieved through a major rise in coal costs, primarily caused by increased railroad transportation rates. The utilities concern about the unfamiliar technology was resolved by the decision by General Electric, shortly followed by Westinghouse, to offer utilities a 'turnkey' plant in which the reactor supplier would take on total responsibility to build and license the plant, at which time it would be turned over to the utility. The reactor supplier also offered to provide for the training of reactor operators and other special nuclear skills needed by the utilities. In 1963, Jersey Central Power & Light Co. made a thorough economic analysis of a General Electric 'turnkey' offer for their Oyster Creek nuclear plant and declared that the plant would be more economical than coal.

This breakthrough led to the rapid expansion of nuclear power, which is the next step in the US story. Before recounting it, a few words should be said about technical developments in the rest of the fuel cycle and the growing importance of nuclear reactor safety regulations during this period.

11.3.4 *Start of Commercial Reprocessing*

In this time frame of nuclear power development, it was accepted that spent fuel would be reprocessed and the residual uranium as well as the plutonium produced in the fuel during operation would be extracted and recycled back into the reactor. Thus, a parallel development of spent fuel reprocessing was pursued, primarily through the national laboratories, with government encouragement of industrial participation in commercial reprocessing. As a result, a reprocessing facility, the Nuclear Fuel Services Plant, was built at West Valley, New York, with a nominal 1 t day^{-1} capacity. The plant successfully commenced operation in April 1966. The plant utilized the Purex solvent extraction process which had been developed for plutonium extraction from spent fuel from the military production reactors.

11.3.5 *Radioactive Waste Storage Demonstration Frustrated*

With the assumption that fuel would be reprocessed, provisions also had to be made for the safe disposal of the residual radioactive fission products from the spent fuel. Thus, the US AEC sponsored a high level radioactive waste storage demonstration. A salt bed site was chosen by the AEC in Kansas and initial work started to demonstrate the storage of high level radwaste. That demonstration project ran into major political opposition and had to be cancelled. It was one of the first significant indications of less than full public acceptance of nuclear power and its associated technology.

In addition to the opposition to the radwaste storage facility which caused that project to be aborted, there were other nascent signs arising in the mid–1960s of reduced public acceptance. As part of the effort to achieve economic competitiveness in nuclear power, a nuclear power plant was proposed in early 1963 to be located in New York City on the East River opposite the UN Building in Manhattan. This proposal was rejected because of public concern about building a nuclear plant in such an area of high population density. A nuclear plant was sited near the San Andreas fault in California but was abandoned after construction had started because of public concern about its adequacy to withstand an earthquake.

11.3.6 *Reactor Safety Regulation Build-up*

The prospects for major expansion of commercial nuclear power led the AEC to accelerate efforts to provide for regulation of these commercial activities. A reactor regulatory branch was set up within the Atomic Energy Commission to carry out this function and a substantial staff built up. This staff superseded the self-regulating activities which had governed the AEC reactors in both the military and the power demonstration programmes. The Advisory Committee for Reactor Safeguards, which had provided an overview for the government's self regulating activities, continued in place to provide an overview to the regulatory branch which now would be the licensing authority for commercial reactors and associated facilities in the back end of the fuel cycle. The National Environmental Policy Act was also passed in 1963 as part of a mounting effort in the United States to reduce environmental pollution. That act had a substantial effect in increasing the level of complexity of nuclear power plant regulation.

The stage was thus set for an expansion of commercial nuclear electric power in the United States. Such an expansion did occur and very rapidly.

11.4 Rapid Expansion of Commercial Nuclear Electric Power Capacity (1968–1975)

11.4.1 *Accomplishments*

11.4.1 Power Plant Construction and Operational Success By 1975, 55 nuclear power plants were in operation and 178 more were planned or under construction. With the exception of the original power demonstration projects, all of these plants were privately financed and fully commercial. The first group of eleven plants following the demonstration programme were turnkey plants ranging in output from 500 to 1000 MWe. The initial impetus for this period of extremely rapid expansion of nuclear power in the United States was the offering of these turnkey nuclear plants by General Electric and Westinghouse. The prices offered made nuclear power competitive economically with coal and the offerings were made in a period of a strong perceived need for additional base load capacity.

The record set by these first plants was indeed impressive. The earliest of them were producing power in not much more than four years after they were authorized. The cost to the utility was in the range $100 to $250 per kilowatt electric, and even counting the losses the reactor manufacturers absorbed, their total power costs were substantially less than equivalent coal plants. They operated with a high availability factor and for the most part, are still maintaining an above average availability factor among all nuclear power plants.

A typical example was the Ginna plant, a pressurized water reactor built on a turnkey contract by Westinghouse Electric for Rochester Gas and Electric. The plant was authorized in 1966, obtained its construction permit in eleven months, and went to power 51 months after that initial authorisation. It has continued to generate 600 MWe with a high availability factor over more than the sixteen years since it went into commercial operation and has paid back its initial investment several times over. An article in *The Energy Daily*, July 10, 1986 entitled 'RG&E: Glorious Ginna' states:

> 'Rochester Gas and Electric's Ginna nuclear plant generated its 50 billionth kilowatt-hour of electricity this week. The 490–megawatt plant went into service in June 1970. In 1985, the plant had an availability factor of 88 per cent. RG&E says the plant last year saved customers more than $50 million when compared to the cost of generating the same amount of power with coal. Over

its lifetime, says the utility, Ginna has saved ratepayers more than $445 million compared to burning coal.'

The success of these turnkey plants gave the utilities confidence to continue expansion of nuclear power generation and to carry out that expansion in the conventional contracting form rather than through a turnkey contract.

11.4.1.2 Industrial and Contracting Structure: A Transition The conventional US-utility industry structure for building a base-load power generation plant is as follows. The utility would authorize the plant and make arrangements for financing it. The utility would engage an architect–engineer and often with the assistance of the architect–engineer, evaluate competitive bids for the nuclear steam supply system and turbine generating system from the reactor and turbine generator manufacturers. The architect–engineer would design the over-all plant and procure the 'balance of plant' equipment and materials. Then, often with a construction management firm, the utility would build the plant. Transition from the turnkey to the conventional approach occurred for two reasons. The reactor manufacturers found that they could not control the costs of construction and implement the changes required to obtain a licence without incurring substantial losses on their fixed price turnkey contract. Utilities and architect engineers also wanted to return to the conventional approach since it had served them successfully in the past and now seemed applicable to nuclear power since familiarity with the technology had been gained.

These conventionally contracted plants took longer to build and cost more, but the 55 that were completed by the end of 1975 continued to show cost advantages over coal as well as with all alternatives except large hydropower plants. They also chalked up an unprecedented safety and environmental record. There were no deaths from radiation during this entire period and the radiation emissions into air and water were approximately 100–times lower than the standards allowed. This outstanding record led to the 'ALARA' principle in which radiation emissions during normal operation and radiation exposure to workers was to be set as low as reasonable achievable (ALARA). Radiation occupational exposure was kept approximately ten-times lower on the average than the standards allowed. Even in the industrial safety area, nuclear plants had a substantially better record than their sister fossil burning plants and other comparable industrial activities in protecting construction and maintenance workers from non-radiation hazards.

11.4.1.3 Expanding Nuclear Power Plant Exports Another element of this rapid expansion for the US reactor manufacturers and architect engineers was an equally rapid expansion of the nuclear power plant export business. Belgium, Brazil, Germany, Japan, Korea, Mexico, the Philippines, the Republic of China, Spain, Sweden, Switzerland, and Yugoslavia also saw the advantages of utilizing nuclear power for electric generation and turned initially to the United States to provide them with the equipment and the technology for their nuclear power programme. Westinghouse and General Electric were the major US reactor manufacturer exporters and Bechtel, Burns and Roe, Ebasco, Gibbs and Hill, and Gilbert Associates were the major US architect–engineers in the overseas markets. In the early 1970s, the French decision to focus on the PWR for their nuclear power expansion programme led to a licensing agreement with Westinghouse to furnish the technology for the French to build their own PWRs.

This tremendous surge of business, both domestically and overseas, provided the US manufacturers and their subcontractors with a business incentive to invest in major new manufacturing facilities. Well over $2 billion was spent for that purpose, so that by 1975 there were not only 55 nuclear power plants operating in the US but many factories in many parts of the country had been built and licensed to produce nuclear power equipment.

11.4.1.4 Progress in the Back-end of the Fuel Cycle This surge of electric generation capacity brought with it the need for major expansion of enrichment capacity. A new complex of diffusion-type enrichment plants were built by the AEC at Paducah, Kentucky and Portsmouth, Ohio, and a variety of measures were taken to upgrade the throughput of the existing diffusion plants at Oak Rridge. Development work was authorized for advanced forms of enrichment with primary emphasis placed on the centrifuge method but with significant effort also on laser enrichment, a more advanced but highly promising method.

The expectation that spent fuel assemblies would be reprocessed and the uranium and plutonium in the spent fuel recycled into the existing light water reactors led to the need to provide for commercial spent fuel reprocessing. A large-scale (5 t day^{-1}) commercial reprocessing plant at Barnwell in South Carolina, designed and built by Allied General Nuclear Services, a joint venture of Allied Chemical and Gulf Oil, was essentially completed by 1975. General Electric also authorized a spent fuel reprocessing plant located at Morris, Illinois, and it too was completed by 1975. However, the plant was not put into 'hot' operation because pre-operational testing introduced concern as to the feasibility of the direct maintenance concept to which the plant was designed.

The development programme for the fabrication of mixed plutonium oxide fuel assemblies for recycle in LWR reactors had been highly successful. Several reloads had been made and operated with mixed plutonium fuel in US reactors. Based on this successful experience, Westinghouse, General Electric, and Exxon authorized the design and construction of major new mixed plutonium-uranium oxide fabrication facilities for recycle fuel.

The one area in the entire fuel cycle which showed relatively little progress was that of large scale demonstration of the disposal of high level radioactive waste from the reprocessing plant. This lack of progress was not considered serious from a technical standpoint since extensive development and testing by the national laboratories had provided methods for the vitrification and encapsulation of high level wastes which would provide for safe storage. In addition, the volume of high level radioactive waste from the commercial programme was small and the military was storing large quantities of high level radioactive waste seemingly successfully, even though the methods were more primitive than those developed for commercial high level wastes. There was, however, concern at the mounting public apprehension of the safety of storing high level radioactive waste and a recognition that that apprehension was growing because of the lack of a large scale demonstration of safe disposal.

11.4.1.5 Development of the Regulatory Framework Another essential step in this rapid expansion of nuclear power was the further development of the regulatory framework to handle this expanded industrial effort. The reactor regulatory branch, formed by the AEC to review licence applications and to grant construction and operating permits, grew rapidly in the late 1960s and early 1970s to handle the heavy licensing workload and to develop a detailed body of technical regulations against which to measure the adequacy of the applicants' designs and operating capabilities.

By 1975 there were about 3000 permanent staff members of what was then called the Nuclear Regulatory Commission (NRC). This commission was formed in 1974 as a separate independent agency by taking away the regulatory authority from the AEC and transferring its regulatory branch to the new Commission. The move was made on the basis that it was inappropriate to have one agency, the AEC, as both a promoter and a regulator of nuclear power. Some of AEC's R and D functions were also transferred, mainly safety R & D for light water reactors. Thus, a research branch to implement the R and D responsibility was established in the NRC along with the reactor regulator branch and an inspection and enforcement branch.

In sum, then, a massive expansion of this

new industry had been successfully effected by the end of 1975. Fifty-five nuclear power plants were in operation in the US and a similar number were operating overseas through US exports and technology transfer. Major new manufacturing facilities had been built and were operating at full tilt. Expanded enrichment facilities were being completed. Fuel fabrication plants for plutonium recycle were being designed. A fully staffed, independent regulatory commission was in operation. Orders were on the books for almost 200 more US nuclear plants for the growth of electricity projected by both industry and government.

11.4.2 *Emerging Problems*

In the latter half of this period of rapid expansion, a series of problems began to emerge. Six of the most significant of these are discussed briefly below.

11.4.2.1 Licensing Changes and Delays In-depth review of the licensing applications in the early 1970s led to much more detailed statements of regulatory requirements which entailed significant design changes. The licensing review process itself became prolonged because of the increasingly detailed nature of the reviews. A significant increase in licensing requirements arose from the Calvert Cliffs case in which the courts invoked a rigid interpretation of the National Environmental Policy Act, which substantially increased the environmental evaluations required and forced a greater level of change during plant construction. Significant delays in the licensing and construction of nuclear plants occurred from then on.

An early casualty of increasing regulation was the West Valley Reprocessing Plant. The plant was shut down for enlargement in 1972, having successfully processed about 620 metric tons of fuel. The plant was never re-licensed and re-started because the cost of modification to meet the new regulations was prohibitive.

This major example was only part of the impact of a growing movement of environmental concern which affected all industry in the United States, but the impact on nuclear power construction was especially severe. These influences also began to show themselves at the local government level with restrictions being placed on the transportation and storage of radioactive materials from nuclear plants. The eleven month interval required to obtain a licence for the Ginna plant in 1970 stretched to three years and more by 1975.

11.4.2.2 Construction Delays Changed licensing requirements and other design changes required major modification of the plants which by this time were well into construction. The need to make these licensing modifications and design changes, combined with delays in getting design information and material to the construction sites, caused significant construction delays. The perceived need for capacity led to attempts to minimize such delays by using a larger workforce, re-ordering construction sequences to accommodate missing design information or materials, and the use of overtime and premium methods to expedite delivery of material and equipment.

11.4.2.3 Equipment Modifications Licensing change required modification of equipment design and systems designs. Outstanding examples of this were the development of more detailed emergency core cooling system requirements and the development of blowdown and seismic pipe supports, and safety equipment qualifications. Equipment modifications had also to be made to reflect initial field operating experience and to remedy incompatibility between nuclear steam supply system and balance of plant interfaces.

11.4.2.4 Construction and Financing Costs The equipment and plant modifications and construction delays each contributed to the mounting capital costs of nuclear power plants. In addition, double digit inflation, which appeared in the early 1970s, raised the costs of every element of material and labour going into the construction job. Many utilities were unable to include construction work in progress in their rate base so that an ever increasing investment had to be financed. The companion to double digit inflation, double digit interest rates, further increased financing costs. Nuclear power was not unique in being the victim of double digit inflation and increasing capital costs. However, the relatively high capital cost of the nuclear power plant made it a much more vulnerable victim.

11.4.2.5 Public Hearing and Judicial Delays The public hearing process had been established in the Atomic Energy Act to assure that there would be appropriate public participation in the decisions involving the utilization of nuclear power. Atomic Safety Licensing Boards were set up to carry out a quasi-judicial process to conduct public hearings before construction and licensing permits would be granted. This process became a growing cause of delays in obtaining construction and operating permits. In addition, injunctions were obtained through the courts which stopped progress in licensing or in construction. The growing influence of such quasi-judicial and judicial actions is attested to in the findings of the Calvert Cliffs case; in the words of the presiding judge:

> 'These cases are only the beginning of what promises to become a flood of new litigation—litigation seeking judicial assistance in protecting our natural environment. Several recently enacted studies attest to the commitment of the Government to control, at long last, the destructive engine of material progress.'

11.4.2.6 Reliability Issues Towards the end of this period of expansion, major field experience was being gained from the plants that had come on-line. Component reliability issues were emerging from this field experience. A significant number of them were from non-nuclear equipment such as the turbine generators and transformers. However, two major ones were unique to the nuclear systems: the corrosion deterioration of steam generators in the PWR systems and the appearance of intergranular stress corrosion cracking in the piping of the BWR systems. These reliability problems, both nuclear and conventional, combined to keep the average nuclear plant availability at a level of less than 70 per cent with capacity factors running, on an average, less than 60 per cent. These levels were substantially lower than the 80 per cent capacity factors assumed in the economic projections when plants were authorized. In addition to the costs of this unavailability, the cost of maintenance and repair were substantial and largely unanticipated.

These emerging problems were to combine with a major unanticipated problem, the oil embargo, to halt summarily the rapid expansion of nuclear power in the United States.

11.5 Cessation of Growth and Increased Public Opposition (1975–1979)

The major new and unanticipated problem was the impact of the OPEC oil embargo of 1974 and the oil cut-off caused by the Iran revolution in 1978. The reduction in the availability of oil in one sense vindicates the use of nuclear power since it substitutes for oil. But this is a long term influence. The short term impact was the tremendous rise in the price of oil which, in turn, created a major worldwide recession causing substantial reduction in industrial activity and its need for electricity. Another significant impact, although not as big, was price elasticity: the rise in the price of electricity caused a drastic effort to conserve electricity.

The combination of these two major factors was to drive the demand for electricity down and change radically the projection of need for electric generating capacity. For the better part of four decades, electricity demand had been increasing at 7 per cent annually, or doubling every ten years. The effect of the oil embargo and cut-offs was to reduce that demand to less than half, meaning that a substantial percentage of the already authorized base load capacity would be in excess of need. As this picture emerged, the US utilities cancelled essentially every nuclear and coal plant which had been authorized and on which only a small investment had occurred. Since the larger percentage of the new capacity was nuclear, a substantially larger number of nuclear plants were cancelled than coal plants.

The expectation that there would be upwards of 250 GW of nuclear generated electricity operating in the US in the mid–1990s suddenly plummeted to an expectation of 160 GW. For those nuclear plants in which a significant investment had already been made, utilities slowed construction or stopped construction temporarily, since the plants would not be needed on the original schedule. These construction delays were a means of reducing the ultimate cost of producing power from the plants but did raise the costs over what would have been incurred if the plants were built and put into use on the original schedule.

The high interest rates which went with the double digit inflation in that period, of course, made construction delays of any kind, whether voluntary or involuntary, extremely expensive. In many areas of the country, utilities were legally proscribed from putting construction work in progress into the rate base so that the consumer could begin to pay for the investment in the plant as it was being built. This meant that with no revenue being received to allay the increasing costs of construction, the utility had to finance the increased plant investment at a higher interest rate. This spiraling of interest costs brought total financing costs to half of the capital cost of the nuclear plant.

With the election of the Carter Administration in 1976, Federal policy took a negative turn with respect to nuclear power. The generation of electricity from nuclear power was put in the category of 'the last resort', a radical change of policy from the Federal promotion and support provided in the 1950s and 1960s. The Carter Administration took steps to stop commercial spent fuel reprocessing, plutonium recycle, and advanced nuclear power generation demonstration projects. It followed that over a period of time the Barnwell reprocessing plant was written off, all commercial plutonium fuel fabrication facilities were closed, and the Clinch River breeder demonstration plant was cancelled. Although this official turn in position had its basic origins in the concern for horizontal nuclear proliferation of nuclear weapons material, public opposition increased in part because the President, a former 'nuclear engineer', seemed to have major reservations about nuclear power.

By 1979, the emerging problems in the nuclear industry, the impact of the oil embargo and cut-offs, and the negative US government policy combined to bring the future expansion of nuclear power to a standstill. No order for a new nuclear power plant has been placed in the US since 1978.

11.6 The Three Mile Island Accident and Its Impact (1979–1986)

In March of 1979, a malfunctioning ion exchanger, combined with a valve misalignment, started a sequence of equipment malfunctions and operator errors which led to loss of core cooling on the Three Mile Island–2 (TMI–2) nuclear power plant near Harrisburg, Pennsylvania, which caused gross melting of the core. The accident was widely heralded by the United States media as a catastrophe, the first time in history that a major industrial accident that caused no loss of life or human injury, was so labelled.

Although the accident was in no sense a catastrophe in human terms, it was a financial catastrophe for the utility owner and the US utilities in general. It caused an immediate reduction in the credit rating of all nuclear utilities and set off a new round of cost increases, particularly for those plants that were still under construction. Operating plant availability was also reduced and the granting of operating permits delayed.

Although the accident was dominantly bad news for the industry, particularly in the short term, there was some good news of potential significance in the long term. The containment system provided to keep the radioactivity from such an accident from entering the atmosphere and harming the public, successfully performed its function and vindicated the early decision made at Shippingport to provide full containment for commercial nuclear power plants.

A special commission, the Kemeny Commission, was established by the President of the United States to investigate the accident. The Commission concluded that the primary causes of the accident were deficiencies in management, both in industry and in the NRC, in operator training, and in the man-machine interfaces in the control room. Contributory causes from deficiences in equipment were also identified, but were judged to be secondary.

During the investigation and for about six months after, encompassing a total of about one year, there was a moratorium on granting any permits for new reactors to go into operation. A major drop in availability occurred for all the reactors of the same basic design as TMI–2, the B&W once-through system, because of the need to shut down for a series of immediate inspections and modifications. All US reactors suffered a reduction in availability, although not as great, for inspection and modifications.

The NRC developed a TMI action plan containing hundreds of pages and identifying a wide variety of changes in all US power plants to incorporate the 'lessons learned' from the accident. Among the key changes were:

1. Control room reviews and modifications, such as the Safety Panel Display System, to improve human factors in the control room.
2. The addition of a senior technical advisor on each shift to be available to provide advice in the event of emergency.
3. The complete overhaul of emergency response procedures.
4. The introduction of hydrogen controls to assure that even with the generation of 75 per cent of the maximum hydrogen that could be produced in a severe accident, a hydrogen explosion large enough to threaten containment integrity would not occur.
5. The institution of off-site emergency planning requirements, including alarms and drills to both alert and train the public, and to carry out evacuation procedures in the event of a future accident.

In addition to implementing the NRC-mandated changes, the utilities undertook several important new initiatives. A Nuclear Safety Analysis Center was formed at the Electric Power Research Institute to set up a system to evaluate all significant operating plant incidents, to identify potential accident precursors, and to disseminate those evaluations to all the nuclear utilities in the United States. The Institute for Nuclear Power Operations (INPO) was formed as a central organization to establish operating standards, to define operator training requirements and accreditation, and to audit the operational effectiveness of all US nuclear utilities. INPO has become today a several hundred man operation headquartered in Atlanta which is performing these vital functions for the nuclear utilities.

Another vital follow-up action was the assurance of safe shutdown of the TMI–2 plant and its eventual clean-up. It is estimated that this will take approximately ten years and over $1 billion to complete. The investor owned utilities, in recognition of the need to complete the clean-up and the value of the R and D results coming from TMI–2, have pledged about $150 million to assist GPU in the clean-up.

In the clean-up process, a vast amount of information has been gained on the consequences of such a severe accident in terms of physical damage to the plant, on the clean-up of massive quantities of contaminated water, and on the decontamination of the containment structures themselves. Of seminal importance has been the follow-up of the finding from the accident that even if containment integrity had been lost, the quantity of airborne radioactive fission products, excluding the noble gases, which would have been released was less than 0.1 per cent of what was expected. Thus, the potential danger to the public was much less than predicted by the prevailing methods which form the basis for regulation.

Another casualty from the accident was the shutdown of the TMI–1 plant, located on the same site. It took over five years for the utility to regain its operating permit so that that plant could return to power. The loss of revenue from the shutdown of both TMI–2 and TMI–1, combined with the expenses of the clean-up and the purchase of substitute power, drove the utility owner, General Public Utilities (GPU), to the verge of bankruptcy.

The cumulative cost impact of the post-TMI changes on top of already rapidly growing construction and financing costs caused a cash flow squeeze on the utilities who still had plants under construction. In addition, quality control problems in construction began to appear. The time it was taking to get a nuclear plant into operation had now stretched out typically to ten years and the capital cost had grown to about \$3000 kWe^{-1}. Some plants had better records than this, but some had worse.

These factors caused a new phenomenon in the US nuclear industry in the 1980s: the cancellation of plants in which major investments of up to \$3 billion had already been made. The justification for these cancellations was that the capacity was not needed and there was some doubt that the investment would be allowed by the Public Utility Commission (PUC) in the rate base even if the plant were completed.

Such drastic measures have been taken for only a few of the plants which were well along in construction but those newer plants which are still under construction are still causing a heavy financial drain on the utility owner. Unfortunately, when the utility finishes the plant, obtains its operating permit, and puts the plant into operation its financial difficulties are not over because the PUC typically does not take timely action to put the investment in the rate base. Even if the PUC takes action, it may authorize only a portion of the plant investment which it considers is justified by the present generation need. This action by the PUCs forms an over-all pattern: When nuclear power was producing power more cheaply, most of the economic benefit went to the consumer. In later days when higher costs have reversed that trend, most of the penalty is going to the stockholder. Although again this trend is not unique to nuclear power plants, it is most severe for them.

In spite of these setbacks, the effect of the many improvements and industry initiatives taken since the TMI–2 accident has begun to show itself on the 'bottom line' of improved availability of US nuclear plants as seen by the following data from 1984–85; the last two years for which a full year's operating statistics are available.

US PWR plants achieved an average capacity factor in 1985 of 64.6 per cent as compared to 60.3 per cent in 1984, and an average availability factor of 70.6 per cent in 1985 versus 65.5 per cent in 1984. US BWR plants recorded an average capacity factor in 1985 of 54.4 per cent as compared to 48.5 per cent in 1984, and an average availability factor of 61.7 per cent in 1985 versus 55.4 per cent in 1984.

Excellent individual records were achieved in 1985. Three plants, Salem 1 of Public Service Electric & Gas, Oconee–1 of Duke Power, and Connecticut Yankee of Northeast Utilities, exceeded a 90 per cent capacity factor. Nineteen plants exceeded an 80 per cent capacity factor as compared to eleven plants in 1984.

Continued improvement in the 'bottom line' is extremely valuable: If 100 US plants improve their availability so as to achieve an 80 per cent average capacity factor, the equivalent of around 30 new 1000 MWe nuclear plants become available at near zero capital cost.

Substantial change also occurred in this period in the activities of the back end of the fuel cycle. The goals of the programme were adjusted significantly because plutonium recycle in LWRs was no longer a viable commercial activity in the US. Thus, major effort was placed on providing for the storage of spent fuel in substantially greater quantities than originally planned for at individual reactor sites. A renewed effort was undertaken to demonstrate the large-scale disposal of high level radioactive waste, where the form of this 'waste' would now be the spent fuel assemblies themselves. The Nuclear Waste Policy Act was passed to establish that the Department of Energy had the responsibility to implement this programme and provide, by the late 1990s, for the transfer of spent fuel assemblies from individual reactor plants to a federal repository, probably an interim, monitored, retrievable storage facility. The spent fuel assemblies would then be conditioned and transhipped to the permanent geologic disposal site. Significant progress was also being made in providing large scale, highly reliable, dry storage for military high level radioactive waste in a stable salt bed in New Mexico.

Companion legislation was passed to provide for the continued storage of low level and intermediate level wastes coming from nuclear plant operations. The legislation, the Low Level Waste Policy Act, called for groups of states to set up regional compacts in a given region to provide a storage facility for nuclear plant low level radwastes as well as for radwaste from hospitals and other industrial activities utilizing radioactive isotopes.

Enrichment also took on a 'new face'. The Nuclear Non-proliferation Act and US government vacillation between restrictive and free natural uranium export policies encouraged other countries in the world to provide their own enrichment capability rather than continuing to rely exclusively on the United States. Thus, a major enrichment facility was built in Europe. The Soviet Union became a supplier of enriched uranium to utilities in the Western world, including the United States. To adapt to this changed world of enriched uranium supply, the United States decided to retrench its diffusion capacity by closing down the old facilities at Oak Ridge and relying on the new ones in Kentucky and Ohio only. It was decided to abandon construction on a major new centrifuge facility in which $2 billion had been invested and focus development on the laser enrichment process for future enrichment capacity. Such steps were taken to make the United States enriched uranium supply cost competitive in the short run through the retrenchments and in the long run through the advanced laser enrichment process which has major potential for cost reduction and reduced power requirements.

The high level of regulatory change following the TMI accident produced an uncertain regulatory regime which did not seem to be stabilizing as the post-TMI action plan and implementation measures taken by the utilities were completed. The Congress therefore undertook a legislative initiative to reform the regulatory process in the form of the Nuclear Power Standardization Bill. This bill would encourage nuclear plant standardization by the industry and regulatory standardization by the NRC.

The advanced reactor development programme was still being pursued, albeit at a slower pace. Peak activity had reached approximately $700 million in R and D and capital expenditures annually but was now down to approximately $200 million with no appreciable capital investment. Major research and test facilities had been built, particularly in support of the breeder programme, and were still in operation. The high temperature gas-cooled reactor programme was receiving approximately 10 per cent of the present funding.

All of the programmes had been reduced in pace, primarily for two reasons:

1. The expansion of nuclear power through the light water reactor programme had halted, making less urgent the introduction of capability to expand the nuclear fuel supply (the breeder) or to expand nuclear power application to a wider industrial sphere (the HTGR).
2. Cost experience with the advanced systems, to date, showed that capital costs were too high to expect introduction of the systems competitively, particularly since additional subsidized plants would have to be built to reduce the technical risks before the private

sector could consider commercialization of the concepts. A major re-emphasis and redirection had been taken in the programmes, therefore, to find economic breakthroughs in the capital costs of the advanced systems.

In sum, by 1986, the impact of TMI seemed to have been absorbed and a vigorous although substantially less ambitious programme of nuclear power development was now continuing in the United States.

11.7 The Chernobyl Accident and Its Impact (1986–to present)

Encouraged by the significant improvements in 1985 operational performance, US industry moved into 1986 with expectations that the major problems besetting the industry were being surmounted. Plant performance was up, legislation had been passed and funding provided to complete the development of both high and low level waste disposal, legislation to streamline the regulatory process was promising to move through the congress, and the Reagan Administration supported nuclear power. Work was under way to design future reactors and to define standardization processes to help reduce their capital costs. A couple of utilities were even quietly evaluating the prospect of building another nuclear plant in the 1990s.

Then the major accident occurred at the Chernobyl plant in the Soviet Union, spewing radiation world-wide, and causing great consternation among the public, particularly in those countries in Western Europe outside the Soviet bloc.

The impact of this accident cannot yet be fully measured in the United States. The present reaction among the public is muted, principally because the level of radiation that reached the confines of the United States was extremely low and harmless. But efforts are underway to obtain congressional action to put further constraints on the nuclear power industry. Local governments have used their role as participants in emergency planning to create obstruction to bringing some of the new plants up to power. Several issues bearing on containment integrity are being raised by the NRC.

The reason for such calls to action are not technically logical because of the vast difference between the Soviet reactor design and its lack of full containment as compared to the light water reactor designs predominantly used in the United States. In fact, even a cursory examination of the two systems vindicates the wisdom of the early US policies to provide full containment on commercial nuclear power plants of the light water reactor type and to separate weapons materials production reactors from commercial electric generating reactors, policies which were diametrically opposite to those adopted by the Soviet Union in the early stages of its programme.

In spite of this lack of a logical foundation, there is substantial emotional reaction to a nuclear power plant accident which has killed tens of people, overexposed thousands, radioactively contaminated substantial areas of land, and dispersed radiation beyond national boundaries. As a result, it is expected that continuing challenges will be made to the adequacy and safety of nuclear power plants in the US.

11.8 US International Nuclear Power Relationships (1945–present)

The initial posture of the United States after the Second World War with respect to international relationships in nuclear technology was to maintain the shield of secrecy on nuclear development and to seek an accord for nuclear disarmament. Proposals were also made that as nuclear disarmament was implemented, the excess bomb material should be destroyed in nuclear power reactors. These initiatives reduced to a low key as the Iron Curtain rose and the 'cold war' commenced between the United States and the Soviet Union. The lack of effectiveness of this posture was revealed as the shield of secrecy was shown inadequate when the Soviet Union exploded its first nuclear weapon and as repeated proposals for nuclear disarmament with verification procedures were spurned by the Soviet Union.

This posture was replaced in 1955 with the much broader initiative for international co-operation launched through the 'Atoms For Peace' programme. This approach led to the formation of the International Atomic Energy Agency under UN auspices and the acceptance by many nations of the Nuclear Proliferation Treaty (NPT) and other related regional treaties. The NPT had as its broad concept, the transfer of nuclear technology for peaceful purposes in exchange for the promise to refrain from the development of nuclear weapons. The US also supported the formation of Euratom to give stimulus in the European community to nuclear power technological advances and the control of fissionable materials. Accompanying agreements were reached in which the US provided enriched uranium for the nuclear power programme in many countries.

These initiatives led to a wide series of industrial licensing agreements and government-to-government technical exchanges. The agreements provided the basis for the development of nuclear power for electricity generation in many countries in the world and led to the export of a large volume of nuclear power plant equipment, services and technology from the United States to countries in Europe and Asia. Legislation governing the export of nuclear technology was established in the first Atomic Energy Act which was amended to reflect the growing export activities in later years.

The Atoms For Peace programme, the associated transfer of technology, and US nuclear export business prospered in the same decade that nuclear power was expanding rapidly in the United States. Around 1975 when that growth was reaching a halt in the US, developments in nuclear export controls began to occur which contributed to halting the growth of US nuclear exports as well. These developments in nuclear export controls occurred because of renewed concern about the horizontal proliferation of nuclear weapons set off by the explosion of a nuclear device by India. Special concern was attributed to the role of nuclear power plants as a source of weapons materials, even though India had used a heavy water research reactor, not an LWR power plant, to generate plutonium for its device.

The international community attempted to address these issues at the instigation of the United States, through the International Nuclear Fuel Cycle Evaluation (INFCE), an international study of the potential of horizontal nuclear proliferation through nuclear power. The studies were detailed, they were engaged in by all countries in nuclear power development and were estimated to have cost over $100 million. It was concluded that nuclear power was only one way in which horizontal proliferation could occur and was a relatively ineffective way. It was also concluded that there were no technical fixes which would assure control and that international institutional measures were needed. A variety of such institutional measures were suggested although essentially none were followed up in subsequent years.

In spite of these findings, the United States chose to act unilaterally on the matter by passing the Nuclear Non-proliferation Act which put severe limitations on the export of nuclear fuel and nuclear power plant equipment and demanded guarantees on safeguards and fuel re-transfer by the importing countries which were difficult for them to accept. These measures substantially reduced the spirit of co-operation in nuclear power between the US and its international friends and was a contributing cause of a major reduction in the export of enriched fuel and nuclear plant equipment.

Although the non-proliferation legislation in the United States has as its objective the stopping of horizontal proliferation, an objective that had been set by the United States back in 1945, that legislation has seemed to set back progress in horizontal proliferation control initiated by the 'Atoms for Peace' programme.

IAEA activities and the NPT are based on multi-national voluntary agreement. The progress had been impressive, having to be achieved against the traditional concepts of international sovereignty. Nations have willingly agreed to constrain weapons development and submit themselves to international inspection to verify their compliance with that commitment. This, in itself, is an historic precedent in the field of international relations and, of course, is the ultimate key to effective international control of weapons.

A recent example of the difficulties created by the US legislation has been the opening up of trade by the People's Republic of China and their interest in nuclear power development. It has taken eight years for an agreement to be reached between the People's Republic of China and the US which satisfies the Nuclear Proliferation Act. The agreement finally permits US firms to legally enter into contracts for nuclear power plant equipment exports to the People's Republic. In that eight year period, firms from Europe and Japan entered the market, established strong positions, obtained contracts, and have initially achieved a dominant position over the US in a market which has a vast potential.

In sum, US international nuclear relationships were dominated by Federal policies which had one constant over-riding objective: to control the horizontal proliferation of nuclear weapons while encouraging the development of nuclear power for peaceful uses throughout the world. Because of the innate conflicts within this objective, US policy oscillated in emphasis over the years: the pendulum swung from unilateral rigidity to multinational co-operation and flexibility and back again. One senses a swing back to flexibility if only because the loss by the US of its once dominant position in nuclear technology will force it.

11.8 The Future Outlook

The difficulties which the US utilities and the US industry have experienced in the nuclear power programme in recent years do not lead one to paint a rosy picture for the future, but it would be quite imprudent to conclude, as some would suggest, that nuclear power is dead in the United States.

There is nothing dead about an industry in which 100 nuclear power plants, representing an investment of over $200 billion, are generating electricity in unit outputs in the range 500–1300 MWe. These plants are generating 16 per cent of the nation's electricity, more electricity than was generated annually in the entire country in the 1940s. There is nothing dead about the supplier industry in the US. These nuclear plants entail an expenditure on the order of $3 billion annually in fuel and plant services, requiring a major supporting industrial structure. There is nothing dead about development programmes that are receiving funding from the utilities at an annual level of over $0.5 billion and from the government at an annual level of over $0.3 billion to assure continued safety and improved availability of present plants, the handling of their radioactive wastes, and the development of future advanced plants.

One must conclude from this that nuclear power has a future in the United States, although a future that will have to be forged in the crucible of today's problems. These problems are causing responses leading to their solution as well as a commitment to strive continually for a higher level of excellence whether it be in design, construction, operation, or maintenance and whether it be in safety, reliability, or cost. Over-all, the nuclear industry is making greater progress than most industries in resolving the problems of a high technology industry in an affluent society which puts a special premium on safety, the environment, and the quality of life.

BIBLIOGRAPHY

Bauman, D. S. (1983). *An analysis of power plant construction lead times*, EPRI Report EA 2880, Palo Alto, California.

Bebbington, W. P. (1976). 'The reprocessing of nuclear fuels', *Scientific American*, **235**, 30.

Behnke, W. B., Jr. (1982). *Economics and technical experience of nuclear power production in the US.* IAEA Conference on Nuclear Power Experience IAEA-CN-42/81, Vienna, Austria, September, 1982.

Benedict, M., Pigford, T., and Levi, H. W. (1981). *Nuclear chemical engineering*, 2nd edition, McGraw Hill, New York.

Blair, C. (1954). *The atomic submarine and Admiral Rickover*, Henry Holt, New York.

Cohen, B. L. (1977). 'The disposal of radioactive wastes from fission products', *Scientific American*, **226**, 6.

Crowley, J. H., and Griffith, J. D. (1982). 'US construction cost rise threatens nuclear option', *Nuclear Engineering International*, **27**, 25.

Fischetti, M. A. (1986). 'The puzzle of Chernobyl,' *IEEE Spectrum*, **23**, No 7.

Glasstone, S., and Jordan, W. H. (1980). *Nuclear power and its environmental effects*, American Nuclear Society, LaGrange Park, Illinois.

Goldschmidt, B. (1982). *The Atomic Complex*, American Nuclear Society, LaGrange Park, Illinois.

Groves, L. R. (1962). *Now it can be told*, Harper & Row, New York.

Kemeny, J. (1979). *Report of the President's Commission on the accident at Three Mile Island*, US Government Printing Office, Washington, D. C.

League of Women Voters (1980). *A nuclear waste primer*, Washington, D.C.

Olander, D. R. (1976). *Fundamental aspects of nuclear fuel elements*, TID 26711.

Rahn, F. J., Adamantiades, A. G., Kenton, J. E., and Braun, C. (1984). *A guide to nuclear power technology*, Wiley Inc., New York.

Rogovin, D. (1980). *Three Mile Island: a report to the commissioners and to the public*, US Government Printing Office, Washington, D.C.

Smyth, H. D. (1946). *Atomic energy for military purposes*, Princeton University Press, Princeton, New Jersey.

Taylor, J. J. (1986). *R&D improvements from operating experience in the United States*, International ENS/ANS Conference, Geneva Switzerland.

U.S. Nuclear Regulatory Commission (1979). *TMI–2 Lessons Learned*, NUREG 0578 Washington, D.C.

Williams, R. C., and Cantalon, P. L. (1984). *The American atom*, University of Pennsylvania Press, Philadelphia, 1984.

Wymer, R. G., and Vondra, B. L. (1981). *Light water reactor nuclear fuel cycle*, CRC Press, Boca Raton, Florida.

Zinn, W. H., Pittman, F. K., and Hogerton, J. F. (1964). *Nuclear power, USA*, McGraw Hill, New York.

Nuclear Power: Policy and Prospects
Edited by P. M. S. Jones

12

France

JACQUES BAUMIER and EVELYNE BERTEL
Commissariat à L'Energie Atomique, Paris

12.1 Introduction

When the Commissariat à L'Energie Atomique (CEA) was created in October 1945 the development of industrial applications of nuclear energy were included in its mandate. Since that date a close interlink between theoretical research and commercial applications has remained a main feature of the French programme for nuclear energy. Although problems were encountered, a coherent policy was developed over time which can be seen as globally satisfactory.

As with any national programme the French one is specific to the domestic context, and cannot be presented as a model but rather as an example of a possible route, that has proved to be rather efficient to adoption of nuclear energy in an industrialized country.

This chapter summarizes briefly the history of the last decades, and underlines some typical facets of the French situation. As far as future forecasts are concerned we ought to be modest, taking account of the moving and rather unpredictable world economic evolution. For this reason, the indications given here are more of a qualitative nature, and are presented in terms of tendencies rather than precise projections.

12.2 Early Development of Nuclear Energy

The first French heavy water moderated research reactor, Zoe, went critical in December 1948. The question during the following years was, for France as well as for other countries, to choose the best reactor type, adapted to its own scientific knowledge and technical know-how.

Although it was not explicitly stated, the French authorities wanted to keep the nuclear weapon option open. With no domestic technology of isotopic enrichment available it was necessary to develop facilities adapted to plutonium production. As it was easier to produce pure graphite than heavy water, it was decided to build in Marcoule the G1 reactor, graphite moderated and air cooled.

The national utility, Electricité de France (EDF), restoring its generation plant after the Second World War, was soon interested in the nuclear option. It was involved from the beginning in the design of devices for heat extraction and electricity production from G1, which went critical and produced electricity the same year (1956), with a power of 2 MWe.

In 1955, the construction of G2 and G3 was started; they were devoted primarily to plutonium production, but EDF was immediately involved with electricity production at the level of 40 MWe for each unit. These reactors were connected to the grid in 1958 and 1960.

This gas cooled reactor type, developed and designed by the French CEA, was adopted by the national authorities and in

1955 EDF was authorized by the government to launch a series of power plants, starting with a 70 MWe unit. This first one, Chinon A1, was connected to the grid in 1965 followed by Chinon A2 (210 MWe) in 1965, Chinon A3 (480 MWe) in 1966, Saint-Laurent A1 (480 MWe) in 1969, Saint-Laurent A2 (515 MWe) in 1971 and then Bugey 1 (540 MWe) in 1972. This last unit, ordered in 1965, closed the series.

By 1961, the first Franco-Belgian PWR (Chooz—300 MWe) had already been ordered by SENA (Société d'Electricité Nucléaire des Ardennes: 50 per cent Société Belge Centre et Sud, 50 per cent EDF) from the AFW consortium (Framatome, Westinghouse Electric Corp and Belgian companies). Framatome, the supplier of nuclear island, was created in 1958 and signed a licensing agreement with Westinghouse at that date.

12.3 From GC Reactors To PWRs: The Reasons for Change

What happened between 1965, when the last gas cooled reactor was ordered, and 1969–1970 when the first 900 MWe-PWRs orders for Fessenheim 1–2 were placed?

During this period a rather complex argument developed between proponents of the different reactor types, namely gas–graphite and light water ones, that could be called *guerre des filières*.

Although President de Gaulle decided in a restricted Ministry Council of December 1967 that a GC reactor would be built at Fessenheim, the light water reactors were finally adopted, mainly for economic and commercial reasons, and recognizing that this reactor type had already experienced industrial development in the leading nuclear country, the United States.

It should be remembered that, while the nuclear programme was on stand-by position, due to the low price of oil, the director general of EDF declared in October 1969: '. . . we have to maintain a nuclear industry until atomic energy becomes competitive. Our engineers must thus pursue some training. For this purpose, the most economic reactor type should be chosen, which is the American light water one . . .'.

The implementation of this option was supported by the development of French isotopic separation techniques. The military plant at Pierrelatte, using the gaseous diffusion process, became operational in 1967 and it was possible to envisage the launching of a civil facility. In addition a prototype nuclear powered submarine had been designed by the end of the 1950s in Cadarache and this reinforced the idea of developing PWRs for civil purposes.

Finally in November 1969, President Pompidou decided in a restricted ministry council to follow EDF's choice and stop the French GC reactors development.

Although it was a rather controversial matter, it is now possible to see that the choice was a good one for several reasons. It would have been difficult, if not impossible, to increase the capacity of the GC units, taking account of safety requirements; cost reduction from scaling-up would thus have been substantially lower than for PWRs. Metallic fuel was required to increase neutronic efficiency in the GCRs, but involved risks of cladding failures due to oxidation by carbon dioxide (CO_2).

The economic incentive, although real, was probably the less significant argument. In fact, theoretically forecast costs appeared to be just slightly higher, by some 10 to 15 per cent, for GC reactors than for PWRs.

The international context, with an industrial development focused upon light water reactors, especially in the United States, argued strongly in favour of their adoption in France to benefit from the general progress that would be achieved worldwide.

12.4 Launching a Large PWR Programme

Although the 1973–74 oil crisis, and the drastic energy price rises it involved for France, was a determinant in the decision then taken to launch a large nuclear programme, the French authorities had already chosen earlier, as we have seen, to develop nuclear electricity production.

As far as PWRs are concerned, beside the Chooz unit (300 MWe) connected to the grid in 1967 and the Tihange one (900 MWe) ordered in 1969, both shared between Belgium and France, six units were ordered before 1974. Following the adoption in 1969 of LWRs for industrial development, Framatome received orders from EDF for the two Fessenheim units and the four Bugey ones, all PWRs of 900 MWe.

By the end of 1973, the nuclear power plants already connected to the French grid, including the breeder prototype Phenix, represented some 1.7 GWe and they had produced during that year 14 TWh, ie 8 per cent of the domestic electricity consumption.

However, 1974 was a milestone for the French nuclear programme. In March 1974 the government, then chaired by M. Messmer, decided to speed nuclear development in order to face the problem of energy and oil dependancy. It should be remembered that French indigenous energy sources accounted, at that time, for less than 25 per cent of national consumption. EDF ordered in 1974 fourteen LWRs; two BWRs in February, which were cancelled later, and 12 PWRs–900 in April. In 1975 the 'Commission pour la Production d'Electricité d'Origine Nucleaire' confirmed this orientation and allowed a rate of ordering of six GWe per year for 1976 and 1977. At the same time, in August 1975, it was decided to cancel previous BWRs orders and, for industrial reasons, to adopt standardized PWRs for all the French nuclear power plants.

Pursuance of a policy of standardized power ratings, which had already been adopted in the construction of classical thermal power plants was applied in the nuclear field to reduce costs, shorten lead times, and increase safety as well as equipment reliability.

While 900 MWe units were ordered and built, design studies began for 1300 MWe units which were better adapted to the size of the French grid and expected to be cheaper.

Additionally since LWRs were considered as an intermediate stage, the long term goal of developing commercial breeders was pursued, with the decision taken in 1975 to build at Creys-Malville a multinational unit of 1200 MWe in co-operation with other European countries.

By the late 1970s and early 1980s, France continued its expanded nuclear programme, with some 5 GWe of nuclear generating capacity ordered every year.

In 1981 the new socialist government had to adapt its nuclear policy to the economic recession, common to most industrialized countries. Although electricity consumption growth rate remained substantial, 5 per cent and 4 per cent respectively in 1980 and 1981, it proved to be lower than the 7 per cent per year expected earlier. This was reflected by a minor reduction in nuclear ordering rates; three units ordered in 1982 and two in 1983, made-up of one PWR–900 and four PWRs–1300; that is to say a yearly increment of some 3 GWe. Two more PWRs–1300 were ordered in 1984, Chooz-B2 and Cattenom–4, and for 1985 EDF had been authorized by the Government to order one unit, Penly–2.

The successive steps of PWRs development in France could be summarized as follows:

1. Six pre-series units (Fessenheim and Bugey).
2. CP1, (Contrat Programme No. 1), eighteen units, PWRs–900.
3. CP2, (Contrat Programme No. 2), ten units, PWRs–900.
4. Class 1300, 20 units PWRs–1300.
5. Class 1400, two units PWRs–1400, N4 type.

12.5 Nuclear Industry Development and Structure

This large programme was sustained by a broad national industrial development involving reactor manufacturing, nuclear island as well as conventional parts of the plant, and fuel cycle services facilities. The French industry was strongly structured around Framatome, the nuclear plant contractor; Alsthom-Atlantique, the supplier of heavy electrical equipment; and Cogema the fuel cycle services main supplier. This 'monopoly type' of structure enabled France to achieve coherence in her global nuclear energy policy and substantial economies of scale. In the same spirit the standardization option proved very efficient in reducing lead and lag times, improving equipment reliability and increasing safety.

This integrated industrial organization includes hundreds of firms beside the main components manufacturers, namely Creusot-Loire, Jeumont-Schneider and Spie-Batignolles, for the nuclear part; Alsthom-Atlantique for steam generators and CGE for electric equipment. This complex has the capacity to construct some six to eight reactor units per year.

From the beginning an important R and D effort was initiated, whose major goal was to develop an original PWR design based on purely French technology. This led Framatome to replace in 1981 its previous licensing agreement with Westinghouse by a co-operation agreement. Moreover, after several years of research the entirely new design of PWR, the N4 model of 1450 MWe, for which Framatome received the first order in 1984, attests to the success of these efforts.

The same strategy was adopted by Alsthom-Atlantique who designed the Arabelle turbine to replace the Brown-Boveri licensed one by 1983.

Although it is almost impossible to review in brief even the main French nuclear firms it is worthwhile mentioning amongst others, Pechiney for its zirconium metallurgy, uranium milling and conversion expertise and Vallourec, which has specialized in the field of nuclear tubes.

The French fuel cycle industry covers the entire process from uranium mining to high level radioactive waste management.

The six milling plants, owned by Cogema, TCM (Total Compagnie Minière) and CFM (Compagnie Française de Mokta) represent a total capacity of some 3900 t of uranium per year.

The Comurhex plant at Pierrelatte is one of the five units converting yellow-cake into hexafluoride in the western world, and the only one able to treat reprocessed uranium.

The multinational gaseous diffusion enrichment facility at Tricastin near Pierrelatte, operated by Eurodif represents, with 10.8 10^6 SWUs per year, some 25 per cent of the enrichment capacity available in WOCA (world outside centrally planned economy area).

Fuel assemblies are manufactured in the two facilities at Romans (FBFC, Société Franco-Belge de Fabrication du Combustible) and Pierrelatte (CFC, Compagnie Française du Combustible). A new design has been developed recently (AFA, Advanced Fuel Assemblies) which improves fuel performance.

Reprocessing of irradiated fuels and high level waste vitrification are achieved at an industrial and commercial level by Cogema at la Hague and Marcoule.

12.6 Technical Problems and Nuclear Power Plants' Availability

The availability of nuclear power plants in service in France went from 58 per cent in 1982 to 68 per cent in 1983 and 75 per cent in 1984. These figures reflect the solutions found to the problems encountered in 1982 with respect to guide tube pins on the control rod assemblies of PWRs–900 and to the dryer-reheater units at Saint-Laurent B1 and B2.

Some tube defects in the turbine moisture separator reheaters have obliged EDF to close down the two PWRs at Saint-Laurent-des-Eaux and delayed their commercial operation. The problems were caused by differential expansion of the tube and the tube support plates in the separator reheaters. Remedial actions were taken for all the on-going units of the series. The gap between the tubes and the support plates was increased to allow more flexibility in structures dilation and the internal structures were strengthened to avoid problems of vibration and corrosion. These modifications carried out while the units were in their final stage of completion involved delays and lowered somewhat the mean availability of the French nuclear power plants in 1982.

Cracking and failures of control rod guide tube support pins were by far the most complex and worrying problem encountered at the French nuclear stations. Such incidents were reported at various PWRs in the world over the first years of the 1980s. In France, the first pin failure occured at Gravelines B1 in January 1982, and was followed by similar incidents at Fessenheim–1 and Bugey–2. These incidents confirmed that pin integrity was being jeopardized in operating units and that an immediate action was necessary to correct any similar defect that might arise in other plants. In September 1982 it was decided to fit improved pins to all French 900 MWe plants in operation and under construction.

Improved pins had to be designed with a modified shape, to reduce stress concentration, and optimized heat treatment, to limit susceptibility to corrosion; the material of the original pins, inconel–750, has been retained. Pin replacement had to take place in 34 units, 21 of which were in operation. This programme, completed by 1985, was achieved during normal refuelling outage periods of the plants concerned, thus avoiding further non-availability. In most cases refurbishment of irradiated assemblies from operating reactors with new pins has been done in a centralized workshop at Pierrelatte, operated by Framatome. Because of the high risk of irradiation exposure, the complexity of the operation, and short time allowed, the French staff had to develop efficient techniques and acquired considerable expertise.

These problems solved, the nuclear plants performed quite well last year. However the availability can obviously be further improved by better maintenance skill and tighter quality control.

The trend is satisfactory as the availability of nuclear units in operation in France is increasing from year to year, as well as the availability of any individual plant which increases with age. Nevertheless continual attention is paid to any cause of non-availability in order to maintain the good results of the past and prevent future difficulties.

12.7 Nuclear Electricity Economics

Economic incentives were, and remain, determinants for nuclear energy development. In France, nuclear power stations proved to be competitive with coal or oil fired plants since the early 1970s and remained so up to now.

As cost evaluations are often controversial it is worthwhile explaining the context of French economic calculations.

In spite of standardization the investment cost of nuclear units has varied in real value since the first plants were built; it depends on the site, the number (first, second, third etc) of the unit on the site, and on other parameters.

Cost increases of some 5 per cent per year in real terms were experienced as a result of more stringent safety requirements as well as of less favourable economic conditions. This means a doubling of the costs between the first 900 MWe units (Fessenheim) connected to the grid in 1977 and the units that will be connected to the grid in 1992, fifteen years later. However this increment is substantially lower than the ones incurred in other countries, Federal Republic of Germany or the United States for example.

Even if international comparisons of investment costs are difficult French costs paid by EDF for its PWRs seem to be somewhat lower than foreign costs for similar types of reactors. This situation is mainly due to the size of the French programme and the effects of standardization allowing to benefit of series manufacture amongst other advantages.

Moreover, the national utility EDF which is at one and the same time industrial architect, owner and operator, is responsible for engineering and insurance. While most foreign utilities order turnkey power stations, EDF takes upon itself most of the risks, thus alleviating the financial charges and own risks of the NSSS manufacturer. Altogether these differences may result in cost reductions reaching more than 10 per cent.

Hereafter we report the costs established by a working party where the Ministry in charge of Industry, EDF, the CEA and the national coal producer (Charbonnages de France) are represented. These costs are expressed in French francs (FF) at the economic conditions prevailing on the first of January 1984 and refer to a plant that will be connected to the grid in 1992.

Investments and operating costs are summarized in Table 12.1.

Fuel costs are the following:

1. Nuclear
   a. Uranium 700 FF kg^{-1} in 1990 with a 2 per cent per year increase thereafter
   b. Conversion 42 FF kg^{-1} Uranium
   c. Enrichment 912 FF SWU^{-1} in 1990
   d. Fabrication 1430 FF kg^{-1} Uranium in 1992 with 1 per cent per year decrease thereafter
   e. Reprocessing, waste management and storage 6150 FF kg^{-1} Uranium
   f. Plutonium credit 100 FF kg^{-1} (assuming Pu recycling in PWRs).
2. Coal
   7.9 cF th^{-1} in 1984 with 1.5 per cent per year increase; which means assuming imported coal with a rate of exchange of 7.4 FF $US\$^{-1}$, although by the end of 1984 the actual rate was 9.5 FF $US\$^{-1}$
3. Fuel oil
   a. Crude oil 29 \$ $barrel^{-1}$ in 1984, increasing at 4 per cent per year thereafter
   b. Heavy products 20 per cent less than crude oil.

Final results are presented in Table 12.2. These costs are calculated for a 9 per cent discount rate and a 25 years lifetime of the plant. Nuclear plants are assumed to operate 6200 h $year^{-1}$, after three years of operation

Table 12.1 Investments and operating costs FF (January 1, 84)

	Nuclear[1]	Coal[2]	Oil[2]
Investments (FF kWe^{-1})	7454	5430	4680
Construction	5362	4316	3720
EDF's engineering	420	316	186
Interest during construction	1390	816	703
Pre-operating expenses	202	82	71
Provision for decommissioning	71	—	—
Operating (FF kWe^{-1} y^{-1})	253	220	187

[1]PWR-1300 MWe, four units on the same site, rhythm of order two units per year
[2]600 MWe units, rhythm of order four units per year

Table 12.2 Comparative generation costs cF (January 1, 84) kWh^{-1}

	Nuclear	Coal	Oil
Investments	12.0	8.2	6.9
Operating	4.3	3.5	3.0
Fuel	6.4	20.9	63.0
TOTAL	22.7	32.6	72.9
(Desulphurization)		(+2.9)	(+8.1)

Table 12.3 Cost of electricity as a function of load factor cF (January 1, 84) kWh^{-1}

Yearly operating duration (hours $year^{-1}$)	Nuclear	Coal	Oil
Nuclear 6200–coal and oil 6600	23	33	73
Nuclear 3200–coal and oil 3400	39	43	81
Nuclear 1635–coal and oil 1700	71	65	100

at a lower power, coal and oil fired ones 6600 h $year^{-1}$. Desulphurization would add 2.9 cF kWh^{-1} and 8.1 cF kWh^{-1} to the coal and oil kWh costs respectively.

Nuclear electricity remains cheaper for higher load factors, with a break-even load factor versus coal of 28.5 per cent at 2500 h $year^{-1}$ operation (Table 12.3).

12.8 Present Status of the Nuclear Programme

By the end of 1984, France had 41 installed nuclear units representing 25,725 MWe; so far, four units (140 MWe) GC type, have been decommissioned. One unit PWR–1300 has been ordered in 1985 and one or two more will be ordered in 1986. Taking account of the six units due to be decommissioned before 1990, it means that the installed capacity will reach some 60 GWe by the early 1990s.

The status of the nuclear programme at 1st January 1984 is the following:

1. Installed capacity—33,248 MWe: 41 units
   31 PWRs–900
   Two PWRs–1300
   One FBR prototype (Phenix)
   Seven old units (mainly GCRs)
2. Under construction—25,725 MWe: 21 units
   Three PWRs–900
   16 PWRs–1300
   One PWR–1400
   One FBR (Superphenix)
3. Firm orders—2565 MWe: two units
4. Optional—1390 MWe: one unit

It is now obvious that the programme launched in 1970 and accelerated in 1974 has been successful in terms of technical and industrial performances as well as reaching the original goal of alleviating France's dependence on imported fuels. The standardization policy has resulted in short lead times, increased safety and equipment reliability and minimization of costs. The average building time, from ordering to coupling to the grid is some 65 months, substantially lower than in most other countries. Operating experience is hightly satisfactory and a new mode is now developed to adapt the output supplied by nuclear plants to the requirements of the grid. Load following and frequency control are now possible with nuclear stations, thus ensuring a proper utilization of the investments engaged in the programme.

During the last decade, from 1973 to 1983, the share of nuclear electricity in the primary energy balance had grown from 2 per cent to 17 per cent. In 1984 the nuclear power plants produced 182 TWh, 59 per cent of the total electricity production, which means equivalent oil savings of some 30 million t or a reduction of more than 70 billions FF of our energy bill.

The nuclear electricity production has allowed a rapid increase of electricity exports which reached 13 TWh in 1983, 25 TWh in 1984 and are expected to remain at the same level or higher until the end of the decade. The share of nuclear in electricity production will continue to increase from 56 per cent in 1984 to some 70–75 per cent by 1990, the remaining part of the electricity production being due to coal fired plants and hydro. Dependence on imports will thus fall from more than 60 per cent in 1984 to less than 50 per cent by 1990.

The evolution of our electricity production system was substantial since 1960, as may be seen in Table 12.4, and nuclear power plants are chiefly responsible for this change.

In addition, over the last 20 years the French nuclear industry has developed its resources, know-how and expertise in most of the disciplines involved in designing, building, fuelling and operating nuclear power plants. The global capacity of this integrated complex has put our country in a leading position in the world. After a period of rapid expansion this industry has now to face a depressed economy involving a lower level of activity both in the domestic and international markets. However, with more nuclear plants coming on stream in France, fuel cycle services and plant maintenance requirements will continue to rise in coming

Table 12.4 Electricity supply and exports

	Nuclear electricity generation (TWh)	Nuclear/total electricity generation (%)	Electricity exports (TWh)
1960	0.13	0.2	0.10
1970	5.15	3.7	0.50
1975	17.45	9.8	−2.50
1980	57.94	23.5	−3.09
1981	99.62	37.7	4.83
1982	103.07	38.7	3.84
1983	136.92	48.3	13.41
1984	170.00	56.0	25.00

years and it is a matter of flexibility to adjust the resource allocation to the future situation. Nuclear exporting is also a promising route to take advantage of domestic experience on the international market and to improve our balance of payments as well as to develop our foreign industrial and commercial activities.

Nuclear electricity is now a vital source of energy in our country and the nuclear industry is a main branch of our industrial system. The challenge will be to maintain and develop the resources in terms of facilities and manpower, even in the present somewhat difficult economic climate, in anticipation of a turnaround, hopefully in the not too distant future.

12.9 Long Term Policy and Breeders

Nuclear power is a reality in France. Its importance has been demonstrated and will increase until the end of the century when nuclear plants will supply some 80 per cent of our electricity. But what are the long term prospects? Of course, the further you go into the future, the more difficult are the forecasts. Nevertheless it is worthwhile trying a projection of the evolution of the economy for the next few decades, and the nuclear power growth it would involve. We have chosen to illustrate possible futures by two contrasted scenarios.

The high case implies a revival of the economy with GDP growth rates close to 5 per cent between 1990 and 2000, and of about 4 per cent between 2000 and 2020.

The low case corresponds to the continuation of the present world economic crisis with GDP growth rates of 1 per cent between 1990 and 2000 and 0.7 per cent between 2000 and 2020.

For both scenarios we assume a 2 per cent GDP growth rate until 1990.

For the low and high case, the elasticity of primary energy consumption growth rate versus GDP growth rate is assumed to be 0.6 per cent; the asymptotic electricity penetration of primary energy use is fixed at a level of 50 per cent; and the nuclear share of electricity is assumed to level up at 80 per cent.

The results in terms of electricity demand growth rate, nuclear installed capacity and uranium requirements are given in Table 12.5. Without the breeder, cumulated uranium requirements from 1990 would reach 380 000 to 650 000 tonnes.

The French known uranium resources, recoverable at costs of up to 130 \$ kg^{-1} uranium are some 120 000 t. Taking account of speculative domestic resources, as well as foreign resources more or less devoted to French supply (French companies shares in mining and milling facilities abroad, fixed contracts etc), this figure would more than double. Our security of supply is thus much better regarding uranium than for oil or even coal. However these resources would be entirely consumed by 2010 or 2020 in the high and low case respectively.

Obviously additional resources could become available at that time horizon

Table 12.5 Future scenarios

		1990	2000	2020
Electricity demand growth rate (%)	I	—	4.3	3.0
	II	—	2.0	1.0
Installed nuclear capacity projection (GWe)	I	56	104	244
	II	56	76	107
Cumulated French uranium requirement (tonnes)	I	—	115 000	650 000
	II	—	80 000	380 000

I High case
II Low case

although the exploration effort has been somewhat reduced in recent years.

Anyhow the importance of nuclear electricity in energy production implies a security of fuel supply of several decades, and a rather early deployment of breeders is thus justified.

Recognizing that the investment cost of breeder reactors is too high for the time being to allow the implementation of a large programme on a national basis, France had favoured international co-operation in breeder development and the creation of ARGO.

We think that fast breeders could reach the break-even point with light water reactors by 2010 if uranium prices rise by a factor of three or four by that date. Before this time a substantial amount of PWR fuel will be reprocessed and the fissile material they contain could be recycled. It seems thus likely that plutonium recycling in PWRs can be envisaged without pre-empting further breeder development. This option is presently being studied and 30 per cent of PWR–900 fuel assemblies could be MOX, (mixed oxide fuel). In that case 70 per cent of the recycled plutonium will be available later on to feed the breeders. This utilization will be compatible with the implementation and feeding of one or two FBRs until the end of the century and four or five more in the first decade of the next century. When the breeders become fully competitive recycling of plutonium in the PWRs will be stopped.

12.10 Conclusion

In France as well as in many other countries, nuclear electricity will have a major impact on the energy supply for several decades. We have now sufficient technical background and industrial power to ensure reliable development of this new energy source. The French industry has committed a large amount of resources in terms of finance, equipment and manpower, to satisfy the requirements of a large national programme. The impacts of the efforts of the past decades could be viewed as highly satisfactory in terms of scientific progress and domestic energy supply ability. The benefits of this deployment will remain substantial for decades provided we are able to adapt our industrial capacities and R and D programmes to a moving global economic environment.

One of the main tasks of the coming years will be to maintain and develop the existing potential. The improvement of the energy market situation gives an opportunity for developing new technologies and reactors with better design and performances. Fast breeders are a key for middle and long term energy supply; the present pause gives an opportunity for delaying their commercial deployment until they reach competitivity and while improving efficiency.

Electricity generated by nuclear power plants is a new industrial and economic reality today from the development of which we have still much to learn and much to gain.

Nuclear Power: Policy and Prospects
Edited by P. M. S. Jones

13

Canada

H. E. Thexton
Nuclear Energy Agency, Paris

13.1 Introduction

In 1984, 37 per cent of the electricity produced in Ontario, Canada's largest and most industrialized province, was produced by nuclear power from twelve reactors at five stations at a cost of about two-thirds of that from contemporary base load coal-fired stations. In the same year in New Brunswick, one of Canada's smaller provinces, nuclear power provided over 30 per cent of the electricity generated at a significant savings relative to coal-fired generation, plus providing export revenue through power sales to neighbouring utilities in the United States.

The reactors which produced these accomplishments were all of the unique design which the Canadians have christened CANDU, for CANada Deuterium Uranium.

What is a CANDU, and what led Canada, a rather small country industrially, to develop a reactor concept so different from the light water reactors (LWRs) adopted for commercial deployment in most industrialized countries? What factors have led to the remarkable technical success of the system, and what future do Canadians see for nuclear power in their country? This chapter will attempt to address all of these questions. To start, let us look at how a CANDU differs from an LWR.

13.2 What is a CANDU?

In any reactor each fission releases two or three neutrons in addition to a substantial amount of energy. In order for the power level of the reactor to remain constant exactly one of these neutrons, on average, must be captured by another fissile atom (usually uranium–235 or plutonium–239) and so cause it to fission, thus maintaining the chain reaction. However, there are a number of ways these neutrons can be lost. Some will be captured by non-fissile uranium–238, converting it to fissile ^{239}Pu but consuming a second neutron to cause another fission. Some neutrons will be captured by the moderator, material surrounding the fuel which slows down neutrons to make it easier for fissile atoms to capture them. Some neutrons will be captured by structural material used to contain the fuel and the moderator or will leak into the shielding surrounding the reactor. The power control systems will also absorb some neutrons without producing new fissions. Thus the first job of a reactor designer is to ensure that the reactor is capable of maintaining a chain reaction in spite of the strong parasitic competition for neutrons.

Two basic approaches have been used by designers. One is to increase the concentration of fissile atoms in the fuel to increase the probability of a fissile capture before a neutron is captured by a non-productive material. The light water reactor designers chose this route and so used 'enriched' uranium with the concentration of ^{235}U increased to about 3 per cent, some four or more times higher a concentration than occurs in nature.

CANDU designers instead chose the other route of using natural uranium (with about 0.71 per cent ^{235}U) and used materials in the core which minimize the capture of neutrons. Thus a CANDU uses zirconium alloys, which absorb far fewer neutrons than steel, for structural components and heavy water, which is essentially transparent to neutrons, rather than ordinary water for a moderator.

Even with this design approach there are not many neutrons to spare in a CANDU reactor. Since the fission products resulting from each fission tend to absorb many neutrons, the fuel must be removed from the reactor before the fission product concentration gets too high. In practice this means that CANDU reactors have to be refuelled on-power to avoid very frequent shutdowns. Light water reactors were able to add enough enrichment to require refuelling only about once per year so their designers chose the shutdown refuelling option.

The core of a natural uranium reactor has to be quite a bit larger in size than an enriched reactor for physics reasons. Early CANDU designers feared that manufacture of such large pressure vessels for high power reactors would prove difficult if not impossible. As a result they chose to contain fuel bundles and the hot, high pressure heavy water coolant inside a large number of individual pressure tubes. This also facilitated on-power refuelling. The moderator, which surrounds the pressure tubes, could then be kept as a separate circuit and held at a low temperature and pressure in a rather simple design of vessel. By comparison, in an LWR the water pumped into the vessel acts as both moderator and coolant. To maintain an acceptable thermal efficiency the water is very hot and it must be kept at a high pressure

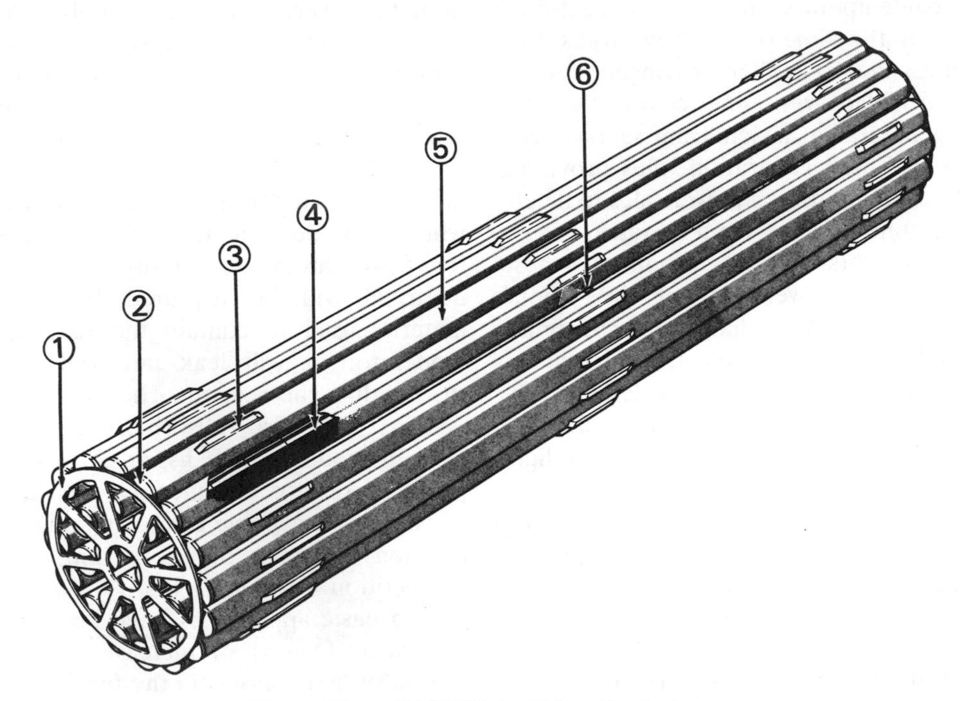

Figure 13.1 CANDU fueld bundle design.

1. Zircaloy end support plate.
2. End caps.
3. Zircaloy bearing pads.
4. Uranium dioxide pellets.
5. Zircaloy fuel sheath.
6. Inter-element spacers.

Figure 13.2 Inspection of CANDU fuel bundles prior to shipment

to prevent the water from boiling in the vessel, or at least to minimize boiling to some practical degree. Thus the vessel must be a very thick walled, complex 'pressure vessel'.

With the pressure tube design it proved convenient to load fuel horizontally and to simply push new fuel bundles into one end of the pressure tubes, taking the used fuel bundles out of the other end at the same time, by means of fuelling machines operating at each end of the reactor. With this arrangement it was practical to utilize short fuel bundles of a very simple design. The bundles were not connected to one another in any way but were simply pushed in to fill the pressure tubes. Also, fresh fuel bundles could be inserted from opposite ends of adjacent channels resulting in a more uniform power distribution over the length of the reactor.

No instrumentation or control assemblies were built into the fuel bundles since it was more convenient to locate these devices in separate assemblies in the moderator vessel where they could operate at low temperature and pressure.

Clearly there are many detailed differences between a CANDU and an LWR but the fundamental features of a CANDU can be summarized as follows:

1. Use of natural uranium in simple fuel bundle assemblies (Figures 13.1 and 13.2);
2. Relatively low fuel burn-up (about one

quarter of that in an LWR) in terms of power produced per kg of fuel put through the reactor.
3. Relatively high energy production per kg of uranium mined (about 20 per cent greater than in an LWR where much of the uranium, including some of the fissile ^{235}U, ends up as 'tails' in the enrichment process.)
4. On-power refuelling.
5. Pressure tubes rather than a pressure vessel.
6. Heavy water rather than light water moderator.
7. Moderator contained in a separate circuit at low temperature and low pressure.

Figure 13.3 shows a schematic view of a CANDU reactor.

13.3 Development of the CANDU

A team of European nuclear scientists were evacuated to Canada in the early days of the Second World War, greatly expanding the small nuclear science programme which Canada's National Research Council had started in the 1930s. This team was given the task of attempting to build a heavy water reactor for plutonium production as part of the allied powers' nuclear weapons programme. This approach was assigned to Canada primarily because some of the Europeans on the team had been experimenting with heavy water before the start of the war. This one factor was responsible, probably more than anything else, for setting Canada on a course which led by logical stages to the CANDU design.

The first Canadian reactor which emerged from this war time programme started operation in September 1945 at the Chalk River Nuclear Laboratories. It was a small, simple unit used for physics tests on heavy water moderated designs. Called ZEEP, for Zero Energy Experimental Pile, it was the first reactor in the world designed and built outside of the United States.

ZEEP's primary purpose was to confirm design parameters for a larger reactor, NRX (National Research Experiment). Although ZEEP and NRX were not finished by the time the war ended, work continued and NRX started operation in 1947. Although it had originally been intended primarily as a plutonium production unit, it proved to be an outstanding research reactor for testing nuclear fuel and materials. Because of the large core and the high density (or 'flux') of neutrons required in the natural uranium fuelled reactor, it was possible to test relatively large size fuel assemblies. This brought researchers from several countries to do experiments in NRX, and later in the more powerful NRU which started up in 1957. Under the agreements for use of the Canadian reactors, Canada benefited from much of information learned by other countries in their fuel development work. One key development which Canada obtained by this route was early access to the use of the zirconium alloy (Zircaloy) developed by Westinghouse for the US navy's submarine reactors. Without this high strength, low neutron capture material, a pressure tube reactor would have been impossible. But this is getting ahead of the story.

In 1952, recognizing the potential for development of a power reactor based on the natural uranium–heavy water principle, Atomic Energy of Canada Limited invited representatives from Canadian utilities and industry to participate in a concept design exercise. They responded, and a study team was established during 1953 and 1954. This early involvement of the utilities proved very worthwhile as it led, from the beginning, to the reactor being designed to their standards for reliable operation and ease of maintenance.

Ontario Hydro, Canada's largest utility at the time, was then nearing the end of developing the province's low cost hydro capacity and was starting to turn to coal-fired power. However, with no coal resources of significance in the province, it realized that this would lead to large expenditures for US coal. It decided that nuclear power offered the chance to provide an alternative generating system which could be fuelled by domestic uranium. It decided, therefore, to proceed,

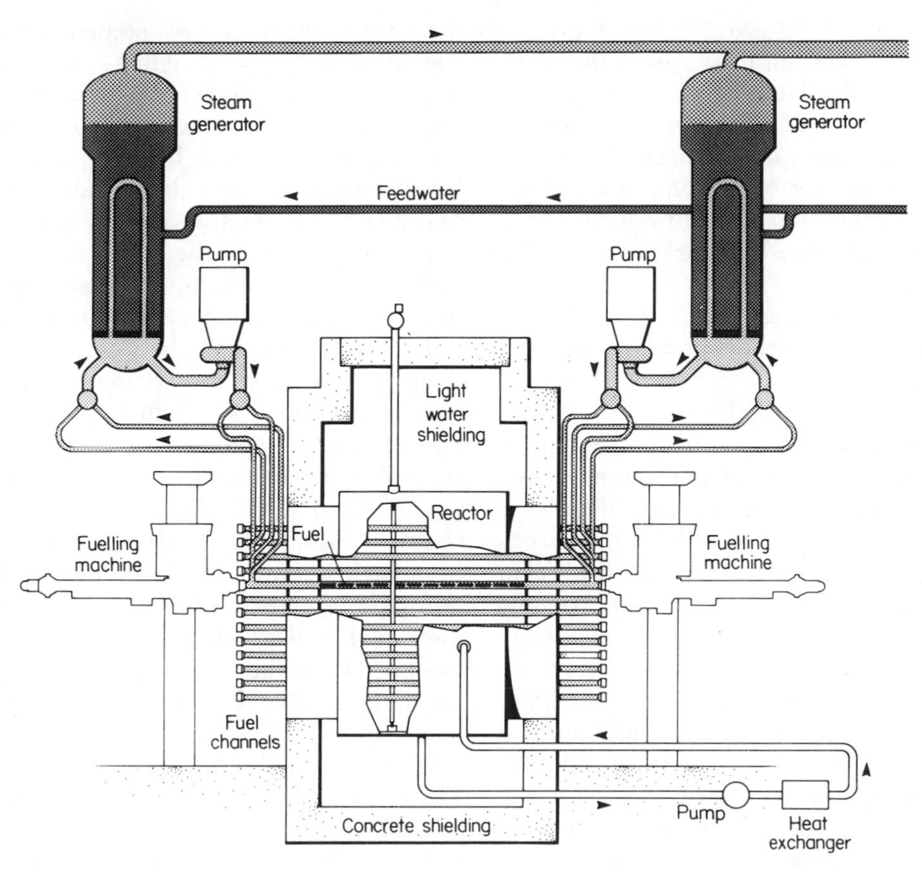

Figure 13.3 Schematic view of a CANDU reactor

in partnership with AECL and Canadian industry, to design and build a 20 MWe nuclear station, to be called NPD (Nuclear Power Demonstration), on the Ottawa River not far from the Chalk River Nuclear Laboratories.

The initial design for NPD was for a vertical core in a steel pressure vessel, very similar to the PWR design being developed in the United States. However, even for this small reactor, the vessel was too large for Canadian industry to build so it had to be ordered off-shore. Although the vessel was under construction, the Canadian scientists and engineers were concerned that it might prove impracticable to build a pressure vessel for a large, commercial reactor in future, again recognizing that the vessel for a natural uranium fuelled reactor would be larger than for an enriched reactor of the same power. It was just at that time that Zircaloy was developed.

A conceptual design for a larger reactor had already shown the advantages of a pressure tube system with a horizontal core. The availability of Zircaloy made this design feasible. AECL and Ontario Hydro therefore decided to stop work on the pressure vessel design and to build instead 'NPD–2' as a demonstration pressure tube reactor with short fuel bundles and several other features which have remained standard conceptual features of all subsequent Canadian power reactors.

NPD–2 (which later reverted in name to simply NPD) started operating in 1962. However there was so much confidence in the system that before it started up the larger 200 MWe Douglas Point plant was committed. This plant was located at what later became the Bruce Nuclear Development on Lake Huron. It was given the name CANDU, which later became the generic name for the Canadian design of reactor.

NPD had its share of learning problems but for the first reactor of the type it did work well, and is continuing to do so still, now over 23 years old. It has a lifetime capacity factor of over 65 per cent.

Douglas Point started up in 1967 and experienced a number of problems with its scaled up systems. However, even before it started up Ontario Hydro had decided to proceed with another ten-fold scale up by building four 500 MWe units in an integrated station at Pickering. The problems at Douglas Point were gradually resolved and the lessons learned were applied to the Pickering design. The idea of a multi-unit station with carefully staged construction and an innovative design to take maximum advantage of common services proved to be technically and economically successful. The four units entered commercial service between 1971 and 1973 and for years set a heady pace in the unofficial international competition for performance as measured by lifetime capacity or load factor (the fraction of electricity actually generated by a plant as a percentage of the amount which would have been generated if it ran at 100 per cent power for 100 per cent of the time since it first generated electricity). At the end of 1983 the four units had a liftime capacity factor of over 80 per cent, compared to a world average for power reactors of about 60 per cent. The economics of the station have also been favourable as will be discussed later.

Douglas Point, having completed its task as a prototype reactor and in need of refurbishment, was permanently shutdown in 1984.

The CANDU system was further developed by AECL and Ontario Hydro after Pickering. In 1977 the first of four new units entered commercial operation at the Bruce Nuclear Generating Station. These units are similar in principle to Pickering although they have a number of detailed differences, such as fewer but larger circulating pumps and steam generators, a new second shutdown system and a new building arrangement. Each unit produced 740 MWe plus about 35 MWe equivalent of process steam for industrial use. They are now being uprated to 848 MWe (equivalent) capacity. In addition, four new 500 MWe units have been built at the Pickering station to the same basic design as the original units though with some new systems and components. They started commercial operation between 1983 and 1986. Another four-unit station is also being built at the Bruce site with three units now operating and one under construction. These are of the same design as the first four units but are rated at 815 MWe. Another Ontario Hydro station, at Darlington, is under construction with four units of 881 MWe each. These units, similar to Bruce but with some evolutionary changes, are scheduled to enter service in the 1988 to 1992 period. They are the last units planned by Ontario Hydro to enter service this century.

In 1984 nuclear power was supplying about 37 per cent of Ontario's electricity. By the time all of its units are operating, totalling over 1400 MWe of capacity, this share will rise to about 65 per cent making Ontario Hydro one of the most nuclear intensive large utilities in the world.

13.4 Other CANDUs

Although Ontario Hydro has been a major focus of the CANDU development AECL has also developed CANDU systems for other customers in Canada and abroad.

While Douglas Point was still being built AECL decided it should develop a back-up design, primarily in case the rate of loss of expensive heavy water from the hot, high pressure primary coolant system proved to be too great a problem. It built a heavy water moderated, boiling light water cooled design it called the CANDU-BLW in a 250 MWe prototype unit for Hydro Quebec at the Gentilly site. This unit had a vertical core but retained the pressure tube design. By allowing the light water coolant to boil in the pressure tubes it was possible to reduce the mass of neutron absorbing water in the core so that the reactor could still use natural uranium fuel. However, the control system required for this design proved to be very complex. The reactor then ran into a series of design and commissioning problems. Since the Pickering reactors, which started operating just after Gentilly–1, ran so well from the beginning, and since heavy water leakage and upgrading did not prove to be a problem, a backup design proved un-needed. Also, Hydro Quebec, with its large water power resources, did not have a major commitment to nuclear power, so the unit was never accorded the priority needed to bring it into full operation. It was permanently shut down in 1977, having run for a total of only a few weeks of full power operation.

AECL's other CANDU design is the 600 MWe unit. This design is similar in principle to the Ontario Hydro units and combines the best elements of the proven Ontario Hydro systems with some technological improvements. Superficially the major difference is that the 600 MWe units are designed as individual units with conventional pressure containment buildings as distinct from the integrated multi-unit Ontario Hydro plants which use shared vacuum containment systems.

Two of the 600 MWe units are in operation in Canada but only one, Point Lepreau in New Brunswick, is being used for continuous base load operation. With a load factor of 89 per cent in its first two years of commercial operation (February, 1983 to February, 1985) its performance has also been outstanding. This has special significance, not only because it demonstrates the dependability of the new design but because it shows that CANDU's good performance is not due only to the skill of a large utility like Ontario Hydro.

Early designs of CANDU reactors were sold to India and Pakistan. Although isolated from on-going Canadian input since the early 1970s for political reasons, India has developed its own CANDU capability and has eight reactors now in operation or under construction.

Later Canada sold reactors of the 600 MWe design to Argentina, Korea and Romania. The Embalsa unit in Argentina and the Wolsung unit in Korea have operated very well. Romania has five CANDU 600 MWe units under construction to come into operation in the late 1980s and early 1990s.

13.5 Technical Problems

Although the CANDU has been a very successful reactor system it has not been without problems.

The first serious problem the Canadian programme ran into was when, in 1952, the NRX reactor suffered a serious accident which resulted in melting of some aluminum structural components and even some uranium metal fuel. This led to serious damage and contamination to the reactor building which required an extensive amount of work, time and money to repair. However, representatives of industries and utilities, which later were to take part in the NPD study team, participated in the NRX rebuilding programme so the lessons learned from this accident led to great improvements to reactor design and safety system design. As a direct result of that very early experience all CANDU reactors have been designed with triplicated systems for measuring critical variables and triggering safety systems. A feature of this approach is that one system can be taken out of service for testing and maintenance while the reactor is at power; this is a contributor to the CANDU's excellent operating performance.

The most serious problems which have been experienced in CANDUs have been with pressure tubes. These are the zirconium alloy tubes which pass horizontally through the core of the reactor and contain the fuel bundles and primary coolant. The zirconium alloys were developed only in the mid–1950s so their behaviour was not as well known by the time they were first used in reactors as were many more conventional materials. However, three particular properties of the materials were recognized early on, though not to the extent to avoid some problems arising later. These properties are:

1. Zirconium alloys tend to 'creep', that is to slowly change dimensions under high temperature, high stress and radiation.
2. Zirconium cannot be welded to steel so mechanical joints must be used between pressure tubes and the connecting steel piping.
3. Zirconium can absorb hydrogen and, under certain circumstances, this can lead to formation of a brittle zirconium-hydride phase.

All three properties have led to problems in some CANDU reactors.

The prototype CANDUs used relatively thick pressure tubes, conservatively designed, and did not run into early problems. However, there is an incentive to keep tube mass low for neutron economy so the first of the Pickering reactors were stressed more highly. It was recognized that this could increase the creep rate of the tubes but the amounts were predicted and allowed for in the design. The diameter increases experienced have been close to predictions but length increases have been substantially higher. This has led to a need to make some structural adjustments to channel piping and supports but, even so, will likely lead to pressure tubes having to be replaced in some reactors before this would be required for other reasons. However, in spite of this phenomenon, tube lifetimes of over 20 years seem likely even in the early reactors. Later ones have modified designs capable of accommodating lifetimes extending to 40 years.

One change to pressure tube design was made quite early in the CANDU history, and that was the type of alloy used. In prototype reactors and in the first two units at Pickering an alloy called Zi-caloy–2 was utilized. All units after that were fitted with a ziconium–2.5 per cent niobium alloy (Zr–2.5 per cent Nb). This alloy was stronger so walls could be about 20 per cent thinner, a further benefit to neutron economy. The alloy was also thought to be less subject to creep, though at the higher stress level the difference has not proved to be important. The change to this alloy, however, did lead to a new and unexpected problem, that of cracking at rolled joints.

Since zirconium alloys cannot be welded to steel a mechanical 'rolled' joint was developed. This joint is similar in concept to that used in joining boiler tubes to end plates. In CANDU reactors grooves are machined on the inside of martensitic stainless steel 'end fittings' which extend out of each end of the reactor. The end of the zirconium alloy pressure tube is fitted inside this machined area and deformed outwards by rotating a system of rollers on a tapered mandrel inside the pressure tube until the tube is squeezed into the machined grooves.

Rolled joints have proven to be simple, reliable and low cost. To date there have been no leaks through the interface of a rolled joint in any of the CANDU power reactors. However, making rolled joints with Zr–2.5 per cent Nb proved to require greater care than with the weaker Zircaloy–2. The problem occurred in the pressure tube just beyond the end fitting. If the rolling tool was allowed to extend into this area it caused some deformation of the unsupported tube, producing very high stresses. Although such stresses apparently quickly relaxed in Zircaloy–2 tubes they were sustained in the stronger Zr–2.5 per cent Nb tubes. Over some months of operation they led to small cracks forming, complicated by hydrogen concentrating at regions of very high stress and precipitating brittle zirconium hydride. Several leaks occured in Pickering units 3 and 4 pressure tubes due to this phenomenon. After detailed inspection, 69 of the 780 tubes in these two reactors were replaced in shutdowns lasting over one year for each unit.

The first of the Bruce reactors had already had rolled joints made by the time the problem arose at Pickering. However, these joints were stress relieved *in situ*, largely eliminating the problem, though cracks in three out of the 480 tubes in this unit were found during 1982 and these tubes were also replaced. The joint rolling procedure was modified to eliminate the problem in subsequent units.

Although some people in the Canadian

nuclear programme wished, when the rolled joint problem was discovered, that they had stayed with Zircaloy–2, it turned out that the move to Zr–2.5 per cent Nb had been a fortunate choice. This became apparent after a potentially serious incident in 1983 when one of the Zircaloy–2 tubes in Pickering unit 2 suddenly ruptured with the failure extending over about half the tube length. This failure turned out to be due to a rather complex set of circumstances, but a key one was that Zircaloy–2 proved to have had a substantially higher rate of hydriding than had originally been expected, and much higher than for Zr–2.5 per cent Nb. The hydrogen level reached in the tube over its twelve years of operation may not have been a problem on its own, but it turned out that there was a second problem in some Pickering tubes.

All CANDU reactor pressure tubes are located inside slightly larger, concentrically located, calandria tubes. A controlled gas annulus between the two tubes insulates the pressure tube, which operates at the primary coolant temperature of nearly 300 °C, from the calandria tube which operates at the moderator temperature of less than 100 °C. The spacing between the two tubes was intended to be maintained by coiled wire spacers placed around the pressure tubes like garters. Two such 'garter spring' spacers were put around each pressure tube, roughly equally spaced along their length. The weight of fuel in the pressure tubes causes the tubes to deflect until they rest on the spacers, locking the spacers into position. However, it turned out that some of the spacers had moved out of position during reactor construction before the fuel had been loaded. With the spacers out of position the pressure tube was able to contact the calandria tube over a considerable length, cooling the outside of the pressure tube.

Hydrogen in solution in zirconium migrates to cold areas and can precipitate to form brittle zirconium hydride. In the ruptured tube, hydride blisters had formed along a considerable length and extended much of the way through the tube wall. When the hydride penetrated to the point of initiating a through wall crack, the rupture was able to run a considerable length before arresting, potentially causing a significant loss of coolant accident. However the calandria tube (a relatively thin walled tube of seam welded Zircaloy–2, but kept cool by its contact with the low temperature reactor moderator) was able to withstand the system pressure and temperature and so contained the primary coolant. Seals at each end of the insulating gas annulus failed, however, so that primary coolant was lost from the system. This alerted the operator who shutdown the reactor by normal means with no emergency shutdown systems being required. Leaking heavy water was recirculated from the sump back into the reactor to maintain cooling so emergency core cooling systems were not called into play. The reactor was cooled down without further incident or significant damage.

Following analysis of the incident Ontario Hydro decided to replace all the pressure tubes in Pickering 1 and 2 with new Zr–2.5 per cent Nb tubes. Although it had been expected that such a large scale retubing exercise would be required in at least the early CANDU reactors at some time, the utility had not yet developed systems and equipment to undertake this job. As a result the work is expected to take three or so years, being completed in 1987.

Although the economic consequences of the pressure tube replacement are significant, there are several positive aspects of the experience. First is the fact that there was no danger to the public arising from the tube rupture. The confidence in the inherent safety of the CANDU system was therefore enhanced by the experience. Secondly the feasibility of replacing the radioactive pressure tubes was demonstrated. Pressure tube replacement will be a key step in refurbishing CANDU reactors which might allow significant extensions to their useful lifetimes, currently assumed to be about 40 years.

A third positive factor is that the properties of zirconium alloy structural components now are very well understood and there is confidence that similar problems will not arise in future CANDU reactors. In this respect the change to Zr–2.5 per cent Nb pressure tubes was most fortunate. Newer CANDU reactors also utilize four garter spring spacers, and those under construction at the time of the tube rupture had the location of all spacers checked and adjusted where necessary before they started up.

In 1985 another pressure tube failure occurred, this one in Bruce unit number 2. This failure was apparently caused by a rare manufacturing fault in the tube. Because it occurred when the reactor was shut down and cold, but pressurized, the resulting 'water hammer' impact caused the calandria tube to also fail and some fuel to be dumped into the moderator. Although the pressure tube and calandria were replaced quickly, startup was delayed until the fuel was recovered. However, this failure was not of major economic consequence.

These events displayed another feature of the Canadian programme that has helped in many less dramatic instances to maintain the high performance of nuclear stations. That is the close co-operation between the utilities, the designers and the R and D support teams. Through the close co-operation of all those involved the cause of failures was determined remarkably quickly and the appropriate actions taken both for the reactors directly affected and for reactors then under construction. This is not to imply that no new problems with pressure tubes could arise, but certainly the level of confidence has been raised that the basic design is sound and that CANDU reactors can be refurbished when needed to have a long and economic lifetime. Utilities take great comfort in knowing that capital intensive generating stations can be refurbished and placed back into economic service after suffering even as major an event as a pressure tube failure.

13.6 Operating Reliability

CANDU reactors have had other equipment problems in addition to the pressure tube problems at Pickering A, described above. However, there have been few significant problems in the sense of impacts on plant operations. In particular boilers, which have been a cause of long outages in many reactors, have caused very little difficulty in CANDUs except for a systematic error in manufacture which damaged tube bundles for the steam generators for several reactors. The fault, which dented a number of tubes during stress relieving, was found before any of the affected boilers were put into service and the tube bundles were replaced. However, startup schedules for several reactors were delayed.

Another concern with heavy water reactors, leakage of heavy water, has not proved to be a serious problem. Some leakage inevitably occurs but most is recovered through room air driers and other means. A make up rate of about 1 per cent of inventory per year has proved to be a realistic target.

On-power refuelling is virtually a necessity for a natural uranium fuelled reactor as explained earlier. However, it has proved a strong virtue for CANDUs as well. It allows the avoidance of long annual refuelling outages which could be expected to take 5–10 per cent of each year's operation. Although the refuelling machines are quite sophisticated they have proved reliable over the years, accounting on average for only about 0.6 per cent outage time.

Fuel has proven remarkably reliable. Of the over 380 000 bundles irradiated in Ontario Hydro's commercial reactors to the end of 1984, only 356 or less than 0.1 per cent have become defective. Defects have led to negligible plant outage time; in recent years it has been zero. This low railure rate, along with the ability to remove defective fuel without shutting down, has contributed to the low background radiation levels in CANDU reactors. This, in turn, has resulted in low rates of exposure to maintenance staff.

A variety of other equipment and systems have caused outage time as well, with the turbines and generators being the leading source. However overall CANDU performance has been excellent, among the best in the world. Table 13.1 lists Ontario Hydro's detailed statistics for outages due to equipment problems for all of its commercial units since they commenced commercial operation.

Table 13.1 Ontario Hydro Experience
Equipment contribution to lifetime[1] incapability to December 31, 1984

Cause of incapability	Incapability (%)			
	Pickering NGS-A	Pickering NGS-B	Bruce NGS-A	Bruce NGS-B
On-power fuelling	0.6	0.1	0.6	0.0
Fuel	0.1	0.0	0.0	0.0
Heat transport pumps	0.2	0.0	0.5	0.0
Pressure tubes	9.1	0.0	1.2	0.0
Boilers (steam generators)	0.3	0.2	1.7	0.0
Turbine and generator	6.6	6.4	4.8	0.3
Instrumentation and control	0.6	1.7	1.3	0.0
Heat exchangers	1.1	4.9	0.1	0.0
Valves	0.4	0.8	0.2	0.2
Other	3.6	1.2	2.8	0.0
Number of units	4	2	4	1
Unit years	50.5	2.6	27.5	0.3
Capability factor	77.4%	84.7%	86.8%	99.5%
Incapability factor	22.6%	15.3%	13.2%	0.5%

[1]Lifetime means since in-service date of each unit

Table 13.2 shows lifetime load factors since first electricity production for these reactors. When weighted for unit capacity and years of service, the average lifetime capacity factor to the end of 1984 for the Ontario Hydro commercial nuclear system is over 80 per cent, even allowing for the outages for pressure tube replacement discussed above.

Operating performance is important for reasons of a utility's system reliability but it also impacts heavily on the economics of nuclear power. This is true for all nuclear stations due to their capital intensity. It is even more the case for the CANDU system which has the extra capital cost of the heavy water inventory. Also the low cost of the CANDU natural uranium fuel cycle results in a lower variable cost component and so further increases the ratio of fixed to variable costs. According to the data provided by Canada to a recent study by the OECD Nuclear Energy Agency, approximately 70 per cent of the levelised cost of generation for a CANDU is due to fixed costs. Most countries building LWRs, reporting in the same study, had fixed costs of about 46–58 per cent.

Table 13.2 Lifetime gross load factors for Ontario Hydro commercial reactors. First electricity to end 1984

Station	Unit	Load factor (%)
Pickering A	1	74
	2	75
	3	79
	4	82
Pickering B	5	75
	6	80
Bruce A	1	83
	2	77
	3	88
	4	88
Bruce B	6	99[1]

[1]Since in-service September 14, 1984

With a high fixed cost ratio it is important for the CANDU reactors to maintain a high load factor in order to spread the fixed costs over a lot of production. The use of on-power refuelling has facilitated this objective but the CANDU designers have considered the need for ease of maintenance in many other ways as well. One example is that in several CANDU stations it is necessary to only reduce power to about 70 per cent, rather than shutting down the unit, to replace a primary pump motor.

As a result of such design features plus on-power refuelling, Ontario Hydro has concluded that CANDUs should be able to maintain a lifetime load factor advantage of about 10 per cent over LWRs operated under identical circumstances.

Another way of looking at the advantage of high load factors is to note that if Ontario Hydro's 14 000 MWe of nuclear capacity in operation or under construction were expected to run at 69 per cent load factor instead of 79 per cent, then to deliver the same amount of electricity annually the utility would have to build an extra 2000 MWe capacity at a cost of some 4000 million Canadian dollars (based on the cost of the Bruce B station as given later). Clearly, the Canadian emphasis to design reactors capable of operating with high load factors has proven to be cost effective.

13.7 CANDU Economics

The cost of generating electricity from CANDU and other reactor types has been given elsewhere in this book using standard assumptions and levelised cost methodologies. However, it is also interesting to review the costs as they impact on Canadian utility customers. This is shown in Figures 13.4 and 13.5, as reported by Ontario Hydro.

Figure 13.4 shows, in dollars of the year, the costs of generation in the utility's two large nuclear stations which are in full operation and in the two most modern coal-fired plants owned by the utility. The cost units are thousanths of Canadian dollars per equivalent net kilowatt hour of electricity delivered to the grid, or m$ kWh−1. The 'equivalent' accounts for the fact that about 12 per cent of the power output of Bruce 'A' is taken off as steam for commercial process heat purposes.

Capital costs are charged by straight line depreciation over 40 years for the nuclear stations and 35 years for the coal units. That is, a fixed depreciation charge of 1/40 (or 1/35) is made each year in addition to the interest charged annually on the total unamortized balance. This procedure leads to a high capital charge in early years, falling later in plant life as the annual interest charges decrease. On the other hand, fuel and operating costs are assumed to rise due to inflation and market forces. For the nuclear

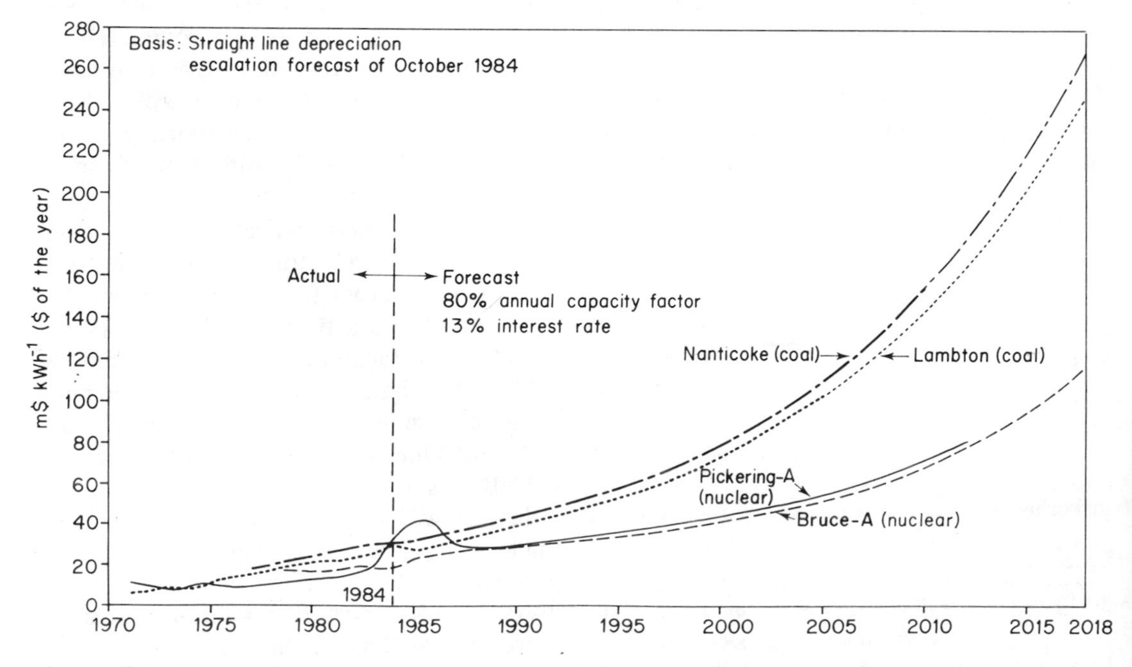

Figure 13.4 Total unit energy cost of Ontario Hydro operating nuclear and coal-fired generating stations

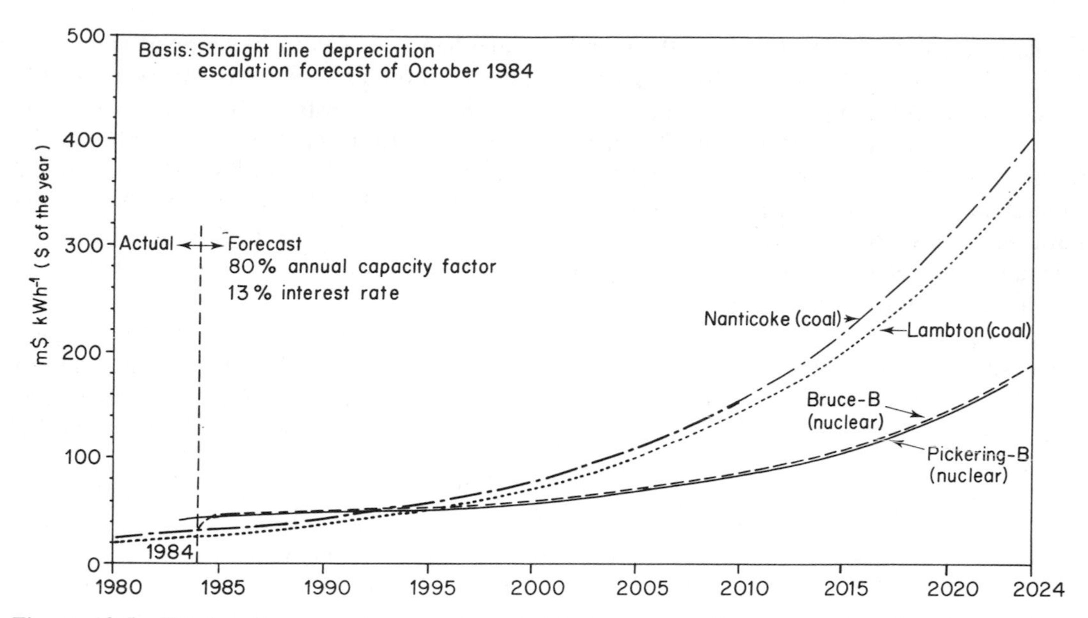

Figure 13.5 Total unit energy cost of Ontario Hydro coal-fired and recent nuclear generating stations

plants the falling interest payments nearly offset the rising variable costs in the early years of nuclear plants until interest payments become quite small and total costs begin to rise. For coal-fired plants the same effect can be seen but the significant rise begins much earlier since the variable costs (primarily the cost of coal) is more dominant.

The curves in Figure 13.4 are based on actual nuclear station data (costs, interest rates and load or capacity factors) up to 1984. For the coal-fired plants up to 1984 the same is true except for load factor which is assumed to be the same in each given year as for the nuclear units. That is, although the coal plants were actually used for intermediate load use they are shown as if base loaded, which they would have been if the nuclear plants had not been built.

Beyond 1984 the capacity factor for nuclear and coal-fired plants is assumed to be 80 per cent and the interest rate is taken as a constant nominal 13 per cent per year.

The anomalous rise in the cost of generation from the Pickering-A plant from 1983 through 1987 illustrates the impact of units 1 and 2 being out-of-service for retubing. However, even with this effect, the curves make it clear that the investment in these nuclear plants is expected to be very favourable over their lifetimes.

Figure 13.5 is similar to Figure 13.4 but shows costs for the newer Pickering B and Bruce B stations which had only some units in operation by the end of 1984. Ontario Hydro has no contemporary coal-fired stations to compare with these new nuclear stations. However, the comparison with the older Lambton and Nanticoke stations is still of interest. It shows that at an 80 per cent load factor electricity will be only slightly more expensive from the nuclear plants in the first decade or so of their lives, after which their advantage is expected to quickly become quite significant. If the comparison had been with contemporary coal-fired plants the trend would have been similar but the cross over would have been a few years earlier.

It is interesting to note that with the remarkably high load factor of 99.6 per cent achieved by the one Bruce B unit in operation in 1984 (as of September 14, 1984) its generating cost was almost as low as that from the older Nanticoke plant.

Tables 13.3, 13.4 and 13.5 show the

detailed calculation for the 1984 data and information about the stations. Note that the nuclear–coal cost comparison in Tables 13.3 and 13.4 are very relevant since the Pickering A and Lambton stations were built about the same time and the Bruce A and Nanticoke plants are also close to being contemporaries.

One other cost factor of particular relevance to Ontario Hydro is the avoided cost of coal. Since Ontario has no indigenous coal resources of significance it imports all of its coal, mostly from the United States but some from Western Canada (Western Canadian coal has to be shipped some 3000 km by train and boat so incurs very high transportation costs). The utility calculates that since 1971 its nuclear programme has allowed it to avoid over $6000 million (Cdn) of coal purchases, over $5000 million of this from the United States, representing a significant balance of payments advantage to Canada.

This section has concentrated on experience in Ontario. However brief mention should be made of experience in two other Canadian provinces with CANDU stations.

Table 13.3 Generating cost comparison for the Pickering A (nuclear) and Lambton (coal-fired) stations during 1984 (1984 Canadian dollars)

	Unit energy cost m$ kWhee^{-1}, (net)	
a. Comparison for operating units only (Pickering A units 1 and 2 did not operate in 1984 due to pressure tube replacement)		
	Pickering A Units 3 and 4 (2 unit-years)	Lambton[1] (4 unit-years)
Interest, depreciation and decommissioning	9.17	1.83
Operation, maintenance and administration	4.81	2.01
Fuelling	3.59	23.01
Heavy water upkeep	0.52	—
Total unit energy cost (TUEC)	18.09	26.85
b. Comparison based on all four units of Pickering A.		
	Pickering A (4 unit-years)	Lambton[2] (4 unit-years)
Total unit energy cost	32.16	30.67
Station data		
	Pickering A (Nuclear)	Lambton (Coal-fired)
Capacity (maximum continuous rating) MWe net	4 × 515	4 × 495
In service	1971–1973	1969–1970
Initial capital cost (M$ Canadian escalated)	746.5	257.0
Specific capital cost ($ kWe^{-1})	362.4	129.8
Economic lifetime (years)	40	35
Depreciation method	Straight line	Straight line
Interest rate (%)	12.4	12.4

[1]Assumes Lambton operated on base load and achieved the same net capacity factor (82.1 per cent) as experienced by Pickering units 3 and 4 in 1984. Lambton's actual capacity factor in 1984 was 54.6 per cent.
[2]Assumes capacity factor of 40.6 per cent for Lambton, as experienced by Pickering A.

Table 13.4 Generating cost comparison for the Bruce A (Nuclear) and Nanticoke (Coal-fired) stations during 1984 (1984 Canadian dollars)

	Unit energy cost m$ kWhee^{-1} (net)	
	Bruce A (4 unit-years)	Nanticoke[1] (8 unit-years)
Interest, depreciation and decommissioning	10.61	3.21
Operation, maintenance and administration	3.45	1.55
Fuelling	4.40	25.47
Heavy water upkeep	0.27	—
Total unit energy cost (TUEC)	18.73	30.23
Station data		
	Bruce A (Nuclear)	Nanticoke (Coal-fired)
Capacity (maximum continuous rating) MWe net	4 × 775[2]	8 × 497
In service	1977–1979	1973–1978
Initial capital cost (M$ Canadian escalated)	1 961.1	872.9
Specific capital cost ($ kWe^{-1})	632.6	219.5
Economic lifetime (years)	40	35
Depreciation method	Straight line	Straight line
Interest rate (%)	12.4	12.4

[1]Assumes Nanticoke also operated as a base loaded station with a net capacity factor of 93.7 per cent. Nanticoke's actual 1984 net capacity factor was 58.1 per cent.
[2]Includes 35 MWe steam per unit produced over and above the turbine–generator capabilities of the unit.

Table 13.5 Generating Costs in 1984 for Pickering B and Bruce B Stations (1984 Canadian dollars)

	Unit energy cost m$ kWhee^{-1} (net)	
	Pickering-B (1.9 unit-years)	Bruce-B (0.3 unit-years)
Interest, depreciation and decommissioning	33.93	24.98
Operation, maintenance and administration	4.16	3.69
Fuelling	4.52	4.13
Heavy water upkeep	0.51	0.11
Total unit energy cost (TUEC)	43.12	32.91
1984 net capacity factor (%)	81.5	99.6
Station data		
Capacity (maximum continuous rating) MWe net	4 × 516	4 × 795
In service	1983–1986[1]	1984–1987[1]
Initial capital cost (M$ Canadian escalated)	3812	6036
Specific capital cost ($ kWe^{-1})	1846.9	1898.1
Economic lifetime (years)	40	40
Depreciation method	Straight line	Straight line
Interest rate (%)	12.4	12.4

[1]Forecast in service date.
Notes: One unit at Pickering B started commercial operation in 1983 and a second in 1984. Two others were still under construction or commissioning.
One unit at Bruce B started commercial operation in 1984. Three others were still under construction or commissioning.

13.7.1 *Quebec*

Hydro Quebec built a 635 MWe CANDU unit which entered commercial operation in 1983. The plant has had a favourable availability factor of 74.4 per cent to 1984 but the province has a very large hydro generating system with significant excess capacity. As a result the nuclear unit, Gentilly–2, has not been used for baseload operation as marginal operating costs of nuclear plants are higher than for hydro.

13.7.2 *New Brunswick*

The New Brunswick Electric Power Commission also brought a 635 MWe CANDU into operation in 1983. The unit, Lepreau–1, has operated remarkably well since it started up, as discussed earlier in this chapter. However, as a single stand-alone unit, and the first nuclear plant built by the utility, its capital costs were significantly higher, per unit of capacity, than the Ontario Hydro stations and its generating costs in 1984 were about 53 mkWh^{-1}. Even so, this cost was competitive with oil-fired and coal-fired generation which provides much of the non-nuclear generation in the province, and on a life cycle basis the plant is expected to show a very significant advantage over the utility's alternatives.

13.8 Future CANDU Options

The Canadian's have studied a number of possible future developments of the CANDU system. These include:

1. 300 MWe CANDU. AECL has designed a 300 MWe unit which is based on standard CANDU components but with a number of design innovations aimed at reducing capital costs. Perhaps the most important of these is the use of a modular building layout which should allow construction time to be shortened significantly, to less than four years. AECL is now actively marketing this design.
2. Enriched fuel. Utilizing slightly enriched fuel (about 1.2 per cent ^{235}U) would increase fuel burnup and slightly reduce fuel costs. With ample uranium resources currently available and fuelling costs being quite low already there is not a strong incentive to develop this option. However, development tests are under way and a commercial scale demonstration may be undertaken during the next few years.
3. Organic cooled CANDU. An experimental organic cooled reactor, WR1, was operated for nearly 20 years at Canada's Whiteshell Nuclear Research Establishment, and an extensive power reactor design study based on this concept was undertaken in the 1970s. The advantage of the system, which still utilizes a low temperature heavy water moderator, is that it can operate with a primary coolant temperature of up to about 400 °C. This results in a significantly higher thermal efficiency and a potentially lower generating cost. The higher temperature could also make this design useful as a process steam producer for industrial projects such as extracting oil from the extensive tar sands formations in western Canada. However, there are no current plans to exploit this system.
4. Thorium fuel cycle. It is possible to use thorium fuel in a CANDU with few changes being necessary to the reactor. Fertile thorium is converted to fissile ^{233}U when it captures a neutron, in the same way that fertile ^{238}U is converted to fissile ^{239}Pu. Although such a cycle may require an external source of fissile material (plutonium or ^{235}U) to supplement the ^{233}U produced, it would require very little uranium supply, perhaps comparable to that of a fast breeder reactor. Since there is a great deal of thorium available in the crust of the earth (about three times as much as uranium), and since the thorium would be recycled, this fuel cycle could be at least as effective as fast breeder reactors in extending the world's fuel resources. However, although some research continues on this option, there

are no plans to develop it commercially at this time since large, low cost uranium reserves favour the current uranium burning fuel cycle.

In summary, there are encouraging options available to keep CANDU designs useful and competitive long into the future. However, there is not a strong economic incentive to move away from the proven heavy water cooled, natural uranium fuelled CANDU at this time. Thus attention will likely be focused on improved construction techniques to reduce construction time and capital cost, and on maintaining the advantage of standardized designs.

13.9 Future Prospects for CANDU Utilization

In spite of the technical and economic success of the CANDU system its future prospects are still somewhat uncertain. Ontario Hydro currently has excess generating capacity and so has no plans for building new units, nuclear or any other type, after completing the Darlington station in the early 1990s. While an upturn in economic growth could lead to a need for new capacity in the mid- to late-1990s there are no immediate prospects for another early order in the province.

Hydro Quebec still has a large amount of hydro power which it can develop in its massive James Bay project so it is not likely to build new nuclear capacity this century.

New Brunswick has been pursuing building of a second 635 MWe unit, Lepreau 2, primarily to provide electricity to the northeast of the United States. Although it has not yet been able to negotiate appropriate contracts for sale of the power, this project continues to be followed closely in both Canada and the United States.

Government of Canada energy studies indicate that of the order of 5000 MWe of additional nuclear capacity should be deployed in Canada by the year 2000, in addition to that already built or under construction, based on expected economic growth and interfuel substitution evaluations. However, electricity generation in Canada is the responsibility of the provinces which establish government owned utilities or license private utilities. To date no provinces other than those already discussed have started significant studies which could lead to early commitments of new nuclear plants. It seems, therefore, that while there may be a demand for further CANDU units in Canada beyond the turn of the century, the industry will have to try to survive until then on export sales and service work. Canada has been a key bidder in all recent international nuclear bidding competitions and there is optimism that these avenues will lead to enough work to maintain the CANDU option. However, it seems likely that the CANDU supply industry will continue to face lean times for some years.

ACKNOWLEDGEMENT The author acknowledges helpful comments of colleagues who reviewed the draft, and particularly the assistance of H. B. Merlin.

REFERENCES

OECD Nuclear Energy Agency (1986). *Projected costs of generating electricity in nuclear and coal-fired power stations for commissioning in 1995*, OECD/NEA.

Ontario Hydro (1985). *Economics of CANDU-PHW 1984*, NGD–10, Ontario Hydro.

Nuclear Power: Policy and Prospects
Edited by P. M. S. Jones

14

The United Kingdom

P.M.S. Jones
Department of Economics, Surrey University

14.1 Introduction

The United Kingdom was one of the first countries to realize the potential implications of nuclear fission for both military and civil use. It was a leader in its civil development and its application for large scale electricity generation. It still leads in the field of gas cooled reactor development and, in collaboration with its European partners, in the development of the liquid metal cooled fast reactor and fusion technology.

Despite the undoubted success of its Magnox programme and its early lead in reactor deployment it has been overtaken by many countries in terms of the contribution nuclear makes to its electricity supplies (see Table 19.1). Its main utility, the Central Electricity Generating Board, has plans to move away from the indigenous gas reactor technology and to adapt and adopt US designed pressurized water reactors; a move not favoured by the smaller South of Scotland Electricity Board.

Despite the unquestioned safety record of its civil nuclear facilities, with over 30 years of power reactor experience and over 20 years of fast reactor operation, there is still a vigorous public debate on the need for and desirability of the nuclear options.

In the paragraphs that follow the history of nuclear development in the UK is traced together with the factors that have influenced policy choices. Finally the present situation and future prospects for nuclear power will be reviewed.

14.2 The Early Years

When the potential of fission energy was beginning to be recognized following the discovery of the phenomenon by Hahn and Strassman in 1939, the UK government, at the instigation of the refugees Frisch and Peierls (1940), set up the Maud Committee to explore the feasibility of developing both power reactors and weapons. The committee's reports confirmed the potential of both and the Directorate of Tube Alloys was set up as a cover for nuclear development in the same year.

The difficulties faced in wartime Britain had led in August, 1940 to discussions with the United States which were followed by the launching of the Manhattan Project by President Roosevelt in December, 1941, initially without UK participation. The Quebec agreement of August, 1943, did establish Anglo-US collaboration and some 40 scientists from Britain joined the American teams. Others, including refugee French scientists from the UK joined civil power research teams in Canada. The intention was that the British scientists would return to the UK at the end of the war to continue their work in collaboration, it was anticipated, with their North American partners.

Following the end of the Pacific war and

the election of a socialist government in Britain, a decision was made to set up an Atomic Energy Research Establishment (AERE) at Harwell. The decision was announced in December 1945 and Dr (later Sir) John Cockcroft, the then director of the atomic energy research establishment in Canada, was appointed to take charge in the following January. AERE like the Armaments Research Department (under Dr W. Penney), and the fissile material production facilities (under Christopher Hinton) were all responsible to the Ministry of Supply and did not have a separate agency status like the organizations formed in the USA and France.

The Minister of Supply was charged under the Atomic Energy Act (November, 1946) with promoting the control and development of atomic energy, to produce use and dispose of it and to carry out research into matters connected therewith. The ministry brought to the task a strong organizational infrastructure with staff experienced in the management of production plants and research establishments.

The climate of the time was still one of wartime austerity with building materials and other resources in short supply. The Fuel and Power Advisory Council (1946) were concerned, *inter alia*, with the post-war restoration of adequate supplies of fuel to the population. Apart from imported oil products for transport, virtually all UK energy needs at the time, including electricity and gas production, were based on indigenous coal (Table 14.1).

In nuclear terms 1946 was a momentous year for the UK which saw not only the passing of the Atomic Energy Act and the creation of the atomic energy establishments at Harwell and Risley, but also of the uranium processing plant at Springfields and the Radiochemical Centre at Amersham. In the same year the US Atomic Energy Act (McMahon Act) ended temporarily US–UK collaboration in atomic energy. The importance to the nation of the coal industry led to its nationalization from January 1st 1947.

14.3 The Formative Years

The initial impetus behind the UK nuclear programme in the immediate post war years remained essentially military with a strong drive to develop a weapons capability. This had, however, a direct impact on the later directions of civil development.

The Harwell laboratories had one experimental reactor (graphite low energy experimental pile, GLEEP) critical in August, 1947, within 18 months of the start of work at the site. A second pile, the British Experimental Pile (BEPO), was operating less than a year later. Use was made of data from the Anglo–Canadian studies and the Harwell radiochemical laboratories were in operation by June 1949 to provide the basic knowledge needed to design and operate large scale production facilities.

A major requirement for the military programme was an assured supply of fissile material and the UK, faced with a choice of costly energy-intensive uranium enrichment or chemically separated plutonium, selected the latter. The production programme managed by Hinton from Risley, which

Table 14.1 Energy consumption in Great Britain 1946 (million tonnes coal equivalent)

	Consumption	Share (%)
Coal (excluding power station use)	147.2	75
Electricity	17.4	9
Gas	7.8	4
Oil (including non-energy use)	(23.4)	12
Total consumption	(195.8)	

Source: Mitchell, B., and Jones, H.G. (1971), *Second abstract of British historical statistics*, Cambridge University Press, Cambridge.

remained a design and contracting centre, necessitated the establishment of the uranium purification and fuel element fabrication plant at Springfields; the design and construction at Windscale of the air cooled graphite moderated reactors (piles), fuelled with natural uranium metal; and the design and construction, also at Windscale, of a fuel processing plant, to recover plutonium and unburnt uranium from the spent fuel. The piles were built in around three years to schedule and cost and went critical in October, 1950 and June, 1951. The Windscale reprocessing plant was operating by the spring of 1950. Britain was thus launched on a nuclear programme firmly based on gas cooled graphite moderated reactor technology, dictated in part by the fact that, unlike the USA, it had no direct access to enrichment.

During the late 1940s and early 1950s the UK was faced with severe economic problems and the world was shaken by the effects of the Cold War, including the Berlin blockade, and by the Korean War. Balance of payments problems were restricting UK economic growth and concern was mounting about the ability to meet energy needs from indigenous coal. Electricity demand was rising in the late 1940s by over 7 per cent p.a. (Table 14.2), and oil imports had risen from 2 million t to 17 million t in only five years. The Ridley Committee (1952) could only foresee continuing fuel shortages and balance of payments problems with little potential to expand coal supply, with a decline in coal exports and with a continuing rise in oil imports.

It was against this background that the decision was taken to design and construct dual purpose piles which would both produce plutonium for a weapons programme and supply electricity to the national network. These new reactors were to be carbon dioxide cooled graphite moderated (Magnox) plants. Meanwhile, a decision had also been taken to proceed with a uranium enrichment plant at Capenhurst, also with fissile material for weapons as its primary objective.

It is perhaps worthy of mention that, following a tradition dating back to the sixteenth century, the UK did not overlook the importance of the environmental implications of energy production. The Fuel and Power Advisory Council (1946) called for smoke abatement and the Radioactive Substances Act (1948) provided for control of such materials in the interests of health and safety.

The arrangement whereby atomic energy was the responsibility of the Ministry of Supply ended with the passage of the Atomic Energy Authority Act of 1954 which set up the UK Atomic Energy Authority (UKAEA) with a chairman and board who were empowered, *inter alia*, to produce, use and dispose of atomic energy and carry out research into any related matters. The chairman reported to the Lord President of the Council who was empowered to give the UKAEA directions after consultation with it, but who did not normally intervene in its conduct of affairs.

The authority was placed under an obligation to see that its activities should cause no hurt to any person or damage to any property and that all radioactive materials were

Table 14.2 The Growth of electricity demand 1946–1951 (10^9kWh = TWh)

	Domestic sector	Commercial	Industry	Other	Total
1946	11.66	3.89	17.63	1.61	34.80
1951	16.94	6.35	25.35	1.87	50.51
% Change 1951–1946	45.3	63.2	43.8	16.1	45.1
% Annual change	7.8	10.3	7.5	3.0	7.7

Source: Mitchell, B., and Jones, H.G. (1971), *Second abstract of British historical Statistics*, Cambridge University Press, Cambridge.

discharged only in accordance with authorization from the appropriate government departments.

At its formation the UKAEA had three groups dealing respectively with research (Harwell), production (Risley, Springfields, Windscale and Capenhurst) and Weapons (Aldermaston). The organization operated in a decentralized manner with a small London Office to serve the chairman and provide links with Whitehall.

14.4 Initial Growth and Euphoria

In August, 1953 work had begun on the new dual purpose reactors at Calder Hall and in April, 1954 the Dounreay Fast Reactor (DFR) had been sanctioned. The newly formed UKAEA was to carry through these projects and the extended reactor programme, which was to result in two additional piles at Calder Hall and four at the Chapel Cross site in Scotland.

Up to this time generation cost considerations had not been important. Such estimates as had been made of electricity production costs from Magnox reactors had shown them to be relatively expensive compared with coal. Thus in 1950 R.V. Moore had estimated that a 90 MW Magnox station would have an investment cost of £100 kW^{-1} and produce electricity at 0.6 d kWh (levelized cost with 4 per cent discount factor and 60 per cent load factor; see Chapter 21), after allowing a sizeable credit for the plutonium produced in the fuel. This figure is equivalent to 2.3 p kWh^{-1} at 1986 prices. Three years later B.L. Goodlet quoted a cost of £227 kW^{-1} for a smaller 35 MW reactor which with a load factor of 80 per cent, an 'optimistic' 20 year life, and an initial fuel cost of £15 000 t^{-1}, would produce electricity at 1 d kWh^{-1} excluding any plutonium credit (3 p kWh^{-1} at 1986 prices).

The first UK nuclear programme was launched in the white paper *A Programme for Nuclear Power* (Cmnd 9389) on the basis of an expectation that peak electricity demand would rise from the 20 GW of 1954 to 55–60 GW by 1975, and the concern that this increase would demand an extra 60 million t of coal p.a. which could not be met from indigenous supplies. Coal output had at the time risen to 220 million t p.a. but the industry was stretched and was facing manpower shortages. An initial move, viewed at the time as temporary, had already been made to equip some stations with dual oil–coal firing.

The 1955 nuclear programme was to construct twelve reactors in three groups of four, totalling 1.5–2 GW, over a ten year period. The first four plants were to be Magnox, but the later stations might be to a more advanced design. The total cost was expected to be in the region of £300 million. One contemporary estimate (Jukes, 1955) put the likely generation cost from these reactors at 0.76 d kWh^{-1} (2.9 p kWh^{-1}, at 1986 prices) made up of 0.36 d capital and 0.40 d fuel charges, excluding any allowance for the plutonium produced. The Central Electricity Authority (1955) also expected costs of around 0.7 d kWh^{-1} for 200 MW plant. These compared with the CEA's estimate of 0.52 d kWh^{-1} from coal plant of similar size. The gap between the fuels was comparable to the net value foreseen for the plutonium produced in the reactors which Jukes had argued to lie in the range of 0.17 d–0.33 d kWh^{-1}, depending on when and how it was used.

This initial programme was undertaken, therefore, with a view to averting a possible energy gap and in the hope that nuclear could roughly break even with coal. There was full recognition of the fact that the cost estimates were based on unproven designs (Calder Hall had yet to be commissioned) and that they could prove optimistic.

Nevertheless there was a conviction that a new industry, the nuclear industry, was about to be born and that there was enormous potential for both domestic and export business.

With government encouragement four design and construction consortia were set up in 1955. These were the Nuclear Energy Company (NEC: AEI-John Thomson), the Atomic Energy Group (AEG: GEC-Simon Carves), the Nuclear Power Plant Company

(NPPC: Parson Reyrolle-Head Wrightson-Macalpine), and the Atomic Power Project Group (APP: English Electric-Babcock and Wilcox-Taylor Woodrow). Each was centred on a large turbo-alternator manufacturer, a boilermaker and a civil engineering firm. A fifth consortium, Atomic Power Constructions Ltd (APC), was formed in 1957.

The year following the white paper saw the Suez crisis which cast doubt on the future security of oil supplies to Europe; growing balance of payment problems, with energy imports to the UK reaching £250 million p.a.; and the triumphant opening of Calder Hall in October. The Magnox reactor designs for the first programme were also showing the feasibility of scaling up the plants to sizes double those envisaged a year earlier, a fact that did not bode well for the infant consortia.

In this climate the trebling of the planned nuclear programme in 1957 (Cmnd 132), only two years after the initial white paper, was scarcely surprising. The new programme was for 5–6 GW total of nuclear plant to be constructed by 1965. Whilst the electricity from the early Magnox stations was then expected to be somewhat more expensive than that from contemporary coal stations, Hinton (1957) expressed his confidence that nuclear stations coming on stream in the early 1960s would be competitive with coal and 'appreciably cheaper' by the 1970s.

The policy of free competition saw all four (existing) industrial consortia tendering for the first three nuclear stations at Bradwell, Berkeley and Hunterston in 1956. The unsuccessful group, APP, was awarded the first of the succeeding batch in 1957 (Hinkley Point), whilst Trawsfynydd went to the fifth consortium in 1961.

The late 1950s was an extremely busy time for the UK nuclear industry. The Atomic Energy Authority was at the centre, providing technical support to the new consortia, producing fuel for reactors and reprocessing spent fuel at Windscale. It had designed and was constructing the reactors at Calder Hall and Chapel Cross. It was developing and constructing advanced prototype reactor designs such as the Dounreay Fast Reactor which went on power in 1959. Work was in progress on nuclear marine propulsion and on fusion. A flourishing business was being built up in radioisotope production for use in research and for applications in medicine, environmental monitoring and industry.

The successful launching of the thermal reactor programme and the confidence it was generating led to the decision to develop the advanced gas reactor which would operate at higher temperatures and higher thermal efficiency than the Magnox plants, using oxide fuels. Early plans to use low neutron absorbing berylium for fuel canning had to be abandoned on technical grounds and the switch to stainless steel, coupled with a desire to get lower capital costs through higher core power densities, led to a move to enriched fuels. This was now possible with the Capenhurst enrichment plant in operation. Construction of the prototype Windscale AGR was begun by the UKAEA in 1958. The AGR was expected to be competitive with light water reactors in both the UK and world markets because of its higher thermal efficiency and the fact that it could be refuelled, unlike the water reactors, on load.

14.5 Problems

The confidence of the late 1950s was shaken by the fire in one of the air-cooled plutonium producing piles at Windscale in 1957. The fire was extinguished with water within two days, but not before significant quantities of radioactive iodine had escaped through the 400 ft stack and filters and been deposited in the surrounding countryside. A ban was imposed on the consumption of locally produced milk for a few weeks and compensation was paid to affected farmers. The pile and its twin were closed down. The accident and its effects are further described in Chapters 6 and 7.

Although the piles with their once through air cooling differed from the civil carbon dioxide cooled designs, it was recognized that special care would be needed to ensure that all nuclear installations were independently vetted. Controls of siting, design and operation were set up under a Nuclear Installation (Licensing and Insurance) Act 1959 which also set up a Nuclear Safety Advisory Committee and the Ministry of Power's Inspectorate of Nuclear Installations to watch over the industry and to act as an independent inspection and licensing authority. Responsibility for the safe operation of installations still rested with the operators themselves.

The reactor export boom which had been hoped for had not materialized; nor did it subsequently. NPPC won a contract for the Latina Magnox station in Italy in 1957 and AEG for the Japanese Tokai-Mura plant in 1959. The developing world market opted for light water reactors (predominantly) based on US designs and the UK won no further turnkey contracts.

In the UK itself the economic break-even of nuclear had been put back by a combination of factors, including increased interest rates (from 4 per cent to 5–6 per cent); a reduction in their expected load factor from 80 per cent to 75 per cent; falling world oil and uranium prices; and reductions in costs of conventional generation through the use of larger plant and increases in thermal efficiency. The interest rate rise hit capital intensive nuclear plant more than coal and the drop in uranium prices made it less likely that plutonium would be used in thermal reactor fuel in the near term, hence reducing its perceived value. Against this the planning life of nuclear plant was increased from 15 to 20 years and higher thermal efficiencies were foreseen so that Jukes (1958) and the Central Electricity Generating Board (Duckworth and Jones, 1958) saw nuclear plant for operation in 1962–63 (0.63–0.70 d kWh^{-1}) being somewhat more costly than coal plant (0.54–0.63 d kWh^{-1}).

By 1960 it was recognized that the short term fuel supply arguments for the switch to nuclear (the gap) were no longer valid and that the reductions in conventional generation costs had weakened the economic case, despite the fulfilment of the predicted downward trend in nuclear costs. A further white paper in this year spread-out the remaining unordered plant programme, with a view to having a steady growth to about 5 GWe by 1968; the lower of the target figures set in the 1957 white paper. The production targets for the coal industry were also cut back by some 25–40 million t.

The declining short term market prospects, the lack of export orders and the increasing size of power plants could not provide the business needed to support the five consortia and the NPPC and NEC merged to form the Nuclear Power Group (NPG), whilst APC and AEG merged to form the United Power Company (UPC); both in 1960.

During this period the Winfrith site was set up and selected as the home for the European Nuclear Energy Agency's prototype high temperature reactor, Dragon, which was subsequently commissioned in 1964. The Culham site was also established for fusion research in 1960. The Dounreay Fast Reactor came to full power in 1963; the same year as the construction of the Steam Generating Heavy Water Reactor (SGHWR) at Winfrith was begun. This represented an alternative thermal reactor design which combined the advantages of the BWR and CANDU reactors and appeared to have better prospects than either. It had a better potential to load follow than the AGR and might be a good candidate for export markets since it could be built readily in a range of sizes (see Chapter 3).

The white paper on the second nuclear programme was published in 1964 (Cmnd 2335) some years before the completion of the first. The plan was to build a further 5 GW of capacity in the period 1970–75, probably based on the prototype Windscale AGR which had operated with high reliability for two years, although water cooled reactor designs were not ruled out.

The three remaining consortia tendered on the basis of AGR designs along with three PWRs and a BWR. After a lengthy technical and economic assessment by CEGB the decision was taken to proceed with the AGR, which was expected to be cheaper in UK conditions, and APC (which survived after the dissolution of UPC) was awarded the first contract for the ill-fated Dungeness B reactor.

Both the decision to proceed with AGR rather than switch to US designed PWRs or BWRs, and the selection of the APC tender were the subject of contemporary and retrospective criticism. Burn (1967, 1978), Henderson (1977) and Williams (1980) have commented at length. What can never be clear is whether the subsequent construction problems would have been averted by any other reactor choice.

There is little question that the AGR had considerable technical attractions over and greater development potential than the LWR. It was on the main line of UK development, had a highly satisfactory prototype in operation and was seen as a step on the way to the still more compact and efficient HTR. Against this the CEGB had some concern over the need to further demonstrate satisfactory graphite and fuel element performance.

In the event five AGR twin-reactor stations were ordered, Dungeness B, Hinkley Point B, Hunterston B, Heysham and Hartlepool. Of these TNPGs Hinkley and Hunterston were the stars. The others were plagued by a range of problems. Those at Dungeness led to the collapse of APC and the transfer of the project to another consortium; the British Nuclear Development Corporation (BNDC). All three stations suffered from technical problems leading to late design changes and experienced major difficulties with labour at the construction sites, so that extensive construction time and cost over-runs were experienced; a problem from which Hinkley B and Hunterston B had also suffered but to a lesser degree (see Table 14.3).

The rapid move from the 30 MWe Windscale prototype AGR to the full scale 600 MW reactors and the pursuit of three parallel designs had led to a number of unforeseen problems: Dungeness had pressure vessel liner and boiler problems, all had vibration problems caused by turbulence in the carbon

Table 14.3 UK Nuclear Power Stations

	Size (MW)	Date start of construction	Original planned date of commissioning	Date of commissioning	Owner
Magnox					
Calder Hall	200	1953	1956	1956	BNFL
Chapelcross	200	1955	1959	1958	BNFL
Berkeley	276	1957	1962	1962	CEGB
Bradwell	245	1957	1962	1962	CEGB
Dungeness A	410	1960	1964	1965	CEGB
Hinkley Point A	430	1958	1963	1965	CEGB
Hunterston A	340	1957	1962	1964	SSEB
Oldbury	416	1962	1966	1967	CEGB
Sizewell A	420	1961	1965	1966	CEGB
Trawsfynydd	390	1959	1964	1965	CEGB
Wylfa	840	1963	1969	1971	CEGB
AGR					
Dungeness B	1200	1966	1970	1986	CEGB
Hartlepool	1320	1968	1974	1986	CEGB
Heysham I	1320	1970	1976	1986	CEGB
Hinkley Point B	1000	1967	1972	1976	CEGB
Hunterston B		1967	1973	1976	SSEB
Heysham II		1980	1986	—	CEGB
Torness		1979	1987	—	SSEB
Prototypes					
Dounreay PFR	250	1966	1974	1974	AEA
Winfrith SGHWR	93			1968	AEA

dioxide coolant; Hartlepool and Heysham needed their boiler closures modified and additional shut down systems.

During the early 1970s there was considerable discussion and debate on future reactor policy with Parliamentary Select Committees and the official Vinter Committee looking at questions of comparative economics, safety and other relevant issues. The latter committee concluded that the choice of systems was finely balanced and that developments in the technologies and in the industrial infrastructure would be important considerations in choice. The government put off making a decision to 1974 to allow further assessment of the merits of AGR, SGHWR and the LWR.

By the end of the review period the CEGB was extremely optimistic about future prospects and Sir Arthur Hawkins, its chairman, had told a House of Commons Select Committee in December 1973 that he saw demand growth in the region of 5 per cent p.a. in the coming decade for which he wished to order nine 1200–1300 MWe PWR reactors in the period 1974–79, with a further nine from 1980–83. The CEGB saw the systems as having world backing and experience (though not at this size). TNPG, the South of Scotland Electricity Board and the UKAEA all favoured the SGHWR which they saw as a proven system with safety advantages over the US PWR.

The early 1970s had seen other developments of importance. The Radiological Protection Act (1970) had set up an independent National Radiological Protection Board to do research and advise government on the effects of radiation and on radiological protection standards. 1970 had seen the Anglo–German–Dutch tripartite agreement on the development and exploitation of the gas centrifuge enrichment process come into force. The old production group of the UKAEA was separated off in 1971 to form the new British Nuclear Fuels company which provided enrichment, fabrication and reprocessing services on a commercial basis. The Radiochemical Centre Ltd, later Amersham International, was similarly created to pursue the sales of radioisotopes and their compounds on a fully commercial basis. 1973 saw the transfer of the Weapons Group from the UKAEA to the Ministry of Defence. The Prototype Fast Reactor (PFR) at Dounreay was operating at low power in early 1974 and the decision was taken a year later to close down the highly successful DFR in 1976 when it had fulfilled its role.

By the time of the 1974 white paper (Cmnd 5695), none of the commercial AGRs had been commissioned although the WAGR had given an extremely satisfactory performance, achieving a load factor of 71 per cent in its first ten years of operation, or 10 per cent higher if deliberate shutdowns of the experimental plant were excluded. The industry had been consolidated with the formation in 1973 of the National Nuclear Corporation, a holding company, with the Nuclear Power Company Ltd (NPC), based on BNDC and TNPG, as its operating arm. Advice was coordinated by the Nuclear Power Advisory Board (NPAB) set up in 1972, on which all parts of the industry were represented. The consequence was the selection of the SGHWR for the next tranche of nuclear power. It had the substantial advantage, it was claimed, that it combined the best US, Canadian and UK technology, could be ordered quickly and was safe and reliable, although this implied no criticism of the safety of other reactor types. The envisaged programme was to order no more than 4 GW over a period of four years. Approval was given for a 4 x 660 MWe station at Sizewell and a 2 x 660 MWe station at Torness in February 1975.

The decision to proceed with the SGHWR had important resource consequences and it was decided on NPAB's advice that the UK could no longer afford to meet the 48 per cent share of the Dragon HTR development costs indefinitely. The European partners could not see their way to increasing their share and the project was closed down in November 1975 despite the fact that the system was still seen as having considerable future potential.

The fact that the UK had failed to achieve

the economic growth rates sought by successive governments in the 1960s and the major disinflationary impact of the 1973 Yom Kippur war, which had led to the formation of OPEC (Organization of Petroleum Exporting Countries) and a sharp rise in oil prices, combined to reduce expectations of energy demand growth. The stillborn SGHWR programme was initially put back and then cancelled in 1976 on the basis of advice from the UKAEA, because its launching costs in the smaller market foreseen were regarded as excessive and because no export market was in prospect.

By this time the Hinkley B and Hunterston B AGRs were ready to enter service, having overcome their teething problems, so that with the exception of the highly successful WAGR, experience was still extremely limited. At the same time some of the major reservations expressed earlier by Sir Alan Cottrell over the demonstration of PWR pressure vessel safety had been satisfied, at least partially, by a detailed technical review by Walter Marshall's Pressure Vessel Safety Working Party.

The then Secretary of State for Energy, Tony Benn, was far from happy with the conflicting advice he was receiving and set up a further review study, the Thermal Reactor Assessment, by the NPC, to assess the merits of the three thermal systems; AGR, PWR and SGHWR. The study, summarized in *Atom* (September, 1977), concluded that there was no longer a case for SGHWR, which would require further development and had no immediate export potential. It favoured the PWR on cost and export potential grounds but saw a need for a further AGR in the period before PWRs could be deployed. At the same time the Nuclear Installations Inspectorate reported that there were no fundamental safety objections to the selection of PWR.

In October, 1977 Benn set up an 'Energy Commission' with representatives from the trade unions, from the nuclear industry and with independent members. This commission favoured the AGR which both CEGB and SSEB now wanted to build. Other strong voices in the industry favoured the PWR and backed the CEGB's wish to switch to this for later reactors.

In the event the government opted (January, 1978) to authorize only the two twin-AGR stations (Heysham and Torness) and to consider the PWR later, when the industry (CEGB) had done the necessary design work. This left the industry in a state of indecision since it appeared that no firm preference had been given to either system. The situation was clarified by the new Conservative administration in 1980, which endorsed the AGR programme and gave the go ahead for the PWR subject to the outcome of a Public Inquiry—the Sizewell Inquiry.

Three other major events had occurred in the latter 1970s. The Department of Energy (1977) published an Energy Policy Review which saw no immediate prospects of an energy or electricity shortfall but anticipated that by the year 2000 demand for both would necessitate a major nuclear contribution. The Advisory Council on Energy R and D (ACORD) also concluded that 'in most perceived views of the future nuclear power will be essential' (Department of Energy, 1976).

The Royal Commission on Environmental Pollution (RCEP, 1976) in its sixth report had taken a somewhat ambivalent stance which praised the industry's safety performance but expressed concerns about the so called 'plutonium economy', about the impacts of terrorism, and about nuclear wastes. On the latter they agreed that they were confident acceptable solutions would be found but felt that 'massive' nuclear deployment should not take place until it had been 'demonstrated beyond reasonable doubt that at least one method exists' for their 'safe isolation for the indefinite future'. Over-all the report was more hostile to the development of the fast reactor (or plutonium fuelled thermal reactors) and reprocessing than to existing thermal reactors and in that it echoed the dominant arguments of the anti-nuclear groups of the time. It completely failed, in this author's opinion, to put the issues in a proper perspective (see Chapter 22 for further discussion on these topics).

The third event of importance was the Windscale Inquiry which examined in detail the proposals of BNFL to build a new 1200 t p.a. thermal oxide fuel reprocessing plant (THORP) at its Sellafield site. The Secretary of State for Energy had become convinced that open public involvement was the best way forward in cases where there was vociferous opposition to development projects. The Inquiry under Mr Justice Parker was set up to give opponents a full hearing. The company put together a strong case based on the expected market, including overseas markets, and the fuel conservation and radioactive waste management benefits. Assurance was given on the general safety of the development for the public and the workforce. The opposition, which included national as well as local groups, argued that the plant was not needed, that it added to radiation risks and that it would increase the likelihood of weapons proliferation or interference with civil liberties.

The inspector's handling of the complex and, up to that time, unique inquiry received wide praise from both sides until his relatively short report recommending approval appeared (*The Windscale Enquiry, 1978*). He concluded that the plant was needed, that it afforded no significant risk through radiation releases or terrorism and that it would lessen the pressure on other states to develop plutonium handling facilities of their own. The report was debated in parliament and approval given for the construction to proceed. The groups opposed to the plant were not pleased.

The other significant event of the late 1970s was not confined to the United Kingdom. President Carter of the United States became concerned that the rapid development of the reprocessing industry and the move to plutonium fuelled reactors might lead to weapons proliferation. He called for a cessation of these activities, which others saw as both legitimate and necessary, and a major international study, The International Fuel Cycle Evaluation, was set up under the auspices of the International Atomic Energy Agency in Vienna. The eight working groups reviewed all facets of nuclear technology, fuel demand and supply and their implications for proliferation, and concluded that there was no problem subject to adherence to the international safeguards regime. The policies being pursued in the UK were not therefore affected (INFCE, 1980).

14.6 The Sizewell Inquiry

Following the Conservative Government's agreement in 1980 the CEGB selected a Westinghouse design for its first PWR and prepared detailed designs and a supporting case in readiness for a public inquiry. This began in January 1983 when it was predicted that it would end within a year. Evidence was taken for over two years and resulted in 11 million sheets of paper with transcripts of evidence being as long as 24 copies of Tolstoy's *War and Peace*. Some 200 witnesses gave evidence; 40 of them from the CEGB itself.

The Board's case was essentially that it expected the PWR to produce electricity somewhat more cheaply than an AGR and significantly more cheaply than coal (or oil) fired plant. It foresaw a need for considerable new capacity in the 1990s and early 2000s, even in the absence of significant demand growth, since a lot of existing fossil and nuclear plant will have reached the end of their useful lives by then. This would justify the construction of a lead station to ensure that the technology was ready and proven even if it involved a small economic penalty. However, on the basis of a wide range of economic and energy demand scenarios, the PWR would be economic even in advance of the need for new or replacement capacity—indeed it would save so much fossil fuel that would otherwise need to be burnt in older inefficient stations, that it would save the CEGB money to build it. Additionally a move to the PWR would provide fuel diversity and provide the UK with an entry into world markets for this popular system.

The exhaustive and detailed case put by CEGB has been subject to detailed analysis and criticism by opponents. They have contested demand projections, all aspects of reactor and fuel cost estimates, fossil fuel price projections (on which the benefits are based), safety arguments and almost all other aspects. Their basic contention has been that the reactor is not needed; if it is needed then it is uneconomic; if it is economic then there are other cheaper systems (AGR, CANDU); if there are not then it is not safe or if it is then there are other safer systems; and in any case proliferation considerations or local impacts should rule out a nuclear choice altogether.

The exchanges in the framework of the inquiry have been of commendably high quality. The degree of detail can be judged by the fact that discussion of such an apparently abstruse topic as international exchange rate movements in the long term, with its associated theoretical modelling, occupied several days.

The South of Scotland Electricity Board, well pleased with its gas cooled reactors, entered the lists and disputed the English Board's claims on comparative reactor economics claiming that a new stretched AGR would be cheaper than the CEGB Sizewell B PWR (Pexton, 1986). It is evident that the two systems have closely similar costs at the present time and both offer considerable savings compared with fossil fuel. The PWR, being a lead station, would appear to have greater potential for future cost reductions however, provided no unanticipated problems arise.

The inquiry system as practised in the UK does provide a mechanism whereby any and all can come and listen or put a point of view. It helps to ensure that all issues are fully aired, but it is not clear that an extended inquiry does much to influence wider public understanding, although the fact that it has been thorough may ensure widespread acceptance of the inspector's findings.

The inspector found in favour of proceeding on the grounds that the reactor was acceptably safe and had a high probability of producing electricity significantly more cheaply than the other options. Following parliamentary debate the Government approved the scheme in March 1987. Construction is proceeding with formal commissioning planned for the spring of 1994.

14.7 The Present Situation

Nuclear power is a well established technology in the UK despite the fact that much of the public debate continues to discuss it as though it was discovered yesterday. Thermal power reactors have been in operation for over 30 years and over 400 reactor years of operating experience have been gained—far more than in any other country including France. The original Magnox stations owned by BNF plc at Calder Hall and Chapel Cross are still operating without problems, as are the first purely civil power stations at Berkeley and Bradwell which were connected to the grid in 1962.

Thc unfortunate use of mild steel bolts to support the core led to downrating (lower power and lower temperature operation) of the Magnox reactors early in their lives to limit corrosion. They have, however, shown themselves capable of giving consistently high availability factors and Hunterston A is among the world leaders with a cumulative lifetime load factor of 84 per cent. Indeed, the experience with Magnox stations has been so good that the Central Electricity Generating Board and the South of Scotland Electricity Board have been confident that they will achieve operational lives of at least 30 years without any major refurbishment. This is significantly greater than the 20 years that was conservatively assumed for amortization purposes when the first civil programme was initiated.

The simplicity of the system is such that it is still seen as a serious contender for situations where smaller highly reliable reactors are required and UK design companies are prepared to tender in appropriate circumstances.

The nuclear share of electricity generated in the UK has risen to over 20 per cent (Table 14.4), equivalent to over 7 per cent of the total UK energy consumption on a primary energy basis. This share is now less than that in most of the UK's major industrial competi-

Table 14.4 Growth of Nuclear Share of Electricity in the UK

Year	Percentage share
1956	0.1
1960	1.6
1965	7.7
1970	9.6
1975	10.5
1980	12.7
1982	16.0
1984	19.5
1985	20.1

Source: Annual Digest of Statistics

tors including France, Germany and Japan; is less than that in Sweden, Finland, Bulgaria and Hungary; and less than that in Japan, Korea and Taiwan. It is however significantly greater than that in the USA, the USSR and Canada. (See Table 19.1.) The current average for the EEC is 35 per cent. The share in the UK will rise further as the existing AGRs at Dungeness, Hartlepool and Heysham I are brought to full power and when the four new reactors at Torness and Heysham II are completed in 1988. By this date some 25 per cent of the UK's electricity will be coming from nuclear sources. Even at this level the nuclear share will be well below the economic optimum which with current estimates of cost and load-duration patterns would require well over 50 per cent of capacity to be nuclear with an even higher share of electricity supply.

The difficulties experienced with the first AGR programme have already been referred to and are illustrated by Figure 14.1. The multiplicity of designs and construction companies, the late design changes and the site labour problems all contributed to lengthy delays and cost over-runs, with only Hinkley B and Hunterston B even moderately near the original schedules. It is a matter for considerable satisfaction that the lessons of the 1970s seem to have been learned so that construction of the latest AGRs has proceeded both to time and cost. The costs of these stations, even so, are still considerably higher than the costs projected for AGRs at the outset of the original programme, due to additional features built in to simplify maintenance in the light of operating experience of the existing plant, and to additional safety features. This real terms cost escalation parallels that which has occurred in all countries, including the United States and France (Moynet, 1985). Some late problems arising from minor design changes have delayed the final commissioning of both reactors.

The performance of Hinkley Pt. B and

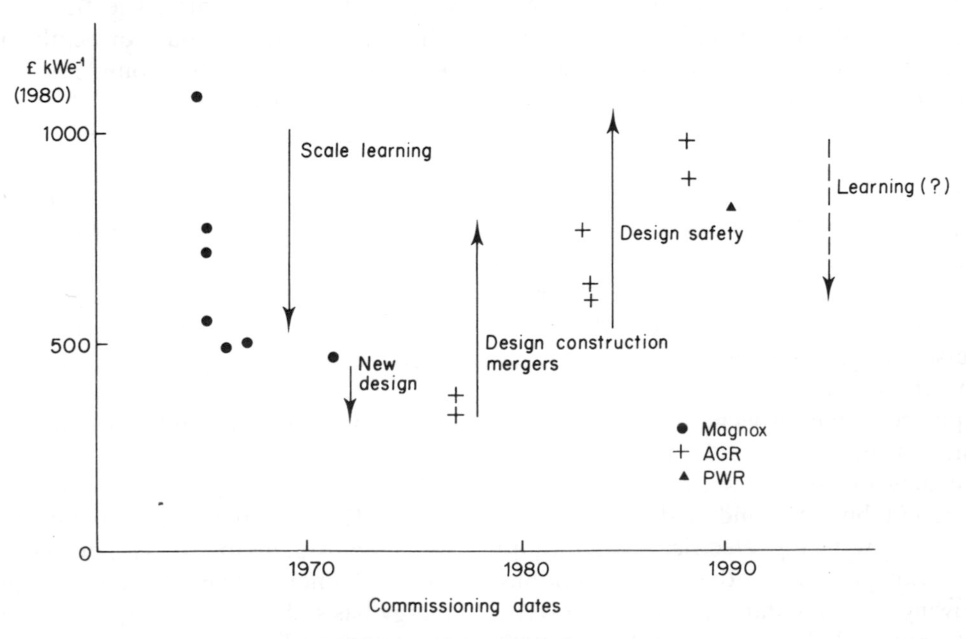

Figure 14.1 Capital costs

Table 14.5 AGR Average Load Factors 1982–1985

Period	Hinkley B 2 × 660 MWe gross %	Hunterston B 2 × 660 MWe gross %
1.8.82–31.7.83	64.2	67.0
1.8.83–31.7.84	72.9	72.6
1.8.84–31.7.85	73.8	76.9

Hunterston B stations has been coming up to expectation with load factors in excess of 70 per cent (Table 14.5). Refuelling is being carried out on part load (30 per cent), a figure which should increase in the future, and the UK's confidence in the AGR concept is beginning to bear fruit.

The Magnox stations cost less to run than either coal or oil stations even when the back-end fuel cycle costs for reprocessing and waste management and disposal are included. The latest published figures (Central Electricity Generating Board, 1985) showed the running costs (fuel plus operations) in March 1984 to be: Magnox, 1.6 p kWh^{-1}; coal, 2.2 p kWh^{-1}; and oil, 3.7 p kWh^{-1}. The availability of the Magnox stations therefore saves considerable sums in fuel bills every year and this was of particular importance during the 1984–85 coal strike when a great deal of oil was being used. During 1984 nuclear output averaged 3.9 terrawatt hours per month which, during the first nine months of the strike would have needed an extra 13 million t of coal or, had it been practicable, 7 million t of oil to replace it. Hypothetically this would have added a further £1 billion to the nation's fuel bills. (At present in normal conditions 74 per cent of UK electricity comes from coal, 6 per cent from oil and 20 per cent from nuclear and hydro. In 1984–85 the oil proportion rose to about 33 per cent.)

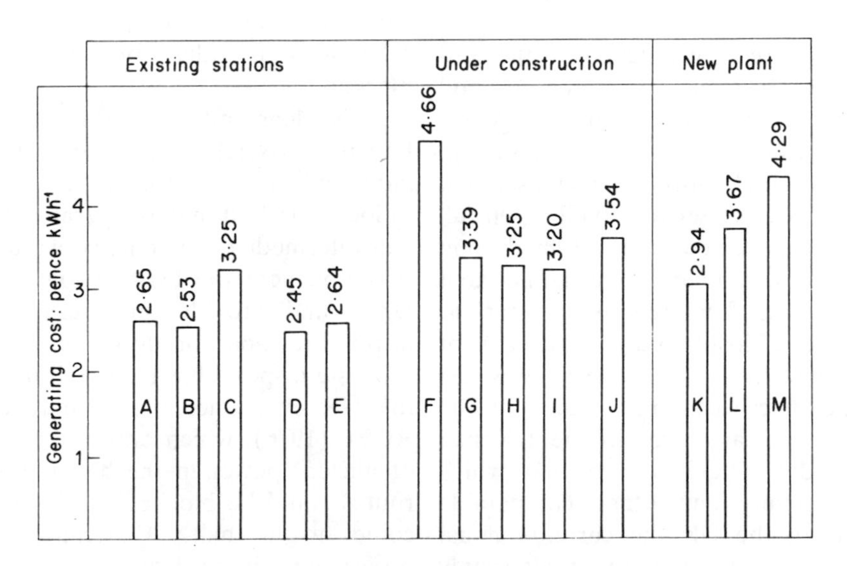

Figure 14.2 CEGB analysis of UK generating costs (March, 1984 money). Estimated lifetime costs at 5 per cent discount rate.
A: Average Magnox B: Old coal stations C: Old oil stations D: Hinkley Point E: Drax A (coal) F: Dungeness B (AGR) G: Hartlepool (AGR) H: Heysham I (AGR) I: Heysham II (AGR) J: Drax B (coal) K: Sizewell B (PWR) L: New AGR M: New Coal

The fuel savings from Magnox will, over the life of the stations, even ignoring the effect of the coal strike, just about provide a 5 per cent return on the initial nuclear investment (see Figure 14.2) on the basis of CEGB expectations for the running costs for contemporary fossil and nuclear plant over the rest of their economic lives. This is a very satisfactory achievement particularly bearing in mind the cost expectations when the original programme was launched which, omitting any plutonium credit, put nuclear electricity costs at some 33 per cent above those from coal fired plant. The projected costs (section 14.4) were also remarkably close to those now experienced (Figure 14.2).

The AGR stations have still lower fuel cycle costs and the savings they achieve will yield an even greater return on their (higher) initial capital costs than Magnox, with the possible exception of Dungeness B where the inordinately long construction period has added greatly to its overall cost in real terms. (Even so the capital costs for Dungeness B are sunk costs and it should pay handsomely from here on to operate it and save fossil fuels.)

The savings are very welcome to the utilities and their customers and provide good justification for the move to nuclear generation. Furthermore, there are more savings to come as the fuel burn-up is increased and the reactors move closer to refuelling on full load. However, even before the 1986 drop in fossil fuel prices, it could not be claimed that the direct saving from the use of nuclear power in place of fossil fired generating plant (expressed in constant money terms) was vast or that the macroeconomic impact of the adoption of nuclear power in the UK had been markedly different from that which would accompany any other comparable investment. In the UK the current annual rate of investment in electricity supply is only about 0.8 per cent of GDP or 4 per cent of gross domestic fixed capital formation, and electricity accounts, on average, for about 4 per cent of industrial costs, so that its effects at this stage are bound to be seen as marginal. With a higher proportion of nuclear electricity the impact on electricity intensive industries such as steel and chemicals could become significant, however, and this could determine whether future production plants are sited in the UK or elsewhere, to benefit from the availability of cheap energy, particularly if the savings are as great as those anticipated for Sizewell B and its successors (Figure 14.2).

A full list of past and present UK power reactors is given at Table 14.3.

In addition to the electricity supply industry the UK possesses design and construction capability for gas cooled and water reactors centred on the National Nuclear Corporation (NNC), and supported by the turbine, boiler making and specialized engineering industries. British Nuclear Fuels plc provide a full range of fuel cycle services from conversion and enrichment to fabrication and reprocessing. Construction of its new 1200 t p.a. capacity oxide fuel reprocessing plant is well under way and it has already sold a large part of the plant's throughput over its first decade of operation to domestic and overseas customers. The centrifuge enrichment process operated by URENCO is fully competitive in world markets.

A Nuclear Industries Radioactive Waste Executive (NIREX) has been set up by the industry, with government encouragement, to locate and ultimately operate sites for low and intermediate radioactive waste disposal, to supplement the existing low level site at Drigg and provide an alternative to sea disposal for appropriate wastes.

Waste disposal has been a difficult issue in the UK ever since the sixth report of the RCEP (1976), which encouraged opponents of nuclear power in the belief that if waste routes could be blocked the industry would come to a halt. Experimental drilling programmes in hard rock were dropped, the consideration of the technically attractive disused anhydrite mines at Billingham abandoned, the sea-dumping of wastes terminated (at least temporarily), and environmentally sound proposals for disposal of encapsulated intermediate level wastes in lined screened trenches given up, in the face of determined opposition. These difficulties have arisen despite the firm support given to all the proposals by the independent Radioactive Waste Management Advisory Committee set up by government in response to the RCEP report. Following the realisation that the costs of new trends, burial sites for low level waste will be no cheaper than the costs of adding this waste to intermediate level waste for cheap burial, the search for shallow sites in the U.K. has been abandoned. The Drigg site is still operational.

On the research side in addition to work on thermal reactors and nuclear safety and waste management, the UK has major collaborative programmes on both fast reactors and fusion. On the latter the Joint European Torus at the UKAEA's Culham site is now operating with deuterium and reaching temperatures high enough for fusion reactions to occur. The experiments are scheduled to move to deuterium–tritium mixtures towards the end of the decade.

The fast reactor programme envisages a series of demonstration reactors within Europe leading to the option of commercial deployment in the post-2000 period when economic or strategic considerations make this the preferred course. In the UK it is already believed that series ordered fast reactors to the latest design with commercial scale fuel plant could compete with PWRs even if their fuel prices remain at current levels. The fast reactor fuel cycle costs are not sensitive to uranium price escalation so any rises not offset by savings in fabrication or enrichment costs will work to the fast reactor's advantage.

Outline planning permission has been sought for a European Demonstration Reprocessing Plant (EDRP) for fast reactor fuels at Dounreay, where the UK's existing fast reactor and its fuel plant are concentrated. This plant is intended to serve the European programme of full scale demonstration reactors and to cut costs of all collaborating countries through the benefits of scale achieved.

14.8 The Future

14.8.1 *Reactor Choice*

Comment on the future is inevitably a mixture of fact and speculation. From the UK's standpoint the most important fact is that electricity demand is growing, and growing at rates higher than those forecast as being most likely by the CEGB and the Department of Energy at the time they submitted evidence to the Sizewell Inquiry in 1983–84. At that time the Department suggested that some 13–28 GWe of nuclear plant might need to be in place by 2000, calling for additions of between 5 GWe and 20 GWe of new plant. Between 2000 and 2010 they suggested a further 19–43 GWe might be needed, or more if strictly economic criteria were followed (Table 14.6). The Cambridge Energy Research Group (Eden and Evans, 1984) has suggested from 4–19

Table 14.6 UK Public Electricity Supply
Possible commissioning of nuclear plant (units GWe net)

Year	1980	1990	Capacity: Constrained build 2000	Constrained build 2010	Economic build 2010
Case X			28.0	65.5	106.2
YU	5.8	10.8	22.6	48.5	75.0
YL			22.6	47.2	66.5
Z			17.6	31.7	44.8
Case A			28.5	65.9	89.1
BU	5.8	10.8	22.6	48.5	61.3
BL			17.6	34.6	52.9
C			12.7	26.4	31.5

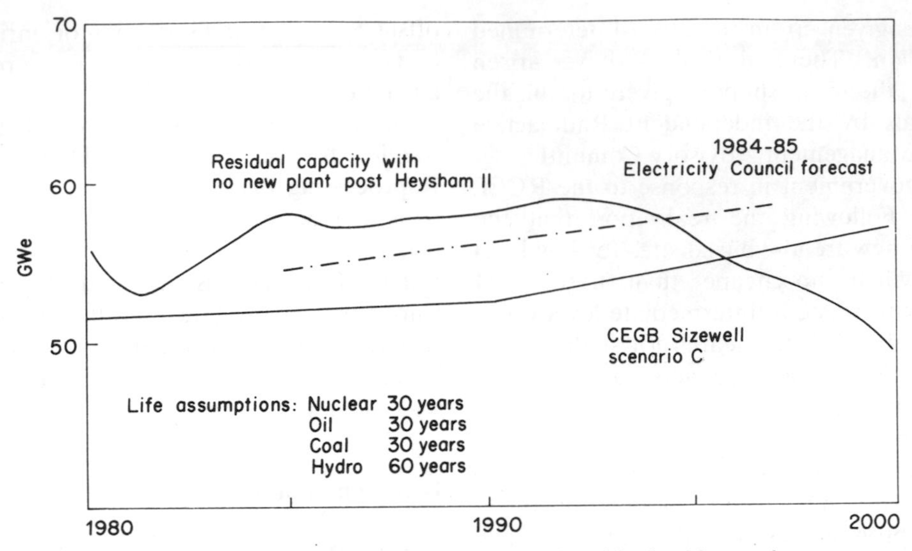

Figure 14.3 UK demand projections and residual grid capacity

GWe of nuclear in addition to Sizewell might be needed by 2000.

Figure 14.3 shows the exisiting capacity on the CEGB grid assuming 30 year lives for Magnox, AGR and oil plant and 40 years for coal plant. The dashed curves show the capacity needed on the basis of the CEGB 'Scenario C', presented at Sizewell, and the more recent Electricity Council projections. The demand incorporates 25 per cent plant margin to allow for uncertainties in winter peak demand, exceptional weather conditions and unplanned plant outages. Without it extensive load shedding could be required with damaging impacts to industry and the domestic consumer. (The CEGB use a 28 per cent planning margin which includes an additional allowance for forecasting errors.)

On this basis new capacity will be needed by CEGB before 1994, or earlier if, for example, a plant had to be shut down sooner. In all some 8–10 Gwe may well be needed by 2000 and, even with no demand growth, up to 20 Gwe could be needed to replace existing capacity by 2005. Some further plant life extension may be possible but not to the extent that would justify deferring Sizewell B or successor stations, whether they be nuclear or coal fuelled.

It was noted in section 14.7 that the earliest date for commissioning Sizewell B is now seen as 1994 so that, contrary to widespread belief, there is no longer a vast surplus of generating capacity in the UK. If Sizewell did not proceed there is virtually no chance of planning and building an alternative in the time. Furthermore there is now an urgent need to order additional plant of some kind to meet the threatened shortfall. If such plant, nuclear or coal, were to be subjected to further full scale inquiries like Sizewell a supply crisis in the mid to late 1900s would become inevitable. Government will have to take a positive line and ensure that decisions are reached timeously. This need not be too difficult since the safety case and economic issues have been examined in great depth and there would be no value in reviewing such aspects again. Future enquiries are therefore likely to be limited to local issues and correspondingly shorter. Potential sites for future nuclear stations have been identified at the exisiting CEGB nuclear sites of Wylfa, Trawsfynydd, Hinkley Point, Dungeness and Sizewell, together with Druridge Bay in Northumberland and Winfrith in Dorset. The latter is already the home of the AEA's SGHWR reactor.

The delay in starting on Sizewell has meant

that even the limited construction experience originally envisaged before ordering further similar plant is no longer feasible. The UK therefore finds itself, once again, in the position of having to order several reactors of a type they have not yet built and demonstrated (although very similar plant will have been operating elsewhere) or reverting to further tranches of AGRs or expensive coal plant. The AGR option would find considerable support but would be at the expense of gaining the replication benefits offered by a series of PWRs.

The decision on the type of future plant to be ordered could be influenced strongly by the outcome of the next general election, which is likely to be the first at which the political parties in the UK have adopted markedly different positions. In the aftermath of the Chernobyl disaster the Labour party has indicated that it might have a nuclear construction moratorium and order only coal plant. The Conservatives, on the other hand, in common with the CEGB, see a continuing programme of nuclear with some replacement coal plant, backed by a vigorous campaign to encourage overall energy conservation. The Alliance parties have yet to agree a policy although the Liberal leader has spoken of continued civil nuclear R and D with a suggestion that renewable energy sources might be looked to for a larger contribution.

The impending capacity shortfall and the need to maintain a dynamic and viable power plant industry to meet the large construction programmes around the turn of the century, will make any radical change of policy difficult. As Torness and Heysham B near completion, the workload on the UK plant construction industry is already declining and pressure from this quarter will grow.

Despite its US design origins and the purchase of the pressure vessel from Framatome, over 90 per cent of the cost of a reactor like the Sizewell PWR is for UK procurement, with work provided for some 10 000 people for a period of seven years in industry and on the construction site, peaking at 15 000. The off-site work is widely distributed through some 24 factories in England, Scotland and Northern Ireland. A modest programme of reactors, be they PWR or AGR, would maintain an industry of some 150 000 people in the UK. An alternative coal based programme would provide fewer construction and manufacturing jobs and it is quite likely that a sizeable proportion of the fuel itself would be imported rather than indigenous, particularly if the new power stations were built on Southern coastal sites to be near the major demand centres, as CEGB has suggested might be the case.

14.8.2 *Scale of the Industry*

The last paragraph has outlined the employment implications of a modest programme of nuclear deployment to the year 2000. In the longer term, even if there were no demand growth, the UK would need to maintain a working stock of some 60 GWe of generating plant. This corresponds to a continuous construction programme averaging 1.5–2 GWe of replacement capacity per annum. In the post–2000 period typical nuclear stations could have a 1.5 GWe capacity so that an ordering rate of one to two new nuclear stations per year would be appropriate, with higher rates if electricity demand grows.

At current costs this would be an investment rate of some £2 billion to £3 billion p.a. which is about 15 per cent of current levels of gross domestic fixed capital formation (GDFC). This compares with some £7 billion per annum currently spent on oil, gas and water supply and is a lower percentage of both GDFC and GDP than the UK was spending on power stations in the 1960s. The total size of the nuclear industry, including the fuel cycle operations of BNF plc would be likely to be in the region of £3 billion to £5 billion p.a. if the UK were firmly committed to nuclear by that time. Since nuclear is, and is expected to remain, the cheapest large scale source of base load electricity, this level of expenditure is less than would be required for alternatives whether based on fossil or renewable sources.

14.8.3 *Nuclear Fuel Supply and Waste Management*

The vigorous and comprehensive fuel cycle services industry in the UK would continue to flourish well into the next century even on the basis of contracted work and existing and planned UK reactors alone. On completion, the THORP reprocessing plant will meet all UK thermal oxide fuel reprocessing needs and provide for reprocessing overseas fuels until the turn of the century. There are no plans at present for reprocessing any UK PWR fuel and no decision on this would be needed until beyond 2000. Further tranches of thermal fuel reprocessing capacity may be needed in the longer term to meet a growing UK nuclear programme, depending on the reactor types selected and the spent fuel management policies being pursued.

The economic centrifuge enrichment process has further development potential and can be expected to remain competitive for a long time to come, even if the laser separation techniques can be brought to a commercial stage.

There are no firm plans in place in the UK for the disposal of intermediate or high level radioactive wastes. Existing and future high level wastes at Sellafield will be vitrified and stored for many decades in surface stores before deep geological disposal. Sites for such disposal will not be needed therefore until well into the next century. The intermediate waste problem is more acute but a policy decision has been taken not to put even those of the wastes which are technically suitable into trenches, and identification of sites for deep repositories is not now foreseen for some time. In the meantime such wastes are stored at existing nuclear sites in suitable surface stores. When the low level waste site at Drigg is filled it is now expected that further LLW will be buried in deep repositories with the intermediate level wastes.

14.8.4 *New Systems*

The fast reactor has been seen as the ultimate fission reactor ever since the start of the UK civil programme and the UK already has some 26 reactor years of operating experience with DFR and PFR. The general nature and expectations of the UK research and development programme were outlined in section 14.7. The slowdown in world economic and energy demand growth in the 1970s, and the slower penetration of nuclear plant into the electricity generation market than had been generally forecast, has led to reduced pressures on uranium supply. This in turn has led to reconsideration of the timing of fast reactor introduction, which was generally seen as a counter to rising uranium prices as known resources were depleted.

In the UK rising uranium prices are not now seen as a necessary precursor (although they would add considerably to the incentives) but, with due caution, the R and D objective is to develop the FR option to the stage where it can be commercially deployed, as and when this becomes attractive economically or strategically. This timing will depend not only on thermal reactor fuel prices but also on reactor choice. The second decade of the next century is the most likely date for large scale introduction but it could be either earlier or later.

Good progress is also being made on fusion through the JET project. Those associated with the programme are optimistic about the prospects of demonstrating the technical feasibility of controlled fusion in relatively few years. However, there is still a long way to go before existing fusion reactor concepts could be brought to a commercial stage, so that even if the costs can be brought to levels comparable to those of fission systems, it is not considered likely that power stations could be built before the third decade of the next century.

At various times in the past the UK has chosen consciously not to pursue civil marine propulsion, high temperature reactors and small reactors for civil use. Research on all these topics has been dropped when the market potential has been assessed to be too limited to merit its continuation. However, these and other systems and fuel cycles are kept under review so that if market or other circumstances changed appropriate action could be taken.

14.9 *Conclusion*

Had it not been for Chernobyl, the UK nuclear industry could reasonably have expected the final years of the twentieth century to be a period of comparative stability with steadier ordering programmes and modest growth as the nuclear share of capacity grew to nearer the economic optimum, followed by a boom of replacement orders post–2000.

Chernobyl, while not altering any of the facts relating to the safety of UK systems or the UK energy situation, has had political impacts that, for the time being at least, make the future far less certain. Events elsewhere could as easily do so again in a more positive fashion. The renewed concern arising from US studies of the greenhouse effect, for example, could bring greater realization of the folly of over-reliance on fossil fuels.

The decisions of the next few years will be crucial since they will determine not only whether adequate generation capacity will be in place in the UK but also whether the nuclear industry can remain viable and grow to meet the even greater demand of the post 2000 period.

REFERENCES

Burn, D. (1967). *The Political Economy of Nuclear Energy*, Institute of Economic Affairs, London.

Burn, D. (1978). *Nuclear Power and the Energy Crisis*, Macmillan, London.

Central Electricity Authority (1955). Cited by Pask, V. A. (1956) in 'The Place of Nuclear Energy in UK Power Development', *British Nuclear Energy Conference Journal*, **1956**, 13–23.

Central Electricity Generating Board (1985). *Analysis of Generation Costs*, CEGB, London.

Department of Energy (1976). *Energy R&D in the UK*, Energy Paper No. 11, D.En., London.

Department of Energy (1977a). *Energy Policy Review*, Energy Paper No. 22, D.En. London.

Department of Energy (1977b). *Government's response to the sixth report of the RCEP*, Cmnd 6820, HMSO, London.

Department of Energy (1982). *Proof of Evidence to the Sizewell B Inquiry*, D.En., London.

Duckworth, J. C., and Jones, E. H. (1958). 'Economic aspects of the U.K. nuclear power programme', *Second Geneva Conf. on Peaceful Uses*, **13,** 575–581.

Eden, R., and Evans, N. (1984). *Electricity's contribution to UK self-sufficiency*, Heinemann, London.

Frisch, O., and Peierls, R. (1940). Appendix 1 to M. Gowing (ed.) *Britain and Atomic Energy 1939–1945*, Macmillan, London.

Fuel and Power Advisory Council, (1946). *Domestic Fuel Policy*, Cmnd 6762, HMSO, London.

Hahn, O., and Strassman, F. (1939). *Naturwiss*, **27,** 11.

Henderson, P. D. (1977). *Two British Errors*, Oxford Economic Papers, pp. 159–205.

Hinton, Sir C. (1957). 'The Future of Nuclear Power', *British Nucl. Energy Conf. Journal*, **1957,** 292–305.

INFCE (1980). *International Nuclear Fuel Cycle Evaluation, Working Group and Summary reports*, IAEA, Vienna.

Jukes, J. A. (1955). 'The Cost of Nuclear Power and the Value of plutonium from early nuclear power stations', *Geneva Conference on the Peaceful Uses of Nuclear Power*, **1,** 322–326.

Jukes, J. A. (1958). 'The economics of nuclear power', *Second Geneva Conference on Peaceful Uses*, **13,** 499–504.

Moynet, G. (1984). 'The Cost of Nuclear Electricity in France', *Rev. General Nucleaire*, **1984, (2),** 141–153.

Pearce, D. W. (1979). *Decision making for energy futures*, Macmillan, London.

Pexton, A. F. (1986): *An up-to-date assessment of the AGR and some comparisons with PWR*, Chartered Mechanical Engineer, pp. 23–27, 31–42.

RCEP (1976). *Nuclear power and the environment*, Cmnd 6618, HMSO, London.

Ridley Committee (1952). *Report of the Committee on National Policy for the Use of Fuel and power resources*, Cmnd 8647, HMSO, London.

The Windscale Inquiry (1978), HMSO, London.

Williams, R. (1980). *The Nuclear Power Decisions*, Croom Helm, London.

Nuclear Power: Policy and Prospects
Edited by P. M. S. Jones

15

The Federal Republic of Germany

Prof Dr DIETER SCHMITT
Emergie-BWL, Universitat-GH-Essen

15.1 Summary Overview

Though only a few decades have passed since the first commercially sized nuclear power plants commenced operation, nuclear energy is rapidly becoming a mature technology. Reactors are demonstrating a high degree of reliability giving high availability and the contribution to the energy needs of the German economy is remarkable. In 1985 about 31 per cent of electricity generated in the FRG was based on nuclear reactors, placing West Germany fifth in terms of world nuclear electricity generation (See Table 19.1). The installed capacity of 17 GWe operated with a load factor of 81 per cent. A further 2800 MWe are starting production in 1986 and three units with the capacity of 4000 MWe are under construction. Eight units with 10 500 MWe are at an advanced stage of planning. In 1985 nuclear based electricity production was 40 per cent greater than that from plants fuelled by (cheap) indigenous lignite, and the plants under construction will lead to nuclear production in 1990 exceeding that produced using hard coal fuel. By the end of the century nuclear capacity could well be doubled and a figure of 32–35 GWe could be regarded as a reasonably central estimate within a possible range of 25–40 GWe unless the public debate in the aftermath of Chernobyl affects future plans.

Even on the basis of this assumption nuclear energy would only be supplying some 16–18 per cent of the FRG's primary energy requirement, i.e. a similar share to natural gas and about 10–20 per cent less than hard coal. It would also be about 50 per cent less than the expected market share of oil for the year 2000. Earlier expectations concerning the size and speed of market penetration may well have to be revised, however.

There are a number of contributory factors in this delay of nuclear penetration in FRG energy markets, even prior to Chernobyl an intense public debate has taken place about the risks of nuclear energy and political indecision has led to major involvement of the courts, leading to delays and increased costs for nuclear energy.

Whilst nuclear has remained the cheapest source for the provision of additional medium and base load electricity capacity, it has faced steep cost increases, which have reduced its economic attractiveness. The growth rate of electricity consumption has also steadily decreased as a result of reduced economic activity, market saturation effects and/or increasing conservation efforts. Added to these factors the energy policy interventions to favour the use of indigenous hard coal, particularly in the electricity sector, has impeded the market penetration of nuclear energy. This will probably not change over the next decade, particularly if the weak energy market persists. If the nuclear debate results in a policy of abandonment the situation would be still worse.

However, nuclear energy represents one of the few technically demonstrated, available, safe and economic options to meet the long-term energy needs of industrial countries like the FRG. Public opposition, at least until Chernobyl, seemed to have lessened as experience with the safe and secure technology with its considerable economic and ecological advantages grew, and as efforts to surmount licensing delays and to close the back-end of the fuel cycle succeeded. In retrospect, the last decades of the twentieth century may appear as the introduction time of nuclear energy, which will reveal its true importance as the growing scarcity of other energy and natural resources becomes apparent. However, environmentalist opposition is calling for all existing facilities to be closed regardless of cost and other impacts; the socialist party favours phasing out nuclear power and abandoning fast reactor and reprocessing R and D; and the other parties are less firmly committed to the nuclear option then they were.

15.2 A Look Backwards

The economic break-through of nuclear energy in the FRG took place in the early 1970s. The first commercially sized nuclear power plants, the Stade 660 MWe PWR and the Wurgassen 670 MWe BWR were commissioned in 1972 and 1973 respectively. Both units had been ordered in 1967 only one year after the first of three medium sized demonstration plants (Gundremmingen A 250 MWe BWR) had commenced operation, and while the two others (Lingen 160 MWe BWR and Obrigheim PWR, originally 300 MWe) were still under construction.

When these first commercially sized nuclear power plants were commissioned even experience with the operation of experimental (prototype) reactors in the FRG was rather small. As Table 15.1 shows the first prototype reactor, a 17 MWe BWR, started operation in 1961. Two others, representing the heavy water and high temperature reactor technologies, had gone into operation in 1965 and 1966. On the other hand, countries like the UK, France and the USA with more advanced nuclear power programmes had a decade of experience showing nuclear to be a reliable and safe electricity generating technology. From today's viewpoint the speed of the early development of nuclear power in the FRG nevertheless looks astonishing, and an indication of the prevailing optimism of the early 1960s when technologists, economists and politicians expected to have at their disposal a mature trouble-free technology, and were convinced that the solution of the few remaining problems would create no significant difficulties.

The FRG was a late entrant to the club of countries using nuclear energy for non-military purposes, political restraints had prevented West Germany entering the field before 1955, well behind countries which had developed their civil programmes from earlier military activities. In common with other countries the West German entry into nuclear energy development was characterized by buoyant optimism, and a large number of scientific institutions and companies entered the field. The government developed the legal and regulatory framework which still operates today. The wide choice concerning the cooling medium, moderator, fuel and fuel assembly, operating temperature, neutron flux levels, led to the first (but not official) atomic energy programme in the FRG (the Eltviller Programme) designating no less than five reactor types to be developed, starting in 1957: the heavy water reactor (HWR), the high temperature reactor (HTR), an organic moderated reactor (OMR), the Magnox type gas-cooled reactor (GCR), and the light water reactor (LWR).

The range of systems considered reflected the variety of ideas, opinions and expectations of the various interest groups engaged in the field of early nuclear development. The HWR was pursued by Siemens, and was a heritage of German nuclear activities during the Second World War. This placed emphasis on independent reactor development with fuel autarky. Krupp and Brown Boveri, both heavy equipment suppliers, developed the

Table 15.1 Nuclear Power Plants in the FRG (March 1985)

Number	Year of commissioning	Year of starting commercial operation (decommissioning	Site	Type	Capacity MWe	Load factor 1984 %	Availability 1984 %
a) Operating							
1	1958	1961	VAK Kahl	BWR	17		
(2	1961	1965 (1984)	MZFR Karlsruhe	DO2PWR	57		
(3	1962	1966 (1979)	Grundremmingen A	BWR	252		
4	1959	1966	AVR Julich	HTR	15		
(5	1964	1968 (1977)	KWL Lingen	BWR	160		
6	1964	1968	KWO Obrigheim	PWR	345	86,0	88,8
(7	1968	1972 (1974)	KKN Niedereichbach	DO2/CO2	100		
8	1966	1973	KNK Karlsruhe	ZH-NA	20		
9	1967	1972	KKW Wurgassen	BWR	670	79,6	80,8
10	1967	1972	KKS Stade	PWR	662	88,4	89,3
11	1969	1975	KK Biblis A	PWR	1204	69,1	70,3
12	1969	1976	KKB Brunsbuttel	BWR	805	78,9	85,7
13	1970	1980	KKP Philipsburg I	BWR	900	83,3	85,2
14	1971	1979	KKU Esensham	PWR	1300	87,7	90,0
15	1971	1976	GKN Neckarwestheim I	PWR	855	84,2	86,7
16	1971	1977	KK Biblis B	PWR	1300	76,7	83,5
17	1971	1979	KKI/I OHU	BWR	907	73,3	82,7
18	1972	1984	KKK Krumel	BWR	1316	87,4	96,9
19	1975	1982	KK Grafenrheinfeld	PWR	1299	89,0	89,8
20	1974	1985	KK Grundremmingen B	BWR	1310	71,6	96,7
21	1974	1985	KK Grundremmingen C	BWR	1310	—	—
22	1975	1985	KKP Philipsburg II	PWR	1349	—	—
23	1975	1985	KKW Grohnde	PWR	1365	—	—
b) Under construction and in phase of commissioning							
1	1970	1985 Plan	THTR Uentrop	HTR	308		
2	1972	1986 Plan	SNR 300 Kalkar	FBR	312		
3	1973	1986 Plan	KK Mulheim-Karlich	BWR	1308		
4	1975	1986 Plan	KK Brokdorf	PWR	1365		
5	1982	1988 Plan	KK N II Neckarwestheim	PWR	1314		
6	1982	1988 Plan	KK L Lingen	PWR	1314		
7	1982	1988 Plan	KK I II OHU	PWR	1370		
c) Planned (advanced stage)							
1			KK Biblis C	PWR	1300		
2			KK Neupotz I	PWR	1350		
3			KK Wyhl	PWR	1350		
4			KK Borken	PWR	1300		
5			KK Hamm	PWR	1300		
6			Pfaffenhofen	PWR	1350		
7			Vahnum A	PWR	1300		
8			Vahnum B	PWR	1300		
d) Total:				34 units,	34 790 MWe		
	of which:	PWR:		21 units,	25 592 MWe		
		BWR:		9 units,	8 543 MWe		
		Others:		4 units,	655 MWe		
		Prototype + demonstration:		6 units,	1 017 MWe		

HTR with its promise of process heat applications in industry. Interatom, then an R and D affiliate of several companies, pursued the OMR, whilst Babcock and Wilcox pursued the GCR with the aim of transferring the Calder Hall experience to German electricity utilities. AEG, one of the traditional power plant constructors in West Germany, offered the LWR, acting as a licensee of the US General Electric Company.

Small prototypes for three of these systems were built in the 1960s and early 1970s: the 17 MWe BWR at Kahl, the 15 MWe HTR at Jülich, and the 57 MWe HWR MZFR at Karlsruhe. The OMR and the GCR lines were dropped in Germany, whilst the 20 MWe KNK, Karlsruhe experimental reactor for breeder development was added in 1973.

Medium sized demonstration plants were commissioned for the four main developments at a later date. As mentioned earlier two BWRs and one PWR started operation in the late 1960s, and the 100 MWe HWR (KKN Niederaichbach) in 1972. The two BWRs as well as the HWR have now been shut-down, but the PWR is still in operation after refurbishment. The demonstration plants for the HTR and FBR systems (the 380 MWe Thorium HTR at Uentrop and the 312 MWe SNR–300 at Kalkar) are heavily delayed and still await completion. Commissioning is now planned for 1986–87.

15.3 The Present Situation

The process of search for the optimum technical solution which characterized the early activities of the nuclear industry was completed in the FRG in the 1960s, and, apart from activities in the fuels of advanced systems like the HTR and FBR, was followed by concentration on a single reactor type, a light water reactor. Since then fifteen commercially sized LWRs with over 16 GWe capacity have commenced operation. Five further units with 6.5 GWe capacity are being commissioned or are under construction and expected to be in commerical operation before 1989, whilst eight additional units with a capacity of over 10 GWe are at an advanced stage of planning. Initially, the reactors were both BWR and PWR, but the PWR has been increasingly favoured. One of the most important factors in the success of the LWR can be seen *ex post* to have been the delayed entry of the FRG into the field of civil nuclear energy. Compared with countries like the US, UK or France, the West German nuclear programmes started in a totally different energy environment. The general energy scarcity of the early 1950s with the doomsday hypothesis about a growing gap between energy requirements and supply, the expectation of rising energy prices and fears for the security of future energy supply, and the resulting call for efforts to increase energy supply at any cost, were things of the past. Once the transport bottle-neck had been overcome in the late 1950s increasing amounts of imported coal entered Europe. The huge Middle East oil reserves with incredibly low almost constant production costs, which had originally been developed to supply United States markets, had sought its outlet elsewhere, including Western Europe, when protectionist measures were adopted in 1957 by the US in the interest of its domestic producers. As a result of the sellers markets which had followed the Second World War a change to a buyers market with energy prices, particularly those of oil, falling in real terms.

Under these circumstances nuclear energy could only expect to enter the West German energy market if and where it seemed likely to be able to compete successfully with other kinds of energy. The choice of the most appropriate nuclear technology for FRG was not affected by a non-civil heritage. West German industry was from the outset expected to make a considerable financial input to the development of nuclear power, although the German government agreed to some subsidization in the face of the perceived necessity to make up for the lag compared with development in the existing nuclear industrial nations. However, up until now only the electrical industry has been prepared to take on this commitment, and this has contributed to the selected reactor strategy. Only supplier's experiences in the field of conventional power plant construction were taken into consideration, and only then if they could offer systems that had proven high availability and a promise of reaching economic break-even within a short time. Even in the early 1960s, the construction of the first demonstration plants and US experience had led to the conclusion that the major increase in unit size, to gain the benefits of economies of scale, would allow economic break-even to be reached. In this environment it reflects a considerable credit on the German public utilities who took the risk of commissioning two 600 MWe units within a year of commissioning the 250 MWe demonstration plant in 1967, and their subsequent decision to commission the world's first 1200 MWe unit only two years later.

On the reactor construction side this left, when viewed realistically, no room for a time-consuming independent reactor development strategy; it favoured construction of power plants under licence and, at best, parallel independent development of the systems. AEG and Siemens (still with a HWR line in parallel to LWR's) on the one side and General Electric and Westinghouse on the other, were prepared to go in this direction. It is, however, questionable whether the US or the US companies would have been fully satisfied with subsequent developments. The US made available long-term loans through Euratom to assist the introduction of nuclear power plants in Europe, and indeed the LWR prevailed in the market but the West German companies soon took over the responsibility for power plant construction in the FRG and developed into strong competitors in world markets. Subsequently the AEG ran into financial difficulties and sold their nuclear activities to Siemens whilst Brown Boveri and Babcock lost 500 million DM on the construction of the 1300 MWe LWR power plant at Mulheim-Karlich and withdrew from the German market. This left the Siemens owned Kraftwerk Union (KWU) as the only supplier of light water reactors. Competition from foreign suppliers has been made more difficult by the sophisticated German regulatory process. Siemens also owns Interatom which is engaged on FBR development (and is interested in HTR technology), and the only other company in the field of nuclear power plant construction is Brown Boveri with its interest in HTE development, although this reactor is still waiting for a commercial breakthrough.

15.4 Nuclear Energy Development

Nuclear energy is the fastest growing energy resource on the West German energy market. In 1985 almost 31 per cent of electricity generation was based on nuclear sources. With 126 TWh of nuclear electricity being produced annually the nuclear output has risen by 150 per cent since 1980 and by a factor of fifteen since 1970. The FRG now ranks fifth in the world in terms of nuclear electricity production (see Table 19.1) and nuclear is as important as lignite in its overall energy supply (Table 15.2).

When the German nuclear industry had closed the gap with other industrialized countries by the end of the 1960s, nuclear energy development entered a period of what could be called euphoric growth. German government, within the first energy programme of 1973, called for a nuclear capacity of 40 GWe or 50 GWe to be installed by 1985. This objective was strengthened by the first oil crisis in 1973, and the recognition of the high dependence of energy users on supplies of oil from a limited number of producing countries. This pointed to a need of urgent restructuring of the West German energy sector to attain a higher degree of security of supply. It is not surprising therefore that licences were issued for ten 1300 MWe power plants in 1973 and 1974. The West German–Brazilian nuclear contract was signed at the same period. A year later contracts were entered into a sale of nuclear power plants in Iran.

The nuclear boom of the mid–1970s abruptly turned into a crisis. A minority of the population had not regarded nuclear energy as absolutely necessary or even desirable. With the first big nuclear power plants going into operation and prospective commercial breakthrough for the technology,

Table 15.2 Primary energy requirements (PER) by kinds of energy (Mtce and percentages)

	1960	1965	1970	1975	1980	1982	1984	1985	2000
Mtce									
Hard coal	128.4	114.4	96.8	66.5	77.1	76.7	79.5	79.2	80–100
Lignite	29.2	30.0	30.6	34.4	39.2	38.4	38.4	36.0	40
Oil	44.4	108.0	178.9	181.0	185.7	159.8	158.5	161.0	130
Natural gas	1.1	3.6	18.5	49.2	64.4	55.1	58.8	59.7	75
Nuclear energy	—	0.0	2.1	7.1	14.3	20.9	30.4	41.3	70–50
Water power/net Imports of Electricity	6.6	6.8	8.4	7.8	7.6	8.1	6.9	6.0	7
Others	1.9	1.7	1.5	1.7	1.9	2.5	3.2	3.5	8
Total	211.5	264.6	336.8	347.7	390.2	361.5	376.5	387.0	400–420
Percentages									
Hard coal	60.7	43.3	28.8	19.1	19.7	21.2	21.2	20.5	19.5–24.3
Lignite	13.8	11.4	9.1	9.9	10.0	10.6	10.2	9.3	9.8
Oil	21.0	40.8	53.1	52.1	47.6	44.2	42.1	41.6	31.7
Natural gas	0.5	1.4	5.5	14.2	16.5	15.2	15.6	15.4	18.3
Nuclear energy	—	0.0	0.6	2.0	3.7	5.8	8.1	10.7	17.0–12.2
Water power/net Imports of Electricity	3.1	2.6	2.5	2.2	2.0	2.3	1.8	1.6	1.7
Others	0.9	0.7	0.4	0.5	0.5	0.7	1.1	0.9	2.0

the anti-nuclear movement succeeded in unifying the opposition under its banner. They occupied sites and adopted legal means of delaying construction of the plant and/or forcing the regulatory authorities to increase safety standards. This led to major delays for the development of nuclear energy and large cost increases. Part of the difficulty arose from a lack of political leadership. A five-year discussion about a national centre for closing the back-end of the fuel cycle in 1979 was terminated by the declaration of the prime minister of Lower Saxony, but the project could not be proceeded with for political reasons. Fears about the risk of nuclear accidents, concern about the possible misuse of plutonium, and concern about terrorism, combined with unforeseen increases in safety requirements provided ammunition for the anti-nuclear groups. The nuclear industry was not helped by President Carter's non-proliferation initiatives which stopped civil reprocessing activities within the US, nor by the Three Mile Island reactor incident of 1978 with the ensuing worldwide press coverage.

Despite the fact that these factors contributed to huge cost increases[1], particularly if extra restrictions were imposed during construction or after the start of plant operation, and the effects of general inflation, nuclear energy remained one of the cheapest sources of electricity, not least because the costs of other types of plant, particularly coal-fired plant, had also increased[2]. Nuclear was also helped by the impact of the second oil price rise associated with the Iran–Iraq war in 1979–80. Nuclear energy has not been able to capitalize on its economic advantage fully because the rates of growth of electricity consumption, which had lain between 7 per cent and 8 per cent p.a. from the early 1970s had declined, initially to about 4 per cent p.a. and then to 2 per cent p.a. (see Table 15.3).

1. Investment costs for 1300 MWe units increased from 650 DM Kw^{-1} to 2200 DM kW^{-1} between the early 1970s and 1980s and are expected to double again in cash terms by the end of the decade.
2. Investment for hard coal-fired plant tripled to 1600 DM kW^{-1} by 1982–83 compared with one decade earlier. The investment for desulphurization and denitrification (mandatory since 1983) is 350–400 DM kW^{-1}, which is similar in magnitude to the total investment for a 350 MWe hard coal plant less than fifteen years ago. In lignite power plants environmental measures called for almost one third of total investment costs.

Table 15.3 Electricity Consumption

	1960	1965	1970	1975	1980	1982	1983	1984	1985
TWh									
Industry	74.7	100.8	131.8	147.4	175.4	166.6	170.6	178.9	183.2
Traffic	3.7	5.6	7.9	8.8	10.6	10.3	10.3	10.6	11.2
Public service sector	4.2	7.2	11.3	17.3	24.1	25.6	26.3	27.4	28.3
Agriculture	1.9	3.3	5.1	6.3	7.1	7.2	7.3	7.4	7.6
Households	12.2	23.8	43.1	67.8	95.5	88.0	90.2	94.1	97.1
Commercial	7.4	12.5	19.4	27.2	34.2	36.6	37.7	39.4	40.8
Net consumption	104.3	153.3	218.6	274.8	336.9	334.3	342.5	375.4	386.4
Total consumption	111.7	163.2	232.5	289.6	351.4	349.3	359.3	394.9	408.6
%									
Industry	66.9	61.8	56.7	50.9	49.9	47.7	47.5	47.8	47.6
Traffic	3.4	3.5	3.4	3.0	3.0	2.9	2.9	2.8	2.9
Public service sector	3.8	4.4	4.9	6.0	6.9	7.3	6.5	7.3	7.4
Agriculture	1.8	2.0	2.2	2.2	2.0	2.1	2.0	2.1	2.0
Households	10.9	14.6	18.5	34.4	24.4	25.2	25.1	25.2	25.3
Commercial	6.6	7.6	8.3	9.4	9.7	10.5	10.5	10.6	10.6
Net consumption	93.4	93.9	94.0	94.9	95.9	95.7	95.3	96.0	95.7
Total consumption	100.0	100.0	100.0	100.0	100.0	100.0	100.0	100.0	100.0
Average growth rates *%/a*									
Industry		6.2	5.5	2.3	3.5	−2.6	2.4	4.9	2.1
Traffic		8.6	7.1	2.2	4.0	−1.4	—	2.9	5.7
Public service sector		11.3	9.4	8.9	6.9	2.1	2.7	4.2	3.3
Agriculture		11.7	9.1	4.3	2.4	0.7	1.4	1.4	2.7
Households		14.3	12.6	2.5	7.1	−4.2	2.5	4.3	3.2
Commercial		11.1	9.2	7.0	4.7	3.4	3.0	4.5	3.6
Net consumption		8.0	7.4	4.7	4.2	−0.4	2.5	4.5	2.7
Total consumption		7.9	7.3	4.5	3.9	0.3	2.9	3.8	3.0

In 1982, for the first time, electricity consumption even decreased. Reduced economic activity, price increases for electricity, the effort put into conservation, saturation effects on the market, and political intervention in electricity pricing aimed at reducing electricity's penetration into heating markets, contributed to the reduction of electricity growth. As capacity planning had been orientated many years on the expectation of a continually growing demand, and because of the long lead times for power plant construction, the capital stock of power plants could not be adjusted, and a temporary over-capacity resulted which led to new investment decisions being deferred.

An even more important factor has been the increasing intervention of energy policy reasons to support the use of indigenous hard coal. Since the late 1950s governments have intervened in energy markets in order to prevent a steep fall in West German hard coal production. This coal has not been competitive with imported coal despite considerable productivity gains (about 250 per cent since 1957), the closure of about 50 per cent of capacity, and the concentration of mining activities in six companies. Taxes and duties on coal and oil products have virtually eliminated them from FRG markets, especially for electricity production. Heavy subsidies, which amounted to more than 50 billion DM over the past ten years, should have improved the market position of indigenous coal. It became more apparent in the late 1970s, however, that even with the help of these massive energy policy interventions domestic hard coal could not maintain even its reduced production levels, especially after coal outlets to the steel industry were hit by the changes that took place in that sector also. Nuclear based electricity threatened the last major market for indigenous coal. Three-quarters of coal mining in FRG is concentrated in the Ruhr region which is still suffering from the economic problems of declining 'old' industries. In order to protect the industry and avoid social problems the government in 1979 encouraged the electricity supply industry to enter contracts for increasing amounts of coal rising from 33 million t year $^{-1}$ to 45 million t by 1995. Imports of coal above an allowance of 5–6 million t per year are only allowed after the contracted indigenous hard coal had been taken up.

The combined effect of decreasing electricity growth rates with the considerable new capacity still under construction, and the requirements to use increasing amounts of indigenous hard coal set limits to the penetration for nuclear energy, even if substitution for oil and gas based capacity are taken into account. The net effect had been an increase in coal-fired power production since the mid–1970s (Table 15.4), despite its uneconomic position when compared with nuclear. No new nuclear plants have been ordered between 1975 and 1982, and recent additions to nuclear capacity (Table 15.1 and Table 15.5) are the result of earlier orders. The prospects for nuclear power appear to have considerably improved, however, since the early 1980s.

15.4 The future: recovering from the crisis

Despite the heavy cost increases during the last decade, nuclear energy remains the cheapest source of the supply of additional medium and base load electricity demand in the FRG into the foreseeable future. Economic hydro-power sources have already been fully exploited. Lignite production can be extended (if at all) only to a limited extent despite the huge reserves and, due to high productivity, the low costs. Additional sites for new lignite power plants at the mine-mouth are not available, and transportation of lignite is quite expensive because of its low calorific content. The most important factor, however, is that lignite has to be mined in large (and deep) open-cast mines within highly populated areas, and this restrains production at a level of almost 40 million t p.a. due to environmental constraints. Oil and natural gas (with the exception of interruptable supplies) have lost competitiveness after the sharp price increases of 1973 and 1979. As a matter of policy new oil-fired plants would not be licensed, and gas would be used for power plants only exceptionally.

Table 15.4 Electricity Generation by Kinds of Energy

	1960	1965	1970	1975	1980	1982	1983	1985
TWh								
Water power	13.0	15.4	17.8	17.1	18.7	19.6	18.9	18.0
Hard coal	62.5	87.5	97.7	74.5	112.4	122.2	132.6	129.0
Lignite	31.0	44.4	59.7	85.2	93.5	94.0	94.9	88.0
Oil	4.0	16.7	36.3	30.0	25.7	17.3	12.6	10.0
Natural gas	5.0	7.5	13.3	60.3	61.0	37.3	36.9	24.0
Nuclear energy	—	—	6.0	21.4	43.7	63.6	65.8	126.0
Others	0.9	1.0	11.8	13.2	13.6	12.8	12.1	13.6
Total	116.4	172.3	242.6	301.8	368.8	366.9	373.8	408.6
%								
Water power	11.2	8.9	7.3	5.6	5.1	5.3	5.1	4.4
Hard coal	53.7	50.8	40.3	24.7	30.5	33.3	35.5	31.6
Lignite	26.6	25.8	24.6	28.2	25.4	25.6	25.4	21.5
Oil	3.4	9.7	15.0	9.9	7.0	4.7	3.4	2.5
Natural gas	4.3	4.4	5.5	20.0	16.5	10.2	9.9	5.9
Nuclear energy	—	—	2.5	7.1	11.8	17.3	17.6	30.8
Others	0.7	0.5	4.9	4.4	3.7	3.5	3.2	3.3

Table 15.5 Power plant capacity by kinds of energy

	1960	1965	1970	1975	1980	1982	1983	1984	1985
GWe									
Water power	3.4	4.1	4.8	5.6	6.5	6.5	6.5	6.6	6.7
Hard coal	17.7	24.4	28.5	28.4	28.6	29.4	30.4	30.5	31.6
Lignite	5.5	7.6	8.8	12.5	14.0	13.9	13.8	13.8	12.9
Nuclear energy	—	0.1	0.9	3.5	9.1	10.4	10.4	11.7	16.5
Oil					14.7	13.9	13.8	13.6	12.1
Natural gas	0.8	4.5	7.7	23.3	13.6	14.7	14.6	14.6	13.0
Others					0.8	0.8	1.1	1.1	1.1
Total	27.5	40.6	50.8	74.4	87.3	89.6	90.5	91.8	93.8
%									
Water power	12.4	10.1	9.4	7.5	7.4	7.3	7.2	7.2	7.1
Hard coal	64.4	60.1	56.1	38.2	32.8	32.8	33.6	33.2	33.7
Lignite	20.0	18.7	17.3	16.8	16.0	15.5	15.3	15.0	13.8
Nuclear energy	—	0.2	1.7	4.7	10.4	11.6	11.5	12.7	17.5
Oil					16.8	15.5	15.2	14.8	12.9
Natural gas	3.0	11.1	15.2	31.3	15.6	16.4	16.1	15.9	13.9
Others					0.9	0.9	1.2	1.2	1.2

The only alternative to nuclear power for electricity generation to meet additional demand and to substitute for oil and gas, is hard coal.

Recent cost calculations have confirmed the economic advantage of nuclear based electricity generation against hard coal, and shown them to be substantial. The cost to generate base load electricity in new nuclear power plants for operation in the early 1990s are about 3 Pf kWh^{-1} (1984 values), or 30 per cent lower than those on the basis of (cheap) imported hard coal (7.5 Pf kWh^{-1}), which in turn are 50 per cent below those for indigenous coal (NEA, 1986).

Nuclear's cost advantage is becoming increasingly apparent. Public utilities relying mainly on coal are incurring the cost of environmental controls, including back fitting of desulphurization and denitrification equipment, and they have had to increase their prices by some 20–30 per cent over a period of five years. Utilities with a high percent of nuclear power have been able to maintain almost constant prices. Differences in price levels and structures between various regions in the FRG which have persisted for many years (with a north-south slope) are being reduced.

Nuclear plants can compete with coal plants using imported coal down to load factors of 3000 to 4000 hours p.a., and for plants using indigenous coal at load factors of 2000 hours p.a. The German mining industry has declared that with the so-called *Jahrhundertvertrag* their objectives with respect to coal burning in electricity generation plants have been fulfilled, and that they are not seeking further outlets within this sector. Nevertheless, it seems unlikely that the contract, as currently designated, will end at the middle of the 1990s, since coal demand in the European steel industry has been reduced so far since the original agreement was signed, and that further closures of coal mining capacity in the FRG seem inevitable.

The German economy like that of the other industrial nations' is recovering from one of the deepest recessions since the 1930s, and with higher economic activity electricity consumption is again beginning to increase. In 1984 it grew by 3.8 per cent and in 1985 by 3.5 per cent. Most experts agree that the growth rates of the 1950s and 1960s are past, but, nevertheless, expect an increase in electricity intensity in the economy with a ratio of growth rate of electricity consumption to that of GDP exceeding unity, mainly because of the attractiveness of the expected economies of electricity for most energy uses.

With assumed GDP growth rates of about 2–2.5 per cent p.a. electricity consumption could increase up to the end of the century by another 25 per cent. This would leave, taken together with the substitution for oil and gas still used in electricity generation, room for fulfilling the energy policy objectives for both coal and the nuclear industry.

The future prospects for nuclear energy therefore look quite favourable. Nuclear power plants have proved their availability and reliability (Table 15.1). German power stations compare well in this respect with the best in the rest of the world; a fact which seems to count heavily with the public. Public opinion is also favourably influenced by the fact that no significant accidents or interruptions of operations have occurred for many years in the country. In consequence the confrontations between anti- and pro-nuclear groups at reactor sites had become rare, and the anti-nuclear movement seemed to have concentrated its attention on facilities to close the back-end of the fuel cycle: i.e. intermediate storage installations for spent fuel elements and the planned reprocessing plant and final waste repository. Efforts to stop or delay the construction of facilities to close the back-end of the fuel cycle have been made through the courts and, as one decision with respect to intermediate storage facilities shows, have been quite effective. Nevertheless, it seemed that the nuclear debate was less heated, and that public attention was turning to other topics such as the siting of missile bases and the environmental damage consequent on the emissions from the fossil fuel burning in power plants, households, industry or cars. Chernobyl has, however, led to renewed questioning of the nuclear option.

There is an expectation within the Federal Republic that costs of some stages of the nuclear fuel cycle will stabilize or decrease (including enrichment and reprocessing), and this will offset some part of the earlier cost increases for nuclear power generation.

The standardization of power plant designs and the mutual acceptance of certificates by regulatory bodies have been brought in by the government since 1981, and have succeeded in streamlining the regulatory process without any reduction of safety or legal protection of the public. As a result, the construction time of nuclear power plants can be expected to be reduced to about six years, that is the original figure, from the eight to ten years that have become common in the late 1970s. Three technically identical 1300 MWe PWRs (the so-called Konvoi-types) were ordered in 1982. All are proceeding according to plan without delays or cost increases. Nuclear power has become subject to rational planning once more in the FRG.

Last but not least, progress has been made with closing the back-end of the fuel cycle. Sizeable intermediate storage facilities for spent fuel elements are being built, transport and storage of these elements has been demonstrated, and the efforts to develop a final storage facility (probably salt-dome in Lower Saxony) are proceeding as scheduled for a commissioning date of 2000.

Public utilities, meanwhile, adopt a range of different options for the back-end of the fuel cycle. Some reprocess in existing or planned plants in France and the UK, some plan reprocessing in a future German reprocessing plant, and some contemplate the direct disposal of spent fuel elements. A fourth possibility has been discussed in the press in which waste or spent fuel might be sent for final disposal to the People's Republic of China, possibly in return for collaboration with nuclear plant. It is too early to evaluate such a proposal since safeguards and the price to be charged for such services would need to be looked at. At the moment these would appear to be three times higher than the costs of the national back-end solutions.

DWK, the German reprocessing company, which is jointly owned by public utilities having nuclear plant, has decided to build a 350 t p.a. reprocessing plant in Bavaria by 1995. A recently published study (PAE, 1985) has shown that direct disposal of spent fuel elements operating on the once through cycle is feasible, safe and that it will not create significant proliferation problems. Furthermore, the study suggests that it will be cheaper than the reprocessing option. Using the once-through route costs, according to the study, are some 40 per cent lower than those which would result from reprocessing in a 700 t heavy metal p.a. plant with sophisticated technology and fairly high plutonium credits. The option would also be cheaper than using foreign reprocessing. DWK plan to build a demonstration plant for spent fuel storage in Lower Saxony.

In summary, there would appear to be good prospects for the reduction or stabilization of the cost of nuclear based electricity generation, and the uncertainties surrounding its future have considerably diminished even though they have not yet completely disappeared.

Nevertheless, even under the quite favourable climate up to May 1986, it has to be realized that expectations with regard to the market penetration of nuclear electricity which were widely held in the 1960s and early 1970s have not been realized, and are now unlikely to be realized within this century. Another doubling of nuclear capacity to 34 GW within the next fifteen years, a prospect that many observers find too optimistic, would increase nuclear electricity generation to about 200 TWh, i.e. about 40 per cent of total electricity production in the year 2000. The market share of nuclear energy under this assumption would have increased to about 17 per cent of total primary energy. This is twice as much as lignite, but only of the same order as natural gas, less than hard coal and only about 50 per cent of the market for oil by the year 2000. It is not likely that there will be significant use of nuclear energy outside the electricity sector within the next fifteen years in the FRG. Nuclear ship propulsion, nuclear based district heating and the use of nuclear heat for industrial processes is not likely to gain a significant foothold on this timescale.

Nuclear energy's failure to penetrate the non-electricity markets arises from the fact that its economics are not yet fully understood nor has the technical feasibility of some of the applications been demonstrated. Nevertheless, consideration has been given within FRG to propulsion, district heating, industrial process steam and medium and high temperature heat, the latter especially in connection with coal gasification.

The progress made so far is small. Plans to continue the development of nuclear ship propulsion in the FRG after a decade (1968–78) of testing the first German nuclear ship, the *Otto Hahn*, have been abandoned for the time being and the reactor has been decommissioned. Plans released in the 1970s to build a combined process heat–electricity generating nuclear power plant on the basis of LWR technology by, and at the site of, a chemical factory were cancelled after regulatory delays and sharp cost increases, setting aside the uncertainty of public acceptance for a nuclear plant within a chemical factory complex. Lack of public acceptability must be regarded, too, as the most important restraint in the FRG for the use of nuclear energy as a basis for district heating. This possibility has never been discussed seriously.

Progress has also been slow with the HTR system despite its broad potential applications and its inherent safety advantages. The prototype HTR (AVR at Julich) has nevertheless run with high availability for almost twenty years and reached temperature levels of more than 900 °C. The first electricity generating HTR demonstration plant (thorium high temperature reactor—THTR 300) has been under construction for a period of fifteen years, and in receipt of subsidies of 3 billion DM from the government (i.e. about 75 per cent of its expected costs). Using the HTR technology for high temperature industrial purposes such as steel or coal gasification still awaits demonstration of its technical feasibility and licensing. It is not certain how long it will take or what kind of special requirements regulatory bodies will impose on these new technologies, or the impacts of these on energy costs. Another difficulty with these applications for nuclear energy is the fact that the unit sizes are small compared with electricity generation plants, and this results in diseconomies of scale. These small sizes correspond to the energy requirements for the specific heat applications. Over-all the economics of introducing nuclear energy into markets other than electricity generation are not only heavily burdened by high and uncertain costs, linked both to the remaining technical and licensing uncertainties, but by low and even decreasing profitability. The markets such as district heating or iron ore reduction would also be damaging to the indigenous hard coal industry, and this would lead to political objections to this expansion of nuclear activity.

The use of nuclear energy for coal gasification has to be seen in the context that the products (substitute natural gas, town gas and synthesis gas) will have to compete with the huge and steadily increasing natural gas reserves being located within Europe, (Netherlands, and the North Sea) and in adjacent regions (the USSR, North and West Africa and Middle East). The situation will be exacerbated by the likely continuance of weak market conditions for oil, which could continue into the 1990s and beyond and be paralleled by low prices for natural gas.

The prospect for non-electric uses of nuclear energy in the FRG would therefore appear rather pessimistic, at least compared with the earlier expectations of the HTR advocates. Further development of this reactor type will depend upon its suitability for electricity generation, and it seems unlikely that the advantages of this type of reactor will be fully realized unless the public utilities are prepared to construct 500 MW units, or if a large industrial electricity producer deploys smaller module HTRs for combined heat and power production. A feasibility study released recently has shown that medium sized HTRs could be competitive for electricity production. However, unless there is a commitment to construct a 500 MW HTR (or a small module by an industrial user) by the end of the 1980s, scheduled for operation in the mid–1990s, it may not be possible to continue the development of this reactor type for non-electric purposes. There is a reasonable prospect that such a decision will be taken, but uncertainty remains.

The most important question for the further development of nuclear energy in the FRG, however, is the final outcome of the debate about nuclear energy after the Chernobyl accident. The majority of the population has been shocked by the potential impact on safety and health, especially of children and pregnant women, of a nuclear accident 1500 km away. In this context it is of minor importance whether the dangers were real or only highly hypothetical, as pointed out by official institutions like the Commission for Protection Against Radiation (*Strahlenschutzkommission*). As a result of a unique press campaign, nourished by the initially limited amount of information available, aided by inconsistent disclosure of information and instructions by official institutions and the contradictory evaluations of 'experts', with possibly, a high degree of misuse of the accident for political purposes, nuclear energy faces serious problems with respect to public acceptance, at least for the moment. A declining proportion of the population wants continuing nuclear development on the basis of the much higher safety requirements in the FRG against the USSR, and the unchanged economic and ecological advantages of this kind of energy as compared to substitutes like fossil-fuels, renewables or conservation. The situation might be different if other countries in Europe or elsewhere changed their nuclear policy.

Environmentalists on the other hand demand an immediate stop of all nuclear activities in the FRG, even if this would result in temporary bigger environmental problems, higher costs to private consumers and industry and/or certain restraints for electricity consumption. The socialist party (as well as the unions) meanwhile, has decided for a stepwise departure from using nuclear energy, with only the timescale and identification of the route with least costs and frictions for society and the economy being questions still to be answered. They are already decided on abandonment of reprocessing and breeder technology. Even within the liberal and conservative parties a majority favours rethinking nuclear policy, regarding this kind of energy as 'transitory' until better solutions to meeting the energy needs of post-industrial societies have been developed.

Even assuming some kind of overshoot at the moment in public and political reactions after the severe accident of Chernobyl, the near to mid-term future of nuclear energy will depend heavily on the results of elections taking place in spring 1987 in the FRG. For many observers even the assumption of nuclear stagnation (23 500 MWe with 150 TWh year^{-1}) after the plants now under construction have been commissioned is potentially too optimistic. The doubts about the political feasibility of beginning reprocessing processes on a large scale have also increased. Last but not least, it seems optimistic to assume that the SNR 300 will be commissioned as long as the present government of Northrhine–Westfalia oppose it.

Virtually all experts, however, expect that in the long run the fossil fuel markets will be characterized by increasing scarcity of fuels, thus opening new opportunities for non exhaustible resources and nuclear energy. The processs may accelerate dramatically if concerns about the near term impact of carbon dioxide on climate hold true. In view of the finite nature of uranium and thorium resources only advanced converter and breeder type reactors (and possibly fusion) can make a very lengthy energy contribution and this is why utilities in the FRG, like those in other industrial nations, are proceeding with the construction of medium sized demonstration plants, particularly for the fast breeder reactor (SNR 300) and the high temperature reactor using thorium fuel (THTR 300). It is also the reason why the West German government provides considerable support to these activities, with both reactors together having received more than 7 billion DM.

The continuation of governmental support for these projects in FRG must be open to question. The costs of the projects have increased dramatically, and for the time being there is a surplus of low cost uranium available which throws into question the need to develop a plutonium strategy. The value of plutonium as a fuel has therefore decreased, whilst the cost of its recovery has tended to increase over the years, and the storage of spent thermal reactor fuel has come to look a more attractive option. How to continue the development of the technological options regarded as necessary in the long-term may become the main question of technology policy.

Future historians may come to look upon the development of nuclear energy in the last decades of the twentieth century as a gigantic failure which absorbed human and capital resources, hindering both technical development in other areas and the timely adjustment of the energy system. On the other hand even the severe problems which nuclear energy faces in the FRG today might ex post be regarded only as temporary disturbance preceding the introduction of nuclear energy as the principal method of generating electricity. In the longer term nuclear power may still penetrate non-electrical applications and increase its share of primary energy supply well beyond the levels that have been anticipated for the turn of the century.

REFERENCES

Nuclear Energy Agency (1986). *Projected Costs of Generating Electricity*. NEA/OECD, Paris. PAE (1984). *Systemstudie Andere Entsorgungstechniken,* Karlsruhe.

Nuclear Power: Policy and Prospects
Edited by P. M. S. Jones

16

Japan

KEVIN HUTTNER and TATSUJIRO SUZUKI
The International Energy Forum (IEF) Tokyo, Japan

16.1 The History of Japan's Nuclear Programme

16.1.1 *Birth of the Programme*

Official interest in nuclear energy in Japan began in 1952 when the US occupation ended and the ban on nuclear research was lifted. This interest increased after President Eisenhower announced his 'Atoms for Peace' project at the end of 1953. The Science Council of Japan, an influential organization composed of experts from a variety of fields, was one of the leading groups involved in the debate on nuclear energy. The council's members were divided in their feelings. Some were against developing nuclear energy unless its peaceful uses were proven completely safe and all nations banned nuclear weapons. Others stressed its potential and called for an active development programme. In the end the council decided to advocate a positive but cautious approach which stressed slow, steady progress based on thorough R and D.

Certain politicians, however, including Yasuhiro Nakasone, the present prime minister, favoured a more active approach. As a result of these politicians' efforts the first nuclear energy budget was allocated by the Japanese Diet in March, 1954. It was hoped the budget would increase national awareness of the possible applications of nuclear power and serve as a catalyst for their development.

The Science Council considered the budget premature and worried particularly that nuclear energy would be used for military purposes. To help prevent this from happening a month later it formally called for an independent civilian development programme open to public scrutiny. (The contents of this proposal became known as the 'three basic principles': independent, civilian, open.) When Japan's Basic Atomic Energy Law specifying the manner in which nuclear energy should be developed in Japan was enacted at the end of 1955, its content closely resembled that of the Science Council's 'three principles'. Thus the ideological framework created by the Science Council had a large influence on the course of nuclear energy development in Japan.

Although a nuclear budget had been allocated, the issue of authority over the new programme remained. The Science Council wanted authority to be vested in an independent commission free from government influence. Members of the government and the conservative Diet parties wanted the government, assisted by a purely advisory body, to have control. In the end a compromise was reached. As a result the Japan Atomic Energy Commission (JAEC) was established in January, 1956. The JAEC serves as an advisory council whose decisions the prime minister must respect. An Atomic Energy Bureau (AEB) was formed to serve as the secretariat to the JAEC. It was located in the same building as the new Science and Technology Agency (STA), an organ charged with the promotion of advanced science and technology. The head of the STA serves simultaneously as the chairman of the JAEC and is an appointee of the prime minister. This decreases the likelihood of the JAEC adopting policies unacceptable to the government. In June the Japan Atomic Energy Research Institute (JAERI) was formed to handle basic nuclear R and D. And the next month the Atomic Fuel Corporation (a public corporation) was established to mine and process uranium. Since their creation these organizations have provided an institutional structure for Japan's nuclear energy programme.

At the same time as this administrative structure was forming Japanese industry was organizing itself to enter the new field. By the end of 1956 five industrial consortia had formed centering on five companies: Mitsubishi, Hitachi, Toshiba, Sumitomo and Fuji. These companies planned to sell machinery to the nine Japanese private utilities that had formed in 1951 after Japan's state owned national utility was dissolved.

16.1.2 *Reactor Importation*

Soon after the first nuclear budget was allocated, a Preparatory Council for the Utilization of Atomic Energy was set up by the government to help guide the nuclear programme until an administrative structure could be formed. The predominant way of thinking in those early days was to develop nuclear power by using domestic resources. In 1954 the Preparatory Council decided a natural uranium heavy water reactor would be the most appropriate reactor to build, since Japan did not have the capability to enrich uranium for LWR fuel. However, in the beginning of 1955 the US decided to offer 100 kg of enriched uranium to foreign countries. In the case of Japan it also offered to pay half the cost of the uranium as an additional inducement. Not long after this, the Preparatory Council announced that in preparation for the construction of a commercial nuclear power plant ten years in the future, Japan should take the following steps:

1. Import a water-boiler type reactor from the US in 1956 (50 kW thermal JRR1).
2. Import a CP–5 type research reactor from the US in 1957 (10 MW thermal JRR2).
3. Build a natural uranium heavy water reactor utilizing domestic materials in 1958 (10 MW thermal JRR3).
4. Import a boiling water reactor power demonstration plant from the US roughly a year later (12 MW JPDR).

All of these reactors were in fact built. The power demonstration reactor produced Japan's first nuclear generated electricity in 1963.

In contrast to the Preparatory Council's long-term approach, the first chairman of the JAEC pushed for the rapid introduction of commercial nuclear power plants from abroad. His efforts were supported by the utilities who needed greater generating capacity in order to satisfy demand from a rapidly expanding economy. Domestic coal was expensive, and the utilities were reluctant to increase their dependence on foreign supplies of fuel.

Some Japanese at that time (including the JAEC chairman) believed Britain's Calder Hall gas-cooled reactor to be the world's most advanced reactor. It seemed particularly suitable for introduction into Japan because it did not require the use of either heavy water or enriched uranium. Interest in the Calder Hall reactor increased when a high ranking official of Britain's nuclear programme announced that the reactor could produce electricity at a cost competitive with coal-fired plants. However, while the utilities were in favour of importing a cost-efficient nuclear reactor as soon as possible, some Japanese scientists were against it. They argued it was premature to import a commercial reactor and that it would affect domestic R and D adversely. Prompted by this opposition, a fact-finding mission was sent abroad by the JAEC to investigate the capabilities of reactors in England, Canada and the USA.

Shortly before the fact-finding mission left Japan at the end of 1956, the JAEC announced its first 'Long Term Plan for the Development and Utilization of Atomic Energy'. This plan called for the development of a fast breeder reactor (FBR) after conducting basic research and constructing experimental reactors. However, as this would take time, the plan also recommended importing commercial reactors to generate electricity until FBRs could be introduced.

When the mission returned it issued a report quite favourable to the Calder Hall reactor. Encouraged by this the JAEC officially recommended the importation of a Calder Hall type reactor with an electrical output of 160 MW. In November of 1957 the Japan Atomic Power Company (JAPCO) was formed to be the owner-operator of the new reactor. Some 80 per cent of JAPCO's capital was provided by private companies (the nine utilities put up half of this money) and 20 per cent was provided by the government. Construction began in early 1960 and was finished in May, 1965.

In the years following the decision to import a reactor from Britain, US nuclear makers appeared to achieve considerable success in the development of light water reactors (LWRs). A Presidential Report issued at the end of 1962 declared that an LWR of 500 MW or more would be economically competitive with thermal (coal or oil) power plants. Another study claimed that new LWRs were far superior economically to the Calder Hall type of reactor. By the end of 1963 five US utilities had placed orders for an LWR.

The flurry of nuclear plant construction in the US helped people in Japan to believe the new LWRs had been proven to be economical. Equally important, the US government had lifted its ban on the supply of enriched uranium (LWRs use enriched uranium as fuel). In 1961 the US informed Japan it was prepared to guarantee a supply of enriched uranium to countries it had nuclear co-operation agreements with. Thus, despite the fact that JAERI's LWR power demonstration plant and JAPCO's own Calder Hall type reactor were still under construction, JAPCO decided in early 1963 to import an LWR from the US. JAPCO asked the two main US nuclear makers, General Electric (GE) and Westinghouse, to submit written estimates for their reactors. While both companies built light water reactors, Westinghouse had developed a pressurized water reactor (PWR) while General Electric (GE) had developed a boiling water reactor (BWR).

In September, 1965 JAPCO announced its choice: it would go with GE's BWR. One reason was GE's aggressive campaign to prove that its reactor was the most economical. However, some people claim the main factor involved was the influence of the Tokyo Electric Power Company (TEPCO), Japan's largest utility. TEPCO was JAPCO's largest stockholder, and TEPCO had a long-standing relationship with GE (TEPCO had purchased equipment and know-how from GE in the past).

Not long after JAPCO made its decision to import an LWR the utilities followed suit. In May, 1966 TEPCO decided to order a BWR from GE. In its self-published company history TEPCO lists the following reasons for its choice: (*Tokyo Denryoku Sanju Rekishi*, 1983)

1. GE had received an order from Spain (which helped to prove GE's capability).
2. GE could draw on experience gained from building JAPCO's BWR.
3. GE and TEPCO had already developed a relationship of co-operation and trust.

TEPCO signed a so-called 'turnkey' contract with GE. This contract placed full responsibility for the project from the beginning of construction to the start of actual operation on GE. A price ceiling clause was also included in the contract.

At roughly the same time, the Kansai Electric Power Company (KEPCO), Japan's second largest utility, announced its decision to import a PWR from Westinghouse. This was a fairly natural decision for KEPCO to make. Its relationship with Westinghouse was similar to TEPCO's relationship with GE.

Also, it wanted to use a reactor different from the one chosen by TEPCO, its rival. Since the PWR had certain favourable technological characteristics and had not been proven inferior in terms of economics or reliability, it was not difficult for KEPCO to justify its decision. Besides, from the utility's point of view, the creation of a competitive supply structure probably seemed preferable to dependence on only one maker.

Following the lead of the two largest utilities, the other utilities soon began to order LWRs. Because it was very difficult to determine which reactor was superior, their decisions tended to be based on factors other than their own technical evaluations. Some decisions were based on the existence of ties of some sort to either TEPCO or KEPCO. One utility chose the BWR mainly because that was the reactor JAPCO had chosen.

It is interesting to note the utilities' attitude toward nuclear power plants then. It appears they viewed LWRs as an 'established' technology, and believed that nuclear power plants could be mastered as quickly as thermal power plants. This view was undoubtedly encouraged by the US makers and by the utilities' desire to reduce their growing dependence on oil. Some Japanese claim this misperception caused domestic R and D to be delayed. That is, since people believed nuclear power plant technology was already well developed, there was little need to press for further development. The accuracy of this claim is difficult to measure. In any case the utilities' attitude toward nuclear power technology helps to explain why they decided to purchase different reactors. Since the technology was already 'established', the difference between a PWR and a BWR appeared to be mainly a difference in brand.

By the end of 1972 Japanese utilities had ordered twelve BWRs and nine PWRs. A pattern emerged as to who would contract for the construction of these reactors. GE was the main contractor for the first BWR introduced, Toshiba and GE shared the contracting for the second, and then either Toshiba or Hitachi became the main contractor. As for the PWR, Westinghouse and Mitsubishi shared the contracting for the first reactor and then Mitsubishi became the main contractor. When a utility decided to order a reactor larger than reactors previously introduced, GE or Westinghouse was usually called in to share the contracting. Even after they took over the main contracting duties the Japanese makers continued to pay royalties to the US companies (and continue to do so today).

16.1.3 Domestic Reactor Development

During the early 1960s the JAEC was in the process of reshaping its nuclear R and D policies. The programme to develop a reactor based on domestic technology had begun fairly slowly. In 1959 JAERI had attempted to construct a reactor with a heterogeneous core. This project did not work out due to various technical difficulties including problems with the coolant. In the next few years reports of advances abroad in the field of nuclear power caused people in Japan to worry about falling behind the times. Feeling pressed by a need to reconsider its reactor development programme, the JAEC set up a Power Reactor Development Council in 1964 to help it do so.

The nature and direction of Japan's present nuclear programme can be traced directly to the recommendations made by the council. Despite its name, the council was not limited to the issue of choosing an appropriate reactor for development. It was expected to examine comprehensively all issues relating to nuclear power development in Japan. For the first time an attempt would be made to outline an integrated strategy. The council, composed of JAEC members and fourteen experts on nuclear power, deliberated for a year and half. During this time a study group was sent to Europe and the US to learn about the state of advanced reactor technology abroad.

In 1966, based on the council's recommendations, the JAEC published its decisions regarding the course of nuclear power development (JAEC, 1966). The JAEC's plan stressed the importance of a domestic reactor to over-all energy policy and reaffirmed FBR development as the primary policy goal. In addition to solving Japan's fuel problems (FBRs consume uranium extremely efficiently), the development of an FBR was expected to upgrade the standard of Japanese technology. For the first time specific steps for advancing FBRs were set: the construction of an experimental sodium-cooled FBR reactor using mixed oxide (MOX) fuel to be followed by a 200–300 MW prototype FBR.

In addition to the FBR the plan also called for the development of an advanced thermal reactor (ATR). This was a very significant decision and had not been an easy one to reach (this decision will be discussed in detail in the ATR section). Also of great significance were sections concerning the establishment of a domestic fuel cycle. Due to expected increases in demand, the importance of securing and using nuclear fuel efficiently was stressed. Accordingly the domestic fabrication of fuel and reprocessing of spent fuel were recommended.

To serve as the main organization for advanced reactor development and the establishment of a domestic fuel cycle, the Power Reactor and Nuclear Fuel Corporation (PNC) was established in 1967 (incorporating the Atomic Fuel Corporation). The government and the private sector agreed to split the cost for developing the first two advanced reactors (the utilities provided two-thirds of the capital from the private sector). The five domestic nuclear makers agreed to work in co-operation with the people at PNC to design and construct these reactors.

The JAEC's plan represented the first time a strong commitment was made to the development of indigenous technology in Japan. This was a break from the pattern of technology importation that had been repeated since the late 1880s. The importance attached to the new programme is clearly reflected by the growth of the government's atomic energy budget in the years before and after the establishment of PNC. From 1959 to 1966 the budget grew a total of 68 per cent; from 1966 to 1971 it jumped 277 per cent (see Figure 16.1).

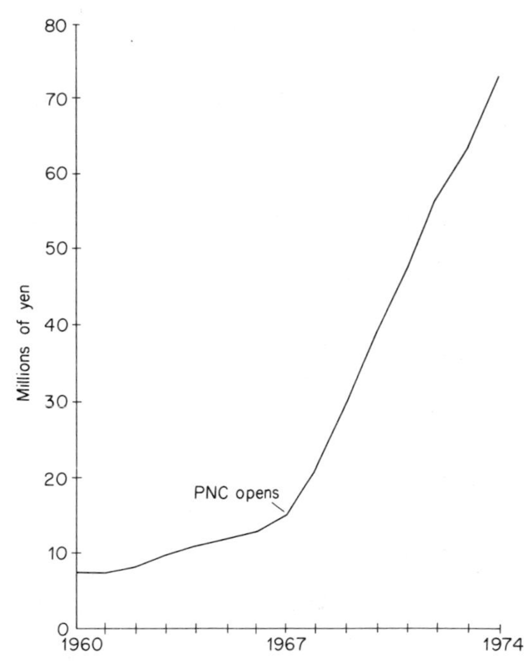

Figure 16.1 History of nuclear budget in Japan. Source: Japan Atomic Industrial Forum

16.1.4 *Effects of Oil Shock*

When the first so-called oil shock occurred in 1973 the Japanese economy was extremely dependent on imported oil. However, this had not always been the case. Actually, Japan's oil dependency had developed over a comparatively short period of time. As late as 1960 only 38 per cent of all energy consumed in Japan was produced from oil (*Sogo Enerugi Tokei*, 1984a). In that year coal was the main source of energy, producing 41.5 per cent of the energy consumed. And significantly, 75 per cent of the coal used came from domestic sources. However, due to the increasing cost of Japanese coal and the availability of inexpensive oil from the Middle East, oil imports grew rapidly when the Japanese economy boomed in the 1960s. Over a thirteen year period the percentage of Japan's energy produced from oil more than doubled. In 1973 it accounted for 78 per cent of the total energy consumed in Japan (*Sogo Enerugi Tokei*, 1984b) (see Figure 16.2).

The tremendous speed of this transition to oil dependence caused the psychological

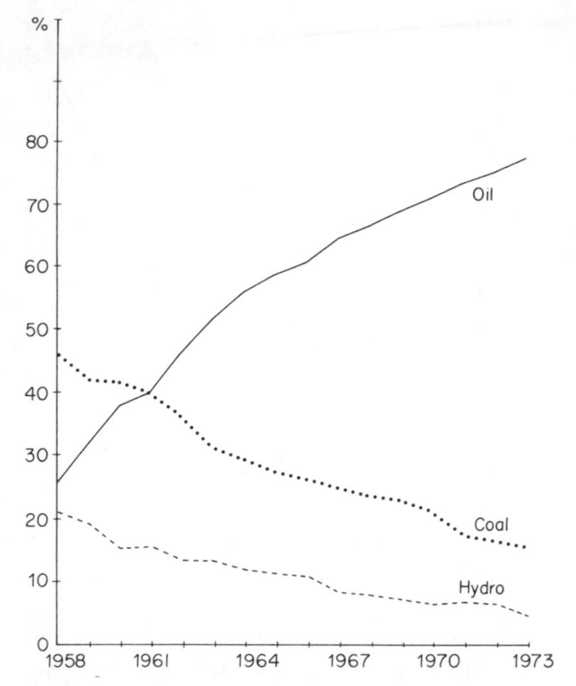

Figure 16.2 Percent of primary energy produced from oil, coal and hydro. Source: Ministry of International Trade and Industry

impact of the oil shock to be particularly great. People were forced to face an unpleasant fact: in roughly a decade their economy had become highly dependent on factors outside their control. Faced with an urgent need to alleviate its dependence on oil, Japan looked increasingly to nuclear power as a means of securing a stable source of energy.

16.1.5 *LWR Improvement and Standardization*

In the early 1970s nuclear plants in Japan were plagued by frequent problems which led to consistently low capacity factors. In 1975 the capacity factor for nuclear plants in Japan was a low 42.2 per cent. It increased to 52.8 per cent in 1976 but fell to 41.8 per cent the following year. Also, operators servicing reactors were being exposed to increasing amounts of radiation. Government officials worried that this poor record would jeopardize their efforts to promote nuclear energy.

The Ministry for International Trade and Industry (MITI), the agency in charge of promoting and co-ordinating industrial activity in Japan, responded by setting up a committee in 1975 to study methods for improving and standardizing LWRs. The main goal was a plan for increasing reactor availability and decreasing operator exposure to radioactivity. Four reactors were selected as objects for the study: 800 and 1100 MW PWRs, and equivalent size BWRs.

Based on the results of the committee's work, the first Improvement and Standardization Plan was carried out from 1975 to 1977. Work on the BWRs began with enlarging their containment vessels. The positioning of various parts and pipes was improved to allow operators more room for repairs and inspections. Additional safety valves and openings for transporting equipment helped to improve operations. The containment vessels of the PWRs were also enlarged. In addition tests were run on high tensile strength steel and pre-stressed concrete to measure their capacity to withstand earthquakes. It was hoped these materials could be used to produce an improved PWR containment vessel.

A second Improvement and Standardization Plan was undertaken from 1978 to 1980. The main goals this time were to improve components and systems, continue work on earthquake-proof designs and expand the scope of standardization to include peripheral equipment. Specifically, counter-measures against stress corrosion cracking were applied to the BWR. Efforts were also made to improve its core and fuel designs. Work on the PWR included improving the reliability of its steam generator. Automatic in-service pipe inspection equipment was introduced on a wide scale and cobalt-free equipment was installed in order to lower radiation exposure. As a result of the above measures, the predicted capacity factor of new plants was placed at 75 per cent, the length of regular inspections reduced from 85 to 70 days and radiation exposure limited to half of previous levels.

In 1978, on the initiative of General Electric, the US maker of BWRs, an international team was formed to do a feasibility study on an advanced version of the BWR (ABWR).

This team was made up of GE, Toshiba and Hitachi (Japan), ASEA-ATOM (Sweden) and Ansaldo Meccanico Nucleare (Italy). After evaluating the results of the feasibility study, TEPCO indicated interest in participating in demonstration tests and further research. In 1981 an agreement was reached between TEPCO, five other Japanese utilities that use BWRs, GE, Toshiba and Hitachi to begin a five year joint ABWR development project in Japan. The main objectives of this project are:

1. Increased safety and reliability.
2. A capacity factor of at least 80 per cent.
3. An 80 per cent reduction of radiation exposure to workers.
4. Reduced operating and construction costs.
5. An ability to practice load following safely and efficiently. (Load following is the altering of electrical output to meet different levels of demand during a given period of time. It will become increasingly important for nuclear power plants in Japan to practice this as the share of electricity produced by them continues to grow.)

To achieve these objectives a number of specific steps will be taken. These include improving reactor core and fuel designs, placing the recirculation pump inside the reactor vessel, etc. The new reactor will be 1300 MW, 200 MW larger than the largest previous model. It is hoped it will be ready for use in the early 1990s.

Later in 1981, to maintain the competitiveness of their reactor, Westinghouse and Mitsubishi announced a joint project to develop an advanced pressurized water reactor (APWR). Utilities in Japan that use PWRs and Bechtel, the huge US engineering and construction company, will also participate in the project. By making major changes in the reactor core and designing a high efficiency fuel, the people involved hope to:

1. Increase the efficiency of fuel consumption.
2. Create an ability to go from 100 per cent output to 50 per cent output (load following) within one hour (at present it takes more than three hours).
3. Allow for the use of uranium enriched to a lower level than uranium in use now.
4. Increase the reliability of the steam generator.

Since the objectives of the ABWR and APWR projects were similar to those of its Improvement and Standardization Plan, MITI decided to help fund them. In effect MITI's plan has been merged with the other two.

On a policy level the over-all goal is the creation of a 'Japanese type reactor'. Although significant improvements were made under the first two Improvement and Standardization Plans, major changes were limited to peripheral equipment. This time more ambitious steps are planned. For the first time the reactor core and fuel assemblies, the most complex parts of a nuclear reactor, will undergo major changes. And unlike the past when US makers were called in to help introduce new (enlarged) reactor models, ABWRs and APWRs will be built completely by Japanese makers from the start. The construction of the first 'Japanese type reactor' will signal the end of reliance on US technology, an event considered long overdue by some people in Japan.

16.2 Present Nuclear Programme

16.2.1 Operating Record

As of March 1986 there are 32 commercial nuclear power plants in operation in Japan. Their combined generating capacity is 24 521 MW. They comprise roughly 15 per cent of the total installed generating capacity and produce roughly 23 per cent of the electricity consumed in Japan. Ten additional reactors are under construction and six more are in the planning stage (see Figure 16.3). Efforts to improve operating records have borne fruit. The over-all capacity factor for nuclear power plants in Japan has risen rapidly over

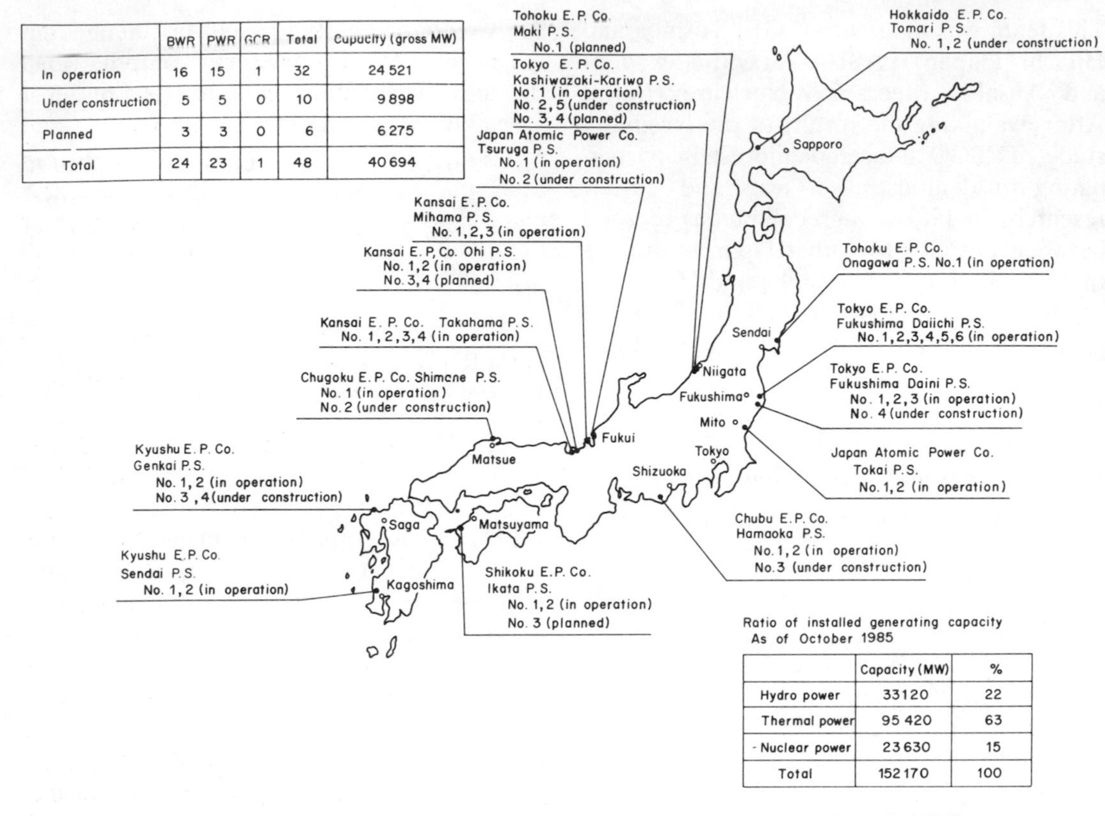

	BWR	PWR	GCR	Total	Capacity (gross MW)
In operation	16	15	1	32	24 521
Under construction	5	5	0	10	9 898
Planned	3	3	0	6	6 275
Total	24	23	1	48	40 694

	Capacity (MW)	%
Hydro power	33 120	22
Thermal power	95 420	63
Nuclear power	23 630	15
Total	152 170	100

Figure 16.3 Commercial nuclear power plant in Japan (As of March 1986)
Source: Japanese Atomic Industrial Forum

the past five years (see Figure 16.4). The 1984 over-all capacity factor (72.3 per cent) was the third highest among the major nuclear countries (behind West Germany and Sweden) and it reached 74.2 per cent in 1985. Three main reasons are usually cited for this vastly improved record:

1. Less time is needed for regular inspections (less regular repair work).
2. Less shutdowns due to improved reliability.
3. Breakdowns are responded to quickly.

Nuclear power continues to be an economical means of producing electricity in Japan. However, the difference in cost between nuclear power and coal, its closest competitor, is shrinking. According to MITI estimates, electricity from nuclear plants was about three yen less per kilowatt-hour (kWh) than electricity from coal plants in 1980. By 1984 this difference had shrunk to one yen (see Figure 16.5). In 1984 the cost of electricity from nuclear plants was roughly 13 yen kWh^{-1} while the cost of coal-fired electricity was roughly 14 yen kWh^{-1}. And when the cost of back end services (disposing of radioactive waste, decommissioning old reactors, etc.) was included nuclear power was estimated to be slightly more expensive than coal power. This trend has been caused by two factors. First, the capital cost of a nuclear plant is a high percentage of its total cost. Since construction costs have risen rapidly in Japan in recent years, the over-all cost of a nuclear power plant has increased more than the cost of a coal plant. Second, the price of imported coal in Japan has gone down in recent years. Nevertheless, nuclear power is expected to continue to be an economical means of producing electricity in Japan. While the cost of nuclear fuel is a relatively small part of a nuclear plant's total cost, the cost of fuel at

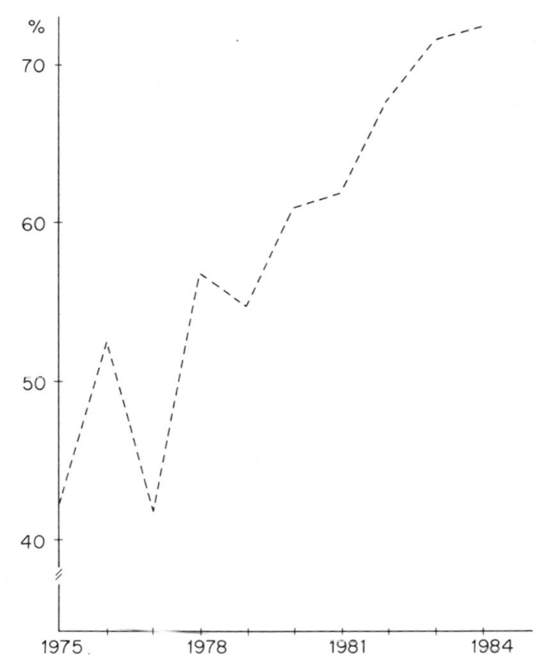

Figure 16.4 Average capacity factor for nuclear plants in Japan. Source: Japan Atomic Industrial Forum

a coal plant is a high percentage of its total cost. Considering that Japan imports almost all of the fuel it uses, a lack of vulnerability to increasing fuel prices should help nuclear power remain competitive. Additionally, active efforts to lower the cost of nuclear power are being made. The applications of newly developed materials are being investigated in hopes of increasing the life span of reactors. Also, a programme to reduce the length of regular inspections by improving the reliability of equipment and streamlining procedures, one that has produced good results so far (see Figure 16.6), will be continued. If nuclear plants continue to spend more time in operation the average cost of the power they produce will go down.

16.2.2 *Programme Administration*

The Japan Atomic Energy Commission (JAEC) has the authority to plan, deliberate and decide on matters concerning the development of nuclear power. The JAEC is composed of a chairman and four other members. Its recommendations are given directly to the prime minister, who is obligated by law to give them due consideration. The Nuclear Safety Commission is responsible for safety matters and also reports to the prime minister. The Atomic Energy Bureau and the Nuclear Safety Bureau of the Science and Technology Agency (STA) serve as the secretariats for the two Commissions. Other divisions of the STA oversee research on new reactor types, nuclear fuel, safety, etc. The minister of the STA serves simultaneously as the chairman of the JAEC. Matters pertaining to the commercial use of nuclear power are administered by MITI. This includes safety examinations, licensing, regulating and promoting. MITI's Advisory

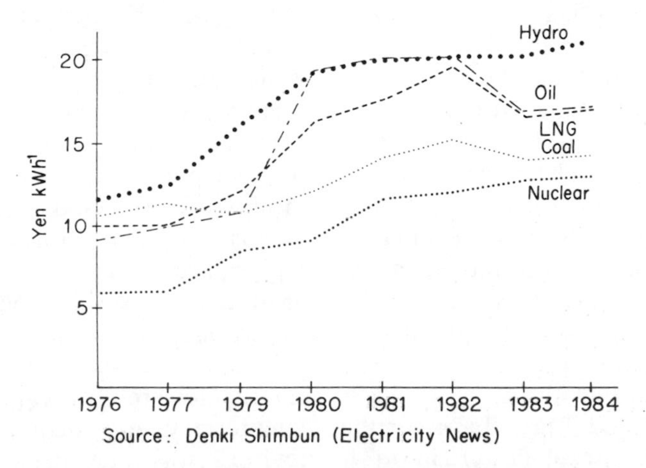

Figure 16.5 Electricity generation cost during the first year of operation. Source: MITI.

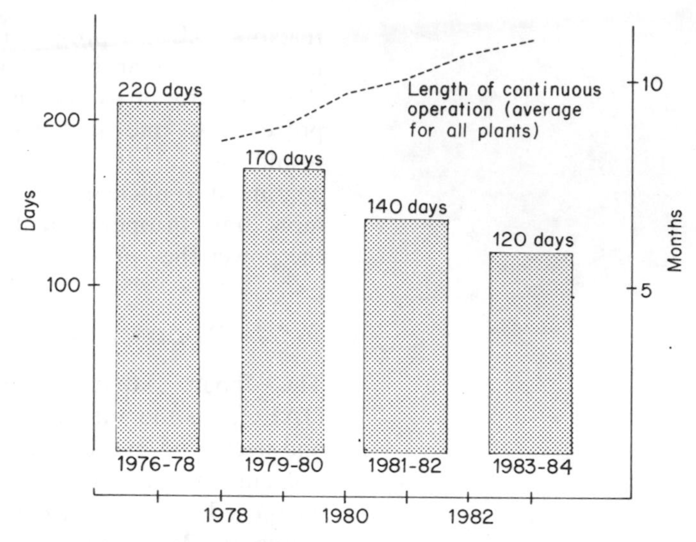

Figure 16.6 Length of regular inspections. Source: Denki Shimbun (Electricity News)

Committee on Energy makes the supply and demand projections used when forming targets for generation capacity. Both MITI and the JAEC set up specialist councils periodically to assist with specific problems.

16.2.3 *Siting, Public Acceptance and Generation Targets*

In the early 1970s people in Japan were becoming increasingly conscious of the pollution created by large industrial projects. This was particularly true in rural areas where power generation plants were built to meet the needs of people living far away. Many of the people living near these plants were fishermen. They worried that their livelihood would be destroyed if the environment was upset. While local opposition was directed at all power generation plants, concern over safety made the siting of nuclear plants particularly difficult. The scarcity of land suitable for sites (Japan is quite mountainous) added to the difficulty. This became an obstacle to the government's programme to promote nuclear power.

In order to deal with the siting problem the government passed 'Three Laws for the Development of Electrical Power' in 1974. The purpose of these laws was to pave the way for new power plants by improving the welfare of the area where they are sited. In accordance with these laws a tax is levied on utilities based on the generating capacity of their new plants. This money is used to finance a special fund. Both the town accepting a new plant and towns in the surrounding area receive money from the fund. They are allowed to use it for building roads, harbours, educational facilities, etc. Limits are placed on the amount of money a given town can receive to prevent a sudden bonanza of funds from disrupting its finances.

Efforts to promote the construction of new power plants have not been limited to financial measures. Beginning around 1977 MITI established a section in a number of its regional offices to promote new power plants. MITI and other government agencies began to hold public hearings where utilities and local residents could exchange views. The idea is to provide information on safety, the importance of electricity and other issues in an attempt to shift local public opinion toward the acceptance of new plants.

In late 1983 the Advisory Committee on Energy lowered its earlier predictions for long-term energy demand. The downward revision was due to a successful energy conservation programme, structural changes

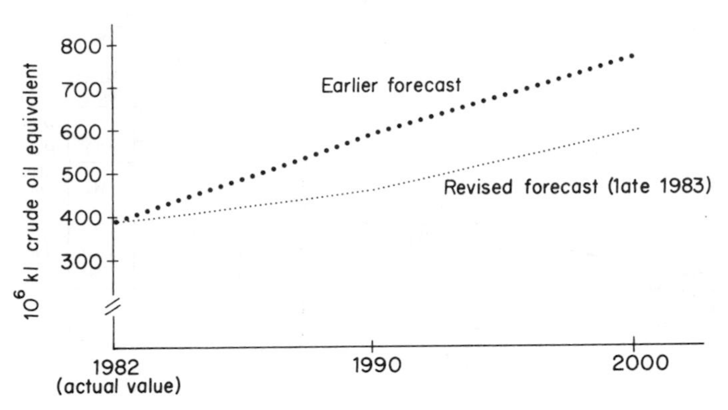

Figure 16.7 Long term energy demand forecast. Source: MITI's Advisory Committee on Energy

in Japanese industry and a lower than expected economic growth rate. In accordance with this revision targets for the generating capacity of nuclear power plants were also lowered (see Figures 16.7 and 16.8). The previous target of 46 000 MW by 1990 was lowered to 34 000 MW, a drop of 26 per cent.

Due in part to siting difficulties, there is little chance the government's targets for power generation capacity will be met in the near future. Still, efforts to promote and gain understanding have produced positive if not spectacular results. Japan's dependence on oil has been reduced. At the peak of its use in 1973 oil supplied 78 per cent of Japan's primary energy. By 1983, despite a 27 per cent increase in total primary energy consumed, this figure shrank to 61.9 per cent. Nuclear power was responsible for producing 42 per cent of the difference (Sogo Enerugi Tokei, 1984b). Naturally government policy supports a continuation of this trend. According to official estimates (which are in fact policy targets) the amount of primary energy obtained from oil should continue to fall (all the way to 42 per cent in 2000) and energy from nuclear power will pick up roughly half of the slack (see Figure 16.9). For more than a decade the number of nuclear power plants in operation has grown steadily. Since the first oil shock in 1973, 22 plants have gone on line. Only twice has a

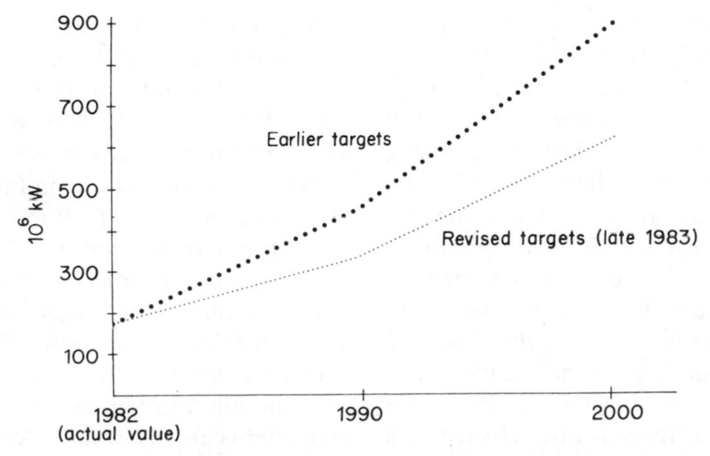

Figure 16.8 Targets for nuclear power generation capacity. Source: MITI's Advisory Committee on Energy

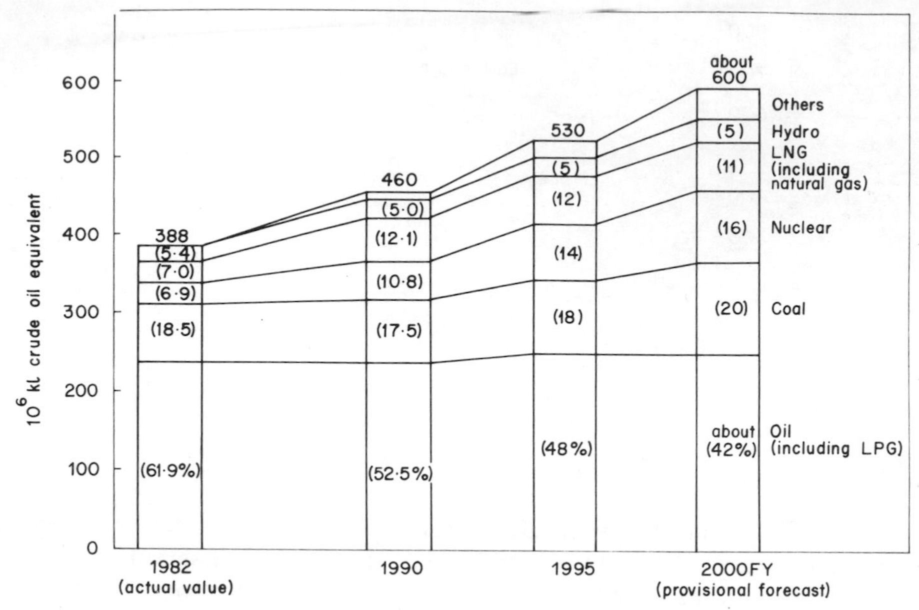

Figure 16.9 Long term supply outlook. Source: MITI's Advisory Committee on Energy, 1983

year gone by without at least one plant being started. Considering the difficulty of siting a new nuclear power plant in other countries during recent years (especially since the accident at Three Mile Island) this is a good record.

16.2.4 *Licensing*

The process of obtaining a licence for a new nuclear power plant can appear complex (see Figure 16.10), but in fact once it begins it almost always progresses fairly smoothly. Although two public hearings must be held, local citizens opposed to a new plant are not legally empowered to interrupt the licensing process. Only once has a licence application been refused after it was filed officially. However, it is not true that local opposition never holds up plans to build a new reactor. Serious negotiations between a utility and local citizens take place before official licence applications are filed. These negotiations can be difficult and time consuming. But once they are concluded there is little chance of a costly delay during construction due to local opposition.

16.2.5 *FBR*

16.2.5.1 The Programme FBR development in Japan is divided into four stages: experimental reactor, prototype reactor, demonstration reactor and commercialization. Japan's experimental FBR (called JOYO) went critical in 1977. In 1982 the reactor was converted into a radiation bed for use as a testing facility. The effects of radiation on fuel and other materials continue to be studied there.

At present Japan's FBR programme is in its second stage. PNC is preparing the site where it will build a prototype 280 MW loop-type FBR reactor (called MONJU). A 280 MW reactor is relatively small; commercial FBRs are expected to be in the range of 1000 MW. The purpose of MONJU is to demonstrate the performance, reliability and safety of Japanese technology before a full-scale reactor is built. The target date for MONJU to reach criticality is 1991. The utilities will help PNC to evaluate designs and supervise construction through a co-operation agreement between PNC and JAPCO. Parts and equipment for the reactor will be designed and manufactured by four companies: Mitsu-

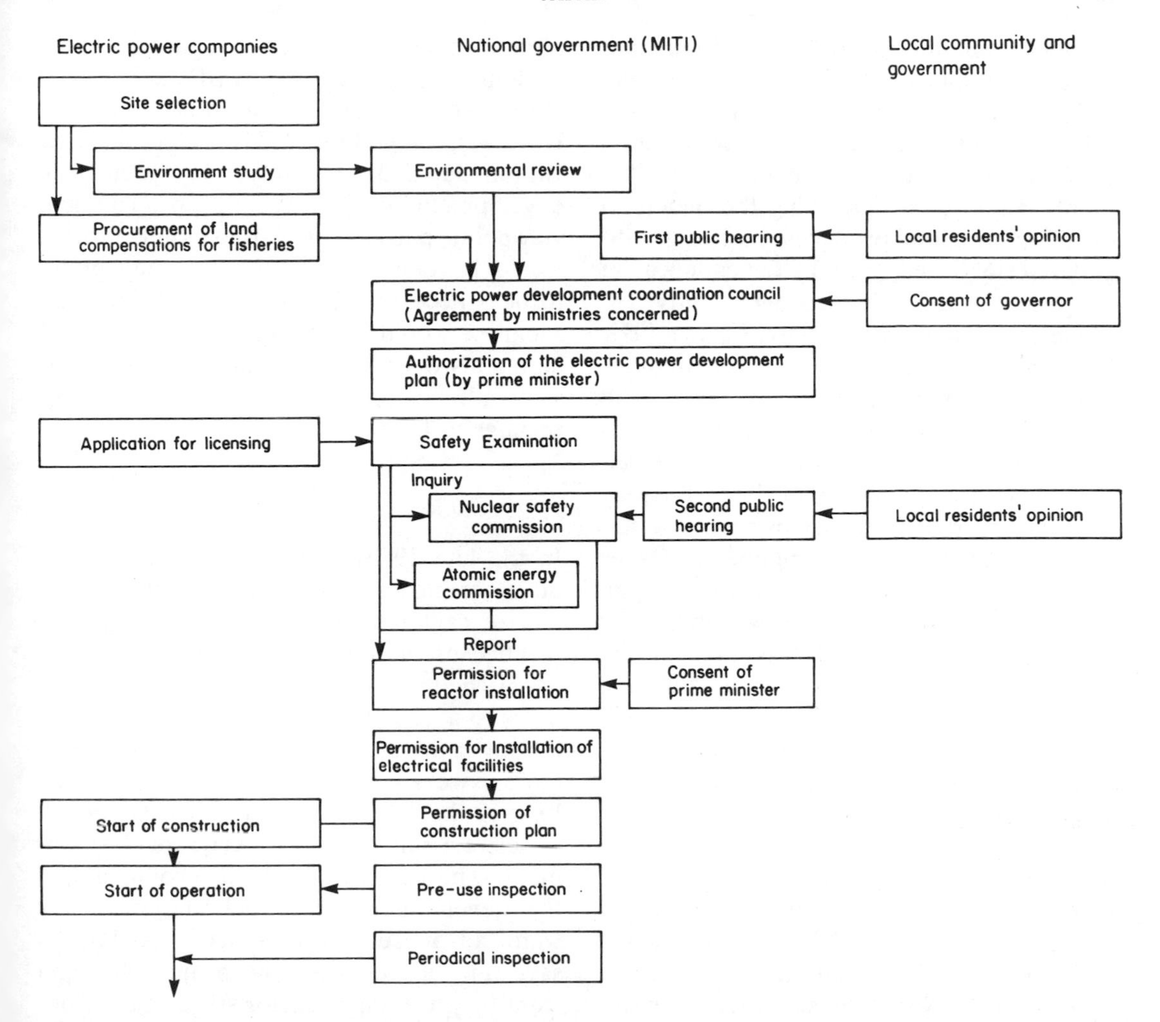

Figure 16.10 Licensing procedure for commercial nuclear power plants. Source: Japan Atomic Industrial Forum

bishi, Hitachi, Toshiba and Fuji Electric.

The next phase of Japan's FBR programme will be to build a 1000 MW demonstration reactor (DFBR). Since predicted construction costs are quite high, the utilities decided to begin a three year cost rationalization study in 1984. PNC is assisting the utilities by conducting related research. The major manufacturers have been asked to come up with their own FBR designs. Mitsubishi is presently working on plans for a more economic loop-type FBR. Hitachi and Toshiba are working on a pool-type FBR. These makers are expected to co-operate with each other while competing to certain extent. It is hoped this arrangement will produce the best design possible. The utilities are scheduled to select a design for the DFBR when the rationalization study ends in 1987.

16.2.5.2 Policy Issues The FBR has always had a strong appeal for people in Japan. They see the FBR as a means of alleviating a fundamental problem: extreme dependence on foreign energy. Japan's lack of natural resources is well known. Nevertheless the figures are striking. According to statistics published by the Agency of Natural Resources and Energy, a full 84 per cent of the energy consumed in Japan in 1983 was produced from imported fuel (*Sogo Enerugi Tokei*, 1984b). This situation fosters a strong awareness of the need to consume fuel efficiently. Experts claim FBRs consume fuel 60 times more efficiently than LWRs. This is mainly because the plutonium produced in FBRs can be used to make new fuel. If a complete fuel cycle (including the means to convert plutonium produced in FBRs into fuel) is established, Japan's dependence on foreign fuel will be reduced significantly.

Due to a relatively late start, Japan's FBR programme is not as advanced as that of other countries. Also, reflecting a world trend, the pace of FBR development in Japan has slowed in recent years. This has been caused by the extreme complexity of the technology involved and rapidly rising costs.

At present there are a number of issues that will hinder FBR development in Japan if they are not resolved. Responsibility for the programme is currently passing from the government to the private sector. Building and operating a commercial scale FBR will not be a simple matter for the utilities. It is important that PNC transfer as much of its experience and know-how as possible to the utilities. The utilities have formed a steering committee to co-ordinate the FBR project. In 1985 they decided that Japan Atomic Power Co. (JAPCO) will serve as the actual owner-operator. This is considered to be only the first step towards the successful management of the DFBR project.

As is true elsewhere, escalating costs are drawing concern. In 1979 the cost of building MONJU was estimated at 400 billion yen. The utilities agreed to pay 60 billion yen of this. However, due to the rising cost of construction materials and a tightening of safety standards, projected costs have risen almost 50 per cent. In 1984 the government asked the utilities to increase the amount of their support to 109 billion yen. Concluding that the cost escalation was unavoidable, the utilities agreed to the government's request in early 1985. As for the DFBR however, the utilities have made it clear they will be extremely reluctant to begin construction if the projected cost exceeds the cost of an LWR by more than 50 per cent.

International co-operation is also an issue of importance. A number of Japanese organizations are doing work on FBRs in co-operation with groups from other nations. However, unlike France, West Germany and Britain, Japan is not involved in a joint FBR development project. From an economic viewpoint participation in a joint project would be beneficial. In fact, it may prove to be a necessity. But if at all possible, the Japanese would like to maintain a degree of autonomy over their programme. How to do so while working with another country remains to be answered.

16.2.6 ATR

In the late 1950s, Japan, like a number of other countries, was trying to develop a reactor capable of using natural uranium as fuel because it did not have access to a supply of enriched uranium. The first reactor worked on was a so-called heterogeneous reactor. However, as mentioned earlier, this project did not work out due to technical problems. In 1962 the JAEC set up a specialist council to select a suitable reactor type for development. The council ended up recommending the development of a fuel-efficient heavy water advanced thermal reactor (ATR). To save time it called for the project to begin from the prototype reactor stage of development. Since ATRs are not as complex as FBRs, it seemed reasonable to skip the experimental stage. And since the ATR was meant to be an 'interim reactor' for use until FBRs are commercialized, it was important to have it ready for use as soon as possible.

The Power Reactor Development Consultation Council (described earlier) was formed not long after the above recommendations were made. To almost all members of the Council the need for developing an FBR was clear, but the need for an ATR programme was not. Members of the Council from the JAEC and JAERI were generally in favour. Certain members from the private sector, particularly the utilities, were opposed. Their reasons can be summarized as follows. First, since we are committed to the FBR, we should concentrate funds and manpower on that reactor. An ATR programme may hinder the development of FBRs. Second, if necessary, introducing a reactor developed abroad is preferable to committing to a domestic ATR programme (*Kinya Ishikawa, Genshiryoku Iinkai no Tatakai*, 1983). This second line of reasoning encompasses a number of ideas. Because huge sums of money and the introduction of new technology would be necessary, an independent programme would be very difficult. And since no conclusion has been reached (in either Japan or abroad) as to which type of ATR is the most appropriate, we should wait until the matter becomes clearer. Moreover, a Japanese ATR might become a hindrance to the introduction of a (superior) ATR developed abroad. This argument carried weight because of Japan's past success at introducing foreign technology. Also, it is probably fair to say that some people lacked confidence in Japan's domestic R and D capabilities at that time.

However, there were some powerful arguments in favour of the ATR. First of all, the ATR is capable of using a variety of fuels (including natural uranium and plutonium) and will therefore alleviate Japan's complete dependence on enriched uranium. Second, since it consumes fuel efficiently, the required amount of natural uranium and enrichment services will decrease. Third, after FBRs are established, the ATR can be used as an efficient supplier of plutonium (a material used to make FBR fuel). Fourth, in case FBRs are 'late' (that is, if their introduction takes more time than expected) the ATR can serve as a fuel-efficient 'interim reactor'. (*Daikibo Kenkyu Kaihatsu ni okeru Kokateki Manejimento*, 1980). These reasons persuaded the council to eventually recommend the development of a 165 MW prototype ATR.

Construction of the prototype reactor (called FUGEN) began at the end of 1970. The reactor attained criticality in the beginning of 1978 and began full operations a year later. Construction took more time than originally expected, but FUGEN has produced a good operating record since it was completed.

After a period of discussion and technical evaluation to allow for the building of a consensus, the JAEC recommended the construction of a demonstration 600 MW ATR in 1982. The Electric Power Development Corporation (EPDC) was placed in charge of the project. The EPDC, a company owned mainly by the government, was formed in 1952 to handle large scale hydroelectric projects. Government plans had called for the transfer of ATR technology from the public sector to the private sector. However, it appears the utilities were not enthusiastic about building the reactor due to its high cost. Thus the EPDC was offered the opportunity to become involved with nuclear power. The EPDC subsequently signed a technical co-operation agreement with PNC, the organization in charge of ATR development in the past.

Now basic blueprints for the new reactor are being drawn up at the EPDC. The beginning of operations is targeted for the early 1990s. In February of 1985 the EPDC announced that estimated construction costs had risen roughly 25 per cent (to 390 billion yen) in comparison with earlier estimates. Accordingly, it asked the utilities to increase the amount of their financial support (they had agreed to bear roughly 36 per cent of the cost). The utilities ended up acceding to the request. However, whether or not the ATR is eventually commercialized will probably depend on the result of efforts to reduce its cost.

16.2.7 *HTR*

High temperature reactors (HTRs) are nuclear reactors designed to produce heat rather than electricity. In Japan electricity comprises only about 30 per cent of the overall energy consumed. Much of the remainder is consumed in the form of process heat. Since in many cases HTRs could be used to supply this heat, official policy places importance on their development and use (*Genshiryoku Nenpo*, 1984a).

Work on HTRs began in Japan in 1969. The pace of this work has picked up in recent years. In 1982 the JAEC announced a specific development schedule calling for the construction of a 50 MW experimental HTR to be completed around 1990. This reactor will be cooled by helium and moderated by graphite. Originally the reactor was designed to produce heat of 1000 °C, but this figure was lowered to 950 °C due to the limitations of current materials technology. In 1986, mainly due to budget constraints the size of the plant was reduced to 30 MWt.

At present detailed plans for the experimental HTR are being reviewed. It is hoped the reactor can be made more economical. Hendel, a large scale demonstration test loop, is being used to test high temperature components under high heat and pressure conditions. Safety testing will be done using a rebuilt experimental critical facility. JAERI is conducting basic research on the technical difficulties of operating a factory using heat from an HTR.

16.2.8 *Nuclear Fuel Cycle*

16.2.8.1 Uranium Prospecting The Atomic Fuel Corporation (AFC) was formed in 1956 to handle the prospecting, mining and refining of uranium in Japan. In the beginning people at AFC hoped to fuel future nuclear reactors with domestic uranium supplemented by foreign supplies. However, by the time AFC was incorporated into the newly formed Power Reactor and Nuclear Fuel Corporation (PNC) in 1967, it was clear that domestic deposits of uranium were very limited. (In 1983 the total known and estimated amount of uranium in Japan was only 0.23 per cent of the world's total known and estimated deposits. *Genshiryoku Nenpo*, 1984b.) Dependence on foreign uranium was accepted as inevitable. Nevertheless, to help secure a stable supply, the decision was made to develop whatever amounts of domestic uranium could be found. Also, Japanese companies were encouraged to become involved in the development of uranium abroad.

At present Japanese utilities have secured a supply of 170 000 t of uranium through contracts with companies in Canada, England, Australia, etc. An additional 20 000 t should be obtained from a joint venture in Niger with France and other countries. This amount (190 000 t) is expected to suffice until the mid–1990s. At that time the need for an increased supply is expected to arise. Since the world uranium market is eventually expected to tighten, Japan plans to increase its efforts to locate and develop uranium.

One possibility being investigated is the capturing of uranium from sea water. Sea water contains a potentially significant amount of uranium. If this uranium could be obtained economically it would be a tremendous boon to Japan. However, due to the existence of many other elements, it is difficult to capture uranium alone from sea water. Thus, the development of a material capable of absorbing uranium selectively will be the key. Testing on this method of obtaining uranium is being done with the support of MITI at a research facility on the coast of Shikoku, the smallest of Japan's four main islands.

16.2.8.2 Enrichment In 1969 a successful trial enrichment of uranium was completed. Still, Japanese enrichment technology was suspected of being behind that of other nations. A specialist committee on enrichment technology was set up by the JAEC to study the matter, and a three year time period to meet specific research goals was set.

By 1972 conditions surrounding the enrichment of uranium had changed. Expectations for nuclear power were growing by leaps and bounds. An increasing number of people believed demand for enriched uranium would exceed world production capacity by 1980. The JAEC responded by setting in motion a project to construct an internationally competitive enrichment plant. In 1976 a 'check and review' was made to evaluate the progress of this programme. The level of enrichment technology attained was judged to be sufficient for building a 50 t SWU year^{-1} (see Chapter 4 for an explanation of SWU) pilot centrifuge enrichment plant. Construction began the following year. The plant started operating on a partial basis in 1979 and by 1981 had become fully operational.

Originally plans called for the construction of a commercial enrichment plant after the pilot plant was completed. However, in the end of 1983 PNC decided to first build a 200 t SWU year^{-1} prototype plant in order to improve the reliability and economics of their technology. The prototype plant is scheduled to begin operating at half capacity in 1987. Full capacity operations should begin in 1988.

Japan's first commercial enrichment plant will be built by the newly established Japan Nuclear Fuel Industries Company, a private company funded chiefly by the nine utilities. This centrifuge plant will have a capacity of 1500 t SWU year^{-1}. It will be located in Aomori Prefecture on the Shimokita Peninsula, the northern most part of Honshu, Japan's largest island. Partial operations at this plant are expected to begin in 1991.

At present Japan is almost completely dependent on foreign enrichment services. Japanese utilities have a long term contract with the US Department of Energy that guarantees a minimum of 600 t SWU of enrichment services per year. Their other source of enriched uranium is Eurodif (Eurodif's enrichment plant is in France). The utilities have contracted with Eurodif for 1000 t SWU year^{-1} between 1983 and 1990. According to the JAEC, the supply of enriched uranium secured under these two contracts should be sufficient until 1995. Official plans call for roughly 30 per cent of the uranium required after that time to be enriched at domestic facilities by the year 2000.

16.2.8.3 Reprocessing In 1961 the JAEC set up a Reprocessing Specialist Committee to investigate the possibility of constructing a domestic reprocessing plant. The following year the committee recommended building a reprocessing plant by the end of the decade based on technology introduced from abroad. This was a break with earlier plans that stressed the development of domestic technology. The change in approach reflected a desire to raise Japanese reprocessing technology to an internationally competitive level as rapidly as possible. In 1963 the Atomic Fuel Corporation hired the Nuclear Chemical Plant Co. of England to help it work on preliminary plans for a reprocessing plant. After delays caused by siting and other problems, the French company Saint Gobain Techniques Nouvelles was chosen in 1966 to draw up detailed plans for a 0.7 t day^{-1} reprocessing plant and to oversee its construction.

By this time the utilities had already begun to order LWRs from the US, and they wanted a reprocessing plant that could handle LWR spent fuel. Because existing reprocessing plants were all designed to handle spent natural uranium fuel, Saint Gobain could not simply copy plans from another plant and use them without changes. And because reprocessing was still a new technology, the plans took more time to complete than expected. They were finally finished in 1969. After the JAEC spent two years examining possible safety problems, construction began in 1971 and was finished by the end of 1974. For the next two years preliminary tests were carried out. These were necessary to train the staff and to confirm the operability of the machinery and the process as a whole.

At the beginning of 1977 when the plant was ready to enter trial operations, a new roadblock suddenly appeared. According to a clause in the US–Japan Agreement on the Peaceful Use of Nuclear Energy, prior consent from the US is necessary to reprocess uranium that was enriched in the US. Invoking this clause, newly elected President Carter caused work at the new plant to stop. In 1974 India had surprised the world by reprocessing spent fuel and using the plutonium obtained therefrom to produce and explode an atomic device. President Carter feared other nations would follow this example and therefore moved to check the spread of reprocessing facilities.

Needless to say, people in Japan were reluctant to abandon a project they had invested so much effort in. Moreover, possessed of new natural resources, Japanese view the efficient use of fuel as a matter of utmost importance. Negotiations on this matter between Japan and the US were held several times over the next nine months. In the end a temporary agreement was reached. The new plant would be allowed to operate for two years provided it did not handle more than 99 t of spent fuel. During this time Japan would take no major action concerning a second reprocessing plant. Also, PNC promised to test co-processing, a new method of reprocessing. It was hoped this method would enable nuclear fuel to be reprocessed without pure plutonium being produced. (It was later determined that co-processing would not prevent a country from obtaining weapons-grade plutonium if it so desired.) In subsequent negotiations limits on the plant were extended two years (until 1981) and 50 t (to 149 t of spent fuel). Japan again agreed to not take major action on a new plant until the designated time period expired.

When Ronald Reagan became President in 1981 the chances of Japan and the US reaching a final agreement seemed to improve. Unlike President Carter, President Reagan was not against the reprocessing of nuclear fuel in Japan. In November of 1981 the temporary agreement on reprocessing in Japan was extended another three years. This time the reprocessing plant was allowed to operate at full capacity. Nevertheless, in contrast to President Reagan's stance, certain members of the US Congress continue to be strongly against the loosening of reprocessing restrictions. Thus it remains uncertain when a final solution will be reached. The Japanse side wants to avoid being affected by the vagaries of US policy. It therefore hopes to sign a comprehensive long-term agreement. On the other hand, the US side does not want to sign a long-term agreement unless the restrictions specified in the US–Japan Nuclear Agreement are strengthened.

Since the reprocessing plant finally opened in 1977 operations have not proceeded smoothly. Various difficulties, including leaks in more than one dissolver, have arisen on a consistent basis. The plant has been shut down numerous times and has yet to operate at full capacity. PNC (the operator of the plant) is not proud of this record. But it hopes to assist future attempts at reprocessing in Japan by applying the knowledge and experience it has gained. For this reason it signed a co-operation agreement with Japan Nuclear Fuel Services in 1982.

Japan Nuclear Fuel Services is a private company formed mainly by the utilities to be the owner-operator of Japan's second reprocessing plant. It plans to build the plant near the new enrichment and low level waste disposal facilities in Aomori Prefecture. Preparatory construction work will begin in 1986 with the start of operations scheduled for 1995. Until the new plant is completed Japan will continue to rely on Britain (BNFL) and France (COGEMA) for the bulk of its reprocessing needs.

16.2.8.4 MOX Fuel The emphasis placed on developing mixed oxide fuel (MOX) in Japan reflects a commitment to develop and utilize the advanced (ATR and FBR) reactors that use it. The development of MOX fuel in Japan got underway at AFC in 1965 when Japan's first fuel development facility was completed. When PNC was formed in 1967 and concrete plans were made to build advanced reactors, the need arose for an additional testing facility. This was completed in 1972. Equipment at the new facility was grouped into two 'lines'—an FBR line and an ATR line. (The FBR and ATR use different types of MOX fuel.)

At first the ATR line produced material used in developmental testing. It also manufactured fuel for use in tests done in Norway and England. In 1974 the ATR line was remodelled and enlarged in order to manufacture fuel for FUGEN. The revamped ATR line succeeded in manufacturing its targeted amount of FUGEN fuel in 1978 slightly ahead of schedule. Meanwhile, based on earlier R and D and co-operative testing done in the US, Britain, France, etc., the FBR line was busy with the production of fuel for JOYO. By 1974 it had manufactured enough fuel for JOYO to reach criticality.

By the end of 1984 a total of 70 t of MOX fuel had been produced in Japan. This placed Japan among the world's leading producers of MOX fuel. At present Japanese facilities have the capacity to manufacture 11 t of MOX fuel yearly (10 t for FUGEN and 1 t for JOYO). A plant to supply MONJU with 5 t year^{-1} of MOX fuel has begun trial operations. It is due to begin full operations in 1988. Construction work will begin soon on a plant capable of supplying the demonstration ATR with 50 t year^{-1} of MOX fuel. The target date for its completion is in the early 1990s.

Until recently Japan had imported all of the plutonium it needed for manufacturing MOX fuel. However, a plant that converts plutonium nitrate (the form plutonium is in after it is recovered from spent fuel) into plutonium suitable for MOX fuel was completed in early 1983. Since then it has been in trial operations. The plutonium obtained at the conversion plant is currently being used to make fuel for FUGEN.

16.2.8.5 Thermal Recycling Official plans to use plutonium in light water reactors go back to the JAEC's Long Term Plan of 1967. The commitment made at that time to actively develop FBRs insured a need for plutonium in the future. Recognizing that FBR commercialization would take time, the JAEC recommended developing thermal recycling technology. Instead of merely storing the plutonium recovered from spent fuel, it made sense to use it in LWRs until FBRs were established.

As the predicted date of FBR commercialization has receded into the future, dealing with excess plutonium has become an important issue. For the time being the only need for plutonium is to make fuel for JOYO and FUGEN. Fuel for MONJU and the demonstration ATR will also contain plutonium, but the amount involved will be comparatively small. At present Japan relies on France and England for its reprocessing needs. A new reprocessing plant is scheduled to open in France in a few years. This means the amount of plutonium obtained from spent fuel and returned to Japan will increase in the early 1990s. And since the new reprocessing plant in Aomori is scheduled to begin operations in 1995, excess plutonium is expected to pile up at a rapid rate from that time on (see Figure 16.11).

Awareness of the need to deal with excess plutonium has spurred efforts to develop thermal recycling technology. In 1982 the Thermal Recycling Specialist Subcommittee of MITI's Advisory Committee on Energy published a plan that calls for full-scale commercial thermal recycling operations to begin in the latter half of the 1990s. The first stage of this plan is the use of small amounts of plutonium in relatively small-scale reactors. JAPCO plans to begin 'burning' plutonium in its 350 MW BWR at Tsuruga later this year. The next stage will be to use plutonium in an 800–1000 MW BWR. This time the amount of plutonium used will be one third of the reactor's fuel. JAPCO believes it can use this much plutonium without making any significant changes to the reactor. As for thermal recycling using PWRs, KEPCO plans to run a small-scale test at its 340 MW PWR at Mihama, and in 1985 it formally announced plans to test plutonium use at two of its large-scale reactors. Other utilities are expected to follow suit in the near future (*Chukan Hokoku*, 1985).

If it proves to be extremely expensive, plans for thermal recycling may have to be changed. But such a decision would be difficult to make. Not all of the benefits of thermal recycling are quantifiable in terms of money.

Another potential obstacle to thermal recycling is political pressure from the US. The US worries that a dangerous precedent would be set if it permitted Japan to use plutonium obtained from uranium enriched in the US. A conceivable method for avoiding this problem would be to use ATRs to consume plutonium in Japan. Japanese ATR technology was developed domestically and has not been transferred abroad. Thus the US might be able to permit Japan to use plutonium in its ATRs without creating a precedent. However, unless ATR costs can be reduced, this method would be even more expensive than recycling plutonium in normal LWRs.

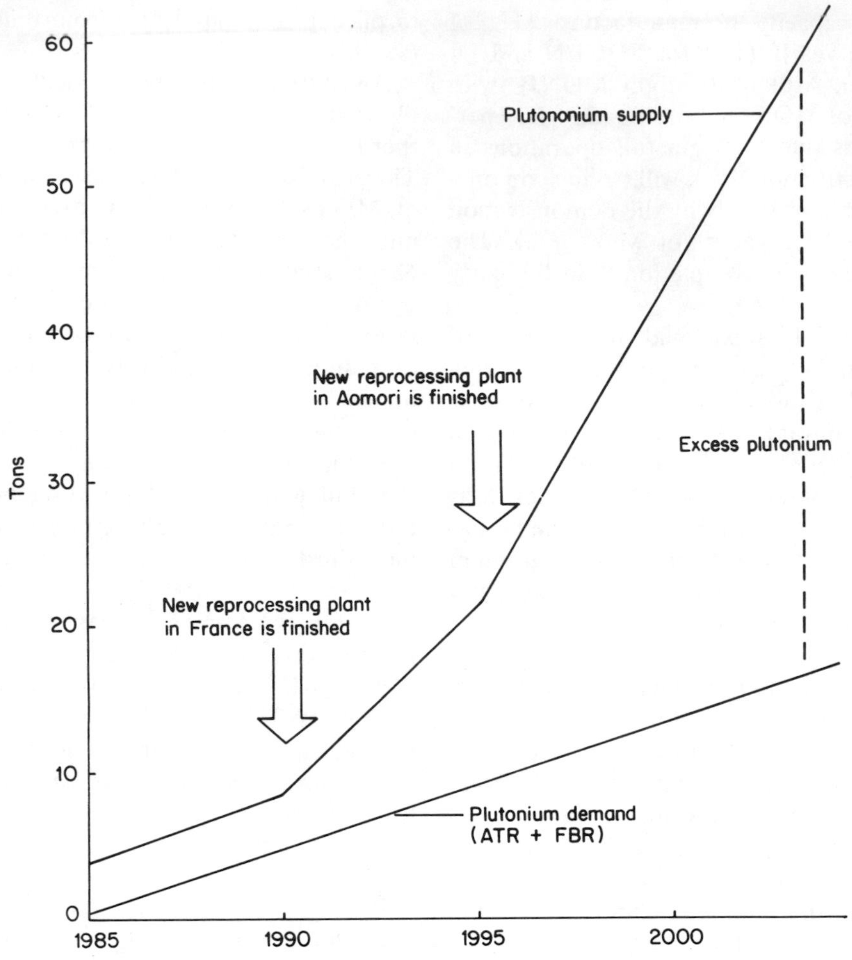

Figure 16.11 Plutonium supply and demand. Source: Nihon keizai Shimbun (Japan Economic Journal). Note: This chart does not reflect actual predictions precisely. It is presented to show a general trend

16.2.8.6 Nuclear Waste Management

(a) *Low Level Waste* Low level waste is currently managed at the locations where it is produced. Most of this waste is first processed in order to reduce its volume. It is then mixed with concrete (or other materials), poured into drums and placed in on-site storage facilities for safekeeping. At the end of 1985 there were approximately 627 000 drums of low level waste in Japan. This number is expected to increase to 950 000 by 1990 and 1 550 000 by the year 2000. Official policy (set by the JAEC in 1976) calls for this waste to be disposed of in both the ground and at the bottom of the sea.

Concerning sea bed disposal, basic research and environmental impact studies are being done. For the time being, however, Japan is unable to carry out on-site testing. In February of 1983 members of the London Treaty (including Japan, the United Kingdom, the US, the USSR, West Germany, France, etc.) agreed to temporarily suspend sea bed waste disposal. An international panel was set up to examine and evaluate disposal technology which reported on technical acceptability in 1985. Japan hopes an acceptable method for sea bed disposal will be found. In the meantime it is striving to gain the understanding of its neighbour countries concerning this issue.

As for ground disposal, the Japan Nuclear Fuel Industries Company is planning to build a storage facility near the commercial enrichment and reprocessing plants in Aomori Prefecture. This facility will be capable of accepting one million 200 l drums over a period of 20 years (and may be enlarged at a later date). It is expected to be completed in 1991 and will cost roughly 100 billion yen. The necessary R and D work for this facility is being done at several places. JAERI is using radioactive isotopes to test the behaviour of different radioactive materials when placed in the ground (shallow burial). The Central Research Institute for Electric Power Industry (CRIEPI), a joint facility run by the utilities, is working to set packing and disposal facility standards. The Radioactive Waste Management Center, a private company funded mainly by the utilities, is working on means for monitoring waste and protecting the environment. This company will also carry out on-site tests for both shallow ground and sea bed waste disposal methods.

At present Japan has no legal standards for classifying levels of radioactivity. This means even extremely low level waste must be stored in the same manner as other radioactive waste. When the decommissioning of reactors (after their 'life span' has been completed) begins in Japan the volume of this type of waste is expected to grow rapidly. Accordingly, the JAEC's Specialist Committee on Waste Disposal proposed that low level waste be divided into three types: low level, very low and waste that does not require special handling. Using this classification system waste management can be rationalized. As time passes and radioactive levels fall, the strict management of waste will become unnecessary.

(b) *High Level Waste* As in many other countries, the deep burial (geologic) method of disposal is considered to be the most promising means of eventually dealing with highly radioactive (high level) waste. The idea is to combine artificial barriers (cannisters, storehouse walls, etc.) with natural barriers (rock strata) deep in the earth to ensure that radiation from waste does not enter the environment. Officials hope to demonstrate the feasibility of this method by the year 2000 (*Genshiryoku Nenpo*, 1984c).

At present high level waste is kept in a storage pool at the Tokai Mura reprocessing plant. In the future it will be vitrified (turned into a form of glass) and then placed in a new storage facility until a more permanent method of disposal is ready for use. Although Japanese vitrification technology lags behind countries such as the US and France, PNC is proceeding with the construction of a pilot vitrification plant that is scheduled to begin operations in 1990. The new facility for storing the vitrified waste should be ready for use by 1992.

REFERENCES

Chukan Hokoku (Interim Report) (1985). Sogo Enerugi Chosakai Genshiryoku Bukai Purutoniumu Risaikuru Shoiinkai (Advisory Committee on Energy, Nuclear Energy Section, Subcommittee on Plutonium Recycling), Tokyo, p. 5.

Daikibo Kenkyu Kaihatsu ni okeru Kokateki Manejimento (1980). (Effective Management in Large Scale Research and Development) Sogo Kenkyu Kaihatsu Kiko (National Institute for Research Advancement), Tokyo, p. 102.

Genshiryoku Nenpo (1984a). (Annual Nuclear Energy Report 1984 Edition) Japan Atomic Energy Commission, 1985, Tokyo, p. 33.

Genshiryoku Nenpo (1984b). (Annual Nuclear Energy Report 1984 Edition) Japan Atomic Energy Commission, 1985, Tokyo, p. 82.

Genshiryoku Nenpo (1984c). (Annual Nuclear Energy Report 1984 Edition) Japan Atomic Energy Commission, 1985, Tokyo, p. 27.

JAEC (1966). *Doryoku Kaihatsu no Kihon Hoshin ni Tsuite* (Concerning the Basic Course of Reactor Development) JAEC, Tokyo.

Kinya Ishikawa, Genshiryoku Iinkai no Tatakai (1983). (Battles of the JAEC), Denryoku Shinposha (New Electricity Information Company), Tokyo, p. 37.

Sogo Enerugi Tokei (1984a). (Comprehensive Energy Statistics 1984 edition) Ministry of International Trade and Industry, 1985, Tokyo, p. 261.

Sogo Enerugi Tokei (1984b). (Comprehensive Energy Statistics 1984 Edition), Ministry of International Trade and Industry 1985, Tokyo, pp. 269, 276.

Tokyo Denryoku Sanju Nenshi (1983). (30 Year History of the Tokyo Electric Power Company), Tokyo Electric Power Company, Tokyo, p. 564.

Nuclear Power: Policy and Prospects
Edited by P. M. S. Jones

17

India

P. K. IYENGAR
Bhabha Atomic Research Centre

17.1 Introduction

The beginning of the Indian nuclear programme was an evolution and not an event. Perhaps because of the experience gained during foreign domination, the nation as a whole had a great will to become self-sufficient and the leaders in science fully shared this national ethos. Again due to the same backdrop, the difficulties in this arduous task did not weigh very heavily on their minds. This ancient country was making a fresh start with the mental vigour of the new. In retrospect, these appear to be important attitudes that influenced the technological choices the country made in developing its nuclear programme.

The pace of evolution of technology from science and its eventual maturing to a full-fledged industry is often determined by compelling reasons like wars, political necessities or just economic compulsions. This evolution, nevertheless, requires enlightened leadership with foresight, ability to muster adequate skills and a spirit of adventure which is essential for progress. Dr H. J. Bhabha provided this leadership in India and mustered the scientific talent in the country to launch a programme for nuclear energy. It was perhaps his association with scientists at Cambridge and other European laboratories in the 1930s, that helped him to develop a clear perspective of frontline science and advanced technologies. More importantly, he realized that these could not just be transplanted into a developing country. These had to be nurtured and the essential component in doing this was to develop expertise and skills in all the sectors of nuclear programme with highly trained manpower—in numbers that are proportionate to the size of the programme.

In March 1944, even before the power of the atom was demonstrated at Hiroshima and Nagasaki, and before the fact that the feasibility of controlled nuclear chain reaction had already been established was known to the world, Bhabha had a plan of action in hand. The foundation of the nuclear programme in India was laid by establishing the Tata Institute of Fundamental Research in 1945, which concentrated on the basic sciences that impinge on development of nuclear energy. This institute not only helped in establishing a scientific base for nuclear research but also in building a nucleus of dedicated scientists. Independence provided the intelligentsia a new opportunity to build an India of their dreams. Jawaharlal Nehru took personal charge of science in India and made it an organized activity with a special place in national life. The Atomic Energy Commission (AEC) was set up in 1948 and it constituted the most prominent Indian scientists of that time. The fact that the charge of the atomic energy programme has always been with the successive Prime Ministers of the country, who were all convinced of its importance to the nation, has had no mean effect on this programme.

Soon after the setting up of the Atomic Energy Commission, the essential elements of the nuclear programme were visualized and new activities were generated. While work in the areas of nuclear physics, electronics and instrumentation was being pursued at TIFR, laboratories for research in chemistry and metallurgy were set up separately. The commission also set up a Rare Mineral Survey Unit in July 1949 which eventually expanded to become the Atomic Minerals Division (AMD). By 1950, this division identified commercially exploitable uranium deposits. Indian Rare Earths Limited (IRE) was established in August 1950 to treat monazite sand for the recovery of rare earths. In April 1953, construction of a thorium plant was started at Trombay to treat the residues after rare earths separation, for the recovery of thorium. This plant was commissioned in August, 1955. As the activities of the commission had expanded substantially, the government of India set up a separate Department of Atomic Energy (DAE) in August 1954. The prime R and D organization of the Department, the Bhabha Atomic Research Centre was established in 1957 where a broad spectrum of research activities of the Department was initiated.

17.2 Bhabha Atomic Research Centre

The activities of this research centre, which was called the Atomic Energy Establishment Trombay, till it was renamed after its founder in 1967, began with the setting up of the Apsara reactor. The charge of enriched uranium was obtained from the United Kingdom. This swimming pool reactor has been used for neutron physics studies and for preparation of radioisotopes ever since it went critical in 1956. It needs no emphasizing that building the first reactor on their own was a tremendous experience for the Indian scientists. It needed inputs from a variety of fields in science and engineering. This not only nucleated groups in all the fields essential for the nuclear energy programme, but also made the community of scientists realize the necessity of multidisciplinary collaboration for its success. The presence of experts in such a broad spectrum of specializations under one roof, as it were, is perhaps one of the most important features of BARC.

This centre went on to build more reactors. As indicated in Table 17.1, CIRUS, a 40 MWt natural uranium heavy water reactor, was built in collaboration with Canada and went critical in 1960. It has rendered excellent service for over two decades. Half of the first load of the fuel elements was made in India and considerable alterations in water purification systems were made at the early stages. It has paid dividends and the reactor vessel has lasted well, for more than 25 years now.

A Zero Energy Reactor for Lattice Investigations on New Assemblies, (ZERLINA) went critical in 1961 and was used for studying heavy water natural uranium systems. This has since been decommissioned in 1983. PURNIMA, a small plutonium

Table 17.1 Research Reactors at Trombay

	Apsara	Cirus	Zerlina[1]	Purnima-I	Purnima-II	Dhruva
Type of reactor	pool type	tank type	tank type	fast	solution	tank type
Fuel	enriched uranium	natural uranium	natural uranium	plutonium	^{233}U as uranyl nitrate	natural uranium
Moderator	light water	heavy water	heavy water	—	light water	heavy water
Coolant	light water	light water	heavy water	air	light water	heavy water
Power level	1 MWt	40 MWt	low	low	low	100 MWt
Maximum neutron flux	1.25×10^{13}	6.7×10^{13}	—	—	—	1.8×10^{14}
Date of criticality	August 4, 1956	July 10, 1960	January 14, 1961	May 18, 1972	May 10, 1984	August 8, 1985

[1]Zerlina was decommissioned in 1983

fuelled fast reactor, was set up to study the physics parameters in 1972. In 1984 PURNIMA-II, another small reactor, went critical. This used ^{233}U as fuel in the form of uranyl nitrate solution. BeO is used as the reflector. Apart from the vital experience gained by making such a variety of reactors, it demonstrates the depth in the expertise generated and the maturity of the Indian programme. This experience also has significant bearing on the capabilities that have been acquired in all the aspects of the fuel cycle.

A large natural uranium fuelled heavy water reactor went critical on August 8, 1985. Dhruva, the largest reactor at Trombay, is designed for 100 MWt. This has been designed and engineered completely indigenously. All the large and complicated components like the calandria and the fuelling machine have been made locally.

R and D work on ore dressing, chemical engineering, unit processes for purification and preparation of UO_2 and U metal, fabrication of fuel rods and other nuclear reactor components, reprocessing and waste management is being presently done at BARC by strong groups which were established early. Today this investment of effort, time and financial resources is paying off. It provides the backbone of the Indian effort in establishing full scale units in all the legs of the fuel cycle and having strong capabilities in each aspect.

Achieving independent capability in these fields is practically impossible without developing strength in basic research and manpower. Strong groups are now established not only in basic sciences like nuclear, neutron and theoretical physics, radiation and radiochemistry and biology but also in applied areas like metallurgy, electronics, instrumentation, chemical engineering, reactor engineering, isotopes production and their utilization.

A special feature of the early era of the Indian nuclear programme was the clarity of vision in planning for the future. For instance, it was visualized that a support of electronics and instrumentation was essential for the growth of nuclear programme. It was also visualized that at the time when the need would be crucial, it would not be forthcoming from the Indian industry for a variety of reasons—ranging from unusual sophistication required to hard economics. BARC was conceived as the cradle in which the R and D efforts in various fields impinging on the programme were to be nurtured. To take care of the needs in electronics, the small seed sown at BARC blossomed into a modern manufacturing unit of Electronics Corporation of India Limited. Similarly, other efforts initiated at BARC were also projected as independent units taking care of major aspects of the Indian nuclear programme. These include the Nuclear Power Board, the Heavy Water Projects and the Nuclear Fuel Complex. Research activities on fast breeder reactors were nucleated and then transferred to an independent laboratory, the Reactor Research Centre at Kalpakkam. Futuristic technologies on which effort has already been initiated at BARC are in the process of being transferred to a Centre for Advanced Technology at Indore. Activities at this Centre will include lasers for fusion research, accelerators and plasma related work. To provide suitable initial momentum, the new units were started with sufficient expertise and manpower provided by BARC.

Since the well defined aim of the nuclear energy programme in India is to develop peaceful uses of atomic energy for the benefit of its populace, utilization of radiations and radioisotopes in medicine, agriculture and industry have been the pursuits of this centre from the early days of its inception. With the advent of ISOPHARM, a newly installed production unit, BARC is presently supplying radiopharmaceuticals for use by over 400 000 patients. A large number of radioimmunoassay kits have been developed for increasing use in the country. A plant for radiation sterilization of medical products has been operating since 1974 and a wide range of radiography equipment and nuclear gauges are being made to meet the needs of the industry. In addition, constant efforts are made to educate the industry and train the necessary manpower. Isotope utilization for hydrology and movement of silt in ports has progressed very well over the years. Radiations have also been used extensively in a variety of applications ranging from wood polymerization to mutation breeding and food preservation.

From the point of view of a developing country the advances in agriculture that are aided by nuclear techniques have very far reaching consequences. BARC has devoted a very substantial effort on improvement of crop plants by induced genetic mutations. Varieties of pulses, wheat, rice, groundnut and jute have been developed which are found to be sturdier and have higher yields. Many of these are already in extensive use and for some, seeds have been released to the farmers. In a large country like India, the economic benefits derived from this development can be considerable. In a country where post crop losses can total over Rs 30 billion, every step in effective preservation can be helpful. Preservation of food by radiation has received considerable attention at BARC. Shelf life of potatoes, onions, and many other products are considerably increased without any adverse effect under local conditions.

Another essential ingredient for establishing high technology on a firm footing was cultivation of manpower of high quality in adequately large numbers and with specializations covering a broad spectrum ranging from physics, chemistry and metallurgy to mechanical, chemical, electrical and computer engineering. A training school for graduate scientists and engineers was started as early as 1957 and about 200 carefully picked graduates in various disciplines undergo a very rigorous training for one year. The course covers, in addition to basic areas, all aspects of nuclear technology which provide the students with a total perspective of nuclear science. The teaching staff is drawn entirely from the working scientists who can also provide better orientation and communicate their practical experience. After completing the training school, the officers have an opportunity to work for their PhD and obtain their degrees from various universities that recognize this centre. This has helped in retaining an academic flavour, which is found very useful for a research centre. Thus BARC, the prime nuclear research organization, has proved to be a launching pad for this country to enter into an era of high technology.

17.3 The Indian Energy Perspective

Four decades ago, in 1947, India emerged as an independent country. It determined to shed its role of a supplier of raw materials which most developing countries are forced to assume in the present world economic system. To develop its industrial capabilities the foremost requirement was power, for which large power projects and bold strategies were needed. These were launched in

Table 17.2 Progress of installed generating capacity in utilities

Year (ending March 31)	Generating capacity (MW)	Share of various modes (%) Hydro	Thermal	Nuclear	Diesel and gas turbines
1950	1712	32.7	58.7	—	8.6
1955	2695	34.9	57.4	—	7.7
1961	4653	41.2	52.4	—	6.4
1966	9027	41.2	48.9	—	5.4
1969	12 957	45.6	51.3	—	3.1
1974	16 664	41.8	51.9	3.9	2.4
1982	32 389	37.6	57.9	2.7	1.8
1985	40 650	35.0	62.3	2.7	—
2000 *planned	100 000	40	50	10	—

the fifties. There has been considerable growth in the installed generating capacity for electricy as shown in Table 17.2. This has risen from 1712 MWe in December, 1950 to more than 40 000 MWe at present. Most of this is contributed by hydro power and thermal power stations based on coal.

The total proven, indicated and inferred, resources of coal in India are 10^{11} t. The present coal production is about 130 million t year^{-1} which is expected to increase to about 500 million t year^{-1} to meet the requirement by 2000. Even this increase would obviously multiply the problems of mining and transport and it is doubtful if this could be achieved without causing new problems in ecology. The oil and natural gas reserves of India are placed at 471 million t and 420 million m^3, respectively. The present policy is to conserve these reserves for petrochemical and transport industry. Hence it is envisaged that the role that would have to be played by nuclear energy will become larger in the foreseeable future, especially as the

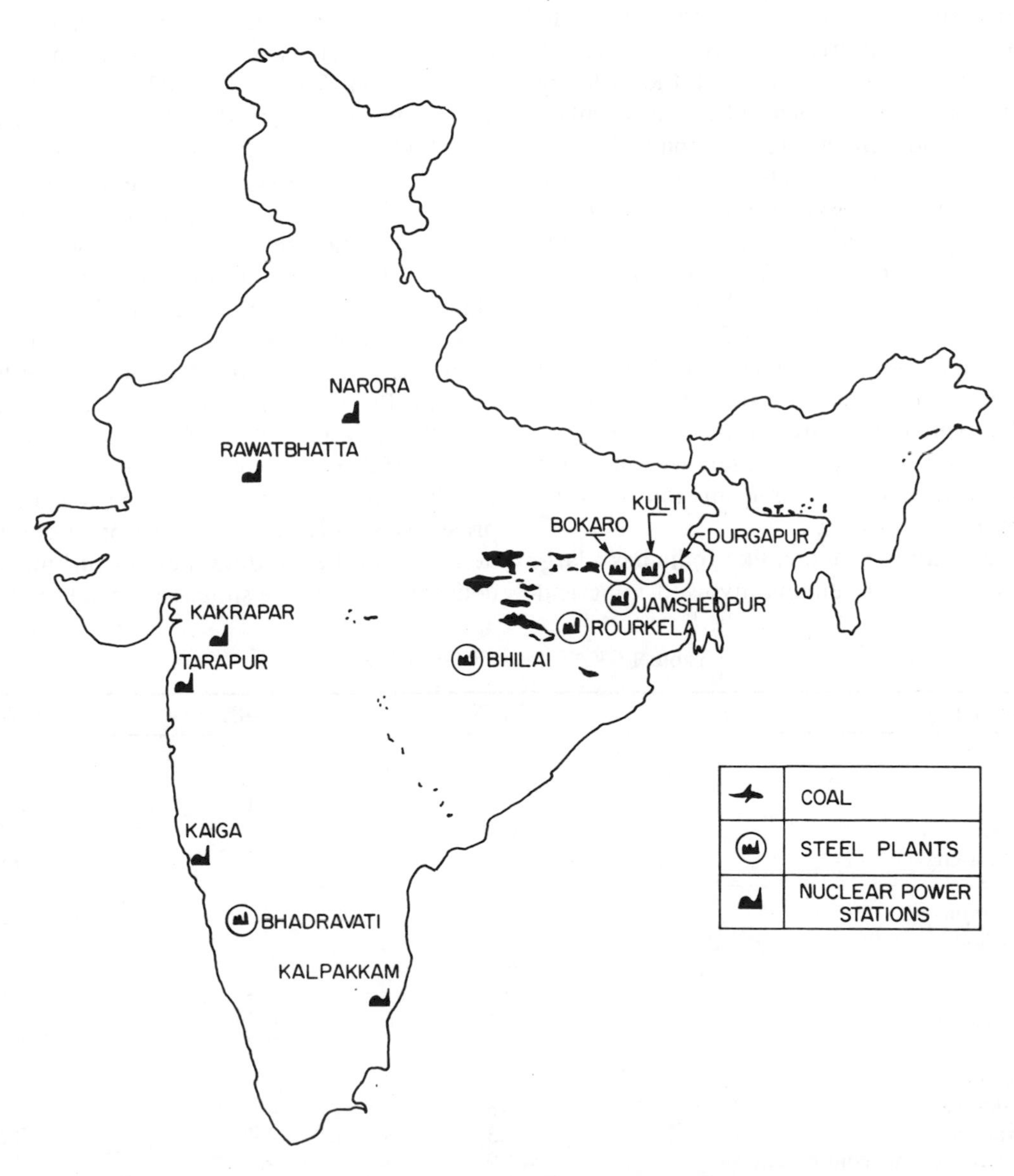

Figure 17.1 Location of resources, large plants and Nuclear Power Plants

other renewable forms of energy still seem far from becoming major commercially viable sources.

As the main reserves of coal are primarily situated in the eastern region, the initial heavy industry was set up in that area. Figure 17.1, for instance, shows the location of the iron and steel plants.

However, India is a large country with a population of 710 million spread in an area of 3.28 million km^2. Industry has been developing in several regions. Other considerations like consumer centres, ports for imports and exports etc. play their due role in developing the location of the industry. Hence the region around Calcutta, Bombay, Ahmedabad, Bangalore etc. have become major consumers of electricity. Therefore, other sources of energy have been brought into play according to need and considerations of economy. The contribution of different sources to the total electricity generation is shown in Table 17.2.

Isolated power stations catering to the needs of their locality have now been joined and extended grids are being developed. This has helped in adapting larger sizes of generating units and EHV transmission to derive economic advantages.

A developing country like India has a large section of its population involved in agriculture. A will to make a significant improvement in the standard of life of the population by rural electrification is served well by establishing large grids. By 1982–83, 71 per cent of the population had electricity available to it and over 53 per cent of the villages have been covered. That electricity production has increased at this rate and not faster is because of economic constraints. This capital intensive sector has accounted for 30 per cent of the total capital investment in the sixth 'five year plan'.

While *per capita* energy consumption is still low, it has increased ten-fold since independence to 145 KWh in 1982–83. About 20 per cent of the power produced goes to agriculture. While this is far from adequate, the modern planners have also started wondering about the wisdom of setting up aims as high as the consumption rates of the most developed nations. *Per capita* consumption in USA is almost two times that in Western Europe but this is not commensurate with the difference in the standard of living. Mrs Indira Gandhi, the late Prime Minister of India said in the World Energy Conference in 1983: 'Thoughtful people have begun to worry whether affluence is not exacting too high a price. We should be good guests on our earth, neither too demanding nor disturbing its delicate balance. We should allow it to renew

Table 17.3 Analysis of Growth Rates

Growth rates	1971–76	1976–81	1971–81
Production			
Coal	6.3	2.9	4.6
Oil	4.1	4.4	4.4
Gas	10.8	—	5.2
Electricity	5.3	6.7	6.0
Consumption:			
By fuel			
Coal	4.5	1.3	2.9
Gas	17.1	3.4	10.0
Electricity	6.5	6.3	6.4
Oil	4.3	7.0	5.5
By sector			
Industry	6.6	4.0	5.3
Transport	2.3	2.4	2.3
Domestic and commerce	1.0	4.4	2.7
Agriculture	11.1	9.1	10.1
Total energy consumption	4.7	3.7	4.2

itself for those who are to follow. Technology must create work, wealth and satisfaction without exhausting the resources of our planet and of the atmospheric layer enveloping it.'

Table 17.3 shows that the maximum growth in the consumption has been in the agriculture sector and this explains how India has today become self sufficient in food. This increase has primarily been due to rapid increase in electrical and diesel driven pumps to lift ground water for irrigation. Some mechanization with tractors and power tillers has also helped. If, however, the energy in the form of chemical fertilizers was included, this consumption growth rate in agriculture sector would be even higher. Increase in the consumption of electricity in the rural areas would directly improve the quality of life of the population of India which lives mainly in villages. As such, the relevant approach is to establish a country-wide grid with which increasing sizes of power plants can be integrated. Nuclear power stations are being set up in the western region presently at sites which are distant from the coal belt.

17.4 Indian Nuclear Energy Programme

The nuclear power programme of India has come of age. The country has five operating nuclear power stations as shown in Table 17.4. The second reactor at Kalpakkam, near Madras, went critical on August 12, 1985 and has been supplying power to the grid. This gives a total nuclear generating capacity of 1330 MWe which forms about 3 per cent of the present electricity generation capacity of the country.

The systems and technology had to be procured from abroad, to begin with, mainly to keep pace with the growing industry and to initiate a viable programme early. When launching on the nuclear power programme, it was necessary to choose the reactor system which would suit India's long-term requirements. The two reactor systems available at that time were the gas cooled reactor (GCR) being set up in the United Kingdom and France and the light water reactor (LWR) being set up in the United States. Another system which was under development in Canada was the pressurized heavy water reactor (PHWR). India made a deliberate choice in favour of PHWR system because of some special features which made this new system particularly attractive for the Indian programme. Some of these are: (a) uranium enrichment is not required, (b) design is amenable to indigenization, (c) uranium requirement per MWe is less, (d) plutonium production per MWe is high, (e) has built-in safety features in the design because of the presence of large quantities of cool moderator and dual shut down system. Since this reactor system was still under development, it was decided to import a nuclear power station based on LWR technology from US in order to gain time as well as experience in the operation and maintenance, and also to get first hand knowledge about the economics of nuclear power production. The construction of India's first nuclear power station started at Tarapur in October 1964 and the two reac-

Table 17.4 Nuclear Power Stations in India

		Type of reactor	Capacity (MWe)	Commenced	Availability factor (%) (in 1984–85)
Tarapur	(TAPS I)	BWR	210	1969	73.90
Tarapur	(TAPS II)	BWR	210	1969	86.78
Rajasthan	(RAPS I)	PHWR	220	1972	—
Rajasthan	(RAPS II)	PHWR	220	1980	67.27
Madras	(MAPS I)	PHWR	235	1983	60.16
Madras	(MAPS II)	PHWR	235	1985	—

tors of 210 MWe capacity each attained criticality in February 1969. Though it was constructed by the General Electric as a turnkey project, Indian scientists and engineers were associated with all aspects of construction and commissioning. Operation and maintenance of this station has been carried out by Indian scientists from the very beginning. This plant has been supplying the cheapest non-hydro power in India, thereby establishing the economic viability of nuclear power. When Canada started the construction of prototype PHWR at Douglas Point, India collaborated with Canada to set up a station at Kota, Rajasthan, based on two PHWRs of 220 MWe capacity each. The indigenous component for the first unit (RAPS-I) was 54 per cent.

The sixth reactor at Kalpakkam, the most recent addition, was completely an indigenous effort with only 10 per cent of the components being imported. The major components were all made locally. An important contribution of the Department of Atomic Energy has been to encourage the industry, by a very close interaction, to start participating in the nuclear programme and hence we have come to a stage where the power reactors are almost completely indigenously built. While there have been problems in some of the reactors set up earlier, they have improved considerably in the last couple of years. MAPP-I has not only performed better than the other reactors set up with foreign collaborations but has done well by any international standards.

Sets of two reactors which are similar in nature (PHWR) and capacity (235 MWe) are at an advanced stage of construction at Narora, Uttar Pradesh and Kakrapar, Gujarat. Hence the present generation capacity of 1330 MWe will be augmented to 2270 MWe.

As the activities have expanded, they have been projected out and are looked after by constituent units under the Department of Atomic Energy. Prospecting for ores is done by the Atomic Minerals Division. The mining and processing is done by the Uranium Corporation of India, fuel fabrication as well as zirconium structurals are the responsibility of the Nuclear Fuel Complex. Design, installation and operation of power reactors are looked after by the Nuclear Power Board. Electronics instruments are manufactured by Electronics Corporation of India Limited.

Beyond this stage it was envisaged that another twelve reactors of 235 MWe and ten reactors of 500 MWe would be set up by 2000 AD. This would raise the generating capacity to 10 000 MWe and at that point of time nuclear power would form 10 per cent of the national electricity generation capacity. The first two reactors in this group will be set up at Kaiga in Karnataka and another two would be added to the present site at Kota, Rajasthan.

On a casual look this seems a large jump. However, when the present capabilities in the individual legs of the fuel cycle are inspected in detail, as will be done presently, it can be demonstrated that this is practical, albeit an ambitious proposition—a proposition which India should be able to tackle more or less independently, leaning on its own sources of strength. In fact, India has always been willing to pay a price for planning a programme which retains the prime parameters for its viability within the limits of its own capability. This was a consideration in opting for 235 MWe reactors. At the present stage of development of the support industry, components like calandrias, steam generators and end-shields for a larger size of the reactors are not yet being manufactured. In addition, there is considerable economic advantage in standardization of sizes and repetitive orders in industry. The existing grids are still small and much larger reactors cannot be integrated with it. Hence for the next one decade the plan is to build 235 MWe reactors. However, work on detailed design for 500 MWe reactors has progressed better than anticipated. It is expected that the first of the 500 MWe reactors would be commissioned sooner than 1995 and the industrial infrastructure and grids would be ready for these larger reactors by that time.

17.5 Fuel Cycle Capabilities

17.5.1 *Raw Material Reserves*

By way of resources, India has 73 000 t of indicated and inferred uranium resources ranging in their U_3O_8 content from 0.07 per cent to 0.015 per cent. These are located mainly in the Singbhum Thrust Belt and many new areas are presently under investigation. Resources which are being presently mined in Jaduguda have 0.067 per cent of U_3O_8. This mine was commissioned in 1967–68. Similar ores of uranium have now been identified in fields near Jamshedpur. For instance the field at Mohuldih has 0.061 per cent to 0.054 per cent U_3O_8. Moreover, there are other locations which have been identified with lower percentages of U_3O_8 but they may still be economical to harness. The location at Turamdih is an example which has 0.031 per cent U_3O_8. However, the overlay is small and the ore is easily accessible by open cast mining. In addition, by-products like copper and magnetite have also been identified and the host rock is a schist from which uranium content can be easily leached out.

The present philosophy is to take into inventory even the ores containing the lower percentages of U_3O_8 but occurring near the high grade ore areas. The seven major deposits provide 56 per cent of the indicated and inferred resources. Of these Narwapahar alone accounts for about one-third. Together with clusters of smaller deposits already identified, these can support a nuclear installed capacity of about 8000 MWe. Uranium from copper tailings will be able to support an additional generating capacity of 2000 MWe. It is hoped that useful quantities of uranium would also become available from phosphate rocks. It, therefore, appears that the known uranium reserves are enough to sustain the envisaged level of the nuclear programme.

The uranium mill at Jaduguda has now been in operation for seventeen years and is presently producing 130 t of U_3O_8. Additional R and D efforts have resulted in recovering magnetite, copper, nickel and molybdenum as by-products. This further helps in economic recoveries being effected from mines with lower percentages of uranium. A new mill is being set up at Bhatin. Two mills for recovery of uranium from copper tailings have already been set up. In the next fifteen years another eighteen mines and mills are expected to be commissioned. These will raise the production to about 1950 t year^{-1} of U_3O_8 by the year 2000.

India also has 320 000 t of thorium reserves in the monazite beach sands of Kerala and Orissa. The technology for processing and purifying nuclear pure thorium has been established. First batch of blanket elements of ThO_2 for the fast reactor has been manufactured. The nuclear strategy for the country has been largely influenced by the necessity of utilizing this fertile material for energy production in the future. This will be discussed in detail at a later stage in this chapter.

17.5.2 *Fuel Fabrication*

The complex technology of fuel fabrication needs high level of skill to make reliable fuel rods. The uranium plant at BARC was commissioned in 1958 and fed the fuel fabrication facility which prepared half the initial metallic fuel charge for CIRUS. Since PHWR power reactors were to use UO_2 fuel, this development was also undertaken. Half the core for RAPS-I was made here but at that stage the required zircaloy tubes were imported.

Research and development in all aspects of fuel fabrication were undertaken from the very early stage of the Indian nuclear programme. Hence all materials like the fuel, the cladding and core structurals were developed at BARC and, when the level of activity was to be brought up to meet the requirements of the power reactors, the Nuclear Fuel Complex at Hyderabad was established. This was sanctioned in 1968 and the constituent plants for UO_2 and zircaloy components were all made operative between

1971 and 1973. Since the two reactors at Tarapur are BWRs using enriched uranium, plants for handling of imported enriched UF_6 and fabrication of fuel rods were also set up. It now meets the entire requirement of the nuclear power programme.

With the long experience in commercial fabrication of fuel rods and zircaloy components, India is confidently prepared to increase its production capacity of uranium and zircaloy from 180 t and 35 t, respectively, to the level required for meeting the targets set for 2000 A.D. By the end of the present financial year the zircaloy product capacity would become 50 t and this would meet the initial charge and replacement requirements for all the six PHWR reactors (upto NAPS-II). To meet the requirements of generation capacity of 10 000 MWe in 2000 A.D, the production targets will have to be 1500 t of U and 250 t of zircaloy. These will be achieved by the expansion programme of the present plant and by setting up one new plant.

Emphasis on R and D back-up has ensured trouble free scale-ups and avoided cost over-runs. Over the years considerable confidence has been gained as Indian fuel rods have been used in reactors and have performed very well. Innovative research is being continued in fuel design, material characterization and basics of fabrication processes. Based on this the present fuel bundle design with nineteen pin cluster will be changed to 22 pin cluster for Narora reactors and beyond. Subsystems for in-pile evaluation of reactor components have been set up. Considerable effort has also been invested in on-line fuelling machines.

17.5.3 *Heavy Water*

Having opted for PHWRs, India has launched an extensive R and D and production programme for meeting the sustained requirements of heavy water, which is a vital input to this system. The most widely used process for heavy water production is based on hydrogen sulphide–water exchange process followed by vacuum distillation. The technology for this process was not available commercially and had to be developed at BARC. A plant of 90 t $year^{-1}$ capacity based on this process has been set up at Kota and this plant has gone into production recently. This highly complex chemical plant was set up completely indigenously and there were delays in its commissioning. However, the experience gained in the process has developed confidence and identified R and D areas which would enable improvements in the future plants like the one being set up at Manuguru (185 t $year^{-1}$). Another process with commercial potential, developed only up to pilot plant level in France, was the monothermal ammonia–hydrogen exchange process. In view of the fact that demand for synthesis gas for this process could be met from several fertilizer plants being built in the country, a bold decision was taken to adopt this process. The plants at Baroda and Tuticorin (45 t $year^{-1}$ capacity each), which are the only commercial plants based on this process in the world, were set up in collaboration with France. However, when problems were encountered in these plants all improvements had to be carried out entirely through indigenous efforts. The two plants are now operating satisfactorily and the production is being optimized. The technology for this complex process has now been mastered and a new plant (90 t $year^{-1}$) is being set up at Thal, which is expected to come on stream by 1986. It has been the experience that operation of ammonia based heavy water plants are often interrupted due to the non-availability of synthesis gas from the fertilizer plants. In order to overcome this problem, studies have been carried out on ammonia–water exchange process to explore the possibility of replenishing deuterium content of ammonia (from ammonia–water exchange) by contacting with water under optimized conditions. Pilot plant studies on this aspect are in progress at Baroda. Success in this experiment would have significant impact on heavy water production in India. Another area where R and D has made significant contribution is upgrading of heavy water. Based on the technology developed at BARC, full scale upgrading plants have been set up at Kalpakkam (electrolysis process) and at Rajasthan (vacuum distillation process).

The programme of 10 000 MWe of PHWR capacity would require about 13 000 t of heavy water. To meet this requirement four more plants beyond Thal and Manuguru, with an aggregate capacity of 970 t year^{-1} are planned to be set up in phase with the rest of the programme.

17.5.4 *Reprocessing*

Along with the setting up of reactors, it was decided in 1958 that a reprocessing plant should also be set up at Trombay to treat the irradiated A1 clad U metal rods discharged from CIRUS and to develop manpower and expertise in this complicated technology. An entirely indigenous plant was commissioned in 1964. The difficulties of handling high radioactivity are compounded by extremely corrosive environment. This leads to a limited life of these plants as preventive maintenance is very tedious. The plant at Trombay was decommissioned in 1973. The facilities have now been rebuilt and refurnished to reprocess additional fuel that would be discharged from Dhruva and the plant again became operational in 1982.

The spent fuel from RAPS is reprocessed at the 100 t capacity plant at Tarapur commissioned in 1982. Another plant of 100 tcapacity is being set up at Kalpakkam to handle fuel from MAPS. A plant with 400 t capacity is planned to become operational by mid–1990s to receive spent fuel from Narora and Kakrapar reactors. It is envisaged that another 400 t capacity plant would have to be suitably located for reactors beyond Kakrapar to bring the total reprocessing capacity to 1000 t by 2000 A.D.

17.5.5 *Waste Management*

While there is a natural phase lag between the front and back end of the nuclear fuel cycle, Indian nuclear programme initiated preparatory R and D effort in all areas early enough. This has generated manpower with 25 years of experience spanning a wide range of expertise. The basic attitude is to concentrate and contain as much radioactivity as possible and discharge only those streams that have activity far below the internationally accepted limits. Continuous review of this important matter over the years has resulted in adopting increasingly stringent measures to minimize discharge in pursuance of the policy that the ultimate goal of radioactive waste management should go far beyond satisfying the prevailing regulations.

The hazard potential of the radioactive wastes generated at different stages of the nuclear fuel cycle like mining and milling, fuel fabrication, reactor operation and finally the reprocessing of spent fuel have been evaluated and analyzed. Work on development of suitable waste treatment methods and safe disposal practices was initiated at BARC, much ahead of our power programme. This lead in time and our technological preparedness have resulted in development of effective management schemes for radioactive waste from all stages of the nuclear fuel cycle.

Due to widely varying environmental features at different nuclear sites in the country, it was necessary to evaluate different characteristics of each site and adopt norms for containment of radioactivity as well as its discharge. For example, while the limits for discharge at coastal locations like Tarapur and Kalpakkam take advantage of large dilution factors available, the levels of discharge from hinterland sites like Rajasthan and Narora are more restrictive. With improved engineering and process design, the release of discharge at Tarapur has been reduced to below the original specifications and during the last few years it was reduced to as low as 6 per cent of the allowed discharge. Radionuclides present in the gaseous form usually get diluted by several orders of magnitude in the atmosphere. Gas cleaning systems are provided as built-in systems for all the off-gas handling in nuclear facilities.

Most of the high and intermediate level waste is generated in reprocessing plants. The first high level waste immobilization plant set up at Tarapur is based on vitrification, incorporating waste in borosilicate matrix. The vitrified radioactive waste needs to be stored under continuously cooled conditions to dissipate decay heat and under constant surveillance to ensure integrity of the vitrified material and its container, until the decay heat reduces significantly. An interim storage facility based on natural convection air cooling by induced draught has been commissioned.

Any ultimate disposal method for the highly active vitrified waste should ensure that the radionuclides will stay isolated from the environment for an extended period of time. Deep geological formations particularly in the Southern Peninsular Shield appear to be promising for long term and even permanent disposal of the vitrified waste. Candidate sites for a repository in peninsular gneisses and granite formations are being investigated.

While radioactive contents of high level wastes produced are very large, their volumes are relatively small. A 1000 MWe nuclear power plant and associated fuel cycle facilities will, for example, generate only about 2 m^3 of vitrified waste per year. The proposed 10 000 MWe programme would generate about 800 m^3 at the end of the designed life of reactors. In as much, the magnitude of the problem is easily manageable and the related fears are really quite unfounded.

17.6 Nuclear Strategy for the Future

Perhaps a good indicator for a sound strategy, like an indicator for a good piece of writing, is whether it seems equally sound in retrospect. One of the most remarkable examples of Dr Bhabha's foresight is the strategy that he enunciated for the future of the Indian nuclear programme. Decades later it seems like the best option for the country when much more information is available to the decision makers.

It has been decided that the second generation reactors, after the PHWRs would be FBRs using Pu–U fuel. Since fast reactors are capable of breeding fissile fuel from fertile materials, they can help grow nuclear power capacity beyond what can be sustained by one-time use of available natural uranium resources. While the initial size and growth rate of the Fast Breeder (FBR) programme would depend on the fissile material inventory from first generation PHWRs, the subsequent growth rate would depend on doubling time. With 10 000 MWe PHWR capacity by 2000, about 4000 kg (3000 kg fissile) of plutonium will be produced every year. Of the various fissile-fertile material combinations, the plutonium–uranium system offers the best breeding ratio at present. Both plutonium and depleted uranium would be available from the spent fuel from the PHWR. On this basis a capacity of around 350 000 MWe can be attained by the latter half of the next century.

Doubling time of around 40 years is achievable with mixed oxide fuel design similar to the ones presently employed in the large FBRs built in the UK, the USA, France and the Soviet Union. Improvement in fuel design and use of mixed carbide fuel can result in increasing breeding ratio and lowering specific fuel inventory. This could shorten the doubling time to fifteen years.

Since this strategy was on the board, and since no drastic changes have come about in it, timely decisions for interphasing the current efforts with future needs has been aided considerably. Work on a fast breeder test reactor (15 MWe) was initiated in 1970 at Kalpakkam. While the effort started with collaboration with France and was based on the design of Rapsodie, Indian engineers were involved from the earliest stage and critical components like the control rod drives, sodium pumps, heat exchangers and steam generators have been manufactured indigenously. This experience will be invaluable in designing a PFBR (500 MWe) which may become operational by 2000 A.D. This is expected to gear the country for setting up commercial reactors within five years of PFBR going critical.

As in all systems facing severe environment, the performance characteristics of the FBR are dependent on the materials used in the core. Materials used in this system would need to withstand fast neutron irradiation and high temperature over extended periods. Their properties and behaviour are to be determined through a sustained materials development programme. Materials of interest include the mixed oxide/mixed carbide fuels together with special grade stainless steel cladding and other structurals which can ensure safe operation of the fuel upto a burn up of 100 000 MWD t^{-1}. Further improvements in cladding and other structural materials within the core would be required to realize higher breeding ratios. Materials for reactor vessels, piping and other circuit components have also to be developed.

Safety of fast reactors has been a matter of debate all over the world. The concern arises mainly from the presence of large mass of highly enriched fuel which can, in an accident, be postulated to undergo very rapid reactivity excursions capable of liberating large amount of energy. Worldwide research has indicated that the accident scenario is in fact not as severe as was thought earlier and the energy liberated in the most severe postulated accident can, in fact, be safely contained. It may be mentioned that the safety debate concerning fast reactors mostly centres around the very low probability postulated events. The designs evolved for fast reactors provide for adequate, reliable and redundant systems to minimize such probability.

As regards operations, it is known now that well designed and operated fast reactors are much cleaner and safer systems. The large quantity of sodium held in the system requires a special consideration from the point of view of sodium–water reaction and sodium fires but has other desirable features. Sufficient experience exists as of today to manage such systems in a safe manner. Continuing R and D effort would further eliminate concern in this area.

The third generation reactors could be based on thorium. While the doubling time in this system may be much longer, this may be compensated by the very large reserves of thorium. Thorium can be bred to ^{233}U which can then be used to fuel the reactors. An early start has been made for developing thorium technology required for this purpose. Commercial FBRs and reactors using ^{233}U bred from Th are still more than two decades away. However, a developing country with limited financial resources and constraints which necessitates self-dependence, needs to initiate R and D on these systems sufficiently in advance. Work on physics of a ^{233}U reactor has been initiated and a small reactor, called PURNIMA II has been set up at Trombay. This reactor uses about 500 g of ^{233}U as uranyl nitrate solution. Obviously, as a reactor this is very small, but as an assignment it entailed several new steps. Thorium was fabricated and put into the reactor. After irradiation, ^{233}U had to be separated from these rods. Because of high activity, suitable glove boxes had to be installed and the entire control system had to be developed. While this is a small beginning, it has generated considerable confidence in our ability to handle several crucial steps like reprocessing. It now seems within the capability of the country to make critical technologies for this available at the appropriate point of time.

Emerging technologies which are likely to play an important role in the production of power in the twenty-first century are those based on nuclear fusion and accelerators. A nucleus of manpower for developing these technologies has been established at BARC and would be expanded for large scale development at the Centre for Advanced Technology (CAT) which is coming up at Indore. Research would be carried out at CAT on the inertial confinement of high temperature plasma generated by lasers. Another promising area of advanced technology relevant to power production is the development of accelerator based neutron sources. India has made a beginning in the accelerator technology by setting up a Variable Energy Cyclotron at Calcutta. Development of accelerator technology would be taken up at CAT and it is hoped that India would soon take significant steps in this technology.

One of the preconditions for a viable

nuclear programme in a developing country is its ability to economically produce nuclear power. The return on investment is one of the main components affecting the cost of nuclear power. Hence, in the Indian context, factors like the rate of interest and the gestation period can contribute even more to the cost than fuel and heavy water put together. Calculations show that the capital cost per kW installed will change approximately from $500 for stations commissioned in the 1970s to $1540 for those to be commissioned in 1992. The corresponding figures in the US nuclear industry are about $500 and $1800 or more.

The figures given for the Indian industry are based on 235 MWe PHWRs. For reasons discussed earlier, India has not yet built larger units. The more developed countries have already established units of 1000–1500 MWe. However, the experience has been that larger units are not necessarily more economic. This is because, multiple and more stringent safety needed for these new systems is often insisted upon as a matter of abundant caution.

Another important component of PHWR cost is the heavy water inventory. It has become an expensive commodity, mainly because of the escalation in capital cost of new installations. Initially India used ammonia process and attached its heavy water plants to fertilizer plants so that ammonia gas could be obtained at very little cost. The experience was that heavy water production became too heavily dependent on the vagaries of the production schedules in the fertilizer plants and the production cycle of heavy water was broken all too often. India is now working on a process to enrich ammonia gas from water and a pilot plant has been set up at Baroda so that the source of deuterium is water. It might well turn out that in future one could decouple from the fertilizer plants and ensure uninterrupted functioning of the plant. The cost of heavy water from newly constructed plants is placed at Rs 6000 kg^{-1}. With an inventory of 255 t in 235 MWe PHWR, with 8 t annual make-up, its contribution to the cost of electricity production comes to 20 per cent.

Another factor which affects the cost very considerably is the scale and commitment to the programme. As the Indian commitment to nuclear energy increases, a larger section of the industry will become involved and find it more economically worthwhile to participate. Notwithstanding this, even at the present pricing the cost of nuclear power generation in India is equal to a similar sized coal fired thermal plant at pithead. If the thermal plant is situated at 800 km from the pithead, it definitely produces more expensive power. This difference will only increase with time–cost of transporting coal is increasing and the coal which will be available will have even higher ash content than that being used presently. At present the cost of nuclear power from Tarapur reactor is approximately 37 paise kWh^{-1}, Rajasthan 38 paise kWh^{-1}, Madras 44 paise kWh^{-1} and the projected cost for those that will be commissioned by 1992 will be 65 paise kWh^{-1}.

The main factor often quoted in favour of thermal plants is the lower initial investment. This, however, is yet another factor which is not quite what it appears to be. It must not be forgotten that even at the stage of MAPP I 90 per cent of the components used were made in India. Effectively, this only ploughs the money back into the economy. In contrast, this is not necessarily true for thermal plants and the back-up industry like mining. This situation is partly because, unlike the field of nuclear power, many of the larger thermal plants are being made with foreign loans and aids. This forces a situation where global tenders are raised and often the equipment has to be imported. This advantage for nuclear plants has strangely been thrust upon India by the various sanctions imposed on it.

These sanctions may have slowed down the nuclear programme but this time has been utilized for establishing capabilities in the entire fuel cycle. In the ultimate analysis, this has not only generated considerable freedom in making choices but has also reduced the vulnerability of the programme in the future. Because for a developing country, even with the best of intentions and efforts in obeying the safeguard clauses, the shroud that safeguards cast can be misinterpreted in many ways and affect the progress of projects.

The Indian experience has been that, quite unlike the conventional sectors of agriculture and industry where collaboration with advanced countries has been readily available to advantage, there are many difficulties in making collaborative efforts in high technology areas like nuclear energy. This is particularly true in specific problem issues which can drastically affect the progress of the entire programme. Problems in heavy water production, as mentioned above, had to be solved independently. The development and testing of mixed oxide fuel for Tarapur was again an issue, perhaps peculiar to a developing country. These and similar issues have decidedly affected the progress of the programme and penalties have had to be paid both in terms of time and cost. However, it is now evident that the Indian nuclear programme has emerged out of these problems with greater strength and resilience. It is also evident that future plans and strategies of a developing country like India must take account of this experience and eventually provide for it by charting a path most suitable for it and developing capabilities that could make it independent, at least in the crucial steps for progress. Perhaps the only successful collaborations in this field are those in which both sides can give equally. The beginning of the Indian nuclear programme was an evolution. In retrospect it seems to have evolved successfully, primarily because the inherent weaknesses were exploited and converted into its strength.

Nuclear Power: Policy and Prospects
Edited by P. M. S. Jones

18

The Developing Countries

IOAN D. STANCESCU
International Energy Consultant, Munich

18.1 Introduction

For better understanding the crucial long-term dilemma of nuclear energy integration in the energy supply of developing countries and the challenging patterns of such an action, some introductory considerations might prove particularly helpful.

These relate to the ambiguous classifications of developing countries, the enormous diversity of their size, population and economic level and structure, their endowment with natural and energy resources, their present and future share in the world population and in world energy consumption and production, etc.

However, in spite of their diversities, these developing countries also present a series of common energy issues, which facilitate the judgement of if, when and how they might envisage adopting nuclear power technologies.

Accordingly, the following two sections of this chapter deal with the mentioned introductory aspects, section 18.4 presents an overview on the past and present status as well as on the prospects of nuclear power evolution in developing countries, section 18.5 underscores the interest for smaller nuclear reactors, section 18.6 examines briefly the launching conditions of a nuclear power programme in developing countries emphasizing the imperative of having nuclear energy optimally embedded in an over-all energy master plan. Section 18.7 exemplifies nuclear power introduction in some developing countries before a short conclusion rounds up the subject.

18.2 Developing Countries

18.2.1 *Ambiguous Definitions and Grouping of Developing Countries*

The group of developing countries is very heterogenous showing great discrepancies in their geographical size, climate, population, natural resources including energy resources, level of socio-economic development, culture and life habits—in brief, the main factors which determine energy consumption both in absolute figures as well as per head.

While the basic concept is clear, its limits are less well established. Depending on interpretation the group of developing countries can include or exclude important sub-groups of countries (IAEA–Stancescu, 1985a). For example:

1. UNCTAD defines the world economic groups of countries as: developed market economy countries, socialist countries of Eastern Europe, socialist countries of Asia and 'developing countries'.
2. Recent IEA and OECD statistics and publications maintain the first group, combine the second and third under CPE, i.e. centrally planned economies, and leave the last as 'developing countries'.

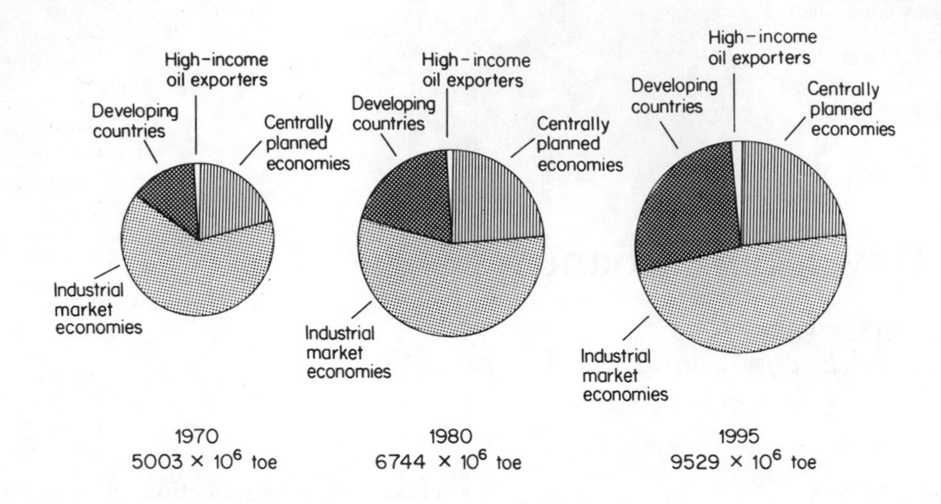

Figure 18.1 Shares of country groups in world commercial primary energy consumption, 1970–1995. Reproduced by permission of The World Bank, 1983

Table 18.1 Commercial Energy Typology of Developing Economies (Reproduced by permission of The World Bank, 1983)

Energy resources or options (relative to country size)	Oil exporters: Large	Oil exporters: Small or medium	Oil importers — Net oil imports as a percentage of primary commercial energy consumption in 1980: 0–25	26–50	51–75	76–100
LIMITED			*Middle income* Lesotho Namibia		*Low income* Burundi Kampuchea Lao, PDR Nepal Rwanda	*Low income* Bhutan Ethiopia Guinea-Bissau Haiti Niger Somalia **Sri Lanka** Togo Upper Volta *Middle income* Barbados Cuba **Dominican Republic** Hong Kong Israel **Jamaica** **Jordan** Lebanon Liberia Mauritania Singapore **Uruguay** **Yemen, AR** Yemen, PDR

MODERATE		*Middle income* *Syria, AR*	*Low income* Zaire *Middle income* Botswana Korea, PDR[2] Vietnam[2] Zambia Zimbabwe[2]	*Low income* Ghana **Pakistan**[1] *Middle income* Brazil[1,2] Chile[1] Guatemala Ivory Coast Mongolia	*Low income* **Bangladesh**[1] Central Afr. Rep. Chad Equatorial Guinea Malawi Mozambique Uganda *Middle income* Costa Rica El Salvador Honduras Korea, Rep. of[2] Paraguay Portugal Turkey[2]	*Low income* Benin Guinea Madagascar Mali Sierra Leone Sudan Tanzania *Middle income* Greece[2] Kenya Morocco Nicaragua Panama Papua New Guinea Philippines Senegal Thailand
SUBSTANTIAL	*Low income* *China*[1,2] *Middle income* *Indonesia*[1] *Iran*[1] *Iraq*[1] *Mexico*[1,2] *Nigeria*[1] *Venezuela*[1]	*Middle income* *Algeria*[1] *Angola* Congo, PR *Ecuador* *Egypt*[1] *Gabon* *Malaysia*[1] *Peru*[1] *Trindad and Tobago*[1] *Tunisia*	*Low income* Burma India[1,2] *Middle income* *Argentina*[1] Cameroon *Colombia*[1,2]	*Low income* Afghanistan[1] *Middle income* Yugoslavia[1,2]	*Middle income* Bolivia[1]	

Note to Table 18.1: Not shown are economies with less than one million population and without production (or prospects of future production) of oil, gas, or coal. The economies included in this table are classified according to their energy resource potential (oil, gas, coal, and primary electricity) that could be economically developable during the next decade. Oil exporters are countries whose official earnings from net oil exports exceed 10 per cent of their total export earnings in 1980–81. *Large oil exporters* refers to those countries that produced more than 70 × 10^6 t during 1980.
[1]Produced one or more × 10^6 t of gas in 1980.
[2]Produced two or more × 10^6 t of coal in 1980.
Economies shown in *italics* produced more than 5 × 10^6 t of oil in 1980.
Economies shown in **bold** print had net energy imports amounting to 30 per cent or more of their merchandise exports in 1980 (information is not available for all countries).

3. IAEA statistics take geographical regions, but when grouping, select 'industrialized countries' and 'developing countries'. The industrialized countries group includes USSR and the German Democratic Republic (GDR) while the developing countries group separates, when split, into the remaining CPE countries as 'developing CPE' subgroup and the other developing countries'. The latter also includes Yugoslavia.
4. Finally, the World Bank (World Bank, 1983) considers the world groups according to Figure 18.1, i.e. distinguishes the industrial market economies, the centrally planned economies (USSR and CPE–Europe), introduces a new subgroup, the high income oil exporters and leaves the remainder in the 'developing countries' group. Table 18.1 lists the developing countries on this basis with further subdivision by commercial energy typology.

The difficulty with the different definitions of the developing countries group is that, while each definition is well suited for its purpose, it becomes difficult or impossible to use comparable statistical data. Whilst totals are identical, explicit breakdowns are not.

For the purpose of this chapter the IAEA classification will be used, since the Energy and Economy Data Bank (EEDB) of this organization, stores the most comprehensive and up to date information on nuclear energy development and the related relevant energy and economic statistics.

18.2.2 *Share of Developing Countries in the World Population*

The world population has been experiencing a dramatically increasing trend and only recently a certain deceleration of its growth rate appears to lead to future manageable figures. From some 205 million in the year 600, the world population increased to over 375 million in the year 1400, 950 million in 1800, reaching 1950 million in the year 1925. Two thirds of this population lived in a northern belt including North America, Europe, USSR, East Asia, etc., the other third in a southern belt comprising Latin America, Africa, South Asia and Oceania. In 1975 the world population reached almost 4.0 billion, equally distributed in the northern and southern belt.

As far as the future evolution of population is concerned, Table 18.2 presents IAEA estimates, both for absolute figures and growth rates. Starting from 1985, the estimates extend over 1990 and 1995 to the year 2000, when some 6 billion inhabitants are expected in the world. In accordance with the IAEA usual statistical presentation the figures are displayed by country groups and finally consolidated for the industrialized and the developing countries (IAEA, 1986).

Owing to higher growth rates the share of the population living in developing countries would increase in relation to the total world population from 78 per cent in 1985, over 79 per cent in 1990, 80 per cent in 1985 to 81 per cent in 2000.

Since population is one of the determining factors of energy demand, both absolute figures and the shift of distribution are important hints for the patterns of future energy—including nuclear–requirements.

The population growth has been accom-

Table 18.2 Estimates of World Population Growth (Reproduced by permission of the International Atomic Energy Authority. IAEA, 1986)

Country group	1985		1990		1995		2000	
	Million inhab-itants	Growth rate %/year⁻¹ 1974–1985	Million inhab-itants	Growth rate %/year⁻¹ 1983–1990	Million inhab-itants	Growth rate %/year⁻¹ 1990–1995	Million inhab-itants	Growth rate %/year⁻¹ 1995–2000
North America	263	0.98	273	0.73	281	0.59	288	0.47
Western Europe	406	0.57	416	0.50	426	0.47	435	0.41
Eastern Europe	419	0.86	435	0.79	450	0.66	462	0.55
Industrialized Pacific	140	0.88	144	0.58	148	0.54	152	0.52
Asia	2536	2.02	2769	1.77	2998	1.61	3220	1.44
Latin America	406	2.49	453	2.20	499	1.98	545	1.76
Africa and Middle East	648	2.79	749	2.94	859	2.77	976	2.59
World total	4818	1.82	5239	1.69	5661	1.56	6078	1.43
Industrialized countries	1067	0.76	1102	0.65	1134	0.57	1162	0.49
Developing countries								
1) In CPE-Europe[1]	99	0.81	103	0.65	106	0.56	108	0.48
2) Others	3652	2.19	4034	2.01	4422	1.85	4808	1.69
3) Total of DC's	3751	2.15	4137	1.98	4528	1.82	4916	1.66

[1]Developing countries in the Centrally Planned Economies (CPE) in Europe: Albania, Bulgaria, Czechoslovakia, Hungary, Poland and Romania.

panied, since the beginning of the century, by a rapidly progressing process of urbanization. In 1900 in the industrialized countries only 30 per cent of the population lived in urban areas, in the developing countries the figure was less that 10 per cent. By 1975 the percentages rose to 65 per cent and 30 per cent respectively. For the year 2000 an urbanization share around 80 per cent is expected in the industrialized countries and of 40 per cent in the developing countries.

The developing world will have around 20 cities of more than 10 million people in the late 1990s for example: Mexico City 31 million, Sâo Paulo 26 million, Shanghai 23 million, Beijing 20 million, Rio de Janeiro 19 million, Greater Bombay 17 million, Jakarta 16.6 million, Seoul 14.2 million, Greater Buenos Aires 12.1 million, Bogota 11.7 million, Delhi 11.7 million, etc. (Konstantinov *et al.*, 1985).

Evidently, the concentration of population in large conurbations in developing countries will require highly centralized energy supply systems, favouring among other solutions adoption of nuclear power.

18.2.3 *Share of Developing Countries in World Energy Consumption and Production*

A rapid overview on the total world commercial energy consumption and its division amongst geographical groups of countries, can be gained from Table 18.3. In addition to these data for 1985 and the future prospects, Table 18.3 displays the share of primary

Table 18.3 Estimates of Total World Energy Consumption (EJ), Percentage used for Electricity Generation, and Percentage Supplied by Nuclear Energy (Reproduced by permission from International Atomic Energy Authority IAEA, 1986)

Country group	1985			1990[2]			1995[2]			2000[2]		
	Total energy consumption	% used for elect. gener.	% supp. by nucl.	Total energy consumption	% used for elect. gener.	% supp. by nucl.	Total energy consumption	% used for elect. gener.	% supp. by nucl.	Total energy consumption	% used for elect. gener.	% supp. by nucl.
North America	81.9	34.2	5.2	92	36	7	98	39	7	105	39	8
				93	37	7	104	39	8	113	41	8
Western Europe	51.9	35.0	10.2	54	37	14	59	38	14	64	40	15
				59	37	13	65	38	14	72	39	16
Industrialized Pacific	19.7	39.4	7.4	21	42	9	23	45	10	26	46	12
				23	41	9	26	42	12	29	44	15
Eastern Europe	77.6	25.8	2.5	88	28	4	98	29	5	107	31	6
				91	28	5	105	30	7	119	31	8
Asia	49.2	17.3	0.9	58	20	1.3	65	24	1.3	73	27	2.2
				63	20	1.2	78	22	1.4	95	25	2.3
Latin America	19.5	22.1	0.4	25	25	0.4	30	29	1.0	35	32	1.4
				26	24	0.5	34	27	0.9	43	30	1.7
Africa and Middle East	19.1	19.4	0.3	24	22	0.3	29	25	0.3	33	27	0.3
				25	22	0.4	33	24	0.5	41	27	0.9
World total	318.8	28.4	4.2	361	30	6	402	32	6	443	34	7
				381	30	6	446	32	7	511	33	7
Industrialized countries	214.7	32.7	5.9	237	34	8	258	36	9	281	38	9
				247	35	8	279	36	9	310	38	10
Developing countries												
In CPE-Europe[1]	14.7	23.3	2.0	17	25	4	18	28	6	19	30	6
				17	25	4	20	28	7	22	30	8
Others	89.4	18.7	0.6	108	22	0.8	125	25	1.0	142	28	1.6
				116	21	0.8	147	24	1.1	180	26	1.9
Total of DCs	104.1	19.3	0.8	124	22	1.2	144	25	1.6	162	28	2.1
				134	22	1.2	166	24	1.7	201	27	2.6

[1]Developing countries in the Centrally Planned Economies (CPE) in Europe: Albania, Bulgaria, Czechoslovakia, Hungary, Poland and Romania.
[2]The top and bottom figures for total energy consumption are low and high estimates, respectively.

energy used for electricity generation and the percentage possibly supplied by nuclear energy.

The share of the developing countries total commercial energy consumption (with 104.1 EJ) in 1985 was around 32.7 per cent of the total world figures of 318.9 EJ. This share is expected to grow to 34.3 per cent in 1990, 35.8 per cent in 1995 and reach between 36.6 and 39.3 per cent in the year 2000. Adding the consumption of non-commercial energy resources, which in some developing countries reach the same order of magnitude as the commercial energy consumption, the share of the developing countries in the total world energy balance would result in correspondingly higher figures.

As far as the endowment of developing countries with indigenous energy resources is concerned, Table 18.1 displays, in the World Bank classification, two basic groups: oil exporters and oil importers. Within each group are three levels in terms of energy resources or options, i.e. limited, moderate and substantial. The oil exporters are further classified as large, small and medium exporters of low and middle income. The oil importers are classified according to the share of net oil imports in their primary commercial energy consumption in 1980—0–25 per cent, 26–50 per cent, 51–75 per cent and 76–100 per cent and are also ranked as low or middle income countries.

The advantage of this classification is that, depending on its position in Table 18.1, the energy problems of a country are easier to identify and, once a diagnosis has been reached, possible solutions and options are more readily discerned.

Figure 18.2 illustrates the relative balance between energy consumption and production in the relevant geographical regions of the world, presented in per head figures (N.N., 1983). For the industrialized countries a decreasing tendency in energy consumption appears after 1973, while in the developing regions the consumption still increases, although at a lower rate than before.

The evolution of the share of primary energy used for electricity generation as displayed in Table 18.3 merits particular attention. From 28.4 per cent worldwide in 1985 this share increases to 34 per cent in the year 2000, whilst for the developing countries it rises from 19 per cent to 27 per cent. The share of nuclear power in the over-all energy balance was still relatively low in 1985, i.e. 4.2 per cent worldwide, 5.9 per cent in the industrialized and only 0.8 per cent in the developing countries. The corresponding figures for the year 2000 read 7 per cent worldwide, 9–10 per cent in industrialized and 2.1–2.6 per cent in the developing countries.

In spite of its numerical format, it must be borne in mind that Table 18.3 as well as the following tables with future projections, represent very rough estimates, having the low and high limits stressing their large uncertainties.

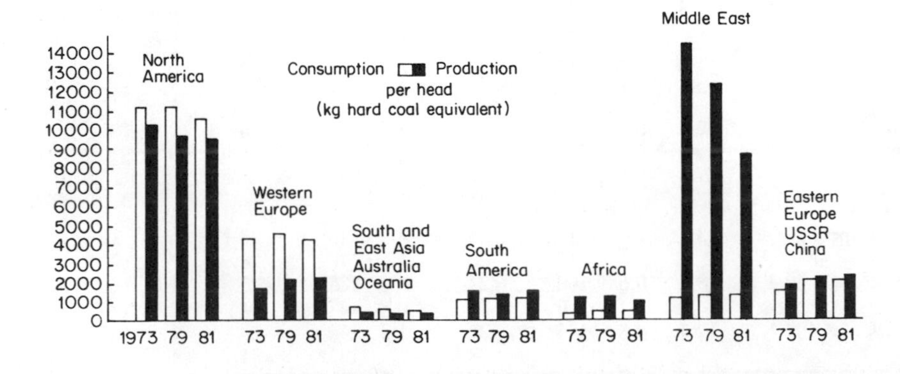

Figure 18.2 Commercial energy consumption and production, per head figures in the world. Reproduced by permission of Esso AG

18.3 Common Energy Issues in Developing Countries

In spite of their tremendous diversity, the developing countries as a group have some broad common energy characteristics and consumption patterns capable of exerting a strong impact on their future energy development. They will be dealt with in general, underscoring however, the characteristic rural-urban dichotomy and some salient problems of energy supply.

18.3.1 *General Aspects*

Enumerated briefly, the following issues might best characterize the present conditions and influence the future evolution of energy patterns in developing countries:

1. Their per head consumption of commercial energy is substantially lower than that of developed countries, in some cases extremely low. While in the industrialized countries the yearly average of the commercial primary energy consumption was around 200 GJ $head^{-1}$ $year^{-1}$ in 1985 (in North America even 311 $head^{-1}$ $year^{-1}$), the average of the developing countries hardly reached 28 GJ $head^{-1}$ $year^{-1}$ with figures of under 10 GJ $head^{-1}$ $year^{-1}$ or even less than 3 GJ $head^{-1}$ $year^{-1}$ in a significant number of countries. The ratio of the *per capita* average energy consumption figures of more than 8:1 embraces some extremes of over-consumption of energy in the first group of countries against substantial under-consumption in the developing regions, especially in their rural and marginal urban areas.
2. In the total energy consumption, the share of non-commercial energy resources, all of which are also renewable resources, is high, in general around 50 per cent reaching 90 per cent or even more sometimes.
3. Size, location and structure of energy consumption develop according to the evolution of the basic human settlements: increasing urbanization tendencies but still a high share of scattered villages.
4. The aforementioned state of under-consumption of commercial energy is even worse than reflected in the per head average figures, since an enormous imbalance exists between the energy consumption of high-income groups and the low-income group, which accounts for approximately two-thirds of the population.
5. The low average per head energy consumption and the substantial under-consumption situation in developing countries denote a considerable potential for increased future demand both for filling the gap of under-consumption and for sustaining the rapid envisaged socio-economic development.
6. Since the great majority of these countries are located in tropical and subtropical areas, energy consumption for space heating is limited to a few regions with temperate climate. This relieves the majority of developing countries of an important component of demand, which in the developed countries reaches a share of more than 40 per cent of the primary energy consumption. However, in the long run, a rapidly growing energy consumption for air-conditioning is expected.
7. Many of the developing countries have serious problems with the depletion of their fuelwood resources, i.e. fighting the effects of the so-called 'second energy crisis', the rapid and massive deforestation with all its additional, major environmental damage and social harm.
8. Most of the developing countries do not have an adequate physical and social infrastructure, i.e. transportation, electric power, medical and sanitary services to cope with their rapid population growth and high rates of economic and energy development.
9. Most of these countries lack trained energy professionals on all managerial levels, as well as adequate institutional framework for over-all energy development.

10. With the exception of a few petroleum exporting countries, the remaining developing countries are net petroleum importers, and the foreign currency payments for petroleum consume a high share of their export revenues.
11. It might be worth mentioning that, in addition to their foreign expenses for petroleum imports and investment financing difficulties, many of the developing countries have heavy internal burdens because of the substantial subsidies governments give, selling energy to low-income consumers at prices well below the real economic cost.
12. In spite of the different financial balances, both petroleum net importing and net exporting countries have the same basic interest in energy conservation since any saved or substituted petroleum means reduced import or increased export of petroleum.
13. In relation to conservation it should be remembered that independent of the degree of their development, the existing energy equipment represents (owing to the high rate of development of these countries) a rapidly decreasing share in the total equipment expected to be in use ten or 20 years later. This means the new equipment will soon become dominant and determine by its concept, design and operating performance, the technical and economic level of the over-all energy system. Therefore, without neglecting the possible improvement of existing equipment, special attention should be given to new energy investments and the chance of providing them with the highest potential for energy conservation, rather than retrofitting them later at higher cost and less efficiency.
14. Last but not least, the developing countries, many of them practically starting from scratch, have the chance of searching for new energy development patterns, avoiding the historical low-efficiency energy evolution in developed countries, and optimizing their own development in the light of the new international and local energy conditions. While in developed countries existing energy wastage has to be eliminated, the developing countries could prevent it from even appearing.

18.3.2 The Rural-Urban Dichotomy

In spite of the accelerated trend in the most populated countries to impressive conurbations, with high population density and a rapidly increasing consumption of commercial energy resources, the majority of the population still lives in rural areas with a low, decentralized energy consumption, mainly based on non-commercial, new and renewable energy resources. This situation leads in the majority of the developing countries to the well known rural–urban dichotomy, which implies a rapidly increasing energy demand, geographically split between urban areas—even in a few immense conurbations—and scattered villages and farms. Accordingly, a double structure of the energy economy develops, both with specific characteristics and different supply solutions.

18.3.3 Salient Patterns of Energy Supply

For the gradually declining share of population—now still over 70 per cent—continuing to live in the rural areas of the developing countries, the energy supply solutions appear to lie in the direction of a decentralized rural energetization (IAEA, 1985a), i.e. a total energy concept based mainly on local non-commercial energy resources, best suited for scattered supply and relatively low and smoothly increasing energy demand.

By contrast, the energy supply of rapidly growing cities and conurbations will develop along similar lines to those in developed countries based on concentrated energy production and corresponding distribution. No doubt, in such structures nuclear energy will have a role to play, as it has already started to do in a few more advanced developing countries.

Compared from the point of view of useful

energy, the imbalance between rural and urban energy supply, remains aggravated by the fact that the first group consumes mainly non-commercial energy resources in conversion processes of rather poor efficiency, while the urban areas benefit from advanced technologies and the advantage of the economy of scale.

However, with extending penetration of electrification in rural areas the imbalance

Table 18.4 Nuclear Power Reactors in Operation and Under Construction (end of 1985) (Reproduced by permission of the International Atomic Energy Authority, IAEA, 1986)

Country	Reactors in operation		Reactors under construction		Nuclear electricity supplied in 1985		Total operating experience to end 1985	
	No. of units	Total MWe	No. of Units	Total MWe	TWeh	% of total	Years	Months
Argentina	2	935	1	692	5.2	(11.3)	14	7
Belgium	8	5486			32.4	59.8	64	1
Brazil	1	626	1	1245	3.2	(1.7)	3	9
Bulgaria	4	1632	2	1906	13.1	31.6	30	6
Canada	16	9776	6	4789	57.1	12.7	151	7
China			1	300				
Cuba			2	816				
Czechoslovakia	5	1980	11	6284	10.9	14.6	22	4
Finland	4	2310			18.0	38.2	27	4
France	43	37533	20	25017	213.1	64.8	338	5
Germany D.R.	5	1694	6	3432	(12.2)	(12.0)	57	5
Germany F.R.	19	16413	6	6585	119.8	31.2	215	4
Hungary	2	825	2	820	6.1	23.6	4	5
India	6	1140	4	880	4.0	(2.2)	54	8
Iran			2	2400				
Italy	3	1273	3	1999	6.7	3.8	69	10
Japan	33	23665	11	9773	152.0	22.7	286	10
Korea R.P.	4	2720	5	4692	13.9	(22.1)	15	5
Mexico			2	1308				
Netherlands	2	508			3.7	6.1	29	9
Pakistan	1	125			0.2	0.9	14	3
Philippines			1	620				
Poland			2	880				
Romania			3	1980				
South Africa	2	1840			5.3	4.2	2	3
Spain	8	5577	2	1920	26.8	24.0	56	10
Sweden	12	9455			55.9	42.3	99	2
Switzerland	5	2882			21.3	39.8	53	10
Taiwan	6	4918			(27.3)	(53.1)	26	1
UK	38	10120	4	2530	53.8	19.3	695	10
USA	93	77804	26	29258	383.7	15.5	954	11
USSR	51	27756	34	31816	(152.0)	(10.3)	531	7
Yugoslavia	1	632			3.9	(5.1)	4	3
Total	374	249625	157	141942	1401.6		3825	3

Note: Figures in parentheses are IAEA estimates.
Construction was cancelled for ten reactors, and suspended for a further eight during 1984

will gradually diminish and in the long run rural areas will join in benefiting from energy of increasingly high quality.

18.4 Electrical Energy and Nuclear Share in Developing Countries

18.4.1 *Past Evolution and Present Status*

As far as the past is concerned, here are a few landmarks in the worldwide evolution of electricity generation:

Year	1930	1940	1950	1970	1980	1983	1985
World electricity production, Twh	270	465	960	4950	8100	8498	9422

The share of the developing countries in the world electricity production increased from some 19 per cent in 1950 to 22.3 per cent in 1985, when the developing countries in CPE–Europe generated 357 TWh, i.e. 3.8 per cent, and the other developing countries contributed 1940 Twh (18.5 per cent) in a total of 2096 TWh (see Table 18.7).

The share of nuclear energy in the world electricity balance was 14.9 per cent in 1985, a combined result of shares of 17.9 per cent in industrialized and 4.2 per cent in developing countries' production.

Table 18.4 presents an overview of the nuclear power reactors in operation at the end of 1985 in the world, grouped by countries with an indication of the number of units, the total installed electric capacity (MWe) and the electricity generated in 1985, the latter expressed in TWh and as a percentage of the total country's electrical energy production. Compared to some very high shares in industrialized countries—led by France with 64.8 per cent, a few developing countries show also remarkable figures, as for example, Bulgaria 31.6 per cent, Republic of Korea 22.1 per cent, etc. Table 18.4 also includes a column indicating the total operating experience to the end of 1984 and two further columns, presenting the number and electrical capacity of the nuclear power reactors under construction at the end of the same year (IAEA, 1986).

Table 18.5 focusses explicitly on the nuclear reactor status as of December 31, 1984 in the developing countries other than the CPE in Eastern Europe (Laue, 1985). The figures were also valid as of 31 December 1985.

The developing countries with on-going nuclear power programmes, as with all countries, have different characteristics. However, from the point of view of their nuclear

Table 18.5 Nuclear Power in Developing Countries[2] (as of December 31, 1984) (Laue, 1985)

Country	Operating		Under Construction	
	No. of Units	Capacity MWe	No. of Units	Capacity MWe
Argentina	2	935	1	692
Brazil	1	626	1	1245
China, P.R.	—	—	1	300
Cuba	—	—	1	408
India	5	1020	5	1100
Iran, Isl. Rep.[1]				
Korea, Rep. of	3	1790	6	5622
Mexico	—	—	2	1308
Pakistan	1	125	—	—
Philippines	—	—	1	621
Taiwan	5	4011	1	907
Yugoslavia	1	632	—	—
Total	18	9139	19	12 203

[1]In the Islamic Republic of Iran, Construction of 4 units has been suspended, but resumption of work on 2 units is under negotiation.
[2]Not including those in the centrally planned economies (CPE) in Europe.
Source: IAEA Power Reactor Information System (PRIS), (13 December 1984)

programmes, some common features can be found.

The group comprising Argentina, Brazil, India, Republic of Korea and Taiwan displays important nuclear programmes with at least one unit in operation and several under construction. In spite of some delays and reappraisals of the programmes, nuclear power development is moving ahead.

The group made up of Bulgaria, Czechoslovakia, Hungary, Cuba and Poland shows a similar approach in the nuclear power programmes, the latter being based on supply and co-operation with the USSR.

Finally, in the third group, Pakistan and Yugoslavia have each one nuclear power plant in operation, while Mexico, the Philippines and Romania have each one under construction. All these countries intend to proceed with nuclear power programmes, but for various reasons none of them have started constructing their second unit yet.

In addition to the above-mentioned countries, there are several other developing countries in different preparatory stages (planning, feasibility studies, acquisition process) which consider, intend to or have decided to go nuclear, but which have not yet launched their first nuclear power project. Among these countries, those which seem to be closest to going nuclear are the People's Republic of China, Egypt and Libya.

Some other developing countries, e.g. Bangladesh, Indonesia, Morocco, Portugal, Peru, Syria and Thailand have stated their intention to introduce nuclear power, but have not yet made definite commitments to nuclear power plant construction, mainly due to financing problems.

Table 18.6 Estimates of Total and Nuclear Electrical Generating Capacity (Reproduced by permission of the International Atomic Energy Authority. IAEA, 1986)

	1985			1990[3]			1995[3]			2000[3]		
	Total elect. (GWe)	Nuclear (GWe)	(%)	Total elect. (GWe)	Nuclear (GWe)	(%)	Total elect. (GWe)	Nuclear (GWe)	(%)	Total elect. (GWe)	Nuclear (GWe)	(%)
North America	784	87.6	11.2	897	118	13	1008	123	12	1068	133	12
				938	127	14	1084	132	12	1214	153	13
Western Europe[1]	502	91.6	18.2	544	124	23	601	141	23	665	158	24
				584	124	21	661	160	24	728	193	27
Industrialized Pacific	208	23.7	11.4	236	31	13	272	36	13	313	48	15
				248	34	14	290	49	17	330	70	21
Eastern Europe	423	34.5	8.2	514	60	12	606	84	14	696	108	16
				538	73	14	658	111	17	775	150	19
Asia	229	8.9	3.9	314	14	4	408	16	4	503	26	5
				330	14	4	458	20	4	607	34	6
Latin America	120	1.6	1.3	169	3	2	227	6	2	290	8	3
				173	3	2	243	6	2	332	11	3
Africa and Middle East[1]	108	1.8	1.7	149	2	1	191	2	1	230	2	1
				155	2	1	215	3	1	286	6	2
World total	2376	249.6	10.5	2823	352	12	3313	407	12	3765	483	13
				2967	378	13	3609	481	13	4271	619	14
Industrialized countries	1817	234.1	12.9	2068	324	16	2345	368	16	2584	423	16
				2181	349	16	2540	429	17	2871	529	18
Developing countries												
In CPE-Europe[2]	86	4.4	5.2	107	11	10	126	17	14	143	23	16
				110	11	10	136	23	17	160	34	21
Others	473	11.1	2.3	649	17	3	842	22	3	1038	36	3
				676	18	3	933	29	3	1239	56	4
Total of DCs	558	15.5	2.8	755	28	4	968	39	4	1180	59	5
				786	29	4	1069	52	5	1400	90	6

[1]Nuclear programmes in Austria and the Islamic Republic of Iran have been interrupted, and the reactors are not included.
[2]Developing countries in the Centrally Planned Economies (CPE) in Europe: Albania, Bulgaria, Czechoslovakia, Hungary, Poland and Romania.
[3]The top and bottom figures for total electric and nuclear capacity are low and high estimates, respectively.

Table 18.7 Estimates of Total Electricity Generation and Contribution by Nuclear Power (Reproduced by permission of the International Atomic Energy Authority, IAEA, 1986)

Country group	1985			1990[2]			1995[2]			2000[2]		
	Total elect. (TWh)	Nuclear[3] (TWh)	(%)	Total elect. (TWh)	Nuclear[3] (TWh)	(%)	Total elect. (TWh)	Nuclear[3] (TWh)	(%)	Total elect. (TWh)	Nuclear[3] (TWh)	(%)
North America	2918	440.8	15.1	3420	655	19	3939	751	19	4271	818	19
				3578	701	20	4236	809	19	4855	941	19
Western Europe	1890	551.4	29.2	2091	763	36	2356	865	37	2661	967	36
				2244	763	34	2591	981	38	2911	1183	41
Industrialized Pacific	807	152.0	18.8	924	192	21	1076	237	22	1252	318	25
				970	206	21	1149	310	27	1320	463	35
Eastern Europe	2088	198.2	9.5	2537	372	15	2993	552	18	3433	710	21
				2654	457	17	3245	729	22	3824	986	26
Asia	884	45.4	5.1	1225	75	6	1611	92	6	2010	166	8
				1290	78	6	1809	118	7	2430	222	9
Latin America	449	8.4	1.9	648	10	2	890	33	4	1160	50	4
				661	13	2	950	33	3	1326	75	6
Africa and Middle East	386	5.3	1.4	552	6	1	735	10	1	921	11	1
				577	11	2	830	16	2	1143	37	3
World total	9422	1401.6	14.9	11399	2075	18	13599	2538	19	15710	3040	19
				11974	2229	19	14810	2997	20	17809	3907	22
Industrialized countries	7325	1313.8	17.9	8503	1917	23	9804	2307	24	10985	2680	24
				8960	2061	23	10619	2695	25	12202	3369	28
Developing countries												
In CPE-Europe[1]	357	30.1	8.4	439	70	16	528	104	20	606	129	21
				454	70	16	568	135	24	681	181	27
Others	1740	57.7	3.3	2457	87	4	3267	128	4	4119	232	6
				2560	97	4	3622	167	5	4925	357	7
Total of DCs	2096	87.9	4.2	2896	158	5	3795	232	6	4725	360	8
				3014	168	6	4190	302	7	5606	538	10

[1]Developing countries in the Centrally Planned Economies (CPE) in Europe: Albania, Bulgaria, Czechoslovakia, Hungary, Poland and Romania.
[2]The top and bottom figures for total electric and nuclear generation are low and high estimates, respectively.
[3]The nuclear generation data presented in this table and the nuclear capacity data presented in Table 18.6 cannot be used to calculate average annual capacity factors for nuclear plants, as Table 18.6 presents year-end capacity and not the effective capacity average over the year.

18.4.2 *Future Projections*

The latest published IAEA estimates of development to the year 2000 are shown in Tables 18.6 and 18.7 (IAEA, 1986).

Table 18.6 presents total projected installed electric capacity figures for 1990, 1995 and the year 2000 and nuclear share in GWe and in percentages, both in a lower and upper bracket, grouped by world regions and summarized for industrialized and developing countries, the latter subdivided in CPE–Europe and other developing countries.

The low–high figures interval includes a large range of possible variations, especially for the second part of the 1990s. However, the tendency of more recent estimates is oriented towards the lower level of the brackets. Table 18.7 presents on the same basis the estimated figures for the yearly total electricity generation (TWh) and the share of nuclear energy in GWh and in percentages.

The share of the installed nuclear electric capacity in developing countries might increase from 15 500 MWe in 1985, i.e. 2.8 per cent of the total world electric capacity to 59 000 to 90 000 MWe, i.e. 5–6 per cent in the year 2000. The corresponding figures for the yearly electricity generation are from 87.9 TWh, i.e. 4.2 per cent in 1985 to possibly 360–538 TWh, i.e. 8–10 per cent in the year 2000. The higher share in the electricity generation is due to the base load operation of the nuclear power plants.

According to Table 18.8, by the year 2000,

Table 18.8 Developing Countries included in Year 2000 Nuclear Power Capacity Projection (Reproduced by permission of the International Atomic Energy Authority. IAEA, 1984)

Countries with nuclear plants in operation or under construction at end of 1983		Additional countries which may consider having nuclear plants in operation by 2000	
Argentina	Mexico	Algeria	Libyan Arab Jamahiriya
Brazil	Pakistan	Bangladesh	Malaysia
Bulgaria	Philippines	Chile	Portugal
China, P.R.	Poland	Egypt	Saudi Arabia
Cuba	Romania	Greece	Syrian Arab Republic
Czechoslovakia	Taiwan, China	Indonesia	Thailand
Hungary	Yugoslavia	Iran. Isl. Rep.[1]	Turkey
India		Iraq	Venezuela
Korea, Republic of		Korea, Dem. Rep.	
Nuclear capacity estimated for 2000: 60–90 GWe in 16 countries		5–15 GWe in five to ten countries	

[1]Construction of four nuclear power units has been stopped, but resumption of work on two units is under negotiation.

the number of developing countries with nuclear power plants in operation or under construction might reach 21–26 countries compared to sixteen at end 1983 (IAEA, 1984a). Figure 18.3 illustrates the corresponding past evolution and the future prospects up to the year 1990, offering in addition, the correlated information for the industrialized countries (Bennett, 1984).

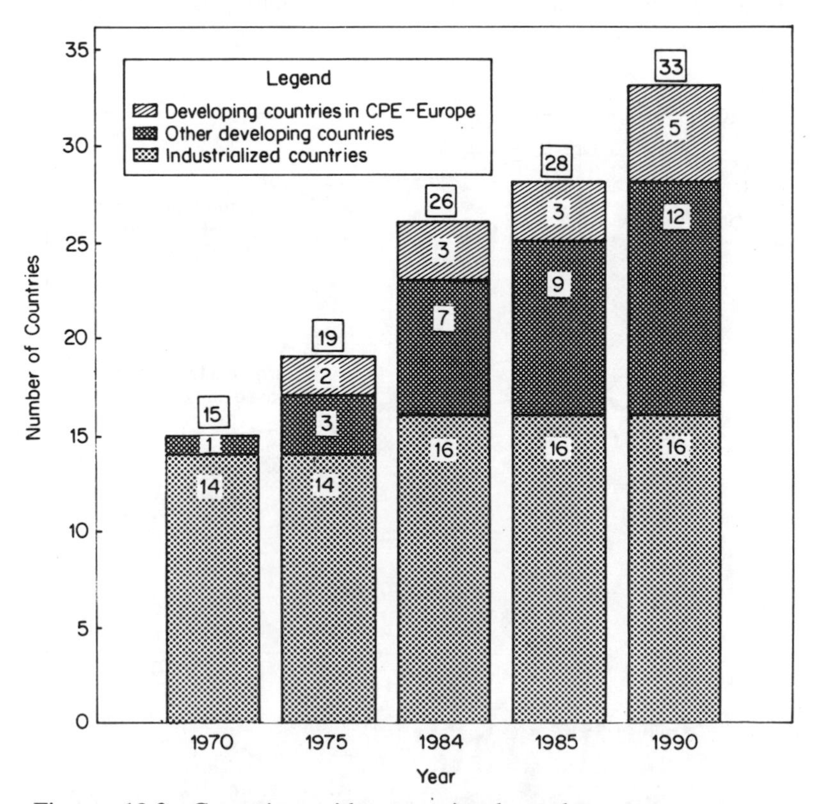

Figure 18.3 Countries with operational nuclear power reactors 1970–1990. Source: Bennett, 1984

18.5 Increasing Interest for Small and Medium Sized Nuclear Power Reactors—SMPRs

The present generation of nuclear power plants has been developed to satisfy principally the needs of the largest market for these plants, which corresponds to the industrialized countries with electric grids that admit the introduction of large units of 600 MWe to 1300 MWe. In contrast, the majority of developing countries have small, weakly interconnected electrical grids that could admit in a first stage of development only SMPRs, i.e. small and medium sized power reactors. Their capacity range is currently 200–500 MWe, for statistical purposes a size limit of 600 MWe has been in use by IAEA for some years (Schmidt, 1984).

Applying the generally used, but also rather simplified, rule-of-thumb that a single power generating unit should not exceed 10 per cent of the total capacity of all plants on the transmission grids, for reasons of stability of the electricity supply, the number of developing countries able to use nuclear power plants increases as the reactor size decreases. The lowest SMPRs in effective operation are presently the 200 MWe standard size used in India, and as highest limits, the USSR standard size 440 MWe power reactor, installed in Cuba and in CPE–Europe countries.

It should, however, be recognized that going nuclear with small reactors will require nearly the same commitments to a high technology as in the large reactor case, namely in respect of manpower, infrastructure, transfer of technology and financing. Nevertheless, in spite of this situation, an increasing number of developing countries would be interested in starting a nuclear power programme, if smaller sized power reactors of reliable design could be offered at acceptable investment costs. Related to the latter, Figure 18.4 displays in addition to cost range of electricity generated by nuclear, coal and oil power plants starting operation in 1990 for large power plants (600–1200 MWe), the kWh cost for electricity produced by SMPRs of 200–400

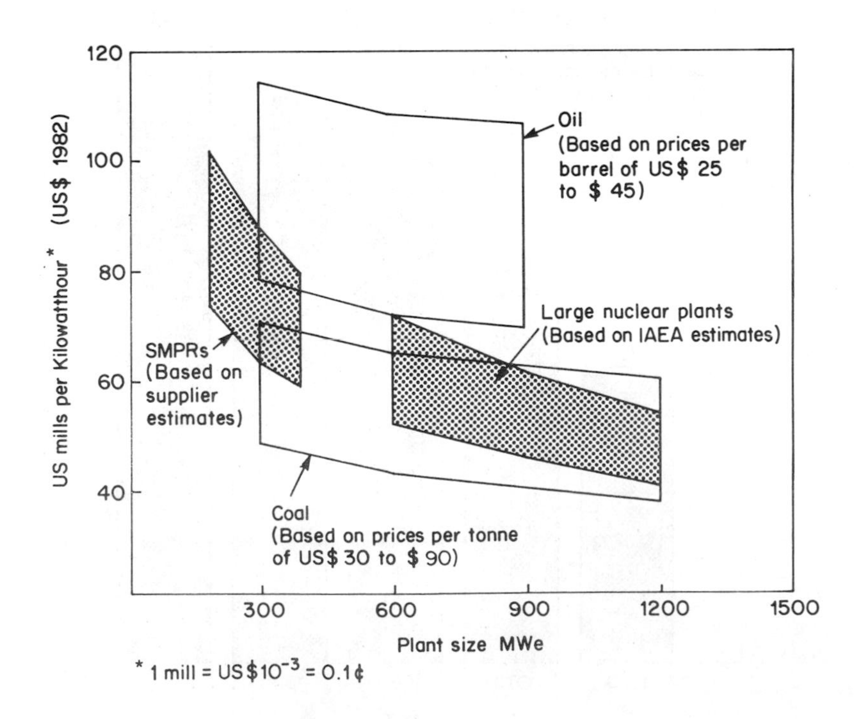

Figure 18.4 Cost ranges of electricity generated by nuclear, coal, and oil power plant starting operation in 1990. Source: Schmidt, 1984

MWe (Schmidt, 1984). Under the IAEA assumptions and the presently large spread in expected capital costs, an expensive SMPR at the 300 MWe level would still compete with oil at current prices and would easily compete at increasing prices. SMPRs in the lower projected cost range, a possibility in the latest cost indications, could be competitive with expensive coal. This could become relevant where local infrastructure and inland transportation costs are added to current world market price levels of coal.

As to competition with hydroelectric plants, the greater flexibility in locating nuclear plants possibly closer to load centres could be an important advantage. While well-located and well-managed hydroelectric plants may be considerably less expensive, recent experience with large hydro plants in countries with poorly developed infrastructures, remote locations, and long transmission lines suggests that SMPRs may be a viable alternative.

For more than 20 years IAEA has tried to promote the industrial production of smaller nuclear power plants primarily for developing countries. However, these efforts have not been successful so far (Konstantinov *et al.*, 1985).

Nowadays, the interest of the supplier industries, faced with diminishing and ever more uncertain home markets and looking for new export opportunities and an emerging interest in some industrialized countries for smaller, standardized nuclear power plants for special situations seem to lead to new developments.

Accordingly, suppliers have begun to invest in updating and readying their SMPR designs. Among the objectives, trends, and philosophies of the updated small-reactor designs, the following appear particularly noteworthy: modularization, a high level of prefabrication–shop fabrication, simplification of process and safety systems (natural circulation, etc.), use of standard components, shorter construction time, emphasis on proven demonstration, enhanced flexibility in site selection, long-term storage of spent fuel (up to 30 years), emphasis on lessons learned, operational services offered for a transition period, etc.

Currently available designs—those which could be offered within a maximum of five years—are summarized in Table 18.9. Others are still under development and review by the respective manufacturers. This relatively large number of designs with a good level of

Table 18.9 Available SMPR plants (Schmidt, 1984)

Country	Company	Type	Power level (MWe)	Ready to bid (years)	Fuel enrichment (%)	Main plant reference
Canada	AECL	Candu	300	0	Natural	600-Plant
France	Framatome	PWR	300	2	4	Pat./Cas.
Germany, F.R.	HRB	HTR	100	0	5–9	AVR
			300			THTR
			500			THTR
	Interatom	HTR	80[1]	0	7.8	AVR
	KWU	PHWR	300	0	Natural	Atucha
Japan	Mitsubishi	PWR	340	0	3	Mihama-1
	Toshiba	BWR	300	0	3	Onagawa-1
			500	0		Hamaoka-1
UK	NNC	Magnox	300	0	Natural	Oldbury
	Rolls Royce	PWR	300	1	3.3	Submarines
USA	GE	BWR	300	4	2–3	600-BWR
	B&W	PWR	90	5	2–4	Otto Hahn
	B&W	PWR	400	5	2–4	Otto Hahn
USSR	Atomen. Exp.	PWR	440	0	4	Many plants

[1]To be offered in two to eight modules.

development and readiness to bid is only a recent phenomenon.

This renewed interest has advanced SMPR design work to such an extent that concrete negotiations could start on most offers. Accordingly, the implementation of SMPRs is becoming a sound proposal from a technical, safety, and economical point of view.

Fortunately the recent developments in SMPR make the nuclear power option feasible to a wide range of developing countries, possibly adding to 20–25 potential candidates before the end of the century (Schmidt, 1984).

18.6 Launching Conditions and Implementation of Nuclear Power Programmes in Developing Countries

Introduction of nuclear power in a developing country is a long, complicated and challenging process, with three distinct phases: a conceptual preparatory phase embracing all basic technical–economic investigations needed for the justification of this very important step; a national infrastructure preparing phase creating the conditions for the launching of a nuclear programme; and finally the implementation phase comprising all project oriented activities leading to the successful commissioning and reliable operation of the first and next nuclear power plants.

18.6.1 *Technical-Economic Conditions*

Thirty years of practical experience and the technical, economic and safety performance of more than 344 nuclear reactors with a total capacity of nearly 220 000 MW operating worldwide in 26 countries at the end of 1984, confirmed nuclear power as a technologically mature and economically competitive option for future electricity supplies.

This option is immediately available on a large industrial scale (Laue, 1985). However, because of the mentioned present unavailability of proven nuclear reactors in a size range of 200–400 MWe, nuclear power cannot immediately be envisaged for developing countries with small, weakly-connected electrical generation and transmission systems.

With this, however, temporary exception, all elements exist for the preparation of the technical–economic investigations and studies, necessary to demonstrate the viability and opportunity of embarking on the introduction of a nuclear power programme. The best study approach for this goal would be the elaboration of a long-term national energy master plan in which nuclear power would appear optimally embedded in the form of a consistent, steadily advancing programme.

In addition to the technical–economic aspects, the study must convincingly demonstrate the necessity of a clear understanding at the decision-making level of the specific aspects of nuclear power, and a thorough knowledge of the tasks and activities to be performed as well as of the requirements, responsibilities, commitments, problems and constraints involved.

A nuclear power programme requires a very large effort (money, resources, manpower, etc.) on a national level over a long period of time. The country has to commit itself to the fulfilment of the requirements and has to establish clear policies to ensure the continuity of the programme of investments and industrial support. In some countries the government's nuclear policy and the commitment to its implementation has been promulgated as a constitutional mandate.

Only a solidly based and long-term nuclear power programme containing a series of nuclear power projects can justify the sizeable effort needed to plan and implement the national infrastructure development and the supporting organizational structures and activities. A single nuclear power plant not integrated into a nuclear programme may become an expensive venture.

These fundamental points have not always been clearly understood in the past, with delays in nuclear power programmes and projects as a result.

Once the conditions are met, the successful launching of a nuclear programme depends, according to the long experience of IAEA in this field, on the creation of the necessary national infrastructure. The main issues involved are the following:

1. An infrastructure of the electricity grid corresponding to the size of the nuclear reactors envisaged.
2. An effective decision making capability with adequate institutional infrastructure.
3. Qualified manpower.
4. Industrial support.
5. Financing.

18.6.2 *National Infrastructure*

18.6.2.1 Infrastructure of the Electricity Grid The interaction of grid characteristics with the design and performances of nuclear power plants is not only one of reciprocal size, although that is the first condition to be met. Problems of structure, reliability and stability of the whole electricity system deserves special attention and study.

In concrete cases substantial deviations from the mentioned 10 per cent rule-of-thumb limitation of the nuclear unit size may prove acceptable, depending to a high degree on the grid concept and the efficiency of the operational behaviour of the control devices.

18.6.2.2 Adequate Institutional Framework In addition to the introductory comments to this section regarding the high commitments required for the introduction of a nuclear programme, an adequate institutional and organizational infrastructure is needed for its practical implementation.

Basicaly, the distribution of tasks, functions and responsibilities between the organizations involved follows patterns similar to those for any other conventional power or industrial programme, but in addition, there will be a regulatory body. The functions of this organization will be the regulation and control of all nuclear facilities, and ensuring that the manufacture, construction and operation of the nuclear plants comply with the safety requirements and the quality standard needed.

Without doubt, the organizational infrastructure is an important prerequisite for launching a nuclear power programme, but setting up such an infrastructure should not constitute *per se* a major constraint for any developing country. Staffing the organizations on the other hand, may constitute a constraint. Even those developing countries that already have substantial nuclear power experience and clearly defined organizational structures are having staffing problems.

18.6.2.3 Qualified Manpower Competent manpower plays a fundamental role for safety and reliability in a nuclear power programme. There can be no compromise on safety; high safety and quality standards must be established and strictly maintained. The required manpower competence and work attitudes can be acquired only through appropriate education. training and experience.

In a country without a nuclear industry, technology is usually acquired from a more advanced country able and willing to transfer it. However, for technology transfer to be successful, the recipient country must be capable of absorbing appropriately the technology, and the key to this is the availability of qualified manpower.

It should be pointed out that in many developing countries the need for scientists and research-oriented personnel, particularly in the nuclear field, is often overestimated while the need for well qualified and experienced practically-oriented engineers, technicians and draftsmen is very much underestimated. Manpower development requires long lead times—ten years or more—and this is frequently not taken into account in programme planning.

The development of an adequate national educational and training infrastructure is the only real way to develop competent local manpower. Any country for which nuclear power is a viable option would have an electric system of reasonable size and a basic industrial infrastructure, and would thus also have certain technical manpower and education and training infrastructure. This will have to be expanded and adjusted in every case to meet the requirements of the nuclear power programme. It is possible and may be necessary to obtain some highly specialized experts and training from abroad, in particular in the early phases of a nuclear power programme, but this can only be utilized in a very limited way and a national manpower development programme is still essential.

IAEA experience shows that all developing countries with on-going nuclear power programmes have found it necessary to invest substantial efforts into manpower development. In fact, it can be said that the degree of success they have been able to obtain in the implementation of their nuclear power programmes and in the achievement of their development goals for national participation and technology transfer, has largely depended on the efforts expended in manpower development and the availability of competent manpower.

In order to support these efforts, the IAEA has offered a substantial contribution to manpower development. In 1975 a training course programme was started which provided more than 1500 participants from more than 50 developing countries with an insight into all aspects of nuclear power programmes. In addition, each year more than 200 fellowships are awarded for on-the-job training in on-going nuclear power projects and related areas.

India is an outstanding example of a developing country where large scale manpower development has been assigned first priority. There is no doubt that without this the country would never have attained the results it has in achieving practical self-sufficiency in the design, construction and operation of its nuclear power plants (see Chapter 17).

18.6.2.4 Industrial Support There are no firm rules regarding the industrial support infrastructure requirements of a country starting on a nuclear power programme, but it has to be recognized that the plants have to be built, the equipment and components have to be installed and tested, and the plants have to be operated and maintained within the country. This means a basic requirement of competent construction and erection firms and of operations and maintenance capabilities. The available industrial infrastructure will probably not have all the technology, know-how, level of quality or the expertise necessary for nuclear power, but these can be acquired.

The national engineering, manufacturing, construction and erection capabilities play an essential role in the promotion and development of the nuclear power programme. These industrial infrastructures should be closely associated with the nuclear power programme to which they provide a pool of necessary skills and human resources. The high quality requirements of nuclear technology call for the enforcement of a strong programme for quality assurance and quality control in particular.

All of the developing countries which have started nuclear power programmes have fairly well developed industrial infrastructures, approaching to varying degrees the levels usually associated with the so-called 'industrialized' countries. They can certainly not be classified as 'least developed' countries.

Experience shows that the higher the level of the industrial infrastructure of a country, the better it was able to meet the challenge of incorporating nuclear power, of absorbing the technology, and of achieving its national participation goals. In this respect, it should be mentioned that, in addition to the level of the available industrial infrastructure of a country, bilateral technical co-operation in transfer of technology has played a major and possibly decisive role.

Experience also shows that up to now no country with a very low level of industrial infrastructure has successfully incorporated nuclear power. This in itself is not sufficient reason to affirm that such countries cannot or should not go nuclear; however, the results of such a case are yet to be seen.

18.6.2.5 Financing The availability of financing on reasonable terms has often been stated to be an over-riding problem for nuclear power introduction in developing countries. The investment related to the gross domestic product (GDP) seems to have been a most important factor. According to studies by the World Bank, electric power expansion investment requirements have remained at about 7–8 per cent of the gross fixed capital formation of developing countries. It is estimated, however, that a shift to higher capital cost plants (including nuclear or hydro) would force an increase in this proportion to about 10–12 per cent. This would correspond to about 1–1.5 per cent of GDP for these countries.

For example, without going into details, the total investment for a 600–900 MWe nuclear power plant will be between 1.5 and 3.0 billion 1984 US$. This is based on investment costs of between 2500 and 3500 US$ kWe^{-1} installed capacity, depending on costs for development of the infrastructure, including the cost of expanding the transmission and distribution system and a possible transition to a high voltage level. Even a power plant in the SMPR range would require a total capital investment of some US$ 1–1.2 billion including interest during construction, and the period of repayment would be some 15–20 years following commissioning of the plant. These are conditions which go beyond normal export credits (Bennett, 1984).

A parallel is often drawn between big hydropower and nuclear projects as the former also are capital intensive and have long project times. There is, however, a fundamental difference as far as export credits are concerned as the civil works for hydro projects are most often financed nationally and the foreign component of equipment is only delivered in the last years of the construction, making the repayment period shorter than for a nuclear plant.

Thus the financing of a nuclear power programme must be seen as a major national effort, which will require long-term arrangements, in order that the impact on the domestic economy could be made acceptable during the long lead time before the savings from the low fuelling cost for nuclear power begin to provide economic benefits. Also, it must be recognized that a nuclear power programme is only one of several development programmes which will compete for available investment funds.

Among the developing countries with ongoing nuclear power programmes, Mexico is possibly the most outstanding example where financing constraints have seriously affected the country's nuclear power programme. The effect of financing constraints has also been causing substantial delays in the programmes of other countries such as Brazil, Romania and Yugoslavia. In fact, there is hardly a country whose nuclear power programme has not been affected negatively by this problem. Bangladesh and Turkey are examples of developing countries where the effective initiation of their first nuclear power project seems to depend mainly on finding a satisfactory financing arrangement.

In this respect, all efforts of Turkey failed to find a possible financing on a leasing basis, i.e. having a first nuclear power plant delivered, constructed and operated for a fifteen year period by the supplier, who would have been paid according to the electricity supplied. Turkey had to turn, therefore, to multilateral negotiations, seeking a possible solution by closer co-operation between industrialized countries sharing supply of components and financing guarantees.

Apparently, financing through special bilateral arrangements between the supplier and buyer (as nearly all imported nuclear power plants in developing countries were financed in the past) is no longer possible in the present economic situation. On the other hand, international financing, e.g. by the World Bank, cannot realistically be expected on a large scale within the foreseeable future, and multilateral financing might be the only solution to the problem. Seed financing from the World Bank of a part of the project could, however, help to obtain other financing more easily.

18.6.3 Nuclear Safety, Waste Management and Nuclear Safeguards

18.6.3.1 *Nuclear Safety and Waste Management* Nuclear safety remains, in developing countries as well as developed countries, of paramount importance both for the personnel working in nuclear plants as well as for the general public. It is, therefore, one of the main tasks given to the regulatory body responsible for full governmental surveillance and control with regard to all problems relevant to safety and environmental protection in the siting, construction, commissioning, operation and decommissioning of nuclear power plants and nuclear facilities within its national boundaries. Accordingly, bilateral assistance on nuclear safety should go far beyond a pure transfer of technology.

Because of the international character of safety problems, international bodies already provide developing countries with invaluable bases for nuclear power plant safety. For example, the IAEA Nuclear Safety Standards Programme (NUSS) has made available an internationally agreed set of five codes of practice and 47 safety guides for thermal nuclear power plants in the field of governmental organizations, siting, design, operation and quality assurance. These documents already play an important role in the transfer of technology, including safety expertise, from developed to developing countries, drawing on advice of the world's leading experts.

Safe disposal of radioactive waste and, in particular, high-level waste has been one of the most important questions frequently raised by the public in industrialized countries. What is now being demanded is tangible evidence, i.e. a demonstration of safe disposal. It would, therefore, be advisable for those developing countries that would like to introduce nuclear power to plan also the back-end of the fuel cycle, including spent fuel storage, waste disposal and decommissioning right from the initial stage, and the back-end costs should be included in the charge for electricity.

18.6.3.2 Nuclear Safeguards Because the peaceful use of nuclear energy is unavoidably accompanied by the production of large quantities of material which could be used for the manufacture of nuclear weapons, supply of equipment and long-term assurance of fuel supply and fuel service are strongly linked to wider international policies. The balance has always been that international supplies must be accompanied by non-proliferation assurances, first bilaterally, later using the IAEA safeguards for verification. The system of agreements and treaties set up to ensure non-proliferation has shown a steady evolution towards a more general non-proliferation regime, now mainly based on the Treaty on the Non-Proliferation of Nuclear Weapons (NPT) and also on the Tlatelolco Treaty for the Prohibition of Nuclear Weapons in Latin America. Both treaties require the conclusion by each State of a safeguards agreement with the IAEA covering all nuclear activities in the state—present and future.

At present, the majority of the developing member states of the IAEA are parties to either the NPT or the Tlatelolco Treaty or both. Considering the generally accepted complementary nature of assurances of nuclear supply and non-proliferation, there is every reason to encourage the remaining few countries operating nuclear power plants or embarking upon a nuclear programme, to accept full scope safeguards, whether under existing or new treaty arrangements, such as regional nuclear-free zones.

Obviously the main constraints to nuclear trade and transfer of technology are political and not technical. Removal of these constraints can be assured only if there is a strong and efficient international safeguards system commanding universal acceptance and respect.

The IAEA safeguards are an important instrument for giving effect to the non-proliferation framework under which transfer of nuclear material, equipment and technology take place. When safeguards are applied to the peaceful nuclear activities of a state, they serve to verify that the state concerned is respecting its commitments under legal agreements. The assurance that such verification gives is in the interest both of the state concerned and of the world community. It adds to mutual trust and hence paves the way for trade and for the transfer of technology.

In this respect a remarkable work is under way in the IAEA's Committee on Assurances of Supply (CAS), seeking to reconcile the interests of buyers and suppliers of nuclear equipment and fuel by assurances for complying with their mutual commitments. However, while progress has been made on some matters, the central issue remains unresolved. CAS continues to provide a forum for this dialogue. Its work will constitute one of IAEA's inputs to the 1987 United Nations Conference on the Promotion of International Co-operation in the Peaceful uses of Nuclear Energy.

18.7 Examples of Nuclear Power in Developing Countries

The overview on the general evolution of integration of nuclear power in the energy systems of developing countries in the preceding sections, merits some additional comments focusing on the practical conditions and achievements in a few, nuclearly more advanced developing countries. However, for space reasons, they have to concentrate on two particularly typical examples, Argentina and Brazil, India having an entire chapter reserved for this purpose (see Chapter 17). The other developing countries with operating nuclear power reactors are only mentioned with a few complementary indications.

18.7.1 *Argentina*

Argentina constitutes an excellent example of a country which, endowed with large uranium reserves and a remarkable industrial potential, decided early to embark on a nuclear power programme with the firm intention of quickly achieving complete nuclear autonomy. Overcoming during 30 years of effort all kinds of difficulties, it now has two power reactors in operation: a 370 MWe heavy-water moderated and cooled reactor known as Atucha I which was connected to the electric grid in 1974, when the latter had 4500 MW installed capacity, and a 648 MWe of Candu type which became operational in 1983 in Embalse Rio Tercero-Cordoba. A 745 MWe PHWR reactor is under construction as Atucha II and is scheduled to be commissioned in 1989. Three other nuclear power stations, each of 600 to 700 MWe, should become operational by the year 2000, the total installed nuclear electrical capacity in the country reaching by that time some 3500 MWe.

The local industrial support developed rapidly, the contribution of the national industry increasing from 33 per cent for Atucha I to 58.3 per cent in Embalse, and 62 per cent for Atucha II. It is scheduled to reach 85 per cent for the nuclear power plant IV, 94 per cent for the plant V and full 100 per cent for the power plant VI. So far as sophisticated local nuclear industry performances are concerned, Argentina exploits her uranium resources, has a uranium enrichment plant, a heavy water production plant, a fuel elements and a heavy components fabrication plant and a strong engineering company ENACE.

The well-conceived and consistently implemented nuclear energy programme in Argentina is one of the most successful in the Third World. However, world concern is clouding this success story, since Argentina continuously refused to sign the nuclear non-proliferation treaty (NPT) and did not ratify the Tlateloclo Treaty, which prohibits the introduction of nuclear weapons in Latin America.

18.7.2 *Brazil*

The first nuclear power plant was commissioned in Brazil in 1982 in Angra dos Reis, 100 km southwest from Rio de Janeiro. The Angra 1, a PWR-unit, has an installed electric capacity of 626 MWe and was supplied on a turnkey basis by Westinghouse. When this contract was signed no plans existed to prepare a general nuclear infrastructure or to start a national industrial participation.

However, a detailed analysis of the Brazilian energy situation in 1974, concluded that in addition to the accelerated harnessing of the considerable hydroelectric potential, a parallel integration of nuclear power would represent the best solution for meeting the rapidly increasing demand for electrical energy to the year 2000. A nuclear power programme which should gradually create in the country the capability of design and construction of nuclear power plants, with the simultaneous development and adaptation of the local industry to the complex nuclear technology, appeared the solution to aim at. In addition, such a programme should be self-sufficient, i.e. including the complete fuel cycle, encompassing enrichment and reprocessing plants.

On this basis, a complex agreement was signed in 1975 with the Federal Republic of Germany, the so-called Bonn–Brasilia Accord. To ensure efficient technology transfer, the foreign partner would not act as a simple consultant but should share a joint responsibility, participating with name and capital in the whole scheme. Several subsidiary joint companies have been created and more than 30 co-operation contracts have, meanwhile, been signed between Brazilian and German industrial companies. This huge and multilateral co-operation endeavour has been the most important ever agreed in the nuclear field.

The nuclear power plants scheduled to be installed in the programme were: Angra 2 of 1250 MWe to become operational (delayed) in 1990, to be followed by seven other reactors of the same PWR type to be commissioned till the year 2000 (postponed from the initial date of 1990).

Regarding the national infrastructure: for Angra 2 and 3, the national industry will have a 55 per cent contribution in the engineering services; the heavy industrial component corporation was inaugurated in 1979, the fuel elements factory in 1982 and the enrichment plant is scheduled to start provisional operation this year.

18.7.3 *Other Developing Countries*

The further developing countries with operational nuclear reactors to be referred to briefly, are the following:

1. India with its five operating reactors totalling 1030 MWe and presenting a particularly successful original and autonomous nuclear power development which is separately dealt with in Chapter 17.
2. Republic of Korea, with three nuclear reactors totalling 1790 MWe in operation and six others (5622 MWe) under construction, is the leading developing country as far as installed nuclear capacity is concerned. The first three nuclear reactors were supplied on turnkey contracts, for the rest an increasing national support is being prepared.
3. Pakistan started early with the 125 MWe nuclear power plant Kanupp generating electricity in 1971–1972.
4. Taiwan had as of December 1984 five nuclear reactors totalling 4011 MWe in operation and a sixth one of 907 MWe under construction. The first two units (PWRs) of 606 MWe and the next two (BWRs) of 985 MWe have been installed in two different power plants. A third power plant will receive the next two 950 MWe units of the PWR type.
5. Yugoslavia has a 632 MWe PWR power reactor in operation at the Krsko plant, supplied on a turnkey basis. However, it is envisaging arrangements for a second unit, that will utilize more extensively the impressive potential of the national industry contribution.
6. CPE–Europe. In the group of the centrally planned developing countries of Europe, USSR supplied nuclear reactors (SMPRs) are predominant. Bulgaria has four units (1632 MWe), Czechoslovakia two units (762 MWe) and Hungary one unit (395 MWe) in operation. The capacity under construction is impressive: eighteen units totalling 9690 MWe, among them two Canadian supplied Candu reactors of 660 MWe each to

Romania. In all these countries, with a strong industrial infrastructure, the national engineering and industrial support contribution is correspondingly very high.

18.8 Conclusion

The role of nuclear power in developing countries is intimately related to the world-wide energy situation and the relationship between energy and socio-economic development. No shortage of energy should be allowed to hamper further economic and industrial development or improvement of the standard of living in these countries.

However, in the last decade the relationship energy–economic growth seems less evident while the relation electricity demand–GDP growth continues to demonstrate a close connection. Since nuclear energy has been so far mainly oriented to electricity supply, although not excluding heat supply as well, the more rapid growth of electricity demand is one important factor favouring its application.

Indeed, while primary energy consumption may double in the developing countries during the period 1983 to 2000, electricity consumption is estimated to increase threefold (Laue, 1985). The accelerated trend to urbanization and great conurbations is one of the main factors in this evolution. Accordingly the share of primary energy consumed for electricity generation in developing countries may increase from the current 19 per cent to around 27 per cent by the year 2000. Comparatively, the shares in industrialized countries are 32 per cent and 38 per cent, which denotes ample room for still further development.

As far as nuclear power is concerned, 40 years of experience demonstrates that it represents a technologically mature, economically competitive, safe and reliable as well as an environmentally acceptable option for electricity supply.

In spite of the described favourable situation, nuclear power cannot be seen as a panacea, or even a viable near-term solution, for the majority of developing countries. This is due to the difficulties in using this technology in small, weakly-connected electrical generation and transmission systems and to the present unavailability of proven nuclear power plants in a size range of 200–400 MWe which could be used in these grids (with the exception of the USSR standardized 440 MWe PWR type, which is still built and exportable).

How narrow the potential market is in this respect, is confirmed by the fact that from the total electricity production of the 161 developing countries in 1981 not including the CPE–Europe—eight of them accounted for 62 per cent of the total production of 1320 TWh. The balance of 38 per cent was distributed among 153 countries with an average of 2.8 TWh per country. The eight major users are China 309 TWh, Brazil 140 TWh, India 117 TWh, Mexico 74 TWh, Yugoslavia 57 TWh, Republic of Korea 43 TWh, Taiwan 42 TWh, Argentina 38 TWh, with a total of 820 TWh. All have on-going nuclear power programmes (Bennett, 1984).

As far as future development is concerned it is expected that from the 1983 position with 24 nuclear reactors with a total capacity of some 11 000 MWe in operation and other 38 totalling 22 800 MWe under construction, the developing world (including CPE–Europe) will reach in the year 2000 an installed electrical capacity between 66 000 and 115 000 MWe, i.e. 5–8 per cent of the total electrical installed MWe. The percentage of nuclear generation in the total electricity production will increase from 2.9 per cent to 8–11 per cent. In terms of countries included in the year 2000 projection, the sixteen developing countries having nuclear power plants currently in operation or under construction may reach 60–90 GWe, with another five to ten additional countries which may join with 5–15 GWe; among the latter are Egypt and the OECD countries: Greece, Portugal and Turkey.

It might, therefore, be concluded that in spite of all difficulties and especially those tied to the preparation of the launching and implementation of a new nuclear programme, nuclear energy will gradually increase its share and contribution to the economic and social progress of the developing world.

REFERENCES

Bennett, L. L. (1984). 'Nuclear Power Status and Trends', paper presented at the IAEA Interregional Training Course on Electric Systems Expansion Planning, September 4–November 1 1984, Argonne National Laboratory, Argonne, Illinois, USA.

Goetzmann, C. A. (1985). 'Design for Low Capital Cost in Small Nuclear Reactors', paper presented at the M.I.T. Conference on Nuclear Power Plant Innovation for the 1990s, January 9–10, Cambridge, USA.

Hirschman, H. (1983). 'Small and Medium-sized Nuclear Power Plants for Developing Countries', paper 2.3–10, 12th Congress of the World Energy Conference, New Delhi.

IAEA (1982). 'Guide-book on the Introduction of Nuclear Power', IAEA, Vienna, p. 349.

IAEA (1986). *Energy, Electricity and Nuclear Power Estimates for the Period up to 2000*, Reference Data Series No. 1, August 1986 edition, IAEA, Vienna,

IAEA (1984b). Nuclear Power: Status and Trends, IAEA, Vienna.

IAEA Stancescu (1985a)–*Energy and Nuclear Power Planning in Developing Countries*—Technical Reports Series No. 245 (Author Stancescu, I. D.), IAEA, Vienna.

IAEA (1986). Nuclear Power Reactors in the World, Reference Data Series No. 2, April 1986 edition, IAEA, Vienna.

Katz, J.E., and Marwah, O.S. (1982). *Nuclear Power in Developing Countries—an Analysis of Decision Making*, Lexington Books, D. C. Heath and Company, Lexington, Massachusetts and Toronto.

Konstantinov, L. V., Laue, H. J., and Bennett, L. L. (1985). 'Nuclear Power and Nuclear Methods Applications in Developing Countries: Status and Prospects', paper presented at the International Conference on Physics and Energy for Development (ICPED) January 26–29, Dhaka, Bangladesh.

Laue, H. J. (1985). 'General Energy Needs for Developing Countries—The Role of Nuclear Power', paper presented at the IAEA Interregional Training Course on Electric System Expansion Planning, January 28–March 29, 1985, Argonne, USA.

Laue, H. J., Bennett, L. L., and Skjoeldebrand, R. (1984). 'Nuclear Power in Developing Countries', *IAEA Bull.*, **26, (1)** 3.

McKenzie, N. C. (1983). 'Financing Nuclear Programmes in Developing Countries', paper IAEA-CN–36/77. Proceedings of an International Conference, September 13–17, 1982, IAEA, Vienna.

N. N. (1983). 'Weltweiter Ruckgang von Verbrauch und Erzeugung von Primarenergien' (Worldwide Decrease of Consumption and Production of Primary Energy), *Brennst.-Warme-Kraft*, **55, (4)** 130.

Schmidt, R. (1984). 'Assessing Prospects for Smaller Reactors', *IAEA Bull.*, **26, (4)** 29.

The World Bank (1983). *The Energy Transition in Developing Countries*, The World Bank, Washington, D.C.

Section IV:
CHALLENGE AND OPPORTUNITY

Nuclear Power: Policy and Prospects
Edited by P. M. S. Jones

19

The Present World Scene

P. M. S. JONES
Department of Economics, Surrey University.

19.1 Introduction

Chapters 11–18 have looked at the history of development of nuclear power and nuclear policy in the world's major western industrial nations and in the developing countries. They have also given consideration to the way forward in thc countrics with which they dealt.

This chapter will look at the over-all world position, at how expectations have been changing and why, and at the general conclusions that can be drawn about the near and longer term future.

19.2 Installed and Planned Nuclear Generating Capacity

At the end of May, 1985, there were 366 reactors operating worldwide with a total capacity of 239 GWe. A further 171 were under construction with a capacity of 155 GWe, whilst 139 were at the planning stage with a capacity of 124 GWe. The detailed breakdown by country is shown in Table 19.1.

The PWR accounts for almost 50 per cent of operating reactors and this high share will approach 60 per cent when all those under construction and planned are completed (Table 19.2). The second most popular type is the BWR although this has been losing ground, so that its over-all share will fall to below 20 per cent. The gas cooled reactors (Magnox and AGR) which have been favoured by the UK; the PHWR developed and exploited by Canada; and the light water cooled graphite moderated system developed and deployed by the USSR each contribute about 7 per cent of the world total. The remaining types, viz. the HTGR and FBR are still at the development stage and have yet to enter the electricity supply market on a significant scale.

The nuclear share of electricity generation has risen over the past 30 years from zero to figures that exceed 40 per cent in six countries (Table 19.1). It is set to rise steadily in most regions of the world as the reactors under construction and planned come into service. In regional terms OECD Europe has the highest nuclear share of generation which is expected to average over 40 per cent by the mid–1990s.

19.3 Forecasts of Capacity

Nuclear power entered the scene in the 1950s, a time when there was great optimism about the future prospects for world economic development. Its low fuel costs and the comparatively low capital costs that its high power density looked likely to offer, appeared to assure it of a major or even dominant role in the supply of electricity.

Forecasts of energy demand produced during the 1960s and 1970s generally assumed a direct link between projected economic growth and projected energy requirements, with electricity gaining a progressively larger share of the total energy, and nuclear a progressively larger share of electricity generation. In consequence very high rates of nuclear capacity growth were expected, particularly in the USA.

The oil crises of 1973 and 1979 had a dramatic effect on world growth, both in the developed and developing countries, and this in turn has led to major reductions in energy and nuclear demand projections; partly due to the lower rates of economic growth and partly due to revised expectations of energy demand for any given level of economic activity. Figures 19.1 and 19.2 illustrate how perceptions of future demand have declined both at the global level and within OECD.

As recently as 1980 the International Nuclear Fuel Cycle Evaluation was projecting that by 2000 there would be 1080–1650 GWe installed worldwide compared with the most

Table 19.1 Nuclear Plants Planned, Under Construction and Operable

	Operable		Under construction		Planned		Nuclear generation in 1984	
Country	Units	MWe	Units	MWe	Units	MWe	TWhe	% of total
Argentina	2	935	1	692	3	1800	4.6	11.4
Austria	1	692	—	—	—	—	—	—
Belgium	7	4479	1	1006	—	—	26.4	50.7
Brazil	1	626	2	2490	6	7470	1.7	0.7
Bulgaria	4	1632	2	1906	4	3812	12.7	28.6
Canada	17	10 037	6	5114	—	—	49.3	11.6
China	—	—	1	288	6	5400	—	—
Cuba	—	—	2	816	—	—	—	—
Czechoslovakia	4	1582	9	4006	3	2676	6.7	8.5
Egypt	—	—	—	—	2	1800	—	—
Finland	4	2310	—	—	—	—	17.8	41.0
France	41	32 993	23	28 355	6	8650	181.8	58.7
German D.R.	5	1694	6	3432	—	—	n.a.	n.a.
Germany F.R.	19	16 414	6	6585	6	7747	92.6	23.2
Hungary	3	1215	1	410	4	1632	3.5	22.2
India	5	1020	6	1110	4	880	4.0	2.6
Iraq	—	—	—	—	1	600	—	—
Israel	—	—	—	—	1	900	—	—
Italy	3	1286	3	1999	10	9500	6.6	3.8
Japan	33	23 264	9	7523	15	13 906	126.1	22.9
Korea	4	2720	5	4692	4	3456	11.8	21.9
Libya	—	—	—	—	2	816	—	—
Mexico	—	—	2	1308	—	—	—	—
Netherlands	2	508	—	—	—	—	3.5	5.8
Pakistan	1	125	—	—	1	900	0.3	1.6
Philippines	—	—	1	620	—	—	—	—
Poland	—	—	2	880	4	2780	—	—
Romania	—	—	3	1980	2	1068	—	—
S. Africa	1	921	1	921	—	—	—	—
Spain	8	5577	7	6518	3	2955	23.1	19.2
Sweden	12	9455	—	—	—	—	48.6	40.6
Switzerland	5	2882	—	—	2	2065	17.3	36.5
Taiwan	6	4918	—	—	4	4120	24.6	47.9
Turkey	—	—	—	—	3	2800	—	—
UK	37	9564	5	3130	2	2500	45.7	17.3
USA	91	75 815	31	35 814	2	2200	325.2	13.5
USSR	49	25 853	36	33 719	38	30 216	131.0	8.8
Yugoslavia	1	632	—	—	1	1000	—	—
Totals	366	239 149	171	155 314	139	123 649	—	—

Reproduced with permission from IAEA Power Reactor Information System and Nuclear Engineering International's *Power Reactor Supplement*, August 1985.

Table 19.2 Reactor Types

Type	Operable	Under construction	Planned	Total
PWR	180	118	108	406
BWR	80	20	12	112
Magnox	34	0	0	34
AGR	9	5	0	14
PHWR	24	15	5	44
LWGR	27	8	8	43
HTGR	3	0	0	3
FBR	6	4	5	15
Other	3	1	1	5
Total	366	171	139	676

recently available OECD estimates of 500–650 GWe, with the regional breakdown shown in Table 19.3. The over-all fall of over 50 per cent in only six years reflects an adjustment of expectations in all parts of the world with OECD Europe least affected.

One consequence of the world economic slowdown was that some countries which had anticipated continued rapid growth had too much capacity in the construction and planning stages for the demand that has subsequently materialized. This has led to rephasing of plans, a slow down in new plant orders and in some countries, notably the USA, the actual cancellation of ordered plants. The resumption of growth during the

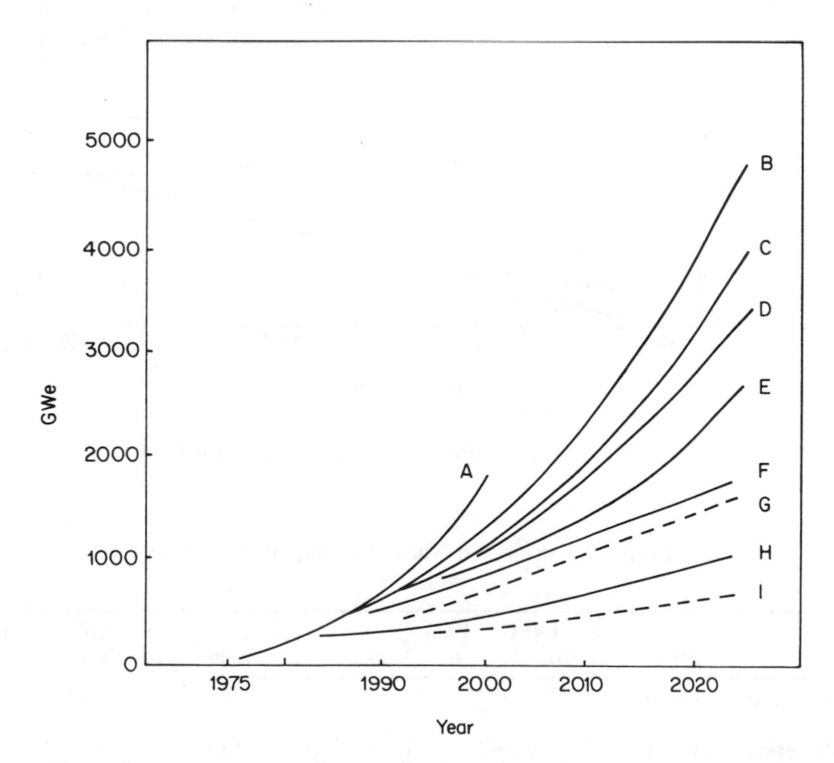

Figure 19.1 Nuclear capacity projections (World). A sample of forecasts of the past decade A – Workshop on Alternative Energy Strategies, 1977
B,F – INFCE, 1980.
C,E – World Energy Conference, 1978
D,H – Jones, AIF/CNA Conference, Quebec, 1981.
G,I – Thornton, NEA, 1987

past few years and the ageing population of fossil and early nuclear plants means that significant numbers of new plant, of some kind, will need to be ordered if capacity shortfalls are not to be experienced in the mid to late 1990s in countries like the USA and the UK.

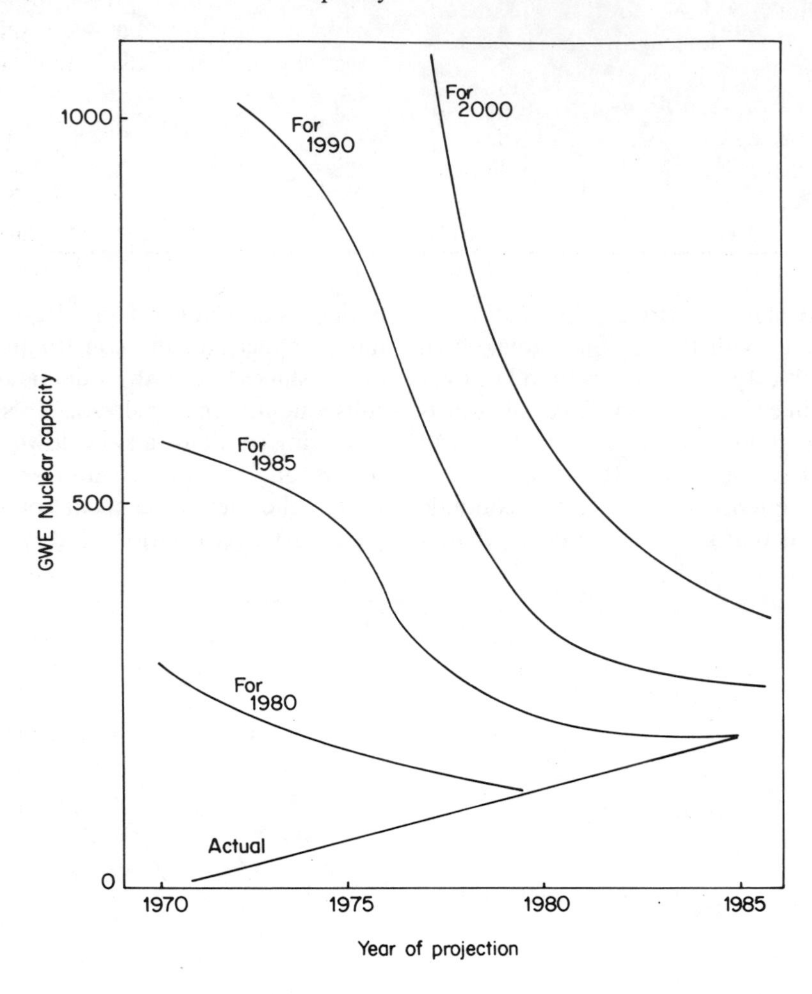

Figure 19.2 Nuclear energy agency forecasts for OECD

Table 19.3 Projected Nuclear Capacity by Region
Net GWe

	INFCE 1980			NEA/OECD (1987)		
	1995	2000	2025	1995	2000	2025
OECD Europe and Pacific	294–391	376–557		188	215–257	325–760
OECD N. America	233–314	307–462	1800–3900	132	150–172	230–390
Developing WOCA[1]	90–100	150–186		27–39	36–71	120–405
Centrally planned economies	169–292	248–447	—	73–89	96–146	200–605
World total	786–1097	1081–1652	—	420–448	497–646	875–2160

[1]World outside the centrally planned economies.
Sources: Nuclear Energy and its Fuel Cycle, NEA/OECD, Paris, 1987.

19.4 Forecasting Methodology

Some digression on forecasting methodology is appropriate at this juncture since the range of options that has to be considered in formulating energy or nuclear policy is closely linked to the likely levels of future demand and the range of uncertainty surrounding estimates of these levels.

There are many approaches to demand forecasting ranging from the simple extrapolation of past trends to complex mathematical modelling exercises. Each has its advantages and disadvantages. The main methods are summarized below.

19.4.1 *Trend Extrapolation*

This method, which was used to some extent in the 1960s, is simplistic in that it assumes implicitly that all factors that have operated in the past will continue to operate in the same way in the future; that the inter-relationships between these factors remain fixed; and that the change in any more fundamental variables (population level, productivity, etc.) continues immutably (Brookes, 1968).

A little more refined approach would take account of foreseeable deviations in any of the factors known to affect energy demand and make such allowances as were judged appropriate.

Whilst such approaches have the merit of simplicity and may give reasonable guidance for the short term, particularly during periods of comparative stability, they lose credibility as soon as a major change in the economic environment takes place, an eventuality that is almost inevitable when looking to the longer term.

19.4.2 *Macro-economic Models*

The econometric approaches seek to link energy demand through a set of empirical relationships to the level of economic activity. They can concentrate at the level of the total economy or can disaggregate the economy into a number of sectors such as domestic, industrial, commercial and transport use. The links may be direct ones between economic activity (e.g. GDP or income per caput) and energy use, or they may be more complex and look at population levels, size of the workforce in employment, and productivity, for example.

Although such methods have the appearance of greater realism than those considered in section 19.4.1 they merely transfer the forecasting problem from one parameter, energy demand, to another, such as GDP; or alternatively to a number of other parameters which contribute to economic activity. These parameters and the inter-relationships between them are just as susceptible to disturbance as energy use itself. Thus, the 1973 and 1979 oil price rises led to a recession that slowed economic growth and contributed to unemployment to an extent totally outside the range that would have been regarded as plausible before 1973.

Furthermore, the relationships between energy use and output or income have changed, due partly to technological progress, partly to the changed price relativity between energy and other commodities, and partly to the influence of public expectations concering fuel availability and prices in the future.

In the 1960s and early 1970s economic growth was widely expected to continue at roughly the rates then prevailing which were high by historic standards (Table 19.4).

Table 19.4 Average annual percentage growth rates of real GDP

Region	Historical growth rates	
	1925–50	1950–75
OECD	2.3	4.2
Developing countries	2.6	5.0
WOCA[1]	2.4	4.3

[1]World outside the centrally planned economies

Indeed, a large part of the world had experienced rates of economic growth exceeding 6 per cent p.a. for extended periods and the world's population was expected to continue growing rapidly, particularly in the developing countries (see Table 18.2).

Since the 1970s there has been a prevailing view that growth either has or will flatten off in the industrial world. This changed perception was partly due to the re-emphasis of the truism that exponential growth can not continue indefinitely without running into supply constraints; a fact recognized in the nineteenth century by Malthus and Jevons. However, the mere recognition of this fact does not give any indication of the time when the levelling-off will occur. Both the level reached and its timing depend on the state of technology. Thus Jevons' concerns over long term energy supply, which focused on coal but attempted to consider possible alternative sources, were invalidated by great improvements in the efficiency of energy use and the unanticipated advent of oil.

Further arguments for long term levelling-off are based on the view that material wealth is reaching saturation in the western industrial nations and that people will not want more, or that wealth is in some way linked to damage to the environment and the population will choose to forego growth in the interests of environmental protection. There is, however, little evidence that the population of the industrial countries are satisfied. There are still great inequalities and few countries will be content with the standards of living they can provide for their most disadvantaged citizens. Conditions in the third world, with two-thirds of the world's population, are nowhere near satisfactory with widespread starvation and disease. On the second point too there is little support. The environmental damage rising from industrial development in the nineteenth and early twentieth centuries was a consequence of prevalent social attitudes and not inevitable. With current awareness, development can be accomplished with minimal environmental impact. If anything, the greatest risk to the world ecology would appear to stem from the excessive use of renewable fuelwood and dung forced on poorer nations by their very poverty (Eckholm, 1975).

Just as the optimistic expectations of the 1960s were invalidated by the unforeseen events of the 1970s, so could the pessimism of the early 1980s prove unfounded. For example, if there are long wave cycles in the level of world economic activity as claimed by Kondratiev (1935). The work of Freeman (1983), Rostow (1978) and others suggests that the observed 50 year cycles could be explained through the process of innovation and consolidation. On the basis of the timing of past cycles one would have anticipated a period of retrenchment in the 1970–1990 period followed by a marked upturn by the end of the century (Ray, 1980). None of the econometric demand models known to the author in use in the western nations or in the international agencies take account of this possibility.

Like trend extrapolation, econometric modelling has to be seen as an approach limited to the short term with an increasingly wide band of error as one moves further into the future (Jones, 1974).

19.4.3 *Disaggregated Analysis*

During the 1970s proponents of low energy futures switched attention to energy conservation and began looking at energy use in a highly disaggregated fashion (Leach, 1979). The approach was not new and had been adopted some years earlier in attempts to explain where all the energy projected to be needed using the (then high growth) macro approaches might be employed (Day and Brookes, 1971). The new proponents looked at individual energy applications and assessed the technical feasibility of reducing the energy consumption needed, for example, for heating homes and offices, for specific manufacturing processes, for transport, etc. This approach then has to be linked to some projected measure of the future number of homes, offices, domestic appliances, cars, etc., and to the future output of the manufacturing processes, to yield estimates of future energy consumption. Account clearly has to be taken of the rate at which the improvements can penetrate the market, taking account of the existing stock of buildings, plant and equipment and of the incentives to implement change (Day *et al.*, 1980).

Such approaches are necessarily country specific and suffer from the inherent weakness that they can only look at existing uses for energy. They are also complicated by the vast number of separate applications that need to be examined for a reasonably comprehensive coverage. In individual areas they probably have greater long term validity than the purely macro-economic methods which can only make crude allowance for technical progress, but they are themselves dependent on potentially inaccurate forecasts of economic activity. On the whole they appear likely to underestimate future demand and it is probably best to view them as complementary to the econometric or trend approaches rather than as an alternative.

19.4.4 *Scenario Methods*

The major upsets of the 1970s demonstrated how vulnerable forecasts based on a narrow range of economic assumptions could be and this led to increased attention being given to the range of alternative futures that might arise depending on particular events or specified developments. In their earlier applications scenarios were often used as a part of normative or goal orientated forecasting in which a desired objective was selected and the means by which it could be reached explored. Thus Lovins (1975) used qualitative world energy scenarios to argue against the need for nuclear power and Leach (1979) to argue that economic growth by 2.5–3 fold could be accomplished in the UK without significant increases in energy consumption.

More usefully scenarios can explore the implications of high and low economic growth, the implications of the election (within a given country) of a right or left wing government, or the effects of a major regional conflict affecting oil supplies, etc.

Those responsible for policy choices or decisions can then set the options they have open to them against the alternative scenarios. They can form a judgement on the preferred option in the light of their perception of the likelihood of the scenarios and of the risks and benefits associated with the outcomes.

The scenario method is no better at predicting the future but, by exploring a wider range of alternatives, it can expose the weakness and strengths of particular energy strategies in terms of their flexibility and vulnerability. The method was used, in the form of alternative economic development paths, by the UK Central Electricity Generating Board in its presentation of its case to the Sizewell Inquiry. The object was to demonstrate that the choice was economically preferable in all of the scenarios considered. The CEGB spent a great deal of effort ensuring that its very detailed scenarios to 2030 were internally consistent and that the economic assumptions underlying the scenarios on such things as fossil fuel prices and dollar–sterling exchange rates were sustainable. Even so the rapid changes in the world energy scene in 1985 and 1986 and the higher than expected growth of electricity had materially altered the position before the outcome of the lengthy Inquiry was known.

19.4.5 *Electricity and Nuclear Shares*

The preceding paragraphs in this section have looked at the question of projecting total energy demand. The share that will be taken by electricity will vary from country to country depending on circumstances. In a market economy the relative prices of competing fuels to the consumer will have a major influence, although the selective or differential imposition of taxes by government may introduce considerable distortions, either accidentally or by design. In centrally planned economy countries or economies that have national energy policies, the share of particular fuels may be set as part of the planning process. Strategic and social considerations may then play a major role.

Goals of energy independence to avoid economic disruption due to external events were popular in the aftermath of the oil crises of the 1970s. The protection of indigenous industries such as coal mining (UK, FRG) or uranium minirg (USA) has also been a major influence in national policy even where centralized planning has not been favoured.

In forecasts where there are no imposed policy constraints the fuel mix can be projected by any of the techniques described earlier; namely trends, econometric or scenario methods, or some combination of these. Additionally recourse can be made to market penetration models. Such approaches have been used by the Nuclear Energy Agency (Thornton, 1982) and the International Institute of Applied Systems Analysis (1981).

19.5 The Uncertain Future

It will be evident from sections 19.3 and 19.4 that there are major problems in trying to predict either national or world energy requirements for any lengthy period into the future. Predicting the level of nuclear capacity is even more problematic. Will the pace of world economic development taper off in coming decades? Will improvements in efficiency of energy use lead to a flattening out or even decline in energy demand? Is the long wave economic cycle effect real and will it lead to an upsurge in economic activity in the 1990s? If so will energy demand follow? Quite credible scenarios can be developed for any of these options.

Over and above this is the question of non-economic factors. Will concern over the environment and the greenhouse effect (see Chapter 21) lead to a massive switch to nuclear power? Will events like TMI and Chernobyl (see Chapters 7 and 22) lead to a nuclear moratorium in some countries? The whole question of public attitudes and how they will change in the future is hardly understood at all (Jones, 1982).

Within this uncertainty there are some inescapable facts: The world population is growing; there are major inequalities of living standards within and between countries; energy is an essential input into manufacturing, agriculture and service activities and to the creation, maintenance and operation of the national infrastructure. The planning and policy implications of these are explored further in Chapter 23.

19.6 The Fuel Cycle and Other Services

The world's nuclear industry relies on uranium supply (covered in Chapter 20) and on enrichment and fuel fabrication services; on the availability of spent fuel storage (interim or longer term); and, ultimately, on reprocessing and/or spent fuel and nuclear waste disposal services.

There are few, if any, problems in the short term. Table 19.5 summarizes the most recent estimates of enrichment and fuel fabrication capacity produced by the Nuclear Energy Agency (Thornton, 1987).

The demand for enrichment services will increase as the capacity of nuclear plant increases, reaching about 39 million SWU p.a. by 1995 in WOCA. At present there are only three major suppliers in WOCA, the US Department of Energy, Eurodif and URENCO, who between them currently expect to have some 36 million SWU p.a. capacity available in 1995; excluding the mothballed 7.7 million SWU p.a. diffusion plant at Oak Ridge, but including expansions by URENCO and in Japan. A further 3 million SWU p.a. is likely to be available from the USSR and small quantities from plants under construction in Brazil and South Africa. These supplies should meet in total the projected WOCA demands, though with little to spare. The present 50 per cent overcapacity will be eliminated.

The time for construction of modular enrichment plant, like URENCO's centrifuge process, is short compared with reactor construction lead times so no problems of supply need arise although most countries will be dependent on external suppliers of enrichment. If uranium prices rose sharply or if higher fuel burn-up were planned there could be an increase in demand for separative work that could lead to pressure on suppliers

Table 19.5 Enrichment and Oxide Fuel Fabrication Capacities and Demand

	Region	Capacity			Annual requirement[1]		
		1985	1990	1995	1985	1990	1995
Enrichment (10^6 SWU year^{-1})	OECD Europe and Pacific	12.5	14.1	16.5	14.0	19.6	24.1
	OECD America[2]	19.5	19.5	19.5	8.3	10.3	11.1
	Developing WOCA	0.0	0.5	0.5	1.5	2.2	3.1
	CPE (export only)	3.0	3.0	3.0	(3.3)	(5.3)	(7.4)
	Total world[3]	35.0	37.1	39.5	22.8	32.1	38.3
Fabrication 10^3 t HM year^{-1}	OECD Europe and Pacific	6.6	7.6	7.2	3.5	4.7	5.6
	OECD America	5.9	5.9	5.9	3.1	4.4	4.2
	Developing WOCA	0.6	0.6	0.6	0.4	0.7	0.9
	CPE	—	—	—	(1.1)	(1.6)	(2.1)
	Total world[3]	13.1	14.1	13.7	7.0	9.8	10.7

[1]Based on 0.25 per cent ^{235}U tails assay.
[2]Excludes US Oak Ridge gaseous diffusion plant which has been on standby since 1985 but which could be brought on line (7.7 SWU year^{-1}) if need be.
[3]Excludes CPE demand and internal supply.

and accelerated expansion of the industry.

Data for the CPE are omitted from Table 19.5 since the plans for future capacity are unknown.

In the longer term the enrichment industry will move away from diffusion which has dominated the field until now, into advanced centrifuge processes and possibly laser isotope separation (see Chapter 4) although the latter is not likely to be available on a sizeable scale before the year 2000. Some expansion of the existing or planned capacity will be needed from around the mid-1990s if projected demands are to be met.

The WOCA requirement for uranium oxide fuel will rise from 7000 to 11 000 t $^{-1}$ year—during the decade 1985–1995 (Table 19.5). The considerable excess of capacity now in place will therefore be absorbed but there should be no supply problems this century—not least because of the short lead times for fabrication plant. Equally there is no problem for gas cooled reactor fuels which are produced in the countries using them.

It is not likely that there will be a major swing to mixed plutonium–uranium oxide fuels during the next decade but some will be needed for the Japanese advanced thermal reactor, for prototype breeder reactors and for use in LWRs. Those countries planning such developments also plan to have their own fabrication plant or are making appropriate contractual arrangements so no major supply problems need be anticipated.

All fuel emerging from reactors has to be stored either at the reactor site or in special facilities. The existing stockpile of spent fuel in OECD at the end of 1984 was about 46 000 t of heavy metal (HM) which will increase to around 122 000 t by 1995 with a further 5000 t in developing WOCA (Thornton, 1987). Total storage capacity is expected to be about 193 000 t (Table 19.6) so that up to 1995 it should be adequate, although not all reactors are adequately provided for and some could run into difficulties.

The construction lead time for storage capacity is not long and there is no need for any imbalance provided there are no major administrative or planning obstacles. In the present climate new reactors would normally include adequate storage as a part of their specification.

The bulk of the reprocessing capacity, installed or planned, is in Europe and Japan. None is anticipated in the USA or Canada in the foreseeable future. The planned capacities are more than adequate to reprocess gas–graphite reactor fuels which are not designed for long term pond storage, and to meet the needs of the mixed oxide fuel (MOX) requirements of fast reactors which are likely to be in place by 1995. (About 1 per

Table 19.6 Back End Services and Requirements

	Region	Capacity			Annual requirements		
		1985	1990	1995	1985	1990	1995
Spent fuel storage 1000 t HM	OECD Europe and Pacific	37.2	54.8	72.6	2.8	3.9	4.7
	OECD America	62.7	94.0	120.2	2.7	4.0	4.3
	Developing WOCA	—	—	—	0.3	0.5	0.7
	Total WOCA[1]	100	150	200	5.8	8.4	9.7
	CPE	—	—	—	0.7	1.2	1.7
	World Total				6.5	9.6	11.4
			A			B	
A. Reprocessing capacity t HM $year^{-1}$	OECD Europe and Pacific	250	3550	5815	2.7	2.7	3.7
	OECD America	0	0	0	0	0	0
	Developing WOCA	100	200	200	—	—	—
B. Plutonium requirements tonnes fissile Pu $year^{-1}$ [2]	Total WOCA	260	3750	6015			
	CPE	—	—	—	3.7	3.7	4.9
	World total	—	—	—			

[1]Rounded and excluding developing WOCA.
[2]Minimum requirement for operating and projected fast reactors.

cent of reprocessed fuel is plutonium—see Chapter 3.)

Some MOX will be wanted for the Japanese ATR and for recycle in LWRs but the quantities are likely to be small and fuel fabrication rather than potential plutonium supply is likely to be the technical constraint, if any. There are already clear plans in this area and countries which will use MOX are in general those which have or will have plutonium stocks from the reprocessing of their spent fuel.

In the longer term there will be a continuing growth in spent fuel requiring storage since reprocessing is not being planned in many countries (e.g. USA, Canada and Sweden). The storage of fuel prior to decisions on some means of disposal, either intact or through reprocessing, will therefore be a growth industry through the year 2000.

As yet there are few clearly defined national plans and schedules for the treatment and disposal of intermediate and high level nuclear wastes. These too will be a growing commitment for the industry in the 1990s and beyond.

The large scale supply of heavy water is concentrated in a limited number of countries but with annual requirements running at about 600 t for new reactors and top-up purposes supplies are more than adequate. Canada alone has an operating capacity of 650 t p.a. with 1200 t that could be brought on line if need be. Only if there were a major expansion of PHWR capacity post 1995 would consideration need to be given to additional capacity.

19.7 The Construction Industry

The nuclear construction industry developed in an environment where rapid growth was expected and expanded rapidly to a size that was not sustainable in the less buoyant markets that developed. In the UK, for example, five design and construction consortia were initially formed which have since been reduced to a single organization, the National Nuclear Corporation. The US industry was geared up to a new ordering rate of tens of GWe $year^{-1}$ together with a potentially flourishing export market. It has seen its hopes dashed with many cancellations and no new domestic orders since 1974. Even the successful French industry, which was taking orders for series of plant at a rate of up to 5 p.a., has seen its future prospects fade as the decline in demand growth has cut new plant orders to around 1 p.a.

At the end of the 1970s the reactor manufacturing capability in the OECD nations had reached some 50–60 new plant commissionings per year. Demand in WOCA had approached this annual level (40 p.a.) around 1972 but has since declined with new orders below 5 GWe $year^{-1}$ in 1984.

Under these circumstances it has been inevitable that the industry has taken steps to reduce its capacity (Walker and Lonroth, 1979). The hard hit US industry has seen three vendors leave the field and the closure of manufacturing facilities. Other major manufacturing countries have seen the scale of operations and workforces cut although work continues on the large number of plants under construction (Table 19.2), and new orders have continued at a rate sufficient to keep design and manufacturing capability ticking over for some time to come.

Nevertheless, apart from Japan and the USSR, which, despite Chernobyl, has declared its intention of pursuing its extensive deployment programme embracing an additional 41 GWe during the next five year plan, the world's pace of construction will slow during the next five years. If the NEA projections (Table 19.3) are correct, then capacity additions in WOCA lie between about 10 GWe $year^{-1}$ and 40 GWe $year^{-1}$ throughout the period 1995–2025, depending on whether their low or high projection is adopted.

The existing world reactor manufacturing industry has certainly more than enough capacity to meet the low projection (France could almost do it alone) but, unless steps are taken to slow the industry's contraction, the upper projection may be beyond what will be left of it by the early 1990s.

To some extent the problem is a reflection of the success of technology. In the 1950s average generation plant capacities were only tens of MWe, compared with the 1000 MWe typical of current plant. The numbers of units needed has therefore fallen enormously with resultant economies of scale, but also with adverse effects on industrial stability.

The overcapacity in reactor manufacture has been paralleled by overcapacity at the front end of the nuclear fuel cycle for uranium supply (Chapter 20), for enrichment and for fuel fabrication. The uranium mining industry has seen many mine closures, particularly in the USA, whilst the USDOE has mothballed a large tranche of diffusion plant, and the UK has closed its diffusion plant in favour of the cheaper advanced centrifuges.

19.8 Research and Development

19.8.1 *Reactors*

The world nuclear slow down has greatly reduced the interest in the development of new thermal reactor systems and concentrated attention on the question of safety assurance and cost control for existing favoured reactor types. Most of the major industrial nations have pursued scale benefits through the move towards larger reactors and turbines—France moving to 1400 MWe, the USSR developing 1800 MWe and 2000 MWe units for the 1990s, the Canadians building multiple tranches of the 880 MWe CANDU which has superseded the earlier 600 MWe design where demand justifies the move.

Over the last few years there has been some counter pressure to develop smaller systems, partly because these appear better matched to the needs of developing countries or countries with small networks, partly because some designs are seen to have potential safety attractions in the smaller sizes (see Chapters 3 and 7), and partly to reduce the financial risk (see Chapter 21). Proponents argue that the lost benefits of scale may be partially or wholly offset by the reduced need for complex and costly engineered safety systems, by the greater replication that would ensue (possibly using factory line methods), and by the greater reliability that some believe the smaller units would offer.

The only 'new' system attracting worldwide interest is the liquid metal cooled fast reactor and its fuel cycle which is being vigorously developed in Europe, the USSR, Japan and India. The US is also enthusiastic for the concept but has been consolidating and redefining its position following the abandonment of the Clinch River LMFBR.

The high temperature gas cooled reactor is being actively developed in Germany and in Japan; both countries which have seen possible uses for industrial process heat. Recent interest has also focused on the attractive safety characteristics of small HTGRs (Klueh, 1986).

In the fuel cycle area a great deal of R& D has concentrated on higher burn-up oxide fuels for the LWR and on cheaper enrichment processes such as the laser and the advanced centrifuge.

Work on reactor safety and on the treatment and disposal of radioactive wastes has also been receiving great attention with the active involvement of the international agencies (NEA, Euratom, IAEA) in the exhange of information and the mounting of co-operative programmes.

Specific points on the directions of development are covered in the technical chapters of Section I and II of this work and in the individual country contributions.

REFERENCES

Brookes, L. G. (1968). 'An S-Curve forecast elec-tricity demand', *Atom*, UKAEA, London.

Day, G. V., and Brookes, L. G. (1971). 'Forecasting the future for nuclear power'. Long Range Planning Society Conference, Shrivenham.

Day, G. V., Inston, H., and Main, K. (1980). *An analysis of the low energy strategy for the UK*, Energy Discussion Paper No.1, UKAEA, London.

Eckholm, E. (1975). *The Other Energy Crisis*, Worldwatch.

Freeman, C. (ed.) (1983). *Long waves in the world economy*, Butterworths, London.

International Fuel Cycle Evaluation (1980). *Fuel and Heavy Water Availability*. Report of Working Group 1, IAEA, Vienna.

International Institute for Applied Systems Analysis (1981). *Energy in a finite world*, Ballinger Pub. Co., London.

Jevons, W. S. (1974). 'The Coal Question', reprinted in *Environment and Change*, **1974**, 373.

Jones, P. M. S. (1977). 'One organization's experience' in *Futures Research*, **1977**, 194–209,

Jones, P. M. S. (1982). 'Discontinuities in Social Attitudes' in B. Twiss (Ed.), *Social forecasting for company planning*, Macmillan, London. pp.77–108.

Klueh, R. (1986). 'Future nuclear reactors—Safety first', *New Scientist*, **1986** (1502) 41–46.

Kondratiev, N. D. (1935). 'Long Waves in Economic Life', *Rev. Economic Stats.;* reproduced in *Lloyds Bank Review*, London, July 1978.

Leach, G. (1979). A low energy strategy for the UK, *International Institute for Energy Development Science Reviews*, London.

Lovins, A. B. (1975). *World Energy Strategies*, Ballinger Pub. Co., London.

Ray, G. F. (1980). 'Innovation as the source of long term economic growth', *Long Range Planning Journal*, **13**, 9.

Rostow, W. W. (1978). *The World Economy*, Macmillan, London.

Thornton, D. (chman), (1982). *Nuclear Energy and its fuel cycle*, OECD/NEA, Paris.

Thornton, D. (chman), (1987). *Nuclear Energy and its fuel cycle*, OECD/NEA. Paris.

Walker, W., and Lonroth, M. (1979). *Viability of the civil nuclear industry*, Royal Inst. of International Affairs, London.

Nuclear Power: Policy and Prospects
Edited by P. M. S. Jones

20

The Potential

P. M. S. Jones
Department of Economics, Surrey University.

20.1 Introduction

The present and near-term future contribution of nuclear power to the world's electricity supplies has been discussed in Chapter 19. This chapter is concerned with the potential for nuclear power in the long term. It looks at two questions; the possible scale of the nuclear contribution and physical factors affecting the rate of its realization.

In the early days of civil nuclear energy development, uranium was regarded as a scarce resource and some commentators saw the nuclear contribution as being strictly limited. This would be true, as will be shown below, if reliance had to be placed on thermal reactors alone. However, with the deployment of fast breeder reactors, the currently known uranium resources are quantitatively comparable to the world's known fossil fuel resources, and they are sufficient to meet all the world's electricity needs for centuries to come.

The following paragraphs set out what is known about the world's uranium and thorium resources and examine the energy contribution they can make, depending on the types of reactor and fuel cycle deployed. The importance of timely development of the mineral resources, reactor systems and fuel cycles is considered.

20.2 The Uranium Resource Base

20.2.1 *Resource Estimates*

As indicated in section 4.3.1.2 uranium is widely distributed in nature with some 10^{12} t dispersed in the upper kilometre of the earth's crust and over 10^9 t in solution in the oceans and seas. The vast bulk of this material is at such low concentrations (about one to two parts per million) that it will never be economically recoverable although, in theory, even the low concentrations in some

Table 20.1 Sources of Uranium

Source	Uranium content	Energy gain[1]	Indicative recovery cost[2] $ kg^{-1}
Conventional ores	~0.2%	15 (IEA, 1975)	20–130
Chatanooga Shales	~0.007%	7 (IEA, 1975)	130–260
Sea-water	0.003 ppm	>6 (Taylor, 1974)	500–1000
Granite	0.002%	2 (Huwlyer, 1975)	very high
Crustal average	203 ppm	≤1	very high

[1]Energy gain equals energy produced from the uranium used as a fuel in a PWR divided by the energy needed to recover the uranium from the source, to enrich it and to manufacture the fuel.
[2]Illustrative costs in 1985 US $ kg^{-1}.

granites (0.002 per cent) or sea-water (0.003 parts per million) could be recovered with a net energy gain if the material was used to fuel thermal reactors (Table 20.1). (Taylor, 1974; INFCE, 1980 a; Institute for Energy Analysis, 1975; Huwyler, *et al.* 1975.)

Estimates of the world's uranium resources are divided into four categories by the Nuclear Energy Agency and International Atomic Energy Agency (Nininger, 1983), depending on the level of confidence surrounding the quantification of the individual deposits.

The best authenticated materials are classifed as reasonably assured resources (RAR). These refer to 'known mineral deposits of such size, grade and configuration' that the uranium from them:

> '. . . could be recovered within the given production cost ranges, with currently proven mining and processing technology. Estimates of tonnage and grade are based on specific sample data and measurements of the deposits and on knowledge of the deposit characteristics.'

Estimated additional resources (EAR), were divided into two sub-categories in the 1983 Red Book (Nininger, 1983). EAR-I contains uranium:

> 'in addition to RAR that is expected to occur, mostly on the basis of direct geological evidence, in extensions of well-explored deposits and in deposits in which geological continuity has been established, but where specific data and measurements and knowledge of the deposits characteristics are considered to be inadequate to classify the resource as RAR.'

EAR-II contains uranium in addition to EAR-I, that is: ' . . . *expected* to occur in deposits *believed to exist* in well-defined geological trends or areas of mineralization with known deposits' (author's italics).

Speculative resources (SR) refer to uranium, in addition to the other categories: ' . . . that is thought to exist mostly on the basis of indirect evidence and geological extrapolation, in deposits discoverable with existing exploration techniques. The location of deposits envisaged in this category could generally be specified only as being somewhere within a given region or geological trend. As the term implies, the existence and size of such resources are highly speculative.'

Each known or inferred deposit in the resource categories is further categorized in terms of the likely costs of uranium recovery which are defined to include direct costs of mining, transporting and processing the ore; the costs of waste treatment; the capital costs of new units and the cost of maintaining non-operating production units, where applicable; and financing costs and overheads including taxes and royalties. Further exploration and development costs are also included where needed for new deposits, but sunk costs are not normally included.

The three cost categories specified by the NEA/IAEA are from zero to \$80 kg^{-1} of uranium (\$0–30 lb^{-1} U_3O_8), \$80–130 kg^{-1} U (\$30–50 lb^{-1} U_3O_8), and \$130–260 kg^{-1} U (\$50–100 lb^{-1} U_3O_8), but it is important to recognize that these costs are not equivalent to the market prices needed to encourage exploitation of the different cost groups.

The relationship between the different categories and their recovery costs is illustrated in Figure 20.1. It has not been customary to attach error bands to the main resource categories but informal estimates have put the uncertainty on RAR at from ±10 per cent to ±20 per cent, with the corresponding range for EAR-I at from ±25 per cent to ±50 per cent. A more formal attempt has been made to attach confidence limits to speculative resource estimates and an over-all figure of around ±50 per cent on the central estimates appears appropriate (see Table 20.3).

In the most recently published NEA–IAEA (Nininger, 1983) assessment for WOCA (the world outside the centrally planned economies) there were some 1.4 million t of uranium in the lowest cost RAR category which is sometimes termed reserves. The \$80 to \$130 kg^{-1} U category contains 0.6 million t, giving a total of 2 million t of RAR.

The same source estimates low and higher

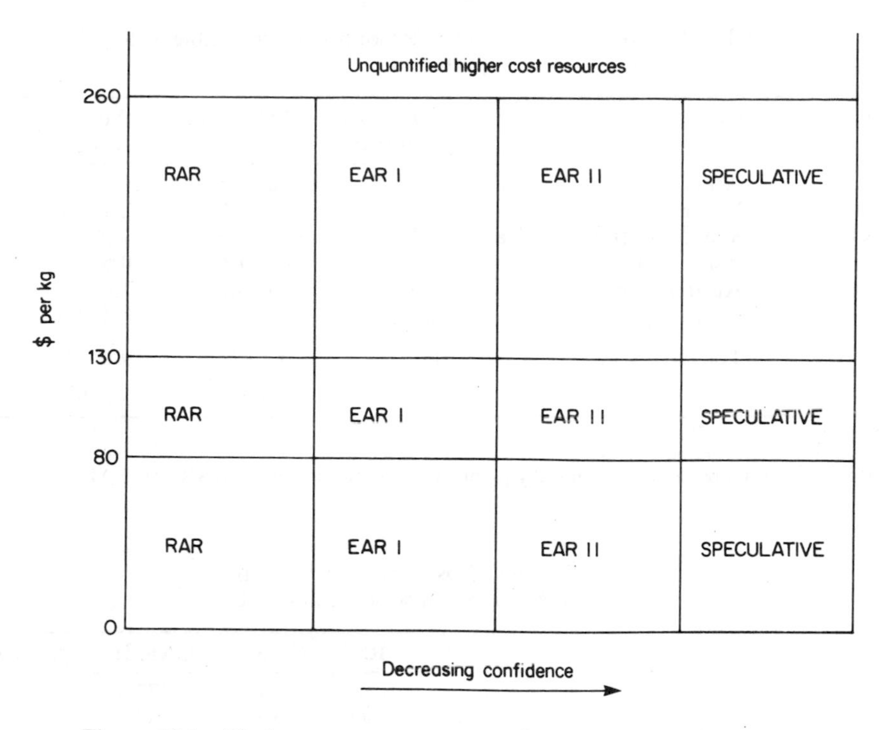

Figure 20.1 Nuclear energy agency uranium resource classification

cost EAR-I resources as 0.9 million t and 0.3 million t respectively, while figures for EAR-II total 0.7 million t and 0.4 million t respectively, although many countries participating in the review failed to provide estimates in this latter category. The figures for RAR and EAR are summarized in Table 20.2.

Speculative resource estimates (Table 20.3) range in total from 6.3 to 16.2 million t with a most likely range of 9.6–12.1 million t.

There is insufficient data available from the Centrally Planned Economy Area (CPEA) to allow similar disaggregation but their total resources including speculative resources are

Table 20.2 Uranium Resource Estimates—1983
(1000 t of uranium)

	RAR		EAR-I		EAR-II	
	<\$80 $kg^{-1}U$	\$80–\$130 $kg^{-1}U$	<\$80 $kg^{-1}U$	\$80–\$130 $kg^{-1}U$	<\$80 $kg^{-1}U$	\$80–\$130 $kg^{-1}U$
USA	131	276	30	52	471	339
Canada	176	9	181	48	179	102
Australia	314	22	369	25	—	—
South Africa	191	122	99	48	—	—
Brazil	163	—	92	—	—	—
Niger	160	—	53	—	—	—
Namibia	119	16	30	23	—	—
France	56	11	27	6	0	12
India	32	11	5	15	—	—
Sweden	2	37	<1	43	—	—
Other (rounded)	100	70	30	50	8	20
Total[1] (adjusted)	1425	575	885	305	(660)	(480)

Source: Nininger, 1983.
[1]RAR and EAR-I adjusted by NEA/IAEA to allow for mining and milling losses not incorporated in some national estimates. EAR-II is not fully reported and totals relate only to reported resources.

Table 20.3 IUREP Speculative Uranium Resources Recoverable at \$130 $kg^{-1}U$ (million t)

Area	Continent	Number of countries	Full range	'Most-likely' Range
	Africa	51	1.3–4.6	2.6–3.5
	N. America	3	1.8–2.9	2.1–2.4
WOCA[1]	S. and Central America	41	0.7–1.8	1.0–1.3
	Asia and Far East	41	0.3–1.6	0.5–0.8
	Australia and Oceania	18	2.0–4.0	3.0–3.5
	W. Europe	22	0.3–1.1	0.4–0.6
	Total WOCA	176	6.3–16.2	9.6–12.1
CPEA[2]		9	3.3–8.4	5.2–6.5

Source: Nininger (1983)
[1]World outside centrally planned economies.
[2]Relates to total potential for centrally planned economies including RAR and EAR.

Table 20.4 High Cost Uranium Resources (1000 t recoverable at \$130–\$260 $kg^{-1}U$)

Country	RAR	EAR-I	EAR-II	Speculative
U.S.A.	241	76	477	595
Canada	69	58	187	—
South Africa	92	357	—	—
Namibia	9	15	—	—
Other (rounded)	8	9	13	26

Source: Nininger (1983)

considered to be in the region of 6 million t. Earlier estimates (WEC, 1980) put their low cost RAR and total EAR at about 0.5 and 1.5 million t respectively.

Estimates of uranium in the highest cost category (\$130–260 kg^{-1} U) are far from complete due to under reporting. The published figures for RAR and EAR-I in WOCA amount to about 0.9 million t with a further 0.7 million t in EAR-II (Table 20.4).

Economics permitting, uranium is already recovered as a by-product of gold mining in South Africa and copper extraction in the USA. Phosphate rock is also a source with some potential which is exploited in the USA, where uranium has been produced as a by-product of phosphoric acid production. Deposits in Morocco are estimated to contain 6 million t at an average tenor of 120 parts per million. Production as a by-product is dependent on the demand for the main product and would not, in general, be economic in its own right.

In summary it can be seen that there is reasonably high confidence that some 3.1 million t of uranium can be made available from known resources, at costs below \$130 kg^{-1} U (\$50 lb^{-1} U_3O_8). A further 1 million t is believed to exist in EAR-II with the possibility of a further 6–16 million t existing in the speculative resource category for WOCA. However, it must be remembered that some of this latter material may not exist, or if it exists it may not be discovered, or if discovered it may not be exploited for political, environmental, geographical or economic reasons. (The 1986 NEA review will have appeared by the time this work is published but the updated figures do not alter the conclusions.)

Thus whilst it was clearly wrong in the early days of considering uranium resources to regard the low cost RAR and EAR as defining the limit of uranium availability, it would be equally imprudent to place high confidence on the availability of 12 million t of uranium currently categorized as EAR-II or speculative resources in WOCA, when selecting future nuclear reactor and fuel cycle strategies.

20.2.2 *Geographical Distribution*

Whilst uranium is widely distributed in the earth's crust, the high grade uranium ores are concentrated in relatively few countries. In WOCA (Table 20.2) some 90 per cent of the principal resource categories (RAR and EAR) are found in nine countries, with about 65 per cent of declared resources in the USA, Canada and Australia.

Europe and Japan, who are rapidly overtaking the United States as the main users of nuclear power, have very small indigenous economically recoverable resources, totalling under 0.3 million t, mainly in France, Sweden and Spain. Even then the 0.08 million t in Sweden is in the Ranstad shale deposit from which production has been banned for environmental reasons.

20.3 Thorium Resources

Thorium, which is a fertile rather than a fissile material (Chapter 4), is also widely distributed in nature and is some three to four times more abundant than uranium in the earth's crustal rocks. Its chemical nature is such, however, that it is more uniformly dispersed and there are fewer concentrated ore bodies (Bowie, 1983). Added to this is the fact that thorium has had few major uses so that deposits have not been sought with comparable vigour to those of uranium.

Table 20.5 Thorium Resources in WOCA recoverable at less than \$80 kg^{-1}Th (1000 t)

Country	RAR	EAR
USA	122	278
Canada	—	136
Australia	19	0.1
Brazil	606	1200
Turkey	380	500
India	320	—
Norway	132	132
Egypt	15	280

Source: Nininger, 1983; Bowie, 1983.

Resources have been categorized in the same way as those of uranium but knowledge of their true extent is far more limited. The most recent resource estimates are given in Table 20.5 together with their geographical distribution.

The total RAR plus EAR could amount to many million tonnes although little of the material is currently worked and the cost categories may not be appropriate to large scale production.

20.4 The Energy Released in Fission

If one atom of uranium–235 undergoes fission it releases on average 3×10^{-11} J. This compares with the 7×10^{-19} J released when one atom of carbon burns in air to produce carbon dioxide. Thus in energy terms the uranium atom (mass 235 units) produced 50 million times the energy, in round terms, of a carbon atom (mass 12 units). The complete fissioning of 1 t of uranium is equivalent, therefore, to burning about 2.5 million t of coal. (Such comparisons are necessarily approximate because the calorific value of coals differs depending on their type and source.) Fully fissioning 1 g of uranium or burning 2.5 t of coal produces heat energy equivalent to 1 MW for one day.

In a thermal reactor the 0.71 per cent of uranium–235 in the natural uranium can undergo fission but the uranium–238 undergoes few fissions at the low neutron energies. One gram of natural uranium burnt in a thermal reactor might therefore be expected to produce at a maximum only 7 kW days of heat energy. However, some of the uranium–238 is converted to fissile plutonium through neutron capture (section 4.2.1) and some of this is burnt *in situ*, whilst the remainder, along with any unburnt uranium–235, is still in the fuel when it is removed from the reactor.

For a once-through LWR 5.8 t of natural uranium are required to produce 1 t of fuel enriched to 3.1 per cent in uranium–235 with 0.2 per cent enrichment tails. When fuel achieving a burn-up of 33 000 MWd t^{-1} is removed from the reactor it contains 0.8 per cent uranium–235, 0.9 per cent plutonium and 3.1 per cent fission products and trans-uranic elements (Jones, 1985). About 3 per cent of the uranium atoms in the fuel have been fissioned; 2.3 per cent through direct fission of uranium–235. This corresponds to 0.5 per cent of the initial uranium. AGRs achieve much the same result.

The 25 per cent of the initial uranium–235 remaining can be re-enriched and recycled (Chapter 4) and, after allowing for some losses, enrichment tails, and the effects of uranium–236, will add some 10 per cent to 15 per cent to the energy recovered from the fresh uranium in its first cycle. The plutonium in the spent fuel, although equivalent in quantity, can do slightly better (because it does not have to be enriched and there are no tails losses) and can add a further 20 per cent to 25 per cent to the energy recovered. Thus the 0.5 per cent of energy extracted from the uranium with a single once-through cycle can be increased by 35 to 40 per cent to around 0.7 per cent with a single recycle or 1 per cent with repeated recycle.

To use the symbolizm developed in section 3.6.1, if ψ neutrons are produced per fission and L of these are absorbed non-productively in reactor materials, then C remain for capture in fertile materials such as uranium–238 or thorium–233, where:

$$C = \psi - 1 - L$$

For a typical light water reactor C, the conversion ratio is found to be about 0.6 so that five fissioning ^{235}U atoms can produce three fissionable plutonium–239 atoms which in turn, when they fission, will produce a similar fraction, to a first approximation. For each initial ^{235}U atom fissioning $(1 + 0.6 + 0.6^2 + 0.6^3 \ldots)$ atoms can be fissioned if the fissionable isotopes are recovered from spent fuel and recycled. This series sums to $1/(1–C)$, i.e. 2.5, so that a light water reactor can burn in theory not the 0.7 per cent initial uranium–235 but up to about 1.7 per cent of the natural uranium feed stock. In practice losses are incurred in reprocessing and some of the plutonium is converted to non-fissile isotopes so that the practical limit for a recycling LWR is nearer to 1 per cent as described above.

Reactors like Magnox or CANDU using natural uranium fuel achieve lower burn-up and contain 0.2 per cent to 0.4 per cent uranium–235 on removal from the reactor. The better neutron economy achieved using heavy water moderator increases the conversion coefficient (C) to 0.8, however. On a single once-through fuel regime about 0.5 per cent of the uranium is consumed in Magnox, the same as with an LWR, of which 0.4 per cent arises from direct fission of uranium–235. In CANDU, with higher fuel burn-up, 0.8 per cent of the initial uranium is consumed. On repeated recycle the total uranium consumption can in theory be increased by up to a factor of five on the initial fissile content (i.e. to 3.5 per cent) but in practice losses, etc., reduce the achievable level to about 2 per cent.

The actual uranium utilization achieved is dependent on the reactor design, the fuel design and the fuel burn-up which all influence the conversion ratio. Adoption of lower enrichment tails and avoidance of materials with high parasitic neutron capture both enhance the fraction of initial uranium that can be burnt. The use of thorium in place of uranium–238 as a fertile material can also improve conversion because the uranium–233 produced yields more neutrons per fission by thermal neutrons that either uranium–235 or plutonium–239 (section 4.5.1). The adoption of thorium-uranium cycles in either CANDU or the HTR can greatly increase the energy recoverable from the initial natural uranium. (Thorium is not itself a fuel and cannot substitute for uranium–235, it can, however, increase the effectiveness with which fission energy available from uranium–235 is utilized.) Table 20.6 gives the quantities of natural uranium needed to fuel different reactor types over their lives. The numbers in the table are only approximate since differences in reactor design, fuel design and fuel use can produce significant variations.

Table 20.6 Uranium Requirements for Reactor Systems
(1000 MWe for 30 years at 70 per cent capacity factor; Uranium tails assay 0.2 per cent)

Fuel Cycle and Reactor Type	Net natural Uranium t	Net fissile plutonium produced t
Magnox		
U once-through	6300	9.8
AGR		
U once-through	4200	3.2
PWR		
U once-through	4400	5.2
Improved U once-through	3700	3.7
U-Pu recycle (s.g.)[1]	2800	2.5
HWR		
Natural U	3700	9.8
Low enriched (1.2%) U	2600	4.4
U–Pu recycle (s.g.)[1]	1800	1.4
Th–^{233}U recycle	1700	0.6
HTR		
U once-through	3200	
Th–^{233}U	1700	small
FBR		
Pu recycle	36	4.5

Sources: INFCE, 1980; Farmer, 1983.
[1]s.g.—reactors only use fissile materials they produce themselves.

In the fast reactor the mixed uranium–plutonium oxide fuels can produce more fissile material in the uranium–238 breeder blanket than they consume, and C, now called the breeding gain, is greater than one. The effect is a net consumption of fertile material which, in theory, could be completely burnt; i.e. 100 per cent of the uranium can be used compared with the 0.4 to 2 per cent in thermal reactors operating once-through or on self generated uranium and plutonium recycle. In practice processing losses reduce the achievable uranium consumption to 50 to 70 per cent.

The energy recoverable from the world's uranium resources can be seen from this to be highly dependent on reactor and fuel cycle choice. The 3.1 million t of uranium recoverable at costs below \$130 kg^{-1} U in the best authenticated resource categories (RAR and EAR-I) are equivalent to about 40 billion t of coal if used in once-through thermal reactors; 75–150 billion t if used in thermal reactors recycling uranium and plutonium; and 4500 billion t if used in fast breeder reactors.

20.5 Resources in Perspective

It can be seen from Tables 20.2 and 20.6 that WOCA's best authenticated uranium resources are only capable of meeting the lifetime requirements of some 750 GWe of thermal reactors operating in the once-through mode, taking their lives as 30 years and their average capacity factors as 70 per cent. The introduction of uranium conserving technologies and/or thorium–uranium fuel cycles could increase this two or three fold. If this were all that were achievable the contribution of nuclear power would indeed be transient. As was seen in the last chapter, WOCA's installed capacity was expected to exceed 1000 GWe by the year 2000 when the INFCE study was performed. More recent estimates (NEA, 1986) suggest that WOCA capacity may not reach half this figure by that date, nevertheless if the known resources were all that could be made available the nuclear contribution achievable with thermal reactors would have a time span measured in decades. If the speculative resources exist, can be located and a reasonable proportion of them brought into production, then a modest thermal nuclear contribution might be sustained for a century or so.

The situation is very different if fast reactors are brought into use. The effective resource base is increased by a factor of around 60 and the better authenticated resources are sufficient to meet lifetime fuel needs of 50 000 GWe of nuclear reactors, even without a contribution from the speculative resources.

As was seen in section 20.4 the WOCA's economically recoverable uranium resources in the well authenticated RAR and EAR-I categories are equivalent when used in breeder reactors to around 4500 billion t of coal, a figure which is comparable to the known geological resources of coal (11 000 billion t) and greater than estimates of the technically and economically recoverable reserves, viz. 700 billion t (World Energy Conference, 1986). It is also considerably greater than the known WOCA resources of oil which amount to 130 billion t coal equivalent (80.4 billion t of oil) and natural gas which are put at 70 billion t coal equivalent (1750 thousand billion ft^3). (British Petroleum, 1983.)

With total WOCA energy consumption running at around 8000 million t of coal per year, expressed as primary energy equivalent, it is clear that the known uranium resources have a vast potential, sufficient in theory to supply even growing world energy demands for many, many, centuries. Furthermore the fuel consumption of fast reactors is so low that they should be able, in the long term, to use even lower grade, higher cost, sources of uranium and these together with any material becoming available from speculative resources increases the time horizon to millenia.

6.6 Future Uranium Supply

6.6.1 *Production from Known Resources*

We have concentrated thus far on the total amount of uranium potentially available but it is the rate at which this can or will be made available that is of greatest importance to the nuclear industry.

The production of uranium in WOCA, having peaked at 36 000 t U in 1959 when it was wanted for weapons manufacture, subsequently declined to below 20 000 t in the mid–1960s. The gradual increase in production in the succeeding decade accelerated rapidly in the mid–1970s when the boom in nuclear plant construction was anticipated following the 1973 oil crisis. Production peaked at 43,900 t in 1981, some 14 000 t more than needed for reactor fuel in that year, and has since fallen back to bring supply more into line with the revised demand expectations.

The Red Book of the OECD Nuclear Energy Agency and the International Atomic Energy Agency has, since 1970, sought to determine current and future anticipated production capacity from existing mines and committed production centres based on known resources. The latest published survey (Nininger, 1983) showed production capacity falling from its 1982 peak of 54 000 t p.a., to around 45 000 t p.a. in 1984, after which a rise of some 600 t p.a. was projected until the mid–1990s, which would bring the level back to the 1982 peak. It is worth noting that output has in the past only exceeded 80 per cent of production capacity for relatively short periods.

Since current designs of LWR and AGR require about 150 t of natural uranium fuel per 1000 MW capacity per year at 70 per cent capacity factor, when operating on the once-through cycle, the anticipated production capacity should be adequate to fuel some 300 GWe of reactors. This is less than the nuclear capacity planned for completion in WOCA by 1990 (Nuclear Energy Agency, 1987) although existing stockpiles of natural and enriched uranium are well able to bridge any shortfall on this timescale.

In addition to the existing and committed production centres plans exist for further expansion in some countries and estimates of possible production from existing, planned and prospective centres, based on known uranium deposits (RAR and EAR-I) have been published (Nininger, 1983). If all of these were to be fully realized then production capacity might rise to a peak of around 75 000 t in 1995, declining progressively thereafter as deposits were worked out.

In view of the extensive mothballing and closure of mines that has taken place in recent years, particularly in the USA, which has from the 1960s been the largest producer, the ability to reach the full projected production must be open to question, despite Australia's entry to the market. (US production has dropped from 17 000 t in 1980 to around 10 000 t p.a. with nine of the 33 mills in operation in 1980 closing. Ten of the remaining operable 24 mills were not producing in 1983.)

20.6.2 *Production from Speculative Resources*

The comparatively pessimistic outlook for future supplies set out in the previous section provides a strong incentive for users to move to more efficient uranium utilization than can be achieved with the once-through LWR. Reactors and fuel cycles with better conversion ratios, spent fuel reprocessing and recycle and the development and adoption of breeder reactors all offer varying degrees of improvement, with the latter, once it has been widely deployed, effectively removing any constraint.

The US Carter administration's wish to slow down or even stop the development and use of reprocessing and plutonium based fuel cycles, led to US efforts to predict supply from speculative resources as a part of the 1979–1980 International Fuel Cycle Evaluation (INFCE, 1980b). To this end intensive effort was devoted to establishing for the first time the levels of speculative resources, and a mathematical model was produced to project the rate at which such resources might be exploited.

The model included assumptions on the time elapsed before significant exploration began; the interval before uranium was discovered; the time required to convert the speculative resource to RAR; the quantity of reserve to be delineated before a decision to open a mine would be taken; and the time then needed to plan, license and open the mine. The nature, grade and size of the undiscovered ore bodies had to be postulated and constraints were applied limiting the rate of expansion of production capacity to match the supplier countries' industrial infrastructure. The model was run with both high and low estimates of speculative uranium resources and with accelerated and delayed development schedules. On the assumptions adopted by its proponents the maxiumum uranium production capability peaked at between 150 000 t and 300 000 t p.a. between 2010 and 2025 (the latter being the arbitrary cut-off date adopted for INFCE). This would be sufficient to sustain a peak capacity (at 70 per cent capacity factor) of 1400–2800 GWe of once-through LWRs.

Whilst recognizing that some production from as yet undiscovered resources is highly likely by the second decade of the next century, other INFCE participants were highly sceptical about the US approach and some of their criticisms were summarized by the present author and European colleagues (INFCE, 1980b). These included the criticisms that the RAR and EAR resources themselves, as then defined, would not be fully recoverable for economic and technical reasons; and that the speculative resources might not all exist or if they existed might not be exploited for political, environmental or economic reasons. The US model also brought into production by 2025 (theoretically) mines containing 38 per cent to 70 per cent of the total speculative resources, contrary to the admonition of the experts on the International Uranium Resources Evaluation Project who had produced the estimates, that they should not be used for planning purposes and that a major part may not be discovered and brought into production on this time scale. The critics concluded that the achievable maximum output was likely to be significantly lower than the theoretical maximum although the extent of the difference was not susceptible to quantification, and was a matter for judgement. It was considered impossible to specify a level of

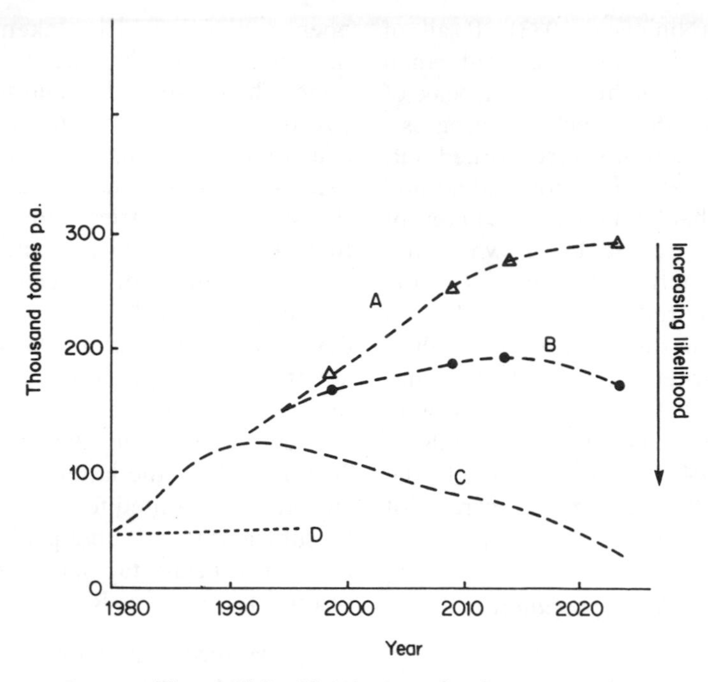

Figure 20.2 Uranium production rate
A,B : Max. output from all resources including speculative (INFCE, 1980)
C : Max. output from known resources of RAR and EAR (INFCE, 1980)
D : Max. output from existing and planned capacity (NEA, 1986)

annual or cumulative uranium production 50 years hence which would find general acceptance.

Figure 20.2 summarizes the eventual INFCE findings on maximum levels of long term supply and the uncertainty over what will be achievable in practice, and superimposes the more recent NEA–IAEA view on production capacity based on known resources. In a free market, of course, supply would only expand at a rate roughly matched to expectations of demand, with market prices moving to bring the two into balance.

20.6.3 *Possibilities and Uncertainties*

The potential long term nuclear contribution, given the development of the appropriate breeder technologies, is certainly as large as that which could be achieved using fossil fuels. There is, however, great uncertainty about the rate at which uranium can or will be made available in the longer term future. The US model's theoretical maximum annual output would be sufficient to meet the annual needs of 1400–2800 GWe of once-through LWRs in WOCA, totals which will not be reached until well beyond the turn of the century. However, unless significant exploration, development and exploitation takes place to bring the as yet undiscovered resources into timely production, supplies might not be adequate even to meet the needs of projected levels of capacity in the year 2000.

From the users point of view the fact that large deposits of uranium may exist is a necessary but not a sufficient condition to satisfy concerns over security of supply. As Tables 20.2 and 20.3 show resources are not spread evenly around the world and major users in Europe and Japan have no significant indigenous resources, whilst other countries like Australia have a large resource base but no nuclear power stations. What matters to the user is not the knowledge that uranium

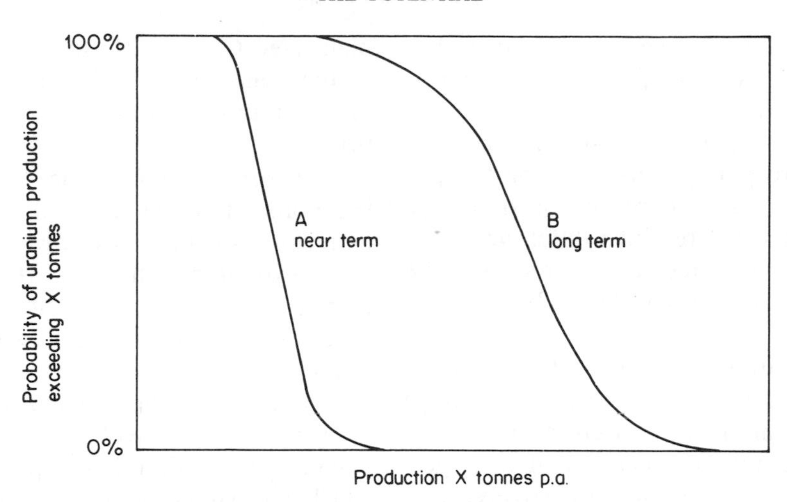

Figure 20.3 Output probability over time

exists or may exist in a potential supplier country, but the confidence that this material will be extracted and made available to him as and when he wants it.

From the users point of view the probability curve for the likely supply of uranium in the future looks something like that shown in Figure 20.3 (Jones, 1980; INFCE, 1980b). In the near term (curve A) in stable economic conditions there may be high confidence of getting total levels of supply approximating to those from existing mines and mills or those nearing completion and there is little room for uncertainty, apart from any political action to curtail supply. In the longer term there may be reasonable confidence that supplies would be higher but there is far greater uncertainty about the total that might be made available. Political or environmental factors may influence whether a supplier nation enters the market or not. Short term economic factors may dissuade new com-

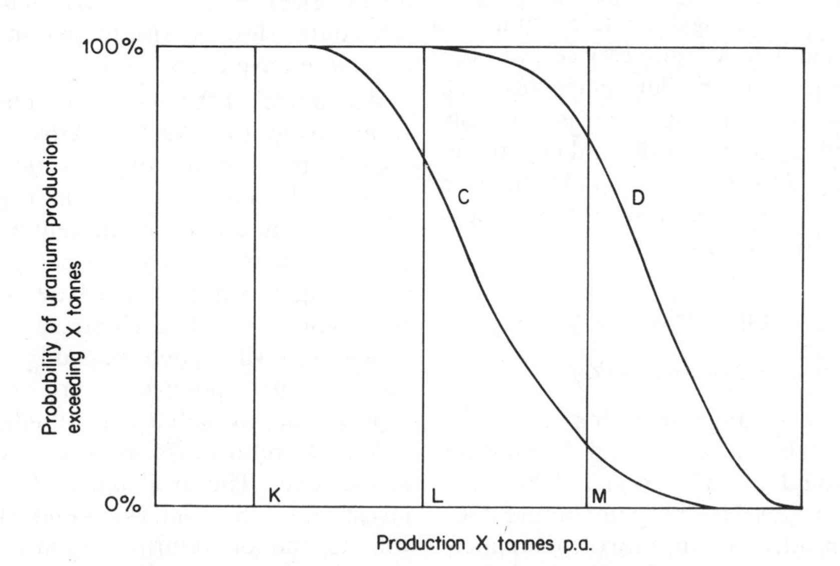

Figure 20.4 Perceptions of output and strategy requirements. C and D represent optimistic and pessimistic expectations about output at some specified date. K,L,M represent needed output for selected reactor strategies

panies from entering the market and lead to later constraints on supply. This situation is represented by curve B.

Figure 20.4 (Jones, 1980) presents the different perceptions of users taking an optimistic (curve D) or more pessimistic (curve C) view of the future. Such differences may arise from their degree of self-sufficiency, the degree to which they have been able to negotiate long term contracts, and their past experience with the reliability of suppliers. If the same users see the estimated world uranium requirements in a given future year for different nuclear reactors and fuel cycle strategies as K, L and M respectively, then both would be confident that requirements would be met for strategy K. The optimistic user would be confident about L but the pessimist would have some doubts. M would be regarded by C as an impracticable strategy whereas D might hope to muddle through.

The point of this illustration is to bring out the importance of perceptions in strategic choice. Some utilities may feel confident that they will always be able to get the fuel they want at an acceptable price regardless of their reactor–fuel cycle choices. Others with less confidence may attempt to secure the future of their fuel supplies by means of physical stockpiles, by entering long term contracts for uranium and enrichment services, or even by taking equity in supplier companies. Still others may feel that such measures only provide medium term security and may therefore favour moves towards the independence from imported fuels offered by the deployment of breeder reactors.

20.7 The Uranium Market

20.7.1 *Theoretical Considerations*

In a free market with perfect competition and perfect foresight the supply and demand for uranium would always be in balance and prices would move to the point where they provided an adequate incentive to suppliers to ensure that this was so, for any given demand level. In practice uranium prices are a sufficiently small component of over-all electricity generation costs in nuclear power plant (see Chapter 21) for the demand side of the equation to be insensitive to modest uranium price movements (the demand is price inelastic).

In theory the supply price would move to the point at which it equalled the cost of extraction of the least favoured resource being exploited at the time (the cost of the marginal supply). If the uranium-resource base were fixed the price would increase steadily as the cheaper deposits were depleted to meet the demands of the nuclear industry. Progressively dearer resources would be worked, at least up to the point at which nuclear power costs themselves were brought up to the cost of the next cheapest means of generating base load electricity on a large scale.

If, on the other hand, the uranium resource base were not fixed, but could be added to through new discoveries, then the market price might temporarily fall if the costs of exploiting the new resources were below the existing marginal supply cost. It would then resume its upward movement from a lower base unless disturbed by further discoveries. The discovery of resources with exploitation costs above the existing margin would not affect prices in the short term but would in due course slow the rise to less than it might otherwise have been.

Any model of the WOCA uranium market would need to reflect anticipated demand growth, the size and exploitation cost spectrum of the resource base, and the rate at which new resources in different exploitation cost categories will be discovered, developed and brought to market. The latter will depend on exploration and development expenditure which itself will be determined by the mining industry's perception of future demand and supply, the availability of funds and the potential returns from other competing investments. The availability of funds from investors or from internal company sources will depend on returns on existing mining activities; not solely in the uranium area.

In reality the uranium market is far from perfect in the economic sense. There is uncertainty about future demand, despite the long lead times for reactor construction (five to ten years). Lead times for resource discovery and exploitation are lengthy (around fifteen years), and the resource base itself is inherently uncertain. Companies may invest too little or too much in exploration, etc., due to faulty perceptions based on imperfect information. For this reason, even in a free market, the movement of uranium prices could be erratic with the present glut, for example, depressing exploration and development, leading to mine closures or inefficient mine working (extracting only the highest grade ore), and contributing to future 'shortages' with high prices, enhanced and excessive exploration, etc., which in turn could lead to a further glut.

Added to this is the fact that the uranium market is not free. Political intervention has occurred on many occasions and in several countries, to establish or protect indigenous mining industries, to support national anti-nuclear weapons proliferation policies and to meet demands of internal anti-nuclear groups.

It is scarcely surprising that under these circumstances there have been widely varying opinions on future uranium prices ranging from the view that there is still a lot of low cost uranium waiting to be found so that prices will be stable (or at worst rise in line with general mining costs), to the view that a significant proportion of speculative resources (or even RAR and EAR) may not be exploited for political or environmental reasons and that disruption of markets is quite likely with major price excursions superimposed on a rising trend, similar to that witnessed in the late 1970s (Figure 20.5).

It is quite easy to build market models

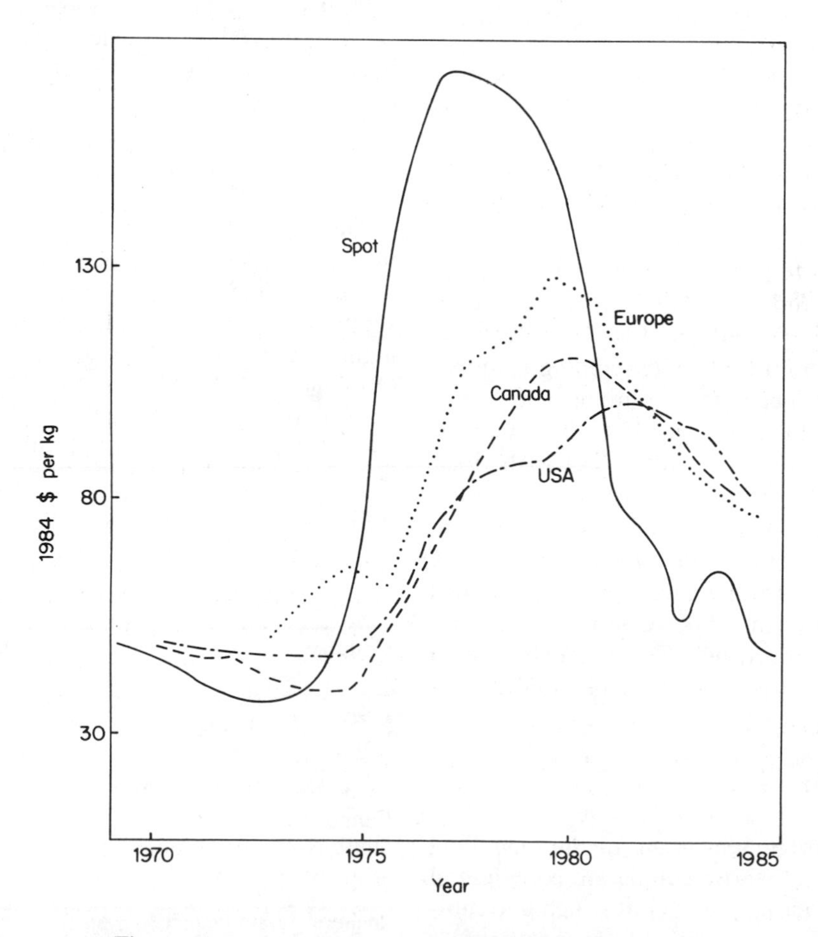

Figure 20.5 Uranium spot prices and delivered prices

Table 20.7 Average Uranium Prices for 1983 Deliveries

	1983 \$ US $kg^{-1}U$
European buyers	80.6
US buyers—foreign origin U	68.1
US buyers—all sources	98.3
Canadian sellers	81.8
Australian sellers	97.3

Source: Jones, 1985

employing plausible assumptions for demand, for lead times, for mining company investment behaviour, and for market disruption through the application of constraints on a portion of the resource base. These can produce widely different price projections to support almost any subjective prejudgement without necessitating the adoption of extreme assumptions. The only merit of such models is that they add weight to the view that there is a genuine wide band of uncertainty about future uranium prices.

In view of these uncertainties a Nuclear Energy Agency expert group (Jones, 1985) adopted a range based on US \$32 lb^{-1} U_3O_8 in January 1984 and rising in real terms at 0 per cent, 2 per cent and 4 per cent p.a. in real terms. This embraced the range of views on possible futures, barring major market disruption, and was centred on 2 per cent p.a. which reflected the long term historic rise in real mining costs (excluding resource depletion effects) for uranium and other minerals, whilst recognizing that technological improvement could offset such rises, and costs of more rigorous environmental control and safety measures might increase them. The upper half of this range was compatible with the long term projections published by the UK Central Electricity Generating Board as part of its evidence to the Sizewell Inquiry (Townsend, 1983).

20.7.2 *Structure of the Real Market*

In practice some 90 per cent of uranium purchases are made under medium or long term contracts whose details are confidential to the contracting parties but which undoubtedly result in a considerable range of prices being paid in any one year. The prices agreed under such contracts reflect the producer's anticipated production costs and allow for cost escalation. They also reflect producer and users views of the present and likely future market conditions. Both parties gain commercial assurance from such contracts for which they both sacrifice some flexibility.

The range of prices at which the bulk of uranium was supplied in 1983 are shown in Table 20.7 to illustrate the degree of variation. As with so many other commodities they are conventionally quoted in US dollars.

The uranium spot market, which accounts for the remaining 10 per cent of sales, has been far more volatile than the contract market and shows a large peak (Figure 20.5) in the late 1970s corresponding to the rapid expansion of nuclear installation plans following the 1973 oil crisis, and the subsequent reduction as world economic growth slowed and nuclear plans were cut back, particularly in the USA. The price decline was accentuated by user utilities, which had over ordered, selling off their inventories and contracted purchases to reduce their short term costs.

The level of existing stockpiles held by producers and utilities is generally believed to correspond to some three to five years supply for currently installed capacity, rather higher than the numbers given in Table 20.8 (Nininger, 1983). Most utilities would probably wish to keep a stockpile of reasonable size to guard against supply disruptions and the costs are not large since the uranium contributes only of the order of 5–10 per cent

Table 20.8 Uranium Stockpiles[2]

Country	Stock t U[1]
USA	60 000
Europe	42 000
S.E. Asia	11 000
Australia	2500
Canada	9000
South Africa	4000
Total	128 500

[1]tonnes natural uranium equivalent
[2]based on estimates by NUKEM

of LWR generation costs and financing charges would therefore amount to around 1 per cent p.a. for each year's stockpile cover. It seems likely that any excess inventory holdings will have been run down before the end of the present decade.

REFERENCES

Bowie, S. H. U. (1983). 'Uranium and thorium raw materials', in W. Marshall, (ed), *Nuclear Power Technology*, Oxford University Press, Oxford, Vol. 2, p.56.

B.P. (1986). *British Petroleum Statistical Review*, B.P., London

Huwyler, S., Ryback, W. and Traube, M. (1975). *Extraction of Uranium from granite,* cited in INFCE, 1980a, p. 186.

INFCE (1980a). *Fuel and heavy water availability*, IAEA, Vienna, p.170.

INFCE (1980b). *Fuel and heavy water availability*, IAEA, Vienna, pp.292–312.

Institute for Energy Analysis (1975). *Net energy from nuclear power*, IEA–75–3.

Jones, P.M.S. (1980). 'Energy choices for the future', *Long Range Planning*, **13**, 18.

Jones, P. M. S. (chmn), (1985). *The economics of the nuclear fuel cycle*, Nuclear Energy Agency/OECD, Paris.

Nininger, R. (chmn), (1983). *Uranium resources*, Nuclear Energy Agency/OECD, Paris; see too later edition (1986). NEFI (1986).

Taylor, K. (1974), *Uranium from seawater*, PAU Report R 14/74, Programmes Analysis Unit, Harwell, Oxon.

Thornton, D. (chmn) (1987). *Nuclear energy and its fuel cycle*, Nuclear Energy Agency/OECD, Paris.

Townsend, M. (1983). *Uranium supplies*, Evidence to Sizewell Inquiry, CEGB/P/7 and CEGB/P/7 (ADD.1), CEGB, London.

WEC (1980). *World energy: Looking ahead to 2020*, IPC Sci. and Tech. Press, London

WEC (1986). *Survey of Energy Resources*, World Energy Conference, London.

White, G. (1986). *The Spot Market and Spot Price*, Uranium Institute 11th Annual Symposium, Uranium Institute, London

Nuclear Power: Policy and Prospects
Edited by P. M. S. Jones

21

The Incentives

P. M. S. JONES.
Department of Economics, Surrey University.

21.1 Introduction

There are four main incentives for the adoption of nuclear power; economic, strategic, environmental and climatic. The greatest stress has generally been placed on its economic attractions (or potential attractions) when compared with alternative means of generating electricity on a large scale from either fossil or renewable resources.

Strategic considerations assumed new significance with the 1973 oil crisis which exposed the vulnerability of the Western industrial economies to disruptions in fuel supply which arose from their heavy dependence on oil. The importance of diversity in the fuel base was also highlighted by the 1984–85 year long miners' strike in the United Kingdom.

The environmental impact of nuclear power differs fundamentally from that of fossil fuel or biomass burning and, whilst radiation emissions are an important consideration (Chapters 6 and 7), there are great attractions in reducing the corrosive gaseous emissions that result from combustion processes. The carbon dioxide produced by fossil fuel and biomass combustion also constitutes a long term threat to the established global climate and means of avoiding its production may assume increasing importance or even urgency in the later years of this century.

These incentives to the development and deployment of nuclear power are examined in the succeeding paragraphs.

21.2 The Economics of Nuclear Power

21.2.1 *The Components of Cost*

The costs of producing electricity have three main components; the capital cost of building the plant; the costs of the fuel to run it, and the costs of labour and non-fuel materials to run it and keep it operational.

The balance of these components differs with the power source. Gas turbines have relatively low capital costs but use expensive fuel, whereas renewable sources relying on wind or water use no fuel and the bulk of their cost is due to the capital needed for construction. Nuclear and coal fired power stations lie between these extremes with nuclear nearer to the renewables and coal nearer to the gas turbine.

Put at its simplest, the cost of producing a unit of electricity will be determined by the fuel and variable operating charges needed for its production, plus the fixed capital and fixed operating charges of the plant divided by the amount of electricity produced. The latter component is reduced if a given plant can be run at higher output, whereas the former component is almost independent of output for any given plant. The inability to start up or close down fuel burning plants instantaneously can, however, affect the over-all efficiency of fuel utilization.

The effect of capacity factor on the unit cost of output is illustrated in Figure 21.1. Capacity factor is defined as the actual energy

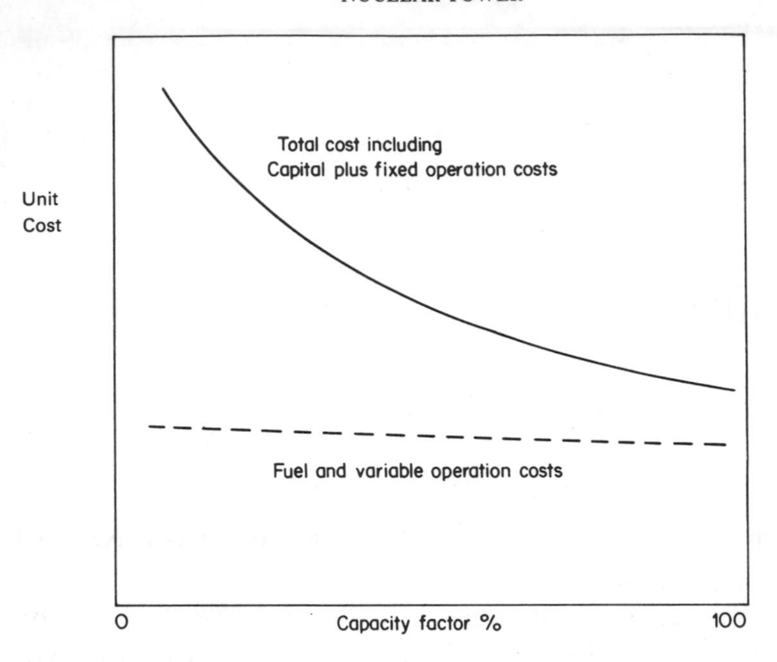

Figure 21.1 Effect of capacity factor on generation cost

output achieved in a given period divided by the output that would have been achieved if the plant had operated at its full design capacity over the whole period. The precise shape of the curve depends on the type of plant and, as will be seen later, total cost curves for different plants cross each other.

Capacity factor (sometimes called load factor) is best defined in terms of power output since the alternative of using plant operating time gives undue weight to periods when the plant is running at below full capacity.

21.2.2 *Defining Costs*

It might appear from the above that the calculation of generation costs is a simple matter. However, even if the costs of building and operating plant and the costs of fuel can be estimated with reasonable confidence, and we return to this aspect later, there are problems. There are many different ways of calculating and presenting unit generation costs using the same input data and these can yield significantly different results.

Perhaps the first question to resolve is whose costs we are concerned with: the costs to the utility or the costs to the nation? The former will deal only in terms of actual or expected charges to the utility for fuels, plant, etc., provided by the contractors or suppliers. The latter might make use of shadow prices to reflect the marginal costs to the national economy of the necessary input factors, or may introduce additional factors to allow for externalities that are not reflected in the market place.

21.2.2.1 National Resource Costs Thus, if a country has balance of payments problems and an increase in the level of imports has a disproportionately large negative impact on the economy, the shadow pricing of imports above their actual market price may be used to bias decisions in favour of domestic activities. Similarly if one supplier is heavily subsidized the true costs to the nation of using his product may be significantly higher than the price he charges for it. Insofar as the subsidy represents a deliberate governmental judgement introduced for welfare, equity, strategic or economic reasons, it may be appropriate to ignore it but, if the level of demand for the product changes radically, the levels of subsidy may diverge from the initial intent.

Alternatively if a supplier is using labour or other factors of production which, in a depressed economy, could not be found alternative employment or use, then it might be appropriate to charge for them at the marginal cost to the economy of their use, rather than at market rates; i.e. gross employment costs net of any welfare or training payments they would otherwise be receiving.

Externalities such as the effects of effluents from fuel and power plants on health, agriculture and the general environment may also have economic impacts which can be reflected in national cost calculations (Jones *et al.*, 1972). Indeed some impacts, such as acid rain, reach beyond national boundaries and this raises the question of where any cut off should be made. For practical reasons governments would usually choose their own borders and deal with international impacts through control of emission levels.

Such exercises in costing are inevitably complex because they can not be limited solely to a single aspect of production or to the immediate market situation; they have to reflect the impact of any choices on future economic development. Thus a government wishing to subsidize or otherwise sustain an uncompetitive industry can only do so by adopting measures which will reduce the profitability or competitiveness of other sectors of the economy. The tax revenues needed to finance a new subsidy come either from other firms, which have to raise prices to maintain viable profit margins, or from consumers, who are then left with lower net incomes and can purchase less of the products of the non-subsidized firms.

The introduction of externalities into economic calculations is also complicated by lack of data about causal relationships between, for example, effluents and the detriments they are believed to produce and by the problems of attaching economic costs to the detriments themselves (Jones, 1979).

Whilst attempts have been made from time to time to conduct electricity generation cost analyses from this wider national perspective and whilst in principle such analyses could assist governments in policy choices, in practice the complexities and uncertainties are such that decisions have been based on the market criteria appropriate to utilities balanced, when governments have wished to intervene, by subjective judgements on the importance to be attached to the wider considerations.

21.2.2.2 Utility Costs Even the question of utility costs is far from simple. Costs are incurred over a period of time commencing with the planning of the plant, continuing through its construction and operation up to its ultimate decommissioning. During this period both costs and power station output will vary from year to year, inflation may change the value of the currency unit during the plant's life, and factor costs may change in real terms.

(a) *Accounting Costs* The accountant who is concerned with covering costs and maintaining the profitability of the utility will adopt an arbitrary basis for allocating capital charges. This may be done by taking the initial capital cost plus any interest incurred on borrowed money during construction and 'paying off' the debt in equal capital repayments spread over an arbitrary number of years. Each year's electricity output then has to meet the capital repayment plus interest charges on the outstanding capital debt. This method, known as straight line depreciation, results in a decreasing capital element in the over all generation cost (Figure 21.2) with an abrupt decline after all the capital has been repaid.

An alternative is to annuitize the capital repayments so that an equal sum to cover capital plus interest is charged each year. This gives a different time profile from the straight line depreciation method but will also show an abrupt change at the end of the selected annuitization period (Figure 21.2) unless the plant is annuitized over its full working life (which is inevitably indeterminate).

Up until quite recently few questioned the use of actual money sums paid out (historic costs) for amortization purposes. The standard accountancy practices developed in an

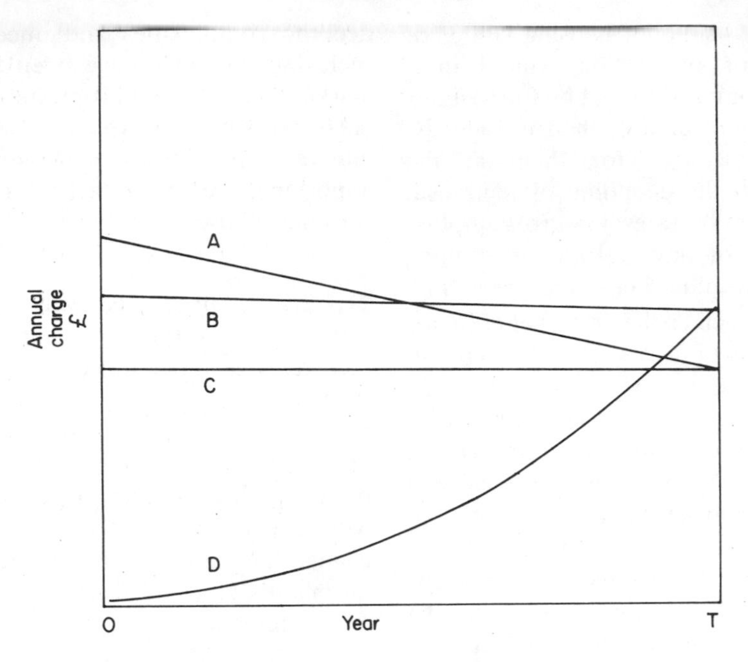

Figure 21.2 Capital charges T = Amortization period For straight line depreciation: C = capital repayment, A = capital + interest For annuitization: D = capital repayment, B = capital + interest

era of low inflation still enabled repayment of borrowings, albeit at the higher interest rates demanded by lenders to offset the effects of inflation, and, if the prices charged by utilities to their customers reflected these historic capital charges, the utility could make a satisfactory profit. However, the alternative view that capital charges should provide the funds to enable replacement of the existing plant at the end of its useful life, thus providing for the continuity of the business, has led to moves for the adoption of current cost accounting in some countries. On this basis the fraction of capital remaining in any year is revalued from its actual cost to its equivalent cost in money of the year in question and this 'new' figure is employed as the basis for capital recovery charges. If this figure is used as a basis for prices to customers the utility will accumulate surpluses and current consumers are effectively providing the capital to finance plant for future users; a situation which many feel unfairly penalizes today's consumers. Perhaps the best that can be said of this approach is that it will bring accountancy costs closer to the long run marginal costs which the economist favours as a basis for pricing in order to optimize the use of national resources.

Both accountancy approaches yield arbitrary annual charges, depending on the time basis selected for amortization, which are independent of electrical output. The apparent unit cost of electricity will vary from year to year depending on the plant performance and if a plant were shut down for extended maintenance its generating costs might appear very high one year and much lower the next. Annual generation costs for existing plant on an accounting basis are therefore of little value for comparing either the worth of past investments or the potential value of future investment in similar plant. Thus situations can be envisaged in which the annual generation costs from contemporary fossil and nuclear stations could show one better one year and the other the next due to performance variations; could show either cheaper depending on the basis adopted for amortization; or could show a clear advantage to nuclear due to use of historic costs despite the fact that nuclear had used more real resources. Utilities in the USA, Canada, UK, France, and elsewhere have published annual generation costs from existing stations and the values have frequently been wrongly used to support arguments for or against particular systems.

The situation is simpler when future plants are being compared on an identical basis in an appraisal. Here projections of annual costs can yield meaningful comparisons although the over-all advantage of one system over the other may involve a trade-off between two time streams of costs with one system 'cheaper' in the near term and the other in the longer term. If the costs are calculated in cash terms allowing for anticipated inflation, as is frequently done in the USA, the visual picture presented may grossly distort the real resource costs of the two options.

(b) *Levelized Costs* By far the best method for simple comparison purposes is the constant money levelized cost approach in which the costs incurred each year from start of planning to final decommissioning are discounted (see next paragraph) to an arbitrary base date and the sum of these discounted costs is divided by the total electrical output expected over the station lifetime also discounted to the same base date. The result, which is independent of the base date chosen, is a levelized cost in constant currency units which, if charged for each unit produced, would exactly meet all costs and provide a return on investment equal to the selected discount rate. The method provides a valid basis for comparing generation plants with different economic lives and effectively annuitizes capital costs over the whole station life at an interest rate equal to the discount rate. By relating capital charges to units produced it eliminates the annual variations and the arbitrariness of accounting lifetimes.

Discounting is a practice developed by economists to take account of two factors. Firstly that even in the absence of inflation £1 in the hand today is preferable to the promise of £1 in several years time, and secondly that the use of resources on a project now means that the opportunity of using them for other purposes is foregone. The opportunity cost (or value of the opportunities foregone) is measured by the return that could have been earned through investment in other projects and this is reflected in financial markets through share dividends. Market interest rates similarly reflect the inducement needed to encourage individuals to forego current consumption for future gain. Other things being equal the two rates should come together for low risk investment, due to the possibility of cross investment, and they thus set a value for the opportunity cost of capital which can be used in discounting calculations.

Thus a sum £y spent today could have earned interest at a compound rate R per cent p.a. to become £x in t years time, where for small values of R $x = y\ (1+R/100)^t$. Conversely a sum £x in t years time only has an equivalent present worth of y, i.e. $x\ (1+R/100)^{-t}$.

If the money spent on plant construction, fuel, operations, etc., in year t is £x, in constant £ units, and the discount rate is R per cent p.a., then the total present worth discounted cost is:

$$\text{P.W.} = \sum_{O}^{L} x\,(1+r)^{-t} \qquad (21.1)$$

where r = R/100 and L is the length of time from the base date over which costs are incurred. (Σ means sum of costs from time t = 0 to t = L).

For a station producing U kWh output in year t the levellized cost per unit of output is C where:

$$C = \sum_{O}^{L} x\,(1+r)^{-t} / \sum_{O}^{L} U(1+r)^{-t} \qquad (21.2)$$

Groups of people acting in concert may be more willing to forego present consumption to achieve future goals than individuals acting alone. There is therefore a social time preference rate which can differ from the opportunity cost rate. Nevertheless it seems reasonable for utilities to adopt the opportunity cost of capital as the basis for discounting in their economic assessments (see Appendix 1).

Levelized costs can also be calculated in cash terms by introducing assumptions about future inflation and price rises. This would give a cost at which the £1 notes spent were recovered but not necessarily the true value of resources invested. It is not therefore favoured for economic appraisal.

(c) *Systems Costs* A completely different approach looks not at a single power station investment but at the effects of alternative investments on the over-all costs of the utility's network of power stations. This systems cost analysis is the correct method for investment appraisal and it will be discussed in some detail later. It has the limitation that it is system specific, that it is linked to the specific investment decision only, and that the results are difficult to compare between countries. For simplicity, therefore, the following paragraphs deal with single station costing before returning to systems cost analysis.

21.2.3 *Single station costs*

21.2.3.1 General Although this work is focused on nuclear power the underlying economics of the different technologies for electricity production are sufficiently different to justify some digression at this point, to enable the reader to understand some of the problems of intercomparison.

For comparison purposes, costs of generation are customarily considered up to the point at which the electricity is fed to the transmission system, i.e. the station bus bar which lies between the high voltage terminals of the generator and the switchyard and distribution equipment. These costs include the costs of buying the site; the construction, operational, maintenance and fuel costs; and the costs of treating and disposing of wastes and of decommissioning the plant at the end of its useful life.

The electricity output from the plant is that fed to the grid, net of any internal use within the station itself. Transmission losses are a characteristic of the grid and not the generation plant and are usually ignored for single station generation cost calculations. The bus bar costs are, however, only sufficient for comparison purposes if the alternative stations have similar transmission needs. This may be the case for large base load fossil or nuclear stations where a common set of possible sites is dictated by cooling water requirements, but it may not be true for those renewable sources which have to be sited at geographically favourable sites which happen to be well removed from demand centres.

21.2.3.2 Variable Renewable Sources Three of the commonly discussed renewable energy sources are inherently variable and unpredictable, viz. wind, waves and sunshine. A fourth, the tides, varies in a more predictable manner with solar and lunar cycles on which are superimposed the effects of wind. With such sources the costs of electricity production depend on the capital costs of the plant and any operational, maintenance and refurbishment costs incurred during its useful life. These operational costs may be quite significant in cases where, by their very nature, the generators have to be placed in hostile environments where they are subject to corrosion, storm damage or fatigue failure. Offshore siting for wave or aerogenerator devices will add to transmission costs and the variability of the individual units will require additional capital expenditure to provide a stable synchronized input to the grid system for the alternating current grids used almost everywhere.

If the capital and total operating costs are known then the present worth lifetime costs can be calculated (equation 1), taking due account of any special transmission and voltage regulation and synchronization equipment needed that is additional to that required for fossil or nuclear stations. Generators are matched to the anticipated weather–operating environment and capital costs are generally quoted per peak kilowatt, i.e. the maximum design output given favourable wind, wave, sun or tide conditions.

The extreme case of the variable source is the aerogenerator which may be designed to close down above specific wind speeds to avoid structural damage, and where its electrical output varies non-linearly with wind speed. It is impossible to know in advance how much energy it will be producing at any given hour of a given day nor can there be any assurance that it could give output when output is wanted. In this it differs from fossil and nuclear stations whose output can be predicted and whose availability at a specified time can be reasonably well assured.

All that can be said about the aerogenerator is that if past weather conditions lead one to expect a certain statistical distribution of wind speeds and directions, then a statistical distribution of electrical output (subject to any high or low wind speed constraints) could be expected and an average potential output could be calculated. Discounting and summing this over the aerogenerator's expected useful life would enable a levelized generation cost to be calculated using equation 2.

However, it will be obvious that an unpredictable aerogenerator producing electricity at a levelized cost of x pence kWh^{-1} is not a complete substitute for a coal fired or nuclear power station producing electricity at the same levelized cost: reliability has a value. One way of reducing the uncertainty over availability would be by using more plant in different locations, another would adopt some means of reversible energy storage. In either case additional costs would be incurred to achieve the improved predictability. Although wind is an extreme example, simplistic comparisons of the costs of electricity from any variable renewable source with that of firm supplies from conventional fossil or nuclear sources is likely to be misleading and a full systems costing approach is needed if meaningful comparisons of investments are to be made (see section 21.2.4).

21.2.3.3 Fossil Fuelled Power Plant In the case of fossil fuelled generation plant the main uncertainty is likely to be associated with the future price of its fuel. Since the life of a fossil fuelled power station may be extended to 40 years or more if this proves economically worthwile, some estimate of the likely long term escalation of coal prices in real (constant money unit) terms is needed. This requires judgement on future world supply and demand, mining and transport costs. Views on prices may differ depending on whether indigenous or imported coal are expected to be used and relative movements in exchange rates can also have a big impact on national perspectives.

Views on fossil-fuel prices can differ widely and rapidly. In the mid–1970s oil prices were generally expected to more than double in real terms by the turn of the century and many models used projections of oil prices as the basis for estimates of future coal prices. This was based on the then plausible view that as oil and gas scarcity forced prices up, coal would move in to fill the 'gap' and would extract the maximum price it could from users; this would be lower than that of oil by a margin reflecting the differences in convenience and end use efficiency. By the early 1980s apart from more optimistic consensus views on oil supply and prices, the link was no longer regarded as realistic. The potential supply of coal, at least in the near to medium term, is large compared with demand and prices will be set by the cost of supply at the margin in a competitive market. The arguments about mining cost, supply levels and world coal prices were very thoroughly examined at the UK Sizewell Inquiry (Turner, 1985) (Layfield, 1987).

The capital and operating costs of coal or oil fired plant are well defined because the technology is established and there is a great deal of experience in their construction and operation. The two elements of uncertainty for the future are concerned with the adoption of anti-pollution measures and the possible introduction of newer technology. This is not the place to go into detail but the introduction of stack scrubbers to remove acid gases can add some 10–20 per cent to the investment cost of new plants and reduce their overall fuel conversion efficiency by some 5 per cent; a total increase of some 15 per cent in unit costs of electricity sent out. The precise costs depend on the quality of the coal used and the desired purity of effluent gases. The costs can be considerably higher for retrofitting control devices to existing plant. (For further details see IEA, 1985.)

The move to more advanced fluidized bed and combined cycle coal burning plant has been under active development but has not yet reached the point of exploitation in large scale power plant. They increase the efficiency of fuel use but have higher capital costs and may affect reliability adversely so that the over-all economic effect is hard to predict at this time.

The costs of disposal of solid wastes from coal fired stations and the costs of decommissioning the station at the end of its life should be included in the over-all generation cost calculation but they are both relatively small in relation to the other costs.

The costs of stations burning organic wastes and biomass can be determined in the same way as for fossil fuelled stations but their design and scale will be different and there may be technical and economic problems associated with the variability of their fuels.

21.3.2.4 Nuclear Costs (Principles)

(a) *General* Although nuclear power has superficial parallels with fossil fuel burning in that it takes a fuel and uses it to produce heat for electricity generation whilst at the same time creating waste products, the differences are considerable. The fuel is not a simple mineral product but it is refined and processed into carefully engineered units to tight specifications. In a thermal reactor only about 1 per cent of the energy content of the fuel is extracted and the spent fuel has to be specially processed and the highly radioactive wastes have to be stored under supervision for long periods prior to ultimate disposal.

The costs of providing a unit of nuclear electricity are therefore spread out in time to a far greater extent than the direct costs of fossil fuel burning (although the indirect environmental costs of the latter may well be felt long after the fuel has been burnt). Table 21.1 illustrates the time range over which costs might be incurred.

(b) *Capital Costs* Since the capital element is the biggest item in nuclear electricity costs it is important that estimates of the costs of new plants are reliable. They might be expected to be reasonably clearly defined in view of the number of plants built world-wide (Chapter 19) but with the exception of France, there has been surprisingly little replication even of LWR designs. Capital costs have risen considerably as the complexity and number of safety systems have increased and this rise, even in France (Moynet, 1985), has more than offset the benefits that would otherwise have been expected from the increases in reactor size.

Cost estimates for new reactors can be and are based on past experience and design information, but there are two opposing viewpoints. The engineer looks in detail at the components of the plant, makes allow-

Table 21.1 Time Leads and Lags for Nuclear Power

Nature of cost	Time over which incurred[1] years	Comment
Construction	0–7	Taken as 6–8 years
New fuel into reactor	5–40	
Spent fuel in reactor pond store	8–44	2 years pond storage
Intermediate fuel storage	11–47	3 years (assumed) prior to reprocessing
Reprocessing of spent fuel	11–47	
Vitrified high level waste storage	11–97	Taken as 50 years
Reactor decommissioning	42–141	Taken as 100 years to Stage III, greenfield site.

[1]Taking start of construction as time zero.

ances for design changes and any escalation or savings in labour or material costs, and comes up with a considered set of costs for the specific plant, incorporating such contingency allowances as his experience dictates. He is aware of the reasons for past cost overruns due, say, to late design changes necessitating retrofitting of new items, or labour or component supply difficulties.

The statistician, on the other hand, who has nothing to work with but data on past costs, may take these and their trends as representative of the whole family of past and future reactors and seek to extrapolate past experience to the future. He will recognize that there are different reactor types and that for any given type a range of factors may affect costs. This might include scale, manufacturer, purchaser, degree of replication, experience of the manufacturer and purchaser, location, principal design characteristics, vintage, and year of start of construction as a proxy for labour and materials cost.

The best known proponent of the statistical approach, Charles Komanoff (Komanoff, 1981, 1982, 1984) has concentrated his attention on US experience and concluded that there are scale and replication benefits but that real costs have also increased in proportion to the square root of the total number of plants constructed: a finding he interpreted as showing society's wish to hold total accident risk levels down. On the basis of this interpretation he felt able to extrapolate real cost escalation into the future to the detriment of nuclear power when compared with coal fired plants, even though they had a similar cost escalation, because there were more of them. Two sets of Komanoff's correlation coefficients are summarized in Table 21.2. His work has been strongly criticized on technical grounds in the US and the UK. As can be seen from Table 21.2 a large number of explanatory variables are introduced, and the values for the coefficients themselves vary despite their apparent high degrees of confidence. Thus in 1981 (Table 21.2) a north-east US location increased costs by 28 per cent, whereas by 1982 for Westinghouse reactors this had risen to 51 per cent, or to 61 per cent by 1983: the negative correlation with architect engineer experience (1981) was replaced by a negative correlation with utility experience in 1982; the TMI incident was claimed to introduce a step increase of 70 per cent. On the basis of Komanoff's Sizewell Evidence figures the projected costs of Calloway and Wolf Creek reactors (under construction in the US) could be either 1–5 per cent cheaper than expectation or 16–22 per cent dearer depending on the version of the regression equations used.

There may be good reasons for expecting continuing real cost escalations if factor costs are known to be rising or if reactors have to be built at progressively less attractive sites with the best sites having been used first. (Neither is applicable in the UK at present, e.g. the reactor construction cost index has risen in line with general inflation for some time.) However, there is no reason for expecting that reactors, which are already built to extremely tight safety standards (Chapters 7 and 8), will have to be designed against ever bigger earthquakes or with ever more replicated and expensive safety measures to guard against increasingly remote failure modes. Where the reasons for past cost escalation are understood (see for example Chapter 14) and steps have been taken to avoid repetition, then there may be high confidence that past trends can be halted or even reversed. Statistical analyses such as those described above certainly raise questions that require consideration but they can not be regarded as having any predictive power unless there is a clear causative relationship underlying them which can reasonably be expected to continue operating in the future in the same manner as in the past.

Capital costs are not solely affected by the costs of labour and materials however; they are also influenced by the time taken to build the plant. An extended construction time prior to completion and operation means that the resources invested have been unproductive (in terms of electricity output) for a longer period and the income foregone on

Table 21.2 Komanoff's Nuclear Capital Cost Regressions
Capital cost in US $kW without interest during construction equals

From Komanoff, 1981 in mid-1979 $ kW^{-1} all PWRs	T-Ratio	Significance level, %	From Komanoff 1984 in 1982 $ kW^{-1} Westinghouse PWRs	T-Ratio	Significance level, %
6.41 ×	2.60	99.2	104.2 ×		
1.28 if north east U.S. ×	6.56	99.9+	1.61 if N.E. U.S. ×	6.23	99.9+
A – E $^{-0.105}$ ×	6.19	99.9+			
MW $^{-0.200}$ ×	2.54	98.5	MW $^{-0.469}$ ×	2.64	98.5
0.903 if multiple ×	2.48	98.2			
1.34 if dangling ×	4.73	99.9+			
1.20 if cooling tower ×	5.22	99.9+			
(Cum.nucl.capacity)$^{0.577}$	13.55	99.9+	(Cum.Nucl.Cap.)$^{0.512}$ ×	5.91	99.9+
			1.3 if post-TMI × or	4.72	99.9+
			0.772 if pre-TMI		
			0.881 y	4.96	99.9+
sample size = 46 units			sample size = 27 units		
r2 = 0.908			r2 = 0.90		
F value = 64.7			F value = 39.1		
A E = architect experience			TMI = Three Mile Island incident		
			y = utilities previous no. of reactors		

possible alternative uses of the resources will be larger. In the discounted levelized cost analysis this is captured by the larger discount factor applied to the delayed station output compared to the construction costs. Accountancy approaches, which in this respect often get carried over into economic assessments, look instead at interest paid on borrowed capital during construction or interest fore-gone on self financed investment. This is added to the basic capital cost to arrive at a total cost for amortization purposes including interest during construction (IDC). The calculation of interest and the assumption that the full capital plus IDC cost is incurred at the station commissioning date, is mathematically the same as discounted cash flow with moneys debited in the year spent provided the interest rate is the same as the discount rate in real terms.

The time taken between start of construction of reactors to their commissioning varies enormously from country to country. France and Japan have consistently built plants in five to six years, whereas the United States (which was initially as good) has slipped, with construction periods extending to as much as eleven years. Capital financing charges with lead times as long as this, particularly with the high interest rates prevailing in the early 1980s, can add considerably to the over-all costs of nuclear electricity. Again judgement and experience have to be taken into account in deciding the likely construction time for new plants.

(c) *Performance* Because nuclear electricity costs are dominated by the fixed initial capital cost, it is important to get the maximum output from any station and to spread the capital charges over as many kWh as possible. All plants are designed with the intention of achieving high availability with minimum 'down-time' for maintenance, safety checks and for refuelling for reactors like LWRs. In practice many existing reactors have not proved as reliable as their designers would wish and average capacity factors for PWRs, for example, have been about 65 per cent rather than the 70–75 per cent originally expected.

Some take the view that the prediction of

the performance of future plants can be based on statistical examination of the behaviour of existing plants. Once again Komanoff (1982, 1984) and others (Surrey and Thomas, 1980) have applied regression analysis techniques with variables including reactor size, reactor age, number of units on site, salt or fresh water cooling, use of cooling towers, learning effects, using calendar year as a proxy, manufacturer, experience of operating utility, country of operation, and a TMI effect.

Part of the problem with performance has been its variability between units. Some reactors of all types have performed with very high capacity factors—thus individual PWRs, BWRs, PHWRs and GCRs have achieved cumulative load factors over their life to date in excess of 80 per cent. Equally there are examples of all four types with cumulative load factors below 50 per cent (Howles, 1986). Furthermore any regression analysis is complicated by the fact that reactor sizes have tended to increase over the years with parallel changes in steam conditions and a move from two coolant loops to three or four loops in the larger (and more recent) PWRs.

As indicated in section 3.4.2 the most important use of data on existing reactors is to identify the causes of unplanned shutdown so that they can be eliminated or at least reduced in future reactors. Thus whilst it may be interesting to know that British Magnox reactors have an average cumulative capacity factor of 58 per cent this has no predictive value for a new Magnox reactor, should one be built, since the constraints applied to limit corrosion damage to mild steel bolts in the reactor core would not be relevant to new designs avoiding this problem.

There is therefore a conflict between the approach of the engineer and that of the advocate of past trends persisting. The engineer seeks to learn from experience, to eliminate identified weaknesses and to make sure that future plant performance is as good as the best of today's plant, or better. Only when performance has reached an economic optimum will the engineer view the statistical analysis of existing plant as giving guidance for the future—and this point has certainly not been reached yet.

There are, of course, technical reasons why some reactor types can be expected to do better than others. Notably gas cooled reactors and pressure tube reactors like CANDU, which are both designed for on-load refuelling, have a potential 5 per cent advantage in availability over PWR and BWR which have to be shut down for this purpose.

(d) *Fuel Costs* There are a number of different nuclear fuel cycles each with its own costs made up of front-end costs incurred before the fuel reaches the reactor and back-end costs incurred after the spent fuel is removed from the reactor. The former include the costs of purchasing uranium, the costs of converting it to pure oxide or to uranium hexafluoride (depending on whether it is to be enriched), the costs in appropriate cases of enrichment, and the costs of fuel fabrication. The latter include costs associated with the storage of spent fuel, its reprocessing, and the treatment, encapsulation and disposal of wastes. Alternatively in the once-through 'cycle' the spent fuel itself may be encapsulated and disposed of intact. In the reprocessing cycle the recovered uranium and plutonium can be used in fresh fuel and some credit can be set against the reprocessing costs. The different processes involved are set out in more detail in Chapter 4. The Magnox reactor uses unenriched metallic fuel, CANDU unenriched uranium oxide, whilst mixed plutonium–uranium oxide (MOX) fuels are used in fast reactors and, at lower plutonium concentrations, might be used commercially in thermal reactors.

As will be seen later many of the stages in the nuclear fuel cycle contribute little to its over-all cost. The important components of cost are uranium purchase, enrichment, fuel fabrication and, where it is practised, reprocessing.

The inevitable and considerable uncertainties about future uranium price movements have been described in section 20.7.1. Although these uncertainties are large and uranium purchase is the largest single component of nuclear fuel cycle costs, the effect on over-all nuclear generating costs is relatively small because, unlike coal fired stations, fuel accounts for only about 30 per cent of the total. Even if uranium prices trebled the costs of LWR electricity would only rise by about 20 per cent, whereas a similar rise in coal prices would double coal based electricity costs.

Developments in enrichment technology with the move to advanced centrifuges and possibly, in the longer term, lasers, are likely to reduce the costs of this stage of the fuel cycle by as much as 50 per cent or more compared with the costs of the energy intensive diffusion process which has provided the bulk of enrichment in the past.

Fuel fabrication is a well-established technology and there is a keenly competitive market so that fabrication costs are unlikely to increase in real terms.

Reprocessing costs have shown steep rises in the past, partly as a result of the move from the reprocessing of metallic fuels to higher burn-up oxide fuels, and partly as a consequence of the implementation of more stringent safety, security and environmental protection measures. The major reprocessors are confident that the technology is well established and that the controls adopted or to be adopted for operating or planned plant are more than adequate. Estimates of the future costs of large scale oxide fuel reprocessing are therefore fairly close to current contractual costs (Jones, 1985) and even these appear to be based on very conservative assumptions about plant life and throughput, so that reductions in cost can not be ruled out for the longer term (Jones, 1987a).

There is considerable variation in the estimates of the costs of spent fuel storage and disposal and reprocessing waste storage and disposal. This arises in part from different design concepts and in part from differences in the scale of operation contemplated. However, adoption of conservatively high values for these back-end costs has little impact on the over-all costs of the nuclear fuel cycle.

There is no established market for the uranium and plutonium recovered from spent fuel by reprocessing and their values would be dependent on their irradiation history, which affects their isotopic content (section 4.4.1), and on the nature and timing of their intended use. Thus uranium purchase and some separative work costs can be saved by taking advantage of the enhanced uranium–235 content of spent PWR fuel. Alternatively, recovered plutonium can be used to achieve the necessary fissile material content of thermal reactor fuel by mixing it with fresh or recycled uranium, and in this way separative work may be avoided completely.

In the absence of a market, the value of plutonium or recovered uranium can only be assessed in terms of the savings they can yield, after subtracting any extra fabrication costs incurred due to the need for additional radiation screening, and allowing for any modification to the fuel specification arising from the presence of neutron absorbing isotopes such as uranium–236. The value of the recovered materials will also be reduced by any extra costs of storage if they are not used immediately, and additional costs may be incurred if the plutonium has come from high burn-up fuel and its post reprocessing storage has been long enough to allow the accumulation of γ-emitting decay products which complicate its subsequent use.

These substitution 'values' will be dependent on the degree of enrichment needed by the MOX fuelled reactor and on its neutron spectrum (section 4.4.1), so that plutonium destined for a fast reactor would have a higher value than material destined for a thermal reactor, both because of its enhanced fission characteristics in the former and because the separative work savings would be larger.

A further factor will arise if the availability of plutonium releases a fuel constraint on the introduction of an otherwise economically attractive energy option. In such a circumstance the 'value' may be related to the whole of the economic gain rendered available: for example, if uranium supply constraints prevented introduction of thermal reactors in place of more expensive coal stations, or fast reactors in place of more expensive thermal systems. On the other hand if uranium supplies were plentiful and there were no foreseeable intention of recycling the uranium or plutonium, then their existence would entail storage costs with no offsetting gains.

(e) *Decommissioning and Waste Management Costs* The costs of decommissioning nuclear facilities are considerably larger than those for non-nuclear plant due to the need to protect employees and members of the public from radiation induced in the structures by neutron activation or by contamination. The estimated costs of such decommissioning and the disposal of decommissioning wastes are included in nuclear generation costs calculations either as part of the investment charges (for the reactor itself) or as part of the fuel cycle (for fuel cycle plant).

Estimates generally put the cost of decommissioning at about 10 per cent of the initial plant cost which, when discounted to allow for the interval before decommissioning is undertaken (see Chapter 10), contributes negligibly to over-all generation costs.

Some claims for very high decommissioning costs estimates have emerged from time to time in the United States (Ryan, 1978) but these have been based on current rather than constant dollar costs which can greatly distort the picture in times of rapid inflation (Jones and Sargeant, 1978). A recent study by NEA (Delaney, 1986) gives undiscounted costs of immediate decommissioning of about 10 per cent of the initial capital cost.

21.3.2.5 Nuclear Cost Comparisons

(a) *Investment Costs* The Nuclear Energy Agency (Jones, 1983, 1986) and UNIPEDE (Moynet, 1983, 1985) have drawn together international information on nuclear and

Table 21.3 Nuclear Capital Costs. Discount rate 5 per cent, January 1984 $ kWe^{-1}

	Reference plant						
Country	Type	Number of units × Mwe (net)	Method of Cooling	Construction cost	Heavy water inventory	Provision for dismantling	Total investment cost
Belgium	PWR	1 × 1300	mixed	895	—	24	1036
Canada Central	CANDU	4 × 881	lake	892	222	5	1295
Eastern	CANDU	1 × 635	sea	1738		6	2015
Finland	PWR	1 × 1000	sea	1022	—	30	1303
France	PWR	2 × 1390	atm.	737	—	30	870
Germany, F.R.	PWR	1 × 1258	atm.	1250	—	10	1429
Italy	PWR	2 × 1000	sea	964	—	28	1131
Japan	LWR	4 × 1100	sea	1202	—	32	1405
Netherlands	PWR	2 × 1000	sea	1017	—	40	1264
Norway	PWR	1 × 1000	sea	1100	—	28	1338
Spain	PWR	1 × 933	sea	1229	—	37	1497
Switzerland	BWR	1 × 950	atm.	1739	—	26	2084
Turkey	CANDU	1 × 665	sea	1650	—	—	1965
	PWR	1 × 986	sea	1300	—	—	1585
United Kingdom[1]	PWR	1 × 1155	sea	1565	—	23	2080
	PWR	1 × 1155	sea	1251	—	23	1492
US Central	PWR	1 × 1200	atm.	2021	—	27	2865
Central[2]	PWR	1 × 1200	atm.	1415	—	27	1860
Eastern	PWR	1 × 1200	atm.	2033	—	27	2884
Rocky Mountain	PWR	1 × 1200	atm.	1967	—	27	2791

Based on Jones, 1986
[1]UK figures for Sizewell and successor PWR
[2]US second central figure for improved construction performance

fossil station capital costs. A significant feature of these costs was their variability (Table 21.3), some of which was due to design, siting, scale, numbers of plants per site and replication differences. This was only a partial explanation, however, and a major contributor to the apparent variability was the failure of monetary exchange rates to reflect accurately the relative factor costs in the different countries. Thus North American and European relative costs can appear to differ enormously depending on the base date adopted for the currency unit, despite the fact that the reactors are unchanged (Jones, 1983, 1986).

This exchange rate problem and the inevitable differences in factor costs (labour, materials, energy) between different countries led the NEA authors to conclude that the only proper comparison between reactor types (or fossil and nuclear stations) was one based on given designs costed with the ground rules and factor costs appropriate to the country making the choice. Attempts to draw absolute comparisons between studies undertaken in different countries were potentially misleading.

For this reason the only quantitative conclusions to come out of the NEA comparative study were those related to the relative costs of coal and nuclear generated electricity.

(b) *The LWR Fuel Cycle* A recently published Nuclear Energy Agency Working Group report (Jones, 1985) examined the costs of the LWR once-through and reprocessing fuel cycles in considerable detail, taking into account the range of views in member states and the prospects for new technology. The group concluded that the discounted levelized cost method is appropriate for calculating fuel cycle costs. It established a set of reference prices for the individual component stages of the fuel cycle and showed in a simple manner how these could be combined to yield an over-all fuel cost inclusive of waste management and fuel plant decommissioning costs.

The over-all fuel cycle costs for the reference case are set out in Table 21.4. The range of costs for each of three separate uranium price profiles adopted is summarized in Table 21.5. The costs are not sensitive to assumptions about reactor life or load factor, to uranium conversion costs, to spent fuel transport or to waste processing, encapsulation and disposal costs. The biggest single component is that of uranium purchase at 40–45 per cent of the over-all fuel cost in the reference case. Other significant items are the enrichment cost (30 per cent), fabrication (10 per cent) and reprocessing (16 per cent net of reference uranium and plutonium credits); see Table 21.4.

The costs of the fuel cycle are influenced by the discount rate and on the leads and lags in the fuel cycle—i.e. the time prior to use of fuel in the reactor that uranium and enrichment services are paid for and the periods selected for storage of spent fuel and/or reprocessing wastes prior to disposal. However, the sensitivity to any of these items is not large within the plausible range adopted by the working group, and the effects of variations are captured within the uncertainty bands given in Table 21.5.

The reference cost for the once-through PWR was 7.8 US mills (January 1984) kWh^{-1} and that for the reprocessing cycle with a five year spent fuel cooling period prior to reprocessing was 8.6 mills kWh^{-1}, both with an over-all uncertainty band of ± 20 per cent (95 per cent confidence limits), of which half was due to the range of price profiles adopted for uranium. The results were shown to be compatible with other recent studies including the NEA's own earlier generation costs study.

The uranium credits allocated in the reprocessing cycle made allowance for its uranium–236 content and it and the plutonium values were based on substitution for uranium–235 in a PWR, with the materials recycled within a few years of separation. The costs of the reprocessing cycle are not sensitive to the precise values of the credits.

The cost of a mixed plutonium–uranium oxide (MOX) fuel cycle is, however, very sensitive to the assumed plutonium price. With plutonium whose costs have already

Table 21.4 Breakdown of PWR Fuel Cycle Component Costs. 5 per cent Discount rate, January 1984 US mills

	Reprocessing cycle		Once-through	
	Mills kWh^{-1}	%	Mills kWh^{-1}	%
Uranium	3.48	40.7	3.48	44.7
Conversion	0.17	2.0	0.17	2.0
Enrichment	2.28	26.6	2.28	29.3
Fuel fabrication	0.88	10.3	0.88	11.3
Subtotal for front-end	6.81	79.6	6.81	87.5
Transportation of spent fuel	0.14	1.6	0.14	1.8
Storage of spent fuel	0.17	2.0	0.65	8.4
Reprocessing/vitrification	2.18	25.5	—	—
SF conditioning/disposal	—	—	0.18	2.3
Waste disposal	0.08	0.9	—	—
Subtotal for back-end	2.57	30.0	0.97	12.5
Uranium credit	−0.54	−6.3	—	—
Plutonium credit	−0.28	−3.3	—	—
Subtotal of credits	−0.82	−9.6	—	—
Total costs	8.56	100	7.78	100

Source: Jones, 1985

been recovered in charges made by existing thermal reactors (and therefore nominally free) it would be some 30 per cent cheaper over-all than the reference fuel cycles in the previous paragraph, whereas with plutonium priced at its thermal reactor substitution value the costs are, by definition, equal to one or other reference cycle cost depending on whether the MOX itself is reprocessed or not.

The NEA working group have stressed that values of the individual process stage costs will differ between countries depending on the scale and timing of their programmes, on contract terms, and on their policies concerning reliance on indigenous materials or services. They will also differ depending on choices adopted for fuel cycle lead and lag times and, in the reprocessing cycle, on the intended subsequent use of recovered fissile materials. The economics of the choice (for 1995 commissioned reactors) between once-through or reprocessing and recycling fuel options is not clear cut and the differences, although they may have significant financial implications for utilities, are sufficiently small (at about 3 per cent of total generating costs) for social, environmental and strategic considerations to weigh heavily in fuel cycle choice.

Table 21.5 Likely Range of Fuel Cycle Costs (PWR). 1984 Mills kWh^{-1}

	Reprocessing cycle	Once-through cycle
Levelized U price \$83 $kg^{-1}U$ (\$32 $lb^{-1}U_3O_8$) (\$32 $lb^{-1}U_3O_8$, escalation 0 per cent p.a.)	6.6–8.7	5.9–7.2
Levelized U price \$121 $kg^{-1}U$ (\$46 $lb^{-1}U_3O_8$) (\$32 $lb^{-1}U_3O_8$, escalation 2 per cent p.a.)	7.5–9.6	7.0–8.3
Levelized U price \$177 $kg^{-1}U$ (\$68 $lb^{-1}U_3O_8$) (\$32 $lb^{-1}U_3O_8$, escalation 4 per cent p.a.)	8.8–10.9	8.6–10.0

Source: Jones, 1985

(c) *Total PWR Generation Costs* Table 21.6 presents the over-all results of an updating of the NEA study (Jones, 1986) re-expressed in January 1985 US dollars and with reactor life extended from the NEA's 1983 reference value of 20 years to their 1985 reference value of 25 years. It will be seen that despite the large variations in national capital costs estimates the nuclear fuel cycle costs are much less variable and are compatible with the 1985 NEA working group's findings. The coal to nuclear costs ratio is also less variable and shows nuclear to have a significant advantage in Europe and Japan on the basis of the coal price escalation anticipated in the contributing countries. This advantage is not sensitive to nuclear fuel cycle cost changes which are expected to be fairly limited anyway (see section 21.3.2.5 (b)), or to load factor assumptions (Table 21.7). They are sensitive to coal prices and nuclear capital costs but the margin is sufficiently great in most cases outside North America to require a 50 per cent or higher increase in nuclear capital costs or a 40 per cent or more drop in projected coal prices before the nuclear advantage is eliminated.

In the United States the situation is less clear. Access to large supplies of cheap coal should enable coal to compete economically in most parts of the country, for new plants commissioned in the mid 1990s, although nuclear retains a small advantage in Eastern areas remote from the coal fields. However, if the US could sort out its regulatory problems and match the construction performance of France or Japan (or even its own best post industrial experience) the cost savings could make nuclear competitive even in the Rocky Mountain region of the USA (Jones, 1986). It is unlikely that this could be achieved for mid 1990s plant but is a realistic goal for the end of the century

The majority of OECD countries use

Table 21.6 Levelized Discounted Electricity Generation Costs. Discount rate 5 per cent, January 1984 mills kWh^{-1}

	Nuclear				Coal				
Country	Investment	O and M	Fuel	Total	Investment	O and M	Fuel	Total	Ratio Coal: Nuclear
Belgium	11.5	3.5	7.3	22.3	8.1	5.1	22.9	36.1	1.62
Canada Central	14.2	2.5	3.6	20.3	9.5	2.8	16.9	29.2	1.44
Finland	14.3	3.6	7.2	25.1	8.1	4.3	21.0	33.4	1.33
France	9.5	4.0	7.3	20.8	7.1[1]	3.2[1]	27.1	37.4	1.80
Germany F.R.	15.6	5.0	7.3	27.9	7.6	5.8	33.4	46.8	1.68
Italy	12.4	3.5	8.4	24.3	6.8[1]	2.8[1]	24.6	34.3	1.41
Japan	15.4	6.1	10.1	31.6	11.6	5.8	25.9	43.3	1.37
Netherlands	14.0	3.5	7.4	24.9	7.5	3.3	21.9	32.7	1.31
Spain	16.3	5.1	7.6	29.0	6.7[1]	3.7[1]	24.0	34.4	1.19
Switzerland	22.8	4.2	8.4	35.5					—
Turkey									
CANDU	21.6	5.9	3.1	30.6					—
PWR	17.3	4.9	5.5	27.7					—
United Kingdom[2]	22.8	5.5	8.8	37.0	12.4[1]	4.5[1]	34.8	51.7	1.40
	16.1	5.5	8.8	30.3	12.4[1]	4.5[1]	34.8	51.7	1.71
United States									
Central	31.8	4.9	7.1	43.8	14.0	4.8	17.4	36.2	0.83
Central[3]	20.6	4.9	7.1	32.6	14.0	4.8	17.4	36.2	1.11
Eastern	32.0	4.5	7.1	43.6	14.2	4.0	28.8	47.0	1.08
Rocky Mountain	31.0	4.6	7.1	42.7	13.8	3.8	15.3	32.9	0.77

Source: Jones, 1986
[1]Without flue gas desulphurization
[2]UK figures for Sizewell and successor PWR
[3]US second set of central data for improved construction performance

discount rates for investment appraisal of around 5 per cent (Jones, 1986) and that value has been adopted here and in the referenced studies. Adoption of a higher discount rate would reduce the nuclear advantage but, of the nuclear countries, only France (9 per cent) and Belgium (8.6 per cent) use values approaching this level, and they continue to show a nuclear advantage of 48 per cent and 34 per cent respectively. (Jones 1987b)

(d) *Scale and Replication* The anticipated nuclear economic advantage as a source of base load electricity will only be fully realized if costs can be kept down or, better still, reduced. Apart from ensuring good site management, streamlining licencing procedures, avoiding design changes involving retrofitting, and avoiding design weaknesses leading to excessive maintenance down-time; sizeable savings of up to 30 per cent of capital costs can be gained from replicating designs and installing parallel units on a single site in an orderly manner. These benefits arise from reduced manufacturing costs, reductions in the costs of common services and spares, and savings resulting from the employment of a stable and experienced construction workforce. Clearly, some of these advantages can also be reflected in reductions of operating costs (Jones, 1986).

Specific nuclear power costs (£ kW^{-1} capacity) can also be reduced by building larger plants. These conventional scale benefits are larger for nuclear than for fossil fired plants (Figures 18.3 and 21.3) and the move towards larger plant sizes in countries with large national grids or utilities has therefore favoured nuclear. However, in countries with smaller grids smaller plants are preferred for security of supply reasons and a general rule of thumb is that individual plants should be no larger than about 10 per cent of the total grid size (see Fig. 21.4).

At smaller sizes nuclear is less competitive but part (or some would argue all) of this disadvantage may be offset by adopting modular factory based construction, reducing construction periods and hence interest charges, better matching of capacity increments to capacity growth (or replacement needs), and in some circumstances

Table 21.7 Break-Even Values of Parameters. Discount rate 5 per cent, January 1984 mills and $

Country	Parameter value for coal–nuclear break-even: Load factor % Common	Load factor % Nuclear (coal equal to 72%)	Levelized coal price Mills kWh^{-1}	Levelized coal price $ GJ^{-1}	Over-all nuclear fuel price Mills kWh^{-1}	Nuclear investment cost/ reference nuclear investment cost
Belgium	8	38	9.3	1.0	21.2	2.19
Canada Central	24	47	8.0	0.9	12.5 g)	1.63
Finland	29	49	12.7	1.4	15.5	1.58
France	12	32	10.5	1.0	23.9	2.75
Germany F.R.	20	38	14.5	1.4	26.2	2.21
Italy	28	44	14.7	1.6	18.4	1.81
Japan	19	47	14.2	1.5	21.7	1.75
Netherlands	33	50	14.1	1.6	15.2	1.56
Spain	48	57	18.6	1.8	13.0	1.33
United Kingdom[1]	32	47	20.1	2.1	23.4	1.64
	13	36	13.4	1.4	30.1	2.33
US						
Central	—	—	25.0	2.4	—	0.76
Central[2]	47	63	13.8	1.3	10.7	1.17
Eastern	61	66	25.4	2.4	10.5	1.11
Rocky Mountain	—	—	25.1	2.4	—	0.68

Source: Jones, 1986
[1]UK values for Sizewell and successor station
[2]US second central values for improved construction performance

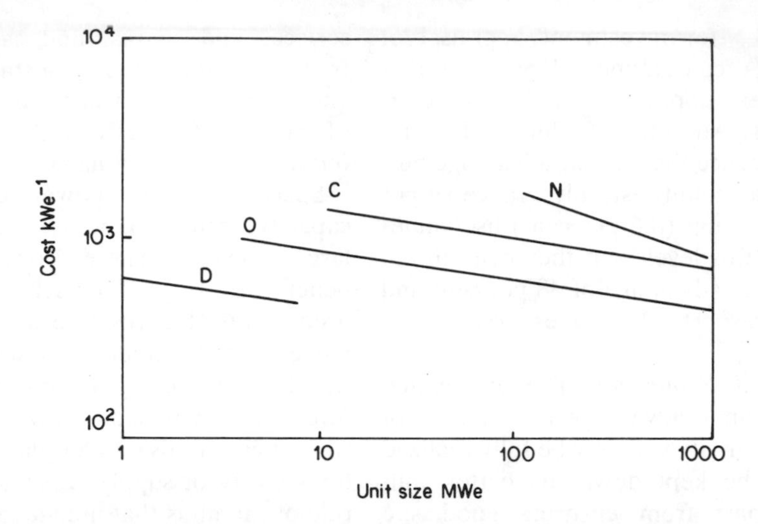

Figure 21.3 Capital cost as a function of scale D : Diesel O : Oil C : Coal N : Nuclear *From Starr and Yu, 1984, reproduced by permission of Chapman and Hall*

reduced financial risk. For this reason the US, which has a number of small utilities, and some other advanced countries have been looking at smaller reactors in the 200 Mwe range for electricity generation. Countries in earlier stages of industrialization also see this as an advantage additional to the capacity constraint imposed by the size of their electricity grid (see Chapters 5 and 18 for further discussion on small reactor types).

(e) *Other Reactor Types* The previous section has touched on small reactor costs. Neither these nor the majority of the alternatives to the PWR have had their economics covered sufficiently well in the literature to permit any generalized conclusions. Perhaps the best that can be said is that PWRs and BWRs appear broadly comparable in over-all generation costs as evidenced by their parallel deployment in the USA and Japan. The AGR and PWR, at its current state of development in the UK, appear broadly comparable in cost, at least to the degree that the

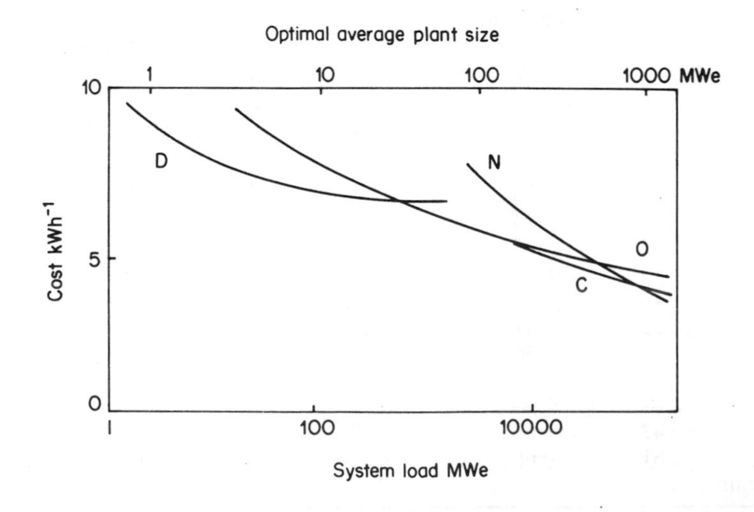

Figure 21.4 Optimum plant size. D : Diesel O : Oil C : Coal N : Nuclear *From Starr and Yu, 1984, reproduced by permission of Chapman and Hall*

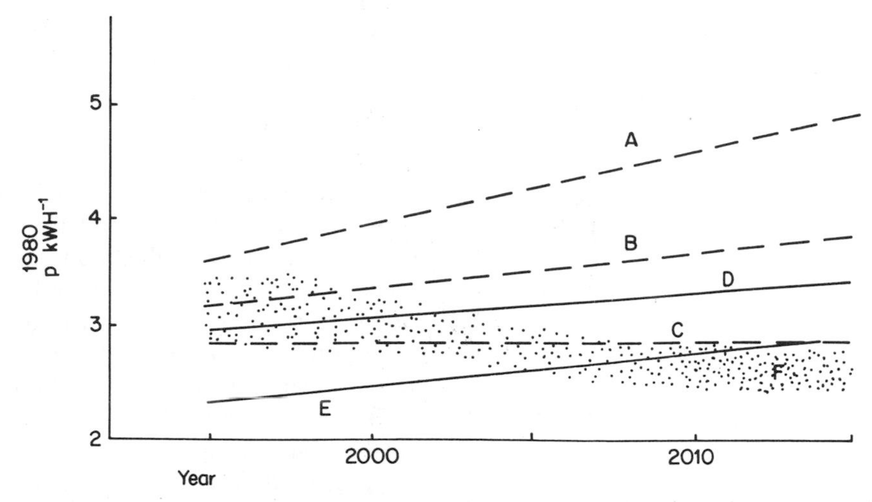

Figure 21.5 Projected fast reactor generation costs. A, B, C : Coal fired plant with 2 per cent, 1 per cent, 0 per cent year[1] coal price escalation
D,E : Range of thermal reactor generation costs
F : Possible range of fast reactor costs Based on Hirsch, 1983.

country's two main utilities can argue about which is best (see Chapter 14).

The strictures in Section 21.2.3.1 against making comparisons between data drawn from different sources limit quantitative statements to the relativities between alternatives where these have been assessed on a common basis.

Thus although the NEA studies (Jones, 1983, 1986) contain data for both CANDU and PWR derived using comparable methodology, it would be wrong to infer from either study that PHWR was a cheaper or dearer option than PWR. The relativities differ considerably with the currency unit and year of comparison for reasons totally unconnected with the reactors themselves (see section 21.2.3.1). The precise relativities to be expected will also depend on a utility's experience with the system, national design and operating requirements, etc., and no clear universally applicable distinction can be drawn at this time even when appropriate data exist for one utility or country. Canadian calculations using a different costing methodology to that recommended here suggest that CANDU would be cheaper than a US PWR constructed in Canada (McConnell and Woodhead, 1984). Comparable calculations by the present author for a single reactor station under UK conditions suggest the reverse would be the case in the UK.

The problems with advanced systems are no less acute. Some comparative data have been published on pool type fast reactors (Farmer, 1984; Hirsch, 1983; Baumier and Duchatelle, 1984) but these systems are still in the rapid learning phase as indicated by the announcement of 20 per cent reductions in specific costs of UK designs. If savings of this magnitude are realized, and other sources also suggest that they may well be (Lindley and Stader, 1986), then the idea that uranium prices would have to rise considerably before fast reactors could compete with LWRs may no longer hold true. A 20 per cent reduction in specific capital costs relative to the figures published by Hirsch or Farmer would reduce their projected generation costs by around 15 per cent to about 2.2p kWh^{-1} (Figure 21.8). This would permit the developed fast reactor with commercial scale fuel plant to compete with UK PWRs even at uranium prices in the region of \$80 kg^{-1}, which is above 1985 spot prices but below the prices agreed under many longer term contracts (see Figure 20.5).

2.2.4 *System Costs*

21.2.4.1 General For the reasons indicated earlier, single station levelized costs give a useful indication of the relative attractiveness

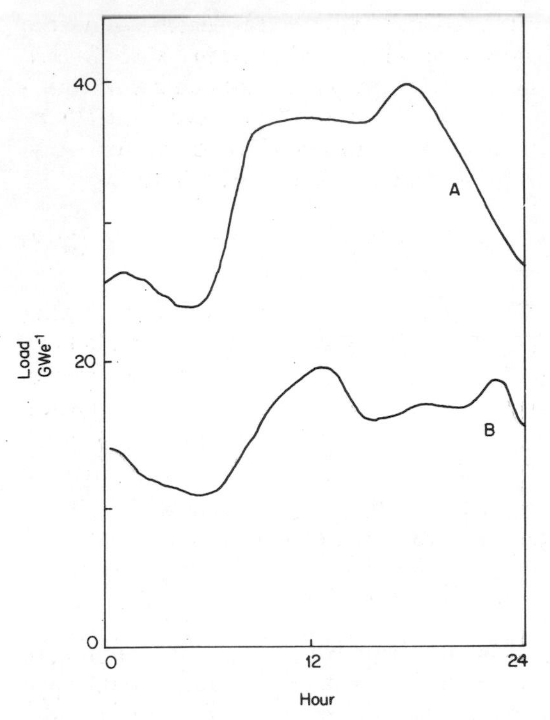

Figure 21.6 Daily load variation (UK) A–Winter's day B–Summer's day. Illustration based on 1983 data

of alternative types of generation station provided they serve a similar function, e.g. supply of base load electricity; and provided they have been derived using data and ground rules appropriate to the utility making the comparison. However, when stations are coupled with a supply grid their particular characteristics may lead to differences in their pattern of use and this can result in stations with similar generation costs (on assumed equal load factors) having quite different impacts on over-all generation systems costs, as the following paragraphs show.

The pattern of electricity demand for winter and summer days in the United Kingdom is illustrated in Figure 21.6. Other countries show similar variations although the timing and height of the demand peaks will differ and the relative winter and summer demand will be strongly influenced by climate. The UK and other high latitude countries use a great deal of electricity for heating in the winter, whereas lower latitude countries may use proportionately more for summer air conditioning. As can be seen from the diagram, there is a base load which continues throughout the day on which peaks are superimposed.

Whatever mix of plant it has, a utility will normally choose to operate the plants with lowest running cost first and bring in plant with progressively higher running costs as demand rises. Plants are ranked in a merit order of running costs, with lowest costs corresponding to highest ranking. Low ranking plants would only be brought in at times of peak demand. In general the utility will call on hydropower or, if it has them, tidal, wind or other fuel-less systems first, followed by nuclear, coal, oil and gas turbine plants in that order.

There are other factors that have to be taken into account. Firstly, fossil and nuclear plant cannot be brought from cold to full power very rapidly so that a sufficient margin of plant has to be kept running (the spinning reserve) to cater for demand increases of which the utility has no advance warning and over which it has no control. Special plant with a short run-up time may be used to deal with peaks and load fluctuations—e.g. pumped water storage schemes or gas turbines. Some limited measure of load control may be practicable over short periods by cutting voltages or adjusting frequencies. but not to the extent that it affects consumers' equipment. The utility will also be conscious of energy losses in the transmission network and location of stations relative to a demand has to be taken into account in operation as well as in initial siting decisions.

A variable source like wind power could, when giving output, displace the highest running cost plant on the system and would therefore tend to save fossil rather than nuclear fuels in the first instance. However, the inertia of large base load plant would limit its ability to follow the rapid variations in net demand after deducting fluctuating wind supply so that a proportion of wind input would give no systems cost saving.

The optimum mix of plant on a self-contained utility network is determined by the shape of the annual load–duration curve

Figure 21.7 Load–duration curve

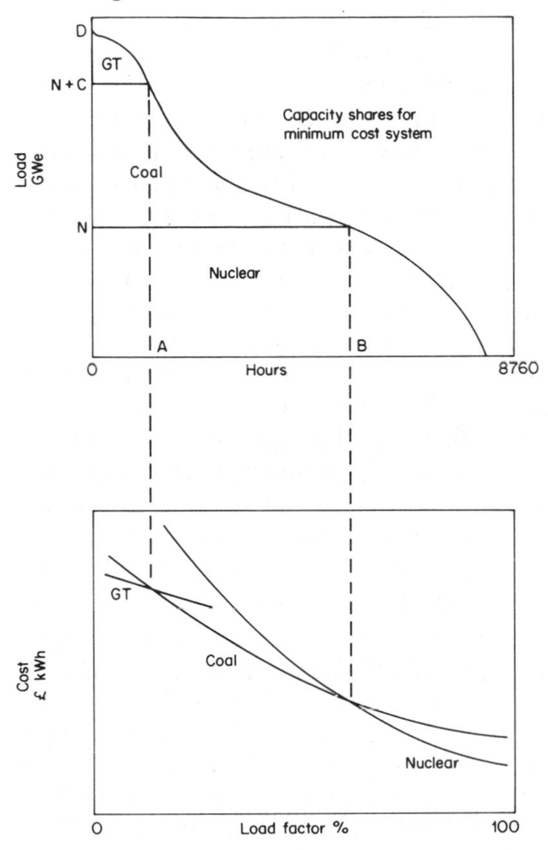

Figure 21.8 Generation cost v. load factor

(Fig. 21.7) and the characteristics of the available plant. Fig. 21.8 illustrates how total generation costs for idealized gas turbine, coal, and nuclear plant vary with load. (For an established network the plants have different ages and different efficiencies so that there will be a range of, say, coal plants with some higher in the merit order than others.) At all except the very lowest loads (in the case illustrated) hydropower is cheapest and would be used to the maximum extent possible. The total supply and its costs are, however, limited by geographical circumstances so that in most developed countries demand greatly exceeds supply from this source.

At loads above 'B' nuclear provides electricity at the lowest overall cost and the optimal nuclear capacity N is determined by the shape of the load–duration curve. Similarly at loads below 'A' gas turbines have the lowest over-all cost while from 'A' to 'B' coal is cheapest. The system maximum demand D would therefore be met most economically by a mix of N, C and G M.W. of nuclear, coal and gas turbine respectively where

$$D = N + C + G \tag{21.3}$$

In practice a utility can only estimate what future system maximum demand is likely to be and it has to take into account the fact that for a variety of reasons some proportion of its plant may not be available at any given time. For this reason a utility will plan to have more capacity on the network than its best estimate of future maximum demand to allow for forecasting errors, for extreme weather conditions and to cater for the uncertainties on the supply side. The surplus of planned capacity over estimated maximum demand is called the planning margin. The magnitude of the planning margin will depend not only on the uncertainties in supply and demand but also on the acceptability of shortfalls in supply. Clearly some risk of shortfall has to be accepted, but once this is specified in quantitative terms the capacity surplus needed to meet the adopted criterion can be deduced, taking into account the statistical uncertainty in projected demand and the availability characteristics of the generating plant (Turvey, 1968).

If there is high confidence that plant will be able to produce electricity when required the planning margin can be small, whereas if the plant is unpredictable much larger capacity surpluses may be required.

Linking power stations through a network provides several benefits. The amount of spare capacity needed to cover breakdowns is reduced since a single spare plant can provide back up for several plants almost as effectively as for one. The larger and more diverse demand covered by the grid should help to smooth load and, in favourable circumstances, reduce peaking. Additional benefits of scale flow from reduced management costs and the reductions in spares holdings and maintenance teams that can be achieved.

Larger linked networks can also accept bigger individual plants without impairing over-all reliability and this can lead to reduced specific plant costs (cost per unit output). This, as can be seen from Figure 21.4 favours nuclear over coal or gas turbines due to the different effects of physical scaling on the different plant types.

21.2.4.2 Investment Analysis The above paragraphs have set out some of the factors that affect the costs of operating different types of plant within an interconnected grid system. In general, decisions on new plant have to be viewed in the context of the existing systems, which may or may not have an optimal plant mix, and the anticipated future changes in demand and fuel prices.

If one predicts demand and its annual and daily pattern of variation one can build a computerized simulation model of the existing system and introduce new plant of different types at different times to see how the over-all costs of running the system will be affected for any range of fuel price projections considered appropriate. These over-all costs include capital expenditure, fuel and operating costs, and can be discounted and summed to give a present worth system cost just as described for the single station earlier. The timing of new capacity additions can also be determined by such a model if the individual plant sizes and availability characteristics and the over-all reliability required of the system have been defined. By iteration it is possible to determine what types of plant should be built and when, in order to minimize the total systems cost—given the particular demand and fuel price assumptions.

One is not generally concerned with planning on such a long time horizon however, and it is only the next plant or next few tranches of plant that are of immediate concern. For this purpose one can assume that the system will add a specified mix of capacity to meet future needs including an appropriate planning margin (24 per cent in the case of the UK's CEGB) and examine the effects on total system cost of making the next plant or plants coal, oil, nuclear, wind, etc., and how the system cost is affected by building the plants sooner or later.

The difference in present worth system cost, with and without the new plant, is the 'net effective cost' (NEC) which in most cases would be positive. The economically preferable (cheapest) alternative to meet the next tranche of a growing demand would be that with the lowest NEC. In rare cases where the system is a long way from the optimum plant mix and there is over-capacity, the 'net effective cost' can be negative—as in the case of the UK Sizewell PWR. This is because the new plant will displace older less efficient plant with high running costs from the system, and the present worth of future fuel and operational savings thus achieved will more than offset the costs of both building and running the new one: i.e. the construction of the new plant actually saves money and reduces future systems costs.

For the alternative of boosting and extending the life of the existing plant on the grid, the net present worth cost of running the system without the new plant is called the net avoidable cost (NAC) which will always be positive. The investment in new plant will be worthwhile if the NEC is smaller (less positive) than the NAC. Table 21.8 presents the NECs provided by UK CEGB for the Sizewell Inquiry.

Similar calculations could have been done for a variable source like an aerogenerator using an appropriate statistical basis for availability and output. If fossil fuel prices are high enough the introduction of aerogenerators (or other variable renewables) might appear economic in terms of potential fuel savings. However, their unpredictability means that the planning margin to allow for the system availability constraint will need to be increased and old thermal plant will need to be maintained as a back-up or, in periods of growing demand, additional tranches of back-up plant would need to be built. For small tranches firm power is close to average power so the effect is small.

21.2.4.3 Treatment of Uncertainty It will be evident that there is unavoidable uncertainty

Table 21.8 Net Effective Cost of New Stations £(1982) kWe^{-1} $year^{-1}$

Item	Sizewell 'B'	Coal station
Capital	91	52
Decommissioning	1	0
O and M	10	9
Fuel	35	126
System saving	−230	−177
Net total	−93	+10

Source: Sizewell Inquiry evidence for no post-Sizewell nuclear plants

surrounding many of the factors that contribute to the relative economic attractiveness of different energy options. However one chooses to make comparisons for future plant, the expectations concerning almost all the inputs affecting costs may turn out, in due course, to have been in error. These include not only the direct costs themselves; i.e. capital, fuel, maintenance and operations; but also availability and plant life. For systems cost studies the demand and load variation forecasts will also be in error to a greater or lesser degree.

There are two main approaches to dealing with this uncertainty in the decision process. The first simply explores the sensitivity of the conclusion to changes in the input assumptions and takes a judgement on whether these are sufficiently large to cast doubt on the conclusion. This approach is illustrated by the example in section 21.2.3.5(c) of this chapter and Table 21.7 where it was seen that the initial assumptions would have needed to be wildly awry before the nuclear advantage in Europe and Japan was eliminated.

The second approach sets values for plausible ranges to the individual parameters and the likelihood of particular outcomes within the ranges. The over-all economic outcome can then be determined as a probability distribution, with a most likely value and a likely range, by performing the appropriate calculations using statistically weighted sampling of the inputs: i.e. taking a capital cost, fuel cost, etc. at random from within the defined ranges and determining the outcome, then repeating the sampling many times in such a way that the distribution of input assumptions matches the defined range and distribution. The approach is tedious but suitable for computerisation using Monte Carlo sampling techniques or, for speed, the latin hypercube method adopted for the Sizewell Inquiry (Hope and Evans, 1983). Care needs to be taken to ensure that linked parameters are not sampled independently.

The output of such analyses can be in the form of confidence ranges (see Table 21.6, Jones, 1985) or in the form of statements such as:-

1. The probability of Sizewell leading to a positive benefit is 60 per cent.
2. If the benefit is negative (with 40 per cent probability), it will lie in the range zero to −£422 million with an expected (mean) value of −£263 million.
3. If the benefit is positive it will lie in the range zero to £17 821 million with an expected value of £4617 million.
4. If Sizewell is built, the over-all expected value of the benefit is £2665 million.

using 1982 money units (Hope and Evans, 1984). (Layfield, 1987, revises these figures).

The recognition of uncertainty and its handling in investment choices is well covered in the standard works on decision analysis.

21.3 Impacts on Environment and Climate

21.3.1 *General*

Under normal operational conditions nuclear reactors and fuel facilities release small readily measurable quantities of radioactivity to the environment. These releases are monitored and controlled to levels where the licensing authorities are satisfied that no harm will come to the public or workers (see Chapter 6) and where such risks as exist are negligible compared with other hazards of everyday living (see Chapter 7).

In the event of major accidents the workers and public could be at risk and for these reasons the designer of nuclear facilities incorporates multiple independent protection barriers with the intention that even in the event of several concurrent system failures the radioactivity would be contained and the likelihood of any incident affecting the public would be negligible. Chapter 7 goes into the question of safety in more detail.

Public confidence worldwide has been badly shaken by occurrences such as those at Three Mile Island and Chernobyl and, in the UK, by the unnecessary release from the Sellafield reprocessing plant in 1984 which culminated in the prosecution of BNF plc in the summer of 1985. Only the Chernobyl disaster has in practice put members of the public at risk and the different media reaction to even minor nuclear events when contrasted with say, the tens of deaths per annum in the UK from domestic gas explosions, the 50 killed in the Bantry Bay tanker explosion, the high death toll (150) from the LPG road tanker crash at San Carlos in Spain, the 123 killed in the Alexander Keilland oil rig collapse or the regular carnage on our roads (5000 p.a. in the UK), is an interesting reflection on our times.

For the purposes of this chapter (but see Chapter 22) consideration is limited to the environmental impacts of routine emissions on the assumption that the designers' objectives with regard to accident avoidance can be met. The routine emissions of radioactivity from nuclear reactors are comparable to those from the radiation contained in emissions from coal fired plant (Commission on Energy and the Environment, 1981); both sources are trivial when compared with natural background radiation (0.2 per cent in the UK), so it is appropriate to focus on other aspects of environmental instrusion.

21.3.2 *Land Use*

Nuclear stations generally require a smaller land area than coal stations of similar capacity. They have no requirement for land for holding coal stocks nor land for holding fly ash or the range of effluents associated with the different desulphurization techniques. They may also benefit from the fact that they need not be sited near coal fields and, if situated on the coast or other situations where direct cooling is feasible, they need no unsightly cooling towers. Typical land areas are given in Table 21.9.

In those countries with mineral resources the land affected by mining operations can also be less, since the quantities of uranium fuel required (Table 21.9) are so much less than the quantities of coal or oil. However, the uranium content of ores is, with some notable exceptions, low; and the actual quantities of ore extracted will need to be some 200–fold higher than the yellow-cake used to produce fuel for the reactor.

The advent of the fast reactor would reduce uranium requirements by up to 100–fold (Chapters 4 and 20) so that it will have great advantage in terms of reduced environmental impacts resulting from mining operations, particularly since the grade of ores worked will inevitably decline over time.

For countries like the UK with indigenous coal and no economically significant uranium resources, the potential environmental gains are even more marked, since the avoidance of increased reliance on coal will also avoid added damage through subsidence, mine spoil tips and the water pollution associated with coal washing (Commission on Energy and the Environment, 1981).

21.3.3 *Visual Intrusion and Noise and Dust*

These are aspects that are of great concern to the public but have impacts that defy meaningful quantification. All that can usefully be said is that the smaller site size and possible avoidance of cooling towers reduces the visual impact of nuclear stations

Table 21.9 Environmental Effects of 1 GWe Power Stations

	Land area hectares	Fuel input t year^{-1}	Wastes t year^{-1}
Nuclear	20	50	Fission products: 1
Coal	70	3 million	CO_2: 7 million SO_2: 120 thousand NO_x: 20 thousand Ash: 0.75 million

compared with equivalent fossil alternatives. The greatly reduced need for fuel and the essential cleanliness of nuclear fuel also lead to minimal impact in terms of noise, traffic, dust, etc., associated with the large scale movement of coal. From the point of view of general amenity nuclear is undoubtedly preferable to coal.

21.3.4 *Other Effluents*

Unlike nuclear power the combustion of fossil fuels is accompanied by the inevitable release to the environment of large quantities of waste products. A 1000 MW coal plant will produce each day the wastes listed in Table 21.9.

Some of these, viz. the sulphur and nitrogen oxides, are causes of current concern as potential contributors to acid rain. Their levels in effluent gases can be reduced, at a price, using a variety of trapping techniques which themselves give rise to considerable quantities of liquid or solid wastes. The widespread damage to forests in Europe and North America is generally attributed to fossil fuel combustion, although the mechanisms and precise causes are not yet understood (Bunyard, 1986). It could, if it progresses, affect local climate and rainfall with unpredictable consequences.

Carbon dioxide is of concern because of fears that in the long term it may affect world climate by raising average temperatures (the greenhouse effect) and consequential impacts on the world's atmospheric circulation and rainfall. Most global climate models predict rises of the order of a few degrees Celsius by the middle of the next century if fossil fuel burning continues at its present rate. (Figure 21.9). Global warming would affect sea levels through reductions in the land based ice masses, desertification through rainfall variations, and the length of the growing seasons. The effect on the world's food supplies and the population at large are hard to predict—even whether it would be beneficial, or detrimental over-all is obscure. The Commission on Energy and the Environment (1981) concluded that:

> 'We consider that it would be premature at this stage to do more than note the potential importance of the possible

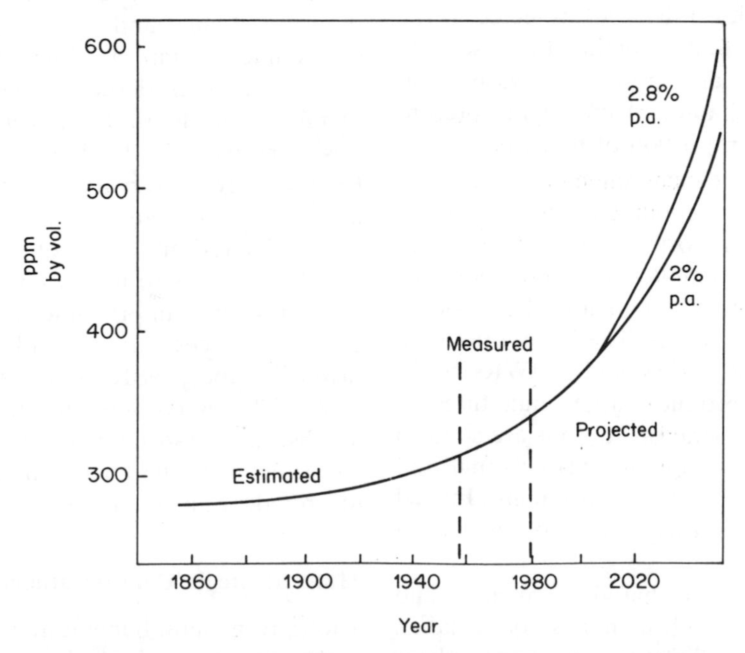

Figure 21.9 Carbon dioxide in the atmosphere. Source: Wigley (1981), 'Uranium and Nuclear Energy', John Wileys & Sons

build-up of carbon dioxide in the atmosphere and that research is being carried out to clarify the problems. No deterioration of the climate due to the build-up of carbon dioxide in the atmosphere has yet been established. If it were deemed necessary in the long run to combat the rise in atmospheric levels, any such action should be international in scope and would need to be based on the outcome of international research.'

Trace emissions from coal burning include such compounds as α-benzpyrene, a well known carcinogen, and small quantities of toxic heavy metals. The impact of such emissions on health is hard to determine, although data given by the Commission suggest that future cancers resulting from α-benzpyrene releases in the UK could lie in the range 0–150 p.a.

None of these pollutants arise from the use of nuclear power and this is one of its great, though so far little appreciated, attractions.

21.3.5 *Thermal Polution*

Since the thermal efficiency of power stations is only in the region of 30–40 per cent significant quantities of heat are rejected to the environment. Some might be used in combined heat and power or waste heat schemes (see Chapter 5) but at the present this is a small proportion of the whole.

The amount of heat rejected to rivers or the sea in direct cooling plants or to the atmosphere in the form of steam when evaporative cooling is employed, is about comparable for fossil plant and advanced gas cooled reactors or fast reactors. The lower operating temperatures of LWRs and PHWRs mean that these do produce larger quantities of waste heat. Nevertheless stations are sited so that the ambient temperature rise is small and not deleterious to the environment. Indeed in some instances it is put to good use in fish farms.

This is not a factor that gives much ground for concern although care has to be taken not to provide conditions inadvertently where small temperature fluctuations can lead to undesirable side effects involving effluents from other industries.

The steam plumes from inland cooling towers can create local microclimatic effects downwind through reductions in sunlight, but even for inland nuclear stations using cooling towers this effect would be broadly comparable to that from fossil stations.

21.3.6 *Environmental Impacts of Other Sources of Bulk Electricity*

The use of gas or oil as an alternative power station fuel has been practised widely in the past and this will continue in the future in some special situations but it will be progressively less economic. Both fuels are preferable to coal in environmental terms. Although some oils produce large quantities of sulphur oxides and the inevitable carbon dioxide, the solid residues are avoided. Gas even avoids the sulphur oxides and apart from carbon dioxide is as benign as nuclear.

Electricity producing renewables which have a 'clean' image are not without problems in their large demand for materials for construction which involves considerable environmental intrusion (see Chapter 8), in their use of land, and in direct ecological and environmental impact. Many of their impacts can only be considered in qualitative terms. Detailed consideration is beyond the scope of this work (see ETSU, 1985; Inhaber, 1981). Even energy conservation through improved house insulation and draught reduction can have deleterious consequences: reduced ventilation leads to increased radon levels in dwellings which in principle could give rise to 1–2 x 10^{-5} excess cancer deaths per gigajoule saved (Inhaber, 1981; Brambley and Gorfien, 1986). This is considerably higher than the deaths that could be associated with routine emissions from the nuclear industry required to provide equivalent energy.

21.4 Strategic Considerations

The third general benefit attached to nuclear is the strategic independence it offers some fuel importing countries. Energy is the driving force of the economy in industrial nations and any threat to their fuel supplies puts the whole nation's wellbeing at risk and, in extremis, threatens local or world peace and stability.

Diversity of fuels and diversity of supplies is one answer, stockpiling another, though the latter can be expensive in terms of money and facilities if significant protection is to be gained.

Nuclear offers the option of a different fuel which is needed in small quantities (physically) and is easy and relatively cheap to stockpile. With the advent of the fast reactor even the present small demands will be reduced so that 1 t per 1000 MWe per year will suffice. Most industrial countries will already have stocks of depleted uranium which would ensure electricity supplies for decades or even centuries. This has been a major consideration in French energy policy although it has been less prominent in countries like the US and UK which are well endowed (at present) with fossil fuel resources.

No less important is the diversity the availability of nuclear can bring within a country. In the absence of a sizeable nuclear generating component the UK government would have had great difficulty in resisting the demands of the coal miners during their year-long strike in 1984–85. The ability to stockpile the relatively small quantities of fresh nuclear fuel needed for reactors makes the electricity supply industry far less vulnerable should strikes occur in the nuclear fuel sector.

21.5 Summary

This chapter has described the incentives for the development and use of nuclear power. It has concentrated on the economic advantages foreseen since this has tended to be the main driving force behind civil nuclear programmes. The second major consideration is the undoubted strategic attraction of this vast energy source.

No less important are its environmental attractions and the crucial role it could play if fears about carbon dioxide impact on the global climate are substantiated by further research.

REFERENCES

Baumier, J., and Duchatelle, L. (1984). 'The case for developing the fast breeder reactor in France', in L. G. Brookes, and H. Motamen, (eds.) *The economics of nuclear energy*, Chapman and Hall, London, pp. 234–254.

Brambley, M. R., and Gorfein, M. (1986). 'Radon and lung cancer', *Energy*, **11**, 589–605.

Bunyard, P. (1986). 'The death of the trees', *The Ecologist*, **1986(1)**, 4–12.

Commission on Energy and the Environment (1981). *Coal and the environment*, HMSO, London.

Crowley, T. H., and Pavlenco, G. F. (1985). 'Main directions of decreasing the capital costs of LMFBRs', IAEA-SM-284/54P, *IAEA Fast Reactor Conference*, Lyon 22–26 July 1985.

Delaney, E. G. (chmn) (1986). *Decommissioning of Nuclear Facilities*, Nuclear Energy Agency/OECD, Paris.

ETSU (1985). *Prospects for the exploitation of the renewable energy technologies in the UK*, ETSU R30 Annex 3, Energy Technology Support Unit, Harwell.

Evans, N., and Hope, C. (1984). *Nuclear Power: Futures, costs and benefits*, Cambridge University Press, Cambridge.

Farmer, A. A. (1984). 'Assessing the economics of the liquid metal fast breeder', in L. G. Brookes, and H. Motamen, (eds.) *The economics of nuclear energy*, Chapman and Hall, London.

Hirsch, Sir P. (1983). 'The fast reactor: perspectives and prospects', *Nuclear Energy*, **22**, 401–415.

Howles, L. (1986). 'Nuclear plant performance', *Nuclear Eng. International*, **June, 1986** 75–79.

IEA (1985). *Electricity in IEA countries*, OECD, Paris.

Inhaber, H. (1981). 'Risk of producing energy and converting it', in *Uranium and nuclear energy*, Uranium Institute, London, pp. 217–230.

Jones, P. M. S., Taylor, K., Clifton, M., and Storey, J. (1972). *A technical and economic appraisal of air pollution in the UK*, PAU Report M20, HMSO, London.

Jones, P. M. S., and Sargeant, J. (1978). 'The Ryan Report', *Atom*, **1978**, 280–281.

Jones, P. M. S. (1979). 'The economics of nuclear power', in C. Sweet, (ed.), *The fast breeder reactor*, Macmillan, London, pp. 143–162.

Jones, P. M. S. (chmn) (1983). *The costs of generating electricity*, Nuclear Energy Agency/OECD, Paris.

Jones, P. M. S. (chmn) (1985). *The economics of the nuclear fuel cycle*, NEA/OECD, Paris.

Jones, P. M. S. (chmn) (1986). *The projected costs of generating electricity*. NEA/OECD, Paris.

Jones, P. M. S. (1987a) *'The Economics of the back-end of the fuel cycle'*, IAEA Conf. on the Back-end of the fuel cycle, IAEA, Vienna, IAEA, Vienna.

Jones, P. M. S. (1987b). *'Costs of Generation'*, Atom, 1987, in press.

Komanoff, C. (1981). *Power plant cost escalation*, Komanoff Energy Associates, New York.

Komanoff, C. (1982). 'The Westinghouse PWR in the United States; cost and performance history', ?

Komanoff, C. (1984). *Proof of evidence to Sizewell Inquiry*, CPRE/P/1.

Layfield, F. (1987). *Repeat of the Inquiry into the Sizewell Reactor,* HMSO, London.

Lindley, R. A., and Stader, J. E. (1986). 'American designers achieve reduced costs', *Nuclear Eng. International*, **31** (**July**), 31–33.

McConnell, L. G., and Woodhead, L. W. (1984). 'The Economics of the CANDU Reactor', in L. G. Brookes, and H. Motamen, (eds.) *The economics of nuclear energy*, Chapman and Hall, London, pp. 255–272.

Moynet, G. (1982). *Generating costs*, Paper to Union International des Producteurs et Distributeurs d'Energie Electrique (UNIPEDE), Congress, Brussels, June 1982.

Moynet, G. (1984). 'The cost of nuclear electricity in France', *Rev. General Nucleaire*, **1984(2)**, 141–153.

Moynet, G. (1985). 'Generating costs', Paper to UNIPEDE Congress, Athens, June 1986.

Ryan, W. (chmn) (1978). *Nuclear Power Costs*, Report of the Congressional Sub-Committee of the Committee on Government Operations.

Starr, C., and Yu, O. (1984). 'The role of centralised energy in national energy systems', in L. G. Brookes, and H. Motamen, (eds.) *'The Economics of Nuclear Energy*', Chapman and Hall, London, pp.30–45.

Surrey, J., and Thomas, S. (1980). 'World-wide nuclear plant performance', *Futures*, **12(1)**, 3–17.

Turner, L. (1985). *Coal's contribution to UK self-sufficiency*, Gower Press, London.

Turvey, R. (1968). *Optimal Pricing and Investment in Electricity Supply*, Allen and Unwin, London.

Nuclear Power: Policy and Prospects
Edited by P. M. S. Jones

22

The Contraints

P. M. S. JONES
Department of Economics, Surrey University.

22.1 Introduction

The potential deployment of nuclear plant is dependent on over-all energy demand, the availability of alternative large scale energy sources and the comparative attractions of the available options. There are no concerns over uranium supply given appropriate technological development. Previous chapters have described the considerable incentives for the use of nuclear in preference to other sources for base load electricity generation in most of the developed world, and in major energy using countries in the process of industrialization (e.g. Korea, Taiwan and India).

There are, however, two very different potential constraints which could result in less use of nuclear power than economic considerations would suggest to be optimal: public concern about the technology and financial regulations which, in some countries, distort the economic picture.

The former of these has been spearheaded (some would say unreasonably stimulated) by a small group of activists, working alone or as members of pressure groups. In Western nations these have steadfastly opposed all nuclear development and disputed virtually every claim from the nuclear industry. Their activities, whether peaceful, as in the UK or North America, or more violent, as on occasion in FRG or on a limited scale in France, attract media coverage which they have exploited to full advantage. The result has been an ongoing 'nuclear debate' the outcome of which, both sides would agree, is important to future generations as well as to ourselves.

Perhaps the first thing to recognize is that neither side in the debate has a monopoly on idealism or concern for man's welfare. The scientists and engineers who have dedicated their lives to the development of nuclear power have seen it as a beneficial addition to the resources at man's disposal and have devoted their best efforts to the economic realization of this resource in a manner that is safe and acceptable. Many who oppose nuclear power do so either because they doubt the engineer's ability to provide adequate environmental safeguards, or because they fear that it will assist the spread of nuclear weapons, or because they dislike modern industrial society and all that goes with it and hanker after something less centralized and on a smaller scale. Others are opposed less idealistically because of the direct impact it may have on their own interests, such as employment in competing industries or the adverse affects of specific construction projects on property values. There is also some suspicion that any protest is exploited by unscrupulous groups in society to further their own political ends.

The central thrust of opposition argument changes over time to capitalize on current events. One would expect that in technical matters such as costs, resource levels and safety which are susceptible to analysis, rational opinion would converge as discussion and study refined the data. Even in these areas, however, lack of trust has acted as a barrier to increased understanding. The arguments are not merely about technical and economic matters: they become political since they relate directly to the future development of society. It is not surprising, therefore that there is a great deal of confusion in the minds of the public, exposed as they are to a shifting blend of emotion and fact.

There are some things on which both sides can agree, although the conclusions drawn would not be the same. The world's population is increasing and will continue to increase at least until the middle of the next century, barring plague, mass starvation, or a world holocaust. It is not reasonable to expect two-thirds of the world's population to exist with living standards greatly below those the industrial nations now regard as necessary, though this is not to say that everyone everywhere should aspire to emulate the life-style of an affluent twentieth century European. The lot of the third world, however, can only be improved by increased economic activity to provide the infrastructure for transport, agriculture, building, hospitals, education, etc., and this, like almost everything man does, requires a plentiful supply of energy at reasonable cost. (This begs the question of the meaning of plentiful and reasonable). In principle there need be no problem since the energy resources potentially available are vast, although there are problems with developing these on the time scale set by the need to replace depleting fossil (oil and gas) reserves.

Views begin to diverge when we examine the future energy problems facing the world. What energy sources can be made available, on what timescale, in what quantity and at what cost? How much energy of what forms (heat, power, etc.) will we need? What are the implications of widespread adoption of any of these energy resources? What are the consequences of failing to provide 'enough' energy?

Perhaps the last of these questions is the easiest to answer, at least in general terms. The consequence of a shortfall in supply will be high energy prices. Depending on the severity of the squeeze the practical consequences could be recession, unemployment, and balance of payment problems. These would be particularly severe in the developing countries and would affect their ability to feed their populations, let alone improve their economic lot. We have seen just this effect following the steep oil price rises of 1973 and 1979. In extreme circumstances with rapidly falling living standards, this sort of crisis could lead to international conflict or internal strife.

Clearly we would wish to avoid such a future and the alternatives open to us are to seek to expand the supply of reasonably priced energy and/or to limit or reduce our demands. In view of world population growth, the need to reduce international disparities and the anticipated future decline in oil and gas supplies, the necessary reduction of demand in industrial nations would appear to require adoption of extensive de-industrialization and totally different life-styles if conservation were to be the sole or dominant route selected. This is a possible political choice for the way forward, but its implications have not been explored or explained by its proponents, nor is there any indication that it would gain much popular support. (See Appendix 2 for a more general discussion.)

This type of argument, however, is not specific to nuclear power and attention will therefore be focused in this chapter on the other issues, viz. safety, environment, proliferation and civil liberties.

The separate issue of finance has a particular significance in the United States and this too will be discussed further below.

22.2 Public Concerns

22.2.1 *Safety*

In a democracy there is little doubt that unless the public at large are satisfied that a technology provides adequate levels of safety it will be rejected. The nuclear industry itself (Chapter 7) is convinced that its technology can be made very safe indeed and that existing facilities throughout the OECD are built and operated in a manner that reduces risks to the population to negligible levels.

Operators of plant have an obligation to ensure that this is the case, and their own business interest together with the interests of their employees in their personal safety provide very direct incentives. Additionally Governments have established formal independent mechanisms to watch over the nuclear industry and to check its designs and operations. In the UK, for example, the Nuclear Installations Inspectorate has to be satisfied with both before it will issue or renew operating licences.

Public confidence in scientists, technologists and industry at large has suffered over the past 20 years from a number of tragic failures, of which the chemical plant at Seveso (Italy), the Union Carbide plant at Bhopal in India and the side effects of the drug thalidomide have had the widest repercussions. Until the Chernobyl disaster it was reasonable for the nuclear industry to point to its excellent record and to regard the fears of major catastrophe raised by some nuclear opponents as lacking in perspective. As indicated in Chapter 7 even the Three Mile Island failure, which has been economically disastrous to the reactor's owner with total losses of billions of US dollars, is unlikely to have led to any excess morbidity or mortality in the local population.

It is difficult writing in the immediate aftermath of the Russian accident to foresee its full implications. Short term deaths from acute effects number over 30 and eventual excess deaths due to radiation effects are likely to be measured in hundreds. Extreme opponents of nuclear will claim that this has fully substantiated their fears and that all nuclear development should cease. The industries of the OECD countries will wish to see if there are lessons to be learned when the full facts are known, but indications so far are that the type of accident and its magnitude were linked directly to the water-cooled graphite moderated pressure tube design which is not in use outside the Soviet Union.

In some countries the difficulty faced by the nuclear industry will be to convince the public and the politicians who represent them that this is true and that the other reactor types are indeed 'safe'. Since the safety cases are highly sophisticated and complex technical documents, the layman can only base his acceptance or rejection of the case on his trust and confidence in those responsible for putting the case and those who formally inspect it. In both France and Canada pride in national achievement and confidence in their procedures seem to go hand in hand. In the US and UK, however, the adversarial nature of litigation or public inquiry, with encouragement from the media, has undoubtedly weakened public confidence in the 'expert' so that even demonstrably true statements may be met with scepticism and disbelief.

Clearly it is incumbent on the industry to ensure that its plants will be safe and to satisfy the public and politicians that this is so. They will also need to be convinced that the environmental and economic advantages are sufficiently large to justify setting aside any residual irrational concerns they may have, since if the benefits are considered to be marginal it would be quite rational to opt for the 'safe' course even when the basis of doubts is recognized to be irrational by those holding them.

If the industry fail to convince the public fairly quickly that Chernobyl was a unique event which is never likely to be repeated, then public opposition will be a major constraint on nuclear development in some industrial nations.

22.3 Waste Management

A second major public concern is with radioactive waste management where there are frequent assertions in the press that this remains an unsolved problem. Chapter 9 shows this to be far from the truth yet there remain concerns in some countries that repositories will not succeed in isolating wastes from the biosphere.

This is one area which ill-founded preconceptions and prejudice has already posed problems for the nuclear undustry, not least in the UK, where local residents, spurred on by pressure groups have resisted even R and D programmes aimed at investigating the geological characteristics of sites. The opposition has been such that government backed plans to investigate a hard rock site in Scotland, plans to look at the use of abandoned anhydrite mine workings in N.E. England and plans to explore the use of clay sites in Southern England for intermediate level waste disposal have been dropped.

Clearly the fear is that experimental sites would be favoured for actual repositories despite the fact that the environmental impact of such repositories would be trivial compared with industrial development or their use for general industrial or domestic wastes.

Equally hard to understand is resistance to the continued disposal of wastes in the deep ocean where the radionuclides, even if all instantly released, would add negligibly to their existing 'natural' levels in sea water.

Again, if public concern is not to add needlessly to nuclear costs the public will have to be convinced of the efficacy of the technical solutions and their inherent safety.

22.4 Radiation

The real concern with safety appears to be linked to the potential effects of large scale exposure of the population to radiation rather than the direct local consequences of accidents. This is illustrated by the passive acceptance of the 50 deaths in the Bantry Bay tanker explosion, the 123 on the Alexander Keilland platform failure, and the 150 in the San Carlos LPG tanker crash, with no calls for the general abandonment of the use or transportation of hydrocarbon fuels, despite the fact that they move in large quantities through busy city streets.

Similarly the concern with wastes is over their hypothesized release to the environment or concentration in food chains in ways that put the public at risk through subsequent radiation exposure.

Radiation appears to be considered by some to be uniquely dangerous even if it poses minimal risk. They doubt the ability of the industry to contain it, they question whether the effects are properly understood and the very basis of the protection standards adopted. In particular fears are expressed for the welfare of future generations and the possible genetic effects of radiation releases.

Chapter 6 looks at this topic in some detail. This is perhaps one area where irrational fears should be most easily exposed though not easily overcome. Radiation is not unique to the nuclear industry: it has always been with us in our environment at far higher levels than those to which the nuclear industry exposes us. We accept the use of medical X-rays and radioisotopes for diagnosis and treatment. We bask in healthy sunlight, and natural springs with radioactive mineral content are still highly regarded. Our environment is full of natural and synthetic agents whose long term effects, if any, are not known although some in large doses are known to be mutagenic or carcinogenic and have the potential for inducing deferred cancers in much the same way as radiation. Unlike many of these substances, radiation has the merit that it is easily measured and its effects are well researched.

Since radiation can be measured and permitted levels of release specified, it is not of itself likely to be the basis of any general moratorium on nuclear activity. However, pressures to reduce radiation emissions can and have added to the costs of the nuclear industry. For example sums spent on the reduction of liquid emissions from the Magnox reprocessing plant at Sellafield (£250 million) might, on a statistical view, save a few lives of dedicated shellfish eaters over the next 10 000 years, whereas the same sum spent on hospital services could have a much larger, more immediate and demonstrable impact.

Public confidence in the industry has not been helped either by the fact that several reductions have been made over the past 30 years in the recommended radiation exposure limits or by the fact that there is no general agreement on an acceptable level of risk for members of the public or workers. What is acceptable appears to differ between industries and countries (see Chapter 7) and a more rational approach in which there is a wider public appreciation of the trade-offs would be highly desirable. So long as each sector of the economy or each issue is treated in isolation such an appreciation will be hard to achieve.

22.5 Weapons Proliferation and Civil Liberties

Everyone is concerned to ensure that public and occupational safety is maintained or improved and everyone who understands the implications of war is anxious to minimize the risks of such an event.

Attitudes to nuclear weapons themselves range from the view that it is only through them (and the balance of terror they produce) that the super powers have stayed at peace since the end of the Second World War, to the view that they are so awful that they should be abandoned unilaterally on moral grounds regardless of the consequences. No one would want to see nuclear weapons finding their way into all the world's national armouries or into the hands of terrorist groups, since either would materially increase the risk that they would be used by someone indifferent to the consequences.

Some opposition to wider deployment of civil nuclear power is based on fears that it could lead to a spread of knowledge and technology that would facilitate the development and manufacture of nuclear weapons by those who do not now possess them. Unfortunately it is already too late (if it was ever conceptually possible) to limit knowledge. It would not be impossibly difficult for any moderately advanced country to develop, at a price, workable weapons, on the basis of published information.

The only solution is acknowledged to be political, with nations voluntarily refraining from weapons development, and demonstrating this by their adherence to the Non-Proliferation Treaty and their acceptance of inspection of their nuclear facilities by the International Atomic Energy Agency safeguards team. This approach provides international assurance that nuclear materials are not being diverted from legitimate civil nuclear operations into clandestine weapons plants. Should such diversion be detected or should undeclared activities come to light, then international political pressure would be needed to dissuade the involved parties from their course of action.

Particular concern has tended to focus on plutonium separation and handling although any fissile material including uranium–235 or thorium–233 could be used to make a fission device.

The world situation would be exactly the same with or without the existence of civil nuclear power programmes and fears about weapons proliferation rationally should not impose a constraint on the deployment of civil nuclear power, provided the international safeguards system continues to function effectively and gets the backing of the countries concerned

The United States, Canada and Australia have imposed additional restrictions on the subsequent transfer, processing and treatment of spent nuclear fuel for which they have supplied the uranium or enrichment, with the objective of ensuring that their materials are not diverted to non-civil uses. Conditions have also been imposed by other suppliers of advanced nuclear technology who expect customers to accept the full range of safeguards inspections.

Some customer nations have not been happy with these restrictions on their freedom of action, particularly in those instances where they have been imposed retrospectively. In an ideal world where all civil nuclear materials and plant were subject to safeguards procedures there would be no need for additional controls and this state may be reached in some world regions if not globally.

The need to prevent diversion of nuclear materials from fuel plant and the need to protect installations from terrorists in the troubled third quarter of this century has led to the adoption of physical monitoring and security measures, including, in appropriate cases, the use of armed guards. This has been seen by some as a threat to the liberty of the individual or the arrival of a police state. This too seems to be a major distortion of the significance or role of nuclear power since a large number of sensitive buildings are now regularly protected in most parts of the western world. These include not only military installations and research establishments but also airports, docks and banks. Many facilities that receive no protection offer far more attractive targets for terrorist actions in terms of their potential impact on the community.

The curtailment of civil nuclear power would have no noticeable impact on the nature or degree of precautions now taken and although the need for such protection may add to nuclear costs, it is not a serious constraint on its deployment.

22.6 Planning Procedures

Most western industrial nations have procedures whereby proposed industrial development can be subjected to scrutiny and its impacts on the community considered before the proposals are put into effect. In general these procedures allow any citizens who feel that they would be adversely affected to voice their concerns.

Since such procedures are well-established they are not of themselves any greater constraint on nuclear development than on any other industrial activity. However antinuclear groups have exploited the system to delay specific plans with great effect. Extreme examples have been seen in the USA and the FRG where the courts have been extensively used to bring about delays of many years.

In the UK hopes of winning wider public support by exhaustive public inquiries culminated in the lengthy Sizewell hearings which will have taken nearly four years before the final decision has been debated and a verdict given.

Such delays, when added to the construction lead times for power stations (five to six years) have had surprisingly little impact due to the general reduction in world economic and electricity demand growth in the early 1980s. However, with more buoyant growth they would make meaningful planning extremely difficult for electricity utilities. There could be a serious risk of energy shortfalls if the situation were to persist; particularly since there is no assurance that any practical alternative to nuclear would have an easier passage.

For these reasons governments have contemplated action to reduce the planning delays. In particular the aim would be to cover generic issues such as reactor safety only once and to limit subsequent discussion to site specific matters and local impacts. The object would not be to stifle opposition but to prevent unreasonable filibustering and repeated re-examination of matters that have been fully aired and on which conclusions have been reached.

Unless governments in those countries most affected succeed in this aim the planning process will prove a constraint and a deterrent to those contemplating investment in new power stations because of the uncertain delays it can introduce.

22.7 Financial Constraints

There are two facets to the financial constraint on nuclear power. The availability of funds and the perceptions of costs. These are dealt with separately.

22.7.1 *Funding*

Like hydroelectric or other capital schemes nuclear power requires a large investment of funds before any revenue can be earned from electricity sales. Whilst this poses no great problem for major utilities in industrial countries, it can be a source of difficulty for smaller utilities, particularly in developing countries, since much of the expenditure has to be on imports of high technology components and the specialized skills needed for plant construction. Additionally special training schemes would be needed to provide a pool of indigenous labour for safe operation and maintenance.

Under some circumstances cash flow and balance of payments considerations may lead a small country to favour fuel using technologies where the costs and revenues are incurred concurrently (in large part) even though these may be more expensive in economic terms.

Up to the present this has not been a major constraint on nuclear power. The grid sizes in the developing countries are too small to accept large nuclear plant and at smaller sizes (<300 MW) coal has been competitive (Chapter 21). This may not be the case in the future and ways are being sought to overcome the barrier.

The 1985 decision by Turkey and Canada to have a joint venture for a CANDU type reactor in the former country has broken new ground. The Canadians are providing the capital which is to be repaid from revenues when the plant is running.

Apart from such special arrangements the customary sources of funding are potentially available, e.g. the World Bank, where nuclear investment would have to stand in competition with other capital schemes.

One particular problem in a free market is the investor's perception of the financial risk he is running both in terms of the technology and its market and in terms of the company or country seeking the capital. The US experience with nuclear construction delays, cost over-runs and retrofitting of modifications, all of which affect economic performance, has to be set against the success of France and Canada (for example) in judging the investment potential. It would not be surprising if banks looked for relatively high rates of return to cover perceived risks, particularly in countries where political stability was not assured. As has been seen in Chapter 21 high interest (discount) rates reduce the economic attractions of nuclear and this too may prove a constraint to its optimal deployment in the future as electricity use in the third world grows.

The same effect could also be a factor in countries like the USA where there are many quite small utilities and investment is dependent on financial markets.

22.7.2 *Rate Fixing*

A problem peculiar to the USA is the regulation of prices charged by public utilities using a long established formula which distorts the true economics. The same formula is used for gas pipelines, town-gas plants, new power stations or a mix of existing older stations.

The price is based on historic capital expenditure plus accumulated interest during construction less recovered depreciation allowances. A specified return on this 'rate base' is derived as the weighted average of interest on debt and a notional return on equity capital allowing for income taxes. The annual capital charge in a given year is the product of the return and the 'rate base' plus the depreciation charge. To this are added labour, materials and fuel costs. In the case of nuclear the initial cost of the core plus interest is spread over the expected plant life. The final price is then the total annual cost divided by expected net electricity output over the year in question.

The use of this formula leads to costs which vary over time, being too high initially compared with the economic levelized costs discussed in Chapter 21. Because US utilities are not allowed to recover costs until the plant is connected to the grid, and because it goes in in large tranches, the phenomenon of rate shock occurs. The introduction of new plant leads to a sudden increase in prices compared with those for the older partly depreciated plant stock, which is exacerbated by the effects of inflation which will have increased the nominal generation costs, even if real costs have remained the same. Nuclear with a high initial capital cost will be more adversely affected than cheaper coal plant.

The net effect is to create a distorted picture of relative costs for the customer which is further exacerbated by the practice of quoting future costs in terms of inflated dollars. These combine to give the customer the impression that generation costs, particularly those of nuclear, are escalating rapidly, and this fuels opposition to further nuclear installation.

As indicated in Chapter 21, nuclear costs have increased in the USA and relative costs, even on a proper economic basis, do not show new nuclear plant with a large advantage over coal unless much better construction schedules and costs can be achieved. Nevertheless the picture is made to appear far bleaker by the rate fixing process.

There is no great incentive to the utilities to change the practices since the formula leads to speedier recovery of their initial investment. Nevertheless the practice does impose an additional psychological constraint on nuclear deployment.

22.8 Summary and Comment

The above paragraphs have identified two main constraints, viz. public opposition and financial, that can act as barriers to nuclear deployment.

It has been seen that most public opposition, where it occurs, is based upon vague fears and a lack of trust in the assurances of experts, officials or politicians. These fears are fanned by anti-nuclear pressure groups who make effective use of the media to further their cause. They have also exploited the planning procedures to delay nuclear development.

The financial constraints, insofar as they distort proper economic development, also involve perceptions of risks and costs.

Fortuitously the slow down in world economic and energy growth since the mid–1970s has meant that the impact of these constraints on nuclear installation has been minimal. However, they have added to the costs of the industry which has sought to meet the concerns.

In the coming decades the resumption of energy demand growth, the need to replace ageing plant and the growing demands of third world countries for electricity will create fresh opportunities. If nuclear is to fill its rightful role as the major source of base load electricity it will be necessary to win and maintain full public confidence in the industry and to establish a meaningful and easily understood basis for the comparison of the options that exist.

REFERENCES

International Nuclear Fuel Cycle Evaluation, (1980). *Final Report of Working Group 4: Reprocessing, plutonium handling, recycle*. International Atomic Energy Agency, Vienna.

Marshall, W. (1978). *Nuclear power and the proliferation issue*. Graham Young Memorial Lecture, University of Glasgow.

Nuclear Energy Agency (1985). *Technical appraisal of the current situation in the field of radioactive waste management*. OECD, Paris.

Pardoe, G.K.C. (chmn) (1984). *Nuclear energy, a professional assessment*, The Watt Committee on Energy, London.

Stauffer, T., (1984). 'The costs of nuclear electricity', in *Uranium and Nuclear Energy*, Uranium Institute, London.

Nuclear Power: Policy and Prospects
Edited by P. M. S. Jones

23

Past, Present and Future

P.M.S. Jones
Department of Economics, Surrey University.

23.1 Introduction

Of all the energy technologies that will be covered in the *World Energy Options Series*, nuclear is unique in terms of the major significance of non-technical factors which have affected and will continue to affect its adoption in many parts of the world. Political factors, in the widest sense, may be of greater importance than technology and economics in determining how nuclear power develops.

In this final chapter the lessons of the earlier chapters will be brought together and their implications for the future considered. The policy choices with their attractions and problems will be outlined, but there can be no recipe for an optimal set of choices that would be appropriate for all countries, or even one country when the full range of possible futures is considered.

23.2 The Past

23.2.1 *Reactor Choice*

We have seen that the initial directions of civil nuclear development were firmly rooted in the events of the Second World War. The United States, the United Kingdom and Canada built on the technologies they had available to them, with the US favouring light water cooled enriched oxide fuelled reactors based on their marine propulsion experience and the availability of large scale enrichment plants; the United Kingdom favouring graphite moderated natural uranium fuelled reactors to overcome their lack of an enrichment and heavy water production capability; and Canada making use of their heavy water technology but also without enrichment plant. The USSR and France, both anxious to have an independent weapon capability, also followed the graphite moderated route initially. Germany (FRG) and Japan, precluded by the peace treaties from weapons development, both opted, after some initial experimentation, for the light water cooled reactor being developed in the USA. This route to a civil nuclear programme was followed by most other western nations including Belgium, Holland, Sweden, Spain and Switzerland. Taiwan and South Korea also bought US technology. Eastern block countries have relied almost exclusively on Soviet developments although some have shown a willingness to consider the adoption of alternative systems, as witnessed by Romania's plans for PHWRs. The latter has also proved attractive to other developing countries. It has been adopted by India and Argentina and is at the planning stage for Turkey.

The LWR has won the major share of world markets, with the BWR and PWR initially sharing the market equally, but with the latter dominating in more recent orders; not least because of the French switch to this system for its main power programmes. Both

France and the Federal Republic have developed the Westinghouse design to suit their own needs and both now compete with the USA in the limited world market for power reactors.

The multiplicity of reactor types which were conceived and, in suitable cases, brought to the prototype stage in the early years of civil development, has therefore narrowed down considerably with only six having been deployed on a significant scale; viz. the PWR and BWR, the PHWR, the graphite moderated Magnox and AGR, and the Russian graphite moderated boiling water cooled RBMK.

23.2.2 *Fuel Cycle Choice*

The initial selection of reactor types was intimately bound up with the choice of fuel cycle, with independence a major consideration for many countries. The USA with sizeable indigenous uranium resources and an enrichment capability was able to capitalize on both. Countries like the UK and France, with neither, opted for natural uranium fuel with chemical reprocessing to recover plutonium, which was seen as the major fuel of the future in view of the limited availability of uranium (as then perceived).

Those countries adopting US water cooled reactors requiring enriched fuel were dependent, initially exclusively, on the US for fuel supplies, and this led to pressures in some countries for the development of indigenous reprocessing and to plans to recycle plutonium as mixed oxide fuel in order to reduce dependence on the monopoly supplier. Alternative sources of enriched fuel from the USSR, UK and URENCO, and France and Eurodif, together with greater knowledge of the world uranium resource base, has reduced US dominance and reduced the perceived urgency for adoption of plutonium based fuel cycles. As explained later this remains a major option for the longer term.

Thorium based fuel cycles, which have been seen as another means of extending the world's fissile material resources, have been considered particularly in connection with the PHWR and high temperature reactors, notably in Canada, India and the FRG. The cycle has not so far been developed to a full commercial scale.

Most European countries and Japan have opted to pursue the route of reprocessing spent fuel to recover plutonium and unused uranium. The resource rich USA and Canada, on the other hand, have adopted a policy of storing spent fuel with the probable intention of disposing of the fuel intact in due course; a policy also formally adopted by Sweden. Following President Carter's efforts to stop reprocessing and plutonium recycle, the US position has changed so that they are no longer firmly committed to the once through cycle, although they have no civil reprocessing capability at this point in time.

23.2.3 *Opposition*

In the 1950s and early 1960s there was no opposition to civil nuclear power although in some western countries small groups opposed to the nuclear deterrent and favouring unilateral nuclear disarmament attracted publicity. The environmental movement which sprang up in some affluent western nations in the mid–1960s was initially concerned more with global ecology and the impacts of modern pesticides and industrial and agricultural practices than with nuclear technology: quite rightly since the use of nuclear power would reduce harmful pollution levels.

However, towards the end of the 1960s a vociferous minority opposition to civil nuclear power began to develop in some countries, though not all. Chapter 22 has looked at some of the arguments they have deployed, often in an opportunistic fashion by seizing on and exploiting any particular concern of the moment. Over the years this opposition has become highly skilled in its use of legal approaches to delaying development (e.g. USA, FRG) and in its use of the media to attract publicity to its cause. In some countries it has become far more professional in its presentation of an argued case and its criticisms of specific nuclear proposals (e.g. UK), whilst in others it has resorted to violent demonstrations in an effort to block schemes (e.g. FRG).

In practice it is doubtful whether it has had a significant impact on the actual level of nuclear deployment. The slow down of the 1980s can be almost exclusively laid at the door of reduced world economic growth following the rise in oil prices and the uncertainty following the Yom Kippur war in 1973 and the start of the Iran–Iraq war in 1979.

Nevertheless, as a consequence of opposition, nuclear power has been a party political issue in some western countries and this had led to the abandonment of Austria's first completed reactor and to a formal decision to limit and ultimately phase out Sweden's large nuclear programme, both following hotly argued referenda. Other countries where political action has had an impact on development include the Netherlands and Australia—the latter more in relation to uranium supply.

In other affected countries such as the USA, FRG, and the UK the main impact has been one of delay in implementation of plans, although so far, because of the economic slow-down, the effect on supply has been unimportant. In the USA in particular these delays have led to major cost escalation which, combined with rate fixing regulations, act as a disincentive to nuclear investment.

Least affected by opposition have been those countries with a clearly perceived need for replacement of existing costly energy sources (France, Japan and India) and a strong indigenous nuclear technology (France, Canada and India). There has also been no obvious opposition in Eastern Europe.

23.2.4 *Non-Proliferation*

In the early years there was a natural desire on the part of the nuclear weapons countries to keep the technology secret. The recognition of the technical impossibility and political undesirability of such a course led to the conferences on peaceful uses and to the Non-Proliferation Treaty. Under the NPT the nuclear powers undertook to make the benefits of the civil technology available to all signatories who in turn promised not to develop weapons. The majority of countries, with a few notable exceptions, became signatories of this or of the regional agreement for South America, the Treaty of Tlatelolco.

By providing for international safeguards, whereby nuclear facilities are inspected by the independent International Atomic Energy Agency, signatories of the NPT make their actions transparent and this allays fears of their neighbours. No signatory of the NPT has breached the agreement. In practice any country with a modern scientific and engineering capability could make nuclear explosive devices independent of their possession of civil power facilities and it has long been recognized that political will is the only assurance one has that they will not do it. The possession of civil nuclear facilities might provide a base of trained personnel but the effect is at worst marginal and, as confirmed by the 1980 Nuclear Fuel Cycle Evaluation, there are no technical means of ensuring that civil materials cannot be used in weapons.

The past attempts to limit the spread of knowledge and skills by maintaining a monopoly on the technology have demonstrably had the opposite effect of encouraging nations to develop an indigenous capability. The NPT through the overt commitment of the parties makes it politically difficult for countries to back-out and would expose any that showed signs of so doing to political and economic pressures which, barring war, are the only deterrents open to its neighbours. It does not seem likely that any scheme better or more effective than NPT could be evolved.

Some supplier countries have sought to supplement the controls on fissile materials by bilateral agreements with users, limiting the latter's rights to transfer or use materials (other than for the original purpose) without specific dispensations. Such restrictions, particularly when imposed retrospectively and unilaterally by a supplier, have not been popular with some countries which see their acceptance of the conditions of NPT as a sufficient assurance of good intent. One obvious consequence would be a move by dissatisfied users to seek alternative sources of supply.

23.2.5 *Cost Escalation*

From the outset nuclear reactors and fuel cycle plant were designed to be safe and economic. Once the technology had been demonstrated and was beginning to enter commercial service the question of safety assurance assumed greater importance since the consequences of the release of a major part of the core inventory of radionuclides were considered to be intolerable.

The decisions reached independently by national authorities in the different countries to provide a very high degree of safety assurance with risks reduced to very low levels, inevitably meant that additional costs were incurred to provide the replicated components, monitoring and control instrumentation and to cover the inspection and testing of the systems. The construction time and labour consequently increased in all countries, contributing further to cost escalation.

Escalation also arose as a result of construction delays, whether associated with opposition tactics as in the USA or site labour problems as in the UK and, not infrequently, from retrofitting or post-design modification when some new problem was identified or some design improvement incorporated. Rising interest rates also added to over-all investment costs. The increases in real terms looked far worse when presented at the prevailing rates of monetary inflation.

Similar escalation took place in fuel reprocessing where progressively more stringent environmental controls have added to costs over the years.

Despite these problems nuclear has been able to gain an advantage, then hold its own against fossil-fuelled power generation in most countries of the world.

23.2.6 *Industrial Development*

The rapid growth of nuclear power necessitated an equally rapid growth of the necessary manufacturing industry and associated service industries. At the same time the unprecedented increase in the size of power plants in the industrial nations, rising from a few tens of MWe to 1000 MWe reduced the numbers of plants needed. Within a relatively few years initial euphoria turned to contraction and consolidation in the manufacturing sector in the UK which had failed to capitalize on its early lead in power reactors, losing out to the US backed LWR technology.

In other countries experience was initially happier but the world recession which began in the 1970s had led to a reduction in demand below that anticipated and consequent overcapacity in both manufacture, fuel supply and fuel services. Major cut backs have occurred, notably in the USA, with its own depressed markets and with the increased international competition for overseas sales of plant and materials.

23.3 The Present

The quantitative aspects of the present world scene have been set out in Chapter 19. Nuclear has become an established part of the world's energy supply with around 400 plants in operation contributing over 15 per cent of the industrial nations' electricity. The nuclear share is set to reach 40 per cent across the European Economic Community by the mid–1990s and already exceeds 50 per cent in both France and Belgium.

As indicated in Chapters 19 and 20 there is considerable overcapacity in many sectors of the nuclear industry. Reactor construction potential is some eight to ten times current rates of new ordering; uranium production capacity is significantly above demand and user stockpiles are still being run down; enrichment and LWR fuel fabrication capacity also exceed requirements by 50 per cent and 100 per cent respectively. Despite the consolidation that has already taken place, with plants and mines being closed down or mothballed, large sectors of the industry are struggling to survive.

With the world economy staging a slow

recovery from the deep depression of the early 1980s and with energy, especially electricity, demand rising, the surplus generation capacity that has arisen in some countries will soon be absorbed and further orders could be expected.

Until recently confidence was growing that the worst was over so far as the nuclear industry was concerned. Developments in the Netherlands, the FRG and Switzerland, for example, offered the promise of renewed progress in those countries. Recent studies have confirmed the economic advantage of nuclear power over coal and oil (Chapter 21) and shown a high degree of consensus on the practicability of safe disposal of nuclear wastes. Although studies on ultrasafe reactor systems were in hand a more rational attitude to the implications of the TMI incident was beginning to prevail with a general recognition that although financially serious the consequences for the population at large were trivial.

Much research and development in the recent past has been directed towards getting the best out of existing reactor systems and enthusiasm for new designs has been tempered by the recognized benefits of replication both for manufacture and licensing.

The world demand slump and the discovery of additional uranium, particularly in Australia, has reduced the pressure on supplies and led to declining prices. It has put back the time where plutonium (or thorium) fuel cycles are needed on resource grounds, although R and D on advanced thermal reactors, high temperature reactors and, especially, fast reactors continues in view of their attractions in the longer term.

The Chernobyl disaster has, however, caused a major upset and the resultant confusion and widespread fission product fall-out has led to calls for a reappraisal of policies in many countries. The Eastern bloc and Soviet Union seem little affected and confident that the basic RBMK reactor design, which is unique to the Soviet Union, is satisfactory. They show no sign of deviating from their planned course. Some other countries have rationally taken the view that their reactors are inherently different to RBMK and that Chernobyl has no direct relevance to their situation or plans. However, in a few countries the disaster has spurred political opposition and led some politicians to adopt an overtly anti-nuclear stance which may lead to nuclear power becoming a significant election issue in the next few years in, for example, the FRG and the UK.

Just as in the aftermath of the TMI incident, uncertainty has been created about the future development of nuclear power in many countries—not because of any material change in the world energy situation (although temporarily low fossil fuel prices create an unreal impression of energy abundance), nor because of any change in the established safety of the vast majority of the world's existing or planned reactors. The uncertainty arises predominantly from public fears and the reaction of politicians and national authorities, fanned in some instances by speculation and alarmist views that tend to be given wide coverage in the media.

The question in such situations is whether decisions will be taken in haste in some countries and whether such decisions will have knock-on effects in other countries? Alternatively, will there be due deliberation and if so what will the outcome be? Whereas the scientists and engineers in the industry tend to look at quantitative risks and costs and compare these with the benefits to society at large, decisions at the political level are often more concerned with balancing the publicly perceived risks and costs against the perceived benefits; and public perceptions can diverge considerably from reality.

23.4 The Future

23.4.1 *Energy Demand*

We have seen in Chapters 18, 19 and Appendix 2 that the growing world population and the unquestioned need to improve the lot of the third world, where population growth is largest, will require a considerable expansion of low cost energy supplies. Equally, despite major efforts to improve the efficiency of energy use, economic growth in the developed countries cannot be decoupled from energy demand growth.

The actual rate of future growth of demand is subject to wide uncertainty, particularly in the long term. The economic growth that might be achieved in a stable world with international co-operation would be quite different to that achieved in a world of conflict—either military or through protectionist trade policies. The actual patterns of socio-economic development will also play a major role in determining long term energy requirements, as will the specific technologies of the future. There are all sorts of ways in which average per capita energy consumption could rise considerably even in advanced countries: e.g. if man sets out to conquer space or the ocean depths; if his demands for minerals require the working of lower grade ores; or if energy is used extensively to provide a more benign environment. Such ambitions are vastly different from those of the third world where energy for heating and cooking are in short supply and almost any welfare improvement demands commercial energy—for construction, transport, agriculture, health, etc.

At the other extreme some industrial nations might decide in the future that their living standards were all they desired and energy demand might then saturate, although this is still some way off in view of the wide disparities in affluence in even the wealthiest nations.

The one thing that is clear is that over-all world energy consumption will rise and that the limited resources of oil and gas will become increasingly costly. Coal could meet the need for many decades but at increasing prices as less economically attractive sources were brought into production. Renewables have significant potential in favoured areas but they are not yet cheap —even hydropower can prove more expensive than fossil-fuel burning. (Lazenby and Jones, 1987)

The only major new source of energy that can compare in scale with the fossil-fuels is nuclear. With the fast reactor or possibly, in the very long term, fusion, it could meet all the world's energy needs for millenia (Chapter 20). Furthermore, the costs of centrally produced nuclear electricity or nuclear heat are comparable to or less than the current costs of fossil energy (Chapter 21) and can be held at present levels long into the future.

Given this fact, and the current status of nuclear as a major contributor to the energy supplies of the majority of the world's industrial nations, it is inconceivable that nuclear will not be a major or even the dominant source of world energy for a long period in the future, whatever its short term vicissitudes.

A far more difficult question, in view of the current uncertainties, is the actual share of energy supply that nuclear will be called upon to provide in the later years of this century and the early years of the next.

The situation for the next decade is largely determined by existing orders and plant under construction because of the long lead times. From then on surplus generation capacity, which has arisen in some countries like the UK and the USA, will have been eliminated by rising electricity demand and the retirement of old fossil-fuelled plant, whilst the nuclear stations built in the 1960s will themselves be reaching the end of their useful lives. Even with low demand growth rates power plant orders will have to increase if there are not to be damaging electricity shortfalls. On economic grounds, provided the licensing and construction delays that have been experienced in some countries can be overcome, the majority of this burst of new orders should be for nuclear plant. Whether it will be will be dependent on the success of the nuclear industry in retaining public confidence or, in countries where this confidence has been lost, regaining it.

23.4.2 *Externalities*

There are several factors that could materially affect nuclear's medium term prospects adversely or favourably.

Any further major accident or release of radioactive materials, whether it were due to physical failure or to human error or negligence, would inevitably add to concern about the technology as a whole. From the public acceptability standpoint one Chernobyl is one too many. The industry will therefore have to maintain its efforts to ensure that design and operational standards are as close to perfection as possible.

On the other hand concerns about competing technologies could assist nuclear. The emissions of acid oxides from fossil-fuel (or biomass) burning plant can be controlled at a price but carbon dioxide emissions can not. As knowledge grows about the likelihood and potential impact of the greenhouse effect there may (or may not) be heightened concern about the widespread combustion of carbonaceous materials. If there were, the only practical option open would be a switch to nuclear for electricity and for the provision of heat, wherever possible.

A further external factor that would speed adoption of nuclear would be the occurrence of major supply problems for fossil-fuels occasioned by war or trade cartels. This is less likely for coal than for oil, but the loss of any major fuel would affect all the others both directly and through its influence on attitudes to security of supply.

23.4.3 *Nuclear Technology*

For the immediate future the range of nuclear technologies available is limited to those that are already being deployed. In the medium to long term the technologies at the prototype or design stage could be available should they prove economically attractive or strategically desirable. These extra options for power generation would include fast reactors, high temperature reactors, advanced thermal reactors and smaller reactors, possibly in one of the 'ultra safe' varieties.

As remarked earlier the resource pressures to move to new high efficiency uranium conserving systems has diminished with the slow down of nuclear construction and the expansion of the world's known uranium reserves. Attention has concentrated on replication and getting the best out of existing plant designs.

The current joint European programme on fast reactors, along with parallel developments in the USSR, Japan and the USA seems certain to bring this technology and its fuel cycle, which have both already been proven, to the stage of commercial availability and scale around, if not before, the turn of the century.

The high temperature reactors have a somewhat less certain development future, although interest in their safety characteristics and their economic potential as power generators should ensure that work continues. The prospects for their use as high temperature heat sources is less clear since the potential markets are not large and the process development costs could be high.

The further development of the Japanese plutonium burning ATR will depend on its economic performance and could rest largely on considerations of fuel supply security specific to Japan. The reactor, if economically successful, could be a new entrant to world export markets in the 1990s particularly if there is a move towards the use of mixed oxide fuels.

Small thermal reactors for power generation or low to medium temperature heat applications are being designed, but prototype demonstrations will be needed and their economics remain unclear. It is unlikely that economic proven designs could be offered much before the year 2000 at the earliest. If development continues and is successful they could find markets in countries with small grid systems (including smaller US utilities) or where capital constraints inhibit the adoption of large scale plant. The safety attractions claimed for some of these systems, including possibly HTRs and fast reactors, may also satisfy a market need, if this remains a point of public concern. It is also conceivable that the lines of development being pursued by India will have attractions in countries faced by similar geographical and infrastructural problems.

On the fuel cycle side the main lines of the technology are already laid down. Higher burn-up uranium oxide fuels and mixed oxide fuels will be available and a range of enrichment technologies will co-exist up to the turn of the century when advanced centrifuges and lasers are likely to be the main technologies in use.

Spent fuel storage, reprocessing and waste management are not likely to look very different although improvements and economies in all the back-end technologies can be expected as further experience is gained.

23.5 Options and Choices

23.5.1 *General*

The utility, agency or government responsible for energy or, more specifically, power supply will continue to be faced with a series of policy choices and have a range of options open. The most important are detailed in the following paragraphs. Some are always applicable, others change with time as circumstances and technology evolve.

23.5.2 *To Buy or Produce*

Assuming that a need for additional electricity supply is foreseen the supplier may have a choice of buying in his supplies from another utility or country or building his own new plant. Thus France and Canada have been looked to in Europe and by the USA respectively as vendors of electricity who can provide a supply at costs that are attractive compared with domestic generation.

The questions that have to be asked by a potential purchaser in each circumstance is whether the supply will be sufficiently secure under all circumstances and whether prices can be expected to remain favourable. Clearly the answers to these questions are dependent on the proportion of one's total supply that is to be bought-in and the practicability of switching to alternative suppliers or indigenous plant if problems arise. Over-dependence on Middle-East oil brought problems in the 1970s and some concern has been expressed about the large scale of gas imports to Western Europe from the USSR, for example.

There is a further possibility which has also been adopted, that of the joint venture. Adjacent countries may decide to invest in new plant to meet both their needs, thus sharing costs and gaining benefits of scale in suitable cases. The China–Hong Kong and Franco–Belgian joint ventures are examples of this approach in the nuclear field.

For the most part these choices are not directly linked to the use of nuclear plant or the particular type of plant although, indirectly, plant choice may affect views on security of supply and/or future prices.

23.5.3 *Nuclear or Non-Nuclear*

Given the decision to build new plant the choice of fuel (or renewable source) will depend on geography, the size of the power grid, and economic and political factors. Assuming that there are no political imperatives, such as protection of an indigenous fuel industry, and that the responsible authority is satisfied on environmental and safety grounds, the main criterion becomes economic.

As described in Chapter 21 nuclear is most attractive in the form of large plants and this demands a large power grid in most circumstances, if reliability of the network is to be maintained. Nuclear would probably not prove attractive at present for countries/utilities with grid capacities much below 10 GWe, although the availability of a smaller economic reactor could reduce this considerably, as evidenced by the experience of India. On the whole, non-industrialized countries with small grids are likely to find coal the more economic choice in the medium term.

There are additional considerations however. A country may choose to build nuclear plant before it is strictly economic to gain diversity of internal supply, to gain experience of nuclear technology which they expect to become economic for their grid at some later date, or to reduce their dependence on fuel suppliers. Environmental considerations may also lead logically to a choice of nuclear in preference to fossil fuel use or renewable sources.

The economic comparison of types of nuclear plant or nuclear versus fossil or renewable is not completely straightforward for reasons set out in Chapter 21. Certainly one can not transfer costs or experience from one country to another, and comparison of costs from different sources is potentially very misleading. Because factor costs, design requirements, taxation, etc. differ from country to country and fuel costs can even vary within countries due to the high transport component, the only meaningful economic comparison requires costing of specific design alternatives in the specific situation and power system they are expected to serve.

Even then it is possible for other financial considerations to affect decisions—for example anticipated future relative currency movements may alter the fuel–capital balance as seen in the user country, or balance of payments concerns may favour a more expensive choice if it has a larger indigenous content.

Even after careful relative assessment the out-turn of an investment is likely to depend as much on construction and operating performance as on basic investment cost.

Clearly there can be no simple universal rule since circumstances differ greatly between and even within countries. All the facts have to be assembled and judgements formed for each specific case.

23.5.4 *Abandonment*

A decision to reject nuclear power regardless of cost is an option that has been taken by Sweden and Austria, the former with plans to phase out existing nuclear plants by 2010. This choice is not one to be taken lightly since plants can not just be shut down without dire consequences for energy supply. Even phasing out has major economic consequences which could include lowered living standards and loss of investment and jobs as industry looks for lower cost energy locations. Clearly this is a political rather than a technical issue but the full consequences will not be apparent for a long time and need to be weighed very carefully before any irrevocable actions are taken.

23.5.5 *Type of Plant*

The economic considerations described in the previous section also apply to plant choice. Further factors influencing choice may be development potential, support services, flexibility of fuel or operation.

Thus utilities will be influenced by the extent of which they can expect support and advice if they run into difficulty with their plant. They may welcome a package which includes assured fuel supply and back-end services. Alternatively, they may see this as a constraint and look for a system offering them a choice of fuel suppliers in the world market. The possibility of indigenous manufacture of fuel has also attracted some customers without an enrichment capability to the PHWR, for example. Others have found the licensing arrangements for the Westinghouse PWR an attractive springboard for developing their own designs and entering world markets.

Increasingly as the nuclear component of a utility's capacity grows there will be interest in the load following characteristics of reactors.

Another technical consideration will be the safety characteristics for particular designs and how these conform to national requirements. The choice here is not merely between reactor types but also between alternative designer's versions of these types. Attitudes are likely to be heavily influenced by any existing operational experience—have the particular versions proved reliable and has their availability been consistently good, for example.

For the immediate future there are only four candidates for new large scale plant; the PWR, the BWR, the AGR and the PHWR. All have shown that they can operate reliably although individual reactors of all types have had their problems in the past. The choice for plants for ordering in the late 1990s may be larger, depending on experience with the prototype FBRs, ATR and HTRS.

Fusion, which has only received passing mention in this volume, is seen as a much longer term technology which is not likely to be at a stage where it is a practicable option much before 2030, even assuming its technical feasibility can be demonstrated and its costs brought to a level comparable with fission power.

23.5.6 *Fuel and Waste Policy*

Some aspects of fuel policy and reactor choice go together. All new commercial systems are using uranium oxide fuel, which is enriched for all but the PHWR. Large users may opt to have their own enrichment and fabrication facilities, but with the existing excess capacity and a competitive world market the gain in security of supply terms is not a major factor.

There is less consensus about the back-end of the fuel cycle where there is a choice for water reactor users to either store their spent fuel or to have it reprocessed. As yet there are no commercial facilities for disposing of either spent fuel or vitrified high activity reprocessing wastes. Gas cooled reactor fuel is not designed for long term pond storage and most of that used to date will need to be reprocessed, although in the future extended dry storage of spent AGR fuel will be a possibility should it be desired.

The choice between reprocessing and storage is partly economic, partly related to the need for plutonium for mixed oxide fuels and partly related to waste management. The economics are sensitive to the price of uranium and reprocessing (Chapter 21) and, like reactor choice, the position will not look the same to all countries.

As the quantities of separated plutonium increase they are faced with the choice of storing it, possibly for later use in fast reactors, or using it in thermal reactor mixed oxide fuel. Once its separation has been paid for there is a significant economic saving to be gained, even at current uranium prices, by using it in LWRs or the ATR, although the size of the benefit depends on the fabrication costs of mixed oxide fuels. Several countries are planning in the medium term to adopt this course, which need not impair progress towards fast reactors.

There will be an expanding need in some regions for spent fuel storage in the coming decades and for waste disposal facilities, although high level vitrified waste or spent fuel disposal are not likely to be practised on a significant scale until after the year 2000.

There remains a wide spectrum of fully satisfactory choices for nuclear waste disposal, and routes have been selected and adopted for low and intermediate level wastes. Few countries have so far seen the need to close options for high level waste disposal by taking decisions on their preferred route. Sweden is the furthest advanced in its planning. At the present time there seems to be little to be lost and much to gain from a policy of storing high level wastes and spent fuel above ground and deferring decisions and minimizing handling until a clearer consensus has emerged. (Jones, 1987)

The possibility of regional repositories has attractions for some countries with small nuclear programmes and/or unsuitable geology. This idea is less welcome to potential 'host' countries but may become less sensitive as confidence in the industry's solutions to the waste problem gain acceptance.

23.5.7 *Development Policy*

Nuclear power has passed from the early phase of rapid change and innovation to a more mature situation in which the policy choices for near term power generation are quite narrowly defined, even though there are still a number of different acceptable options.

The breeder reactor has been demonstrated technically and has many attractive safety features. It will be an essential part of any long term nuclear contribution to world energy supply and since its economics are looking increasingly likely to be close to those for existing thermal reactors, even at current thermal fuel prices, there would seem to be every justification for pressing ahead with its development and no need for any intermediate uranium conserving technologies which would require special development. That is not to say that economies in the fuel cycles for existing reactor types, through higher burn-up or use of MOX, for example, would not be worthwhile. The deployment of the fast reactor will ultimately benefit existing thermal reactors by helping to keep their fuel costs down.

Countries not pursuing a fast reactor route may still find it economically worthwhile to pursue advanced fuel cycles—for example using low enriched uranium or thorium in PHWRs. Such choices will depend on national circumstances.

In view of the political uncertainty there may also be merit in the continued exploration of the range of 'ultrasafe' systems and those which have particularly attractive safety characteristics, including again the FBR, the HTR and small water cooled systems. The economics of such systems certainly require further evaluation to test the claims that are appearing in the literature.

Inevitably further development will be expected on waste management, reactor decommissioning and on ways to improve the economics and reliability of systems.

These lines of development are being pursued in the major industrial nations. The results will be leading to the next stage of nuclear power for deployment in the post 2000 period.

REFERENCES

Lazenby, J. B. C. and Jones, P. M. S. (1987). *Hydropower in West Africa*, Energy Policy, in press.

Jones, P. M. S. (1987), *The Economics of the back-end of the fuel cycle*, IAEA Symposium on the back-end of nuclear fuel cycle, May, 1987, IAEA, Vienna.

APPENDICES

Nuclear Power: Policy and Prospects
Edited by P. M. S. Jones

Appendix 1

Discounting

P. M. S. Jones
Department of Economics, Surrey University.

Discounting is a procedure used to facilitate comparison of investment options where either or both the investment and income are spread out in time. The concept is based on the simple observation that under most circumstances a sum of money in the hand is worth more to the holder than the same sum in real terms in the future, because the holder can use the money in the intervening period.

The question is how much more it is worth to have the sum sooner. In an ideal non-inflationary world (and we work here in real or constant money terms to avoid complication) one view would be that the holder would be indifferent to a choice between having $\$x$ now and $\$x(1+i)^t$ in t years time, where i is the real interest rate he could earn expressed as a fraction. That is, the holder would accept the discounted amount $x(1+i)^{-t}$ now as the equivalent of x in t years and the present value of a series of earnings over time is the sum of these appropriately discounted amounts, viz. $\Sigma x(1+i)^{-t}$. In this example i is called the discount rate and it has been set equal to the rate of real interest the investor believes he could earn on his money. If there is a mixture of income and expenditure the discounted difference is the net present value.

Opportunity Cost of Capital (Social Opportunity Cost)

If purely financial criteria are adopted the discount rate used in assessing a project would be the opportunity cost of the capital being invested, i.e. the return that could be earned on the most productive alternative investments open to the investing organization (utility, government, etc.). This can be assessed objectively by examining the range of investment options or, for policy reasons, a particular target return may be set which projects must reach to be acceptable.

Such a target can, for example, be used to ensure that public sector investment is no less productive than private sector investment by putting the discount rate equal to the return expected on new private sector investment at the margin, i.e. the borderline of acceptability to private sector investors. The net present value for projects discounted at this marginal rate would be negative for projects with lower returns and positive for projects with higher returns. The former would be rejected.

Adoption of the average return on new private sector investment as the cut off criterion would ensure that public sector investment gave higher returns than average private sector investment. For new investment the average return must exceed the marginal rate, and this average on new investment is also likely to be higher than the average return on established investment where competition has forced profits down.

The choice of a minimum target return (i.e. discount rate) depends on the investor and can be set higher or lower if he wishes, although this would imply in either case that he is foregoing real income for reasons of his own. Thus a government may set a high target figure as one means of constraining public expenditure or a low one to encourage it for political reasons.

Risk

Some projects have a higher risk of failing than others. Thus investment in a mineral prospecting company may prove completely fruitless and result in loss of the capital invested, whereas investment in an established industry may yield highly predictable and regular returns. The rational investor looks for a high potential return from high risk ventures to offset the potential loss, and in a portfolio of such ventures would aim to do no worse, on average, than the lower return he might earn on the low risk options.

Some companies incorporate risk allowance into their assessments by adjusting their target discount rates, with higher rates for higher risks. This approach mixes risk and time preference criteria and is better avoided. Probabilities of occurrence can be attached separately to costs and benefits, or better still probability distributions, and expected costs or benefits (i.e. probability x cost or benefit) for any year can be determined, discounted and summed to arrive at an expected net present value (npv) or a probability distribution for npv, which is far more informative than applying arbitrarily raised discount rates. The latter approach incorporates and masks implicit assumptions about risk levels which are better kept explicit in investment appraisal.

The return on low risk investments can be taken as a proxy for the appropriate target rate for expected returns from all projects, at least for investors who have no positive risk aversion.

Social Time Preference

Another approach to discounting, particularly for public sector appraisals, comes from the concept of social time preference (stp). This too can be viewed in financial terms or in terms of public willingness to forego present benefits in return for quantitatively larger future benefits.

It is generally argued that stp discount rates are lower than rates based on the opportunity cost of capital, either because society as a whole is prepared to take a longer term less selfish view than the individual, or because members of society attach positive value to intangible benefits not included in simple financial assessment. (If this were not so, few in a modern society would have children and incur the costs of bringing them up and educating them.)

Establishing the stp discount rate is complicated by this question of value attached to benefits which may themselves be intangible and whose perceived value may itself change with time. The man deferring consumption from today until tomorrow is not necessarily showing negative time preference; he is attaching higher value to tomorrow's consumption than today's, for whatever reason. A country investing in flue gas desulphurization today to obtain future environmental benefits is saying something about the value it attaches to environment and to time preference, and the fact that it does it today rather than yesterday says something about changing real or perceived values.

It is hard to conceive of a society which would not be prepared to spend more in real terms to gain a positive social or economic benefit now rather than the same benefit in ten, twenty or one hundred years' time, i.e. they attach less present value to deferred benefits. However, individuals in that society might see greater advantage in bringing forward benefits within their lifetime and less concern about timing thereafter, i.e. their time preference rate might decline as the time horizon receded, or even show some discontinuity. The average time preference shown by society would then be made up of the time preferences of individuals which could have a wide range.

In practice there does not seem to be a good way of determining the stp rate. Some (Feldstein) argue that it is equal to the individual's marginal borrowing–lending rate; but some are prepared to borrow at very high interest rates to bring consumption forward, while others lend at monetary interest rates that lead to losses in real terms.

In the circumstances, average real returns accepted on established private investment may be as good a measure as any of the average social time preference rate.

Intergenerational Equity

On ethical rather than economic grounds it is sometimes argued that the present generation attaches too little importance to the welfare of future generations and that very low, or even zero, discount rates should be employed in assessment work.

Thus Collard (1979) reviews the literature on discounting in relation to long term projects. His main concern is that discounting prejudges the effective time horizon cut-off beyond which it is unimportant when costs–benefits are incurred and that it conflates the questions of discounting and generational weighting. Like the other economists he cites, he favours lower discount rates for long term projects.

He proposes summing benefit with conventional discounting for the individual beneficiary, say at birth, and then attaching weights to people born in different years and deducing over-all benefits by summing the weighted individual npvs. The present value of \$1 in year t is given by:

$$\sum_{b=0}^{b=t} n_{bt}\, w_t (1+i)_{b-t}$$

where b is the individual's birth date, w_t the weighting function, n_{bt} the proportion of individuals born in year b and alive at t and i the conventional discount rate.

If greater weights are attached to future generations, then the effective discount rate will decrease with time.

Mishan (1975), who also objects to applying discounting to situations, where benefits (and costs) accrue to different generations, does so largely on the question of valuation of social worth for the future, which he argues is impossible. However, if one believes that future generations will be wealthier, more knowledgeable and more advanced technically, then the costs to them of dealing with a problem (like radioactive waste) will appear less than it does to us at this time, so that discounting using current perceptions of cost probably over-weights their potential concern rather than underweights it.

This argument would not apply if one believed that future generations will be less well off than at present, nor could it be applied to resources of increasing scarcity with rising real value, or where values are changing because of changing public perceptions. However, it is not obvious that either situation invalidates the practice of discounting. In the former a belief in the inevitability of declining prosperity would logically lead to negative real discount rates and, in the latter, proper use of the future value of the resource or cost of the detriment in real terms would eliminate the problem. The latter may not be easy to do in practice as observed by Mishan, but it is more straightforward than arbitrarily adjusting discount rates.

Non-Monetary Aspects

A utility can reasonably restrict its concerns to the monetary costs and benefits of its actions which are conducted within legislative or other constraints (which may change with time). The setting of constraints is a matter for government and its regulatory agencies. Company actions are also governed, however, by enlightened self interest—they generally wish to be good neighbours and certainly wish to avoid actions for damages under civil law. Nevertheless they can deal in costs of achieving their objectives within these constraints.

Governments are often expected to take a wider view and to include social impacts in their decision making. Some are quantifiable, albeit imprecisely, such as effects on employment; pollution damage to buildings, crops, livestock and public health (Pearce, 1976; Jones, 1974). Costs can be attached to some of these impacts, though not without argument.

Other impacts are more difficult to quantify and/or set value to—visual intrusion, long term climatic effects of CO_2, weapon proliferation implications, social consequences of energy shortages (Jones, 1980). At the present state of theory and knowledge it is best to identify such impacts separately and not seek to produce an integrated 'cost' which obscures the value judgements built into assessing the importance of the individual impacts.

There seems to be no valid reason for not discounting such non-monetary impacts in assessment, since one is not discounting the effect but the value attached to it, i.e. the sum which would be regarded as exactly compensating for the detriment at the future date. This sum is no different conceptually from the future income earned by an investment.

Discounting and Nuclear Energy

Nuclear power stations have high initial capital and low fuel costs. The main competing option of coal fired plant has lower initial costs and higher fuel costs. Discounting therefore works to the disadvantage of nuclear power by decreasing the relative weight attached to expenditures incurred in the future (fuel) compared with those now (capital).

However, there are also some long term future costs attached to the decommissioning of nuclear plant and the storage and disposal of radioactive waste which have to be included in full comparative costings, and these are also reduced in significance by discounting.

Jeffery (1983) criticizes the application of discounting to long term costs and argues that the concept of discounting was intended to weight short term costs against deferred benefits, not vice versa. He therefore contends that deferred costs should be discounted at a negative discount rate. His logic, if generally applied, would encourage meeting all costs as soon as possible and that is a demonstrable nonsense—it is clearly pointless building repositories for high level waste, for example, decades before the wastes are produced. He singles out nuclear power and wrongly asserts that it is unique in carrying long term costs. The effects of fossil-fuel burning, for example, are just as long term, e.g. atmospheric emissions, subsidence, carbon dioxide build up.

Thus his two claims 'that normal discounting for long term nuclear costs is inappropriate (as) can be seen from the fact that it makes delay appear profitable', and that 'this is not a question of altering the present discounting procedure in general but of an exceptional procedure for a unique problem', are not sustainable. Delay in expenditure is profitable in the real world—who would not prefer to defer paying his bills if he could use the money to earn income during the delay.

From a different perspective the *Rapport du Groupe de Travail sur la Gestion des Combustibles Irradies* (1984) commented that discounting removes practically all value from deferred profits from R and D as well as significantly reducing the current equivalent expenditures on long term expenditure on waste management. They argued that new evaluation methods should be developed, such as calculating the costs of electricity over time for a self financing utility, to see how they would evolve.

This approach has attractions but it is only applicable to a network as a whole, its answers depend on no fewer assumptions than discounting, and it still leaves the question of how one trades off cost differences attached to different strategies. One, for example, might lead to relatively cheaper power now and another relatively cheaper power in the future.

The UK National Radiological Protection Board (1982), in its examination of the cost to attach to health detriments in the optimization of radiological protection, opted to recommend use of social time preference rates (0–3 per cent p.a.) for discounting public detriment and commercial rates for occupational detriment (5 per cent). This distinction appeared to be based on the concept that any detriment to the public is a social rather than financial cost while workers are recompensed through their wage for any risks they incur and the cost is therefore financial. This is totally different to the equity issues raised in the other studies. Their most recent advice has argued that the choice of discount rate for radiological protection assessment does not influence the general conclusions (National Radiological Protection Board, 1986), a view not shared by this author.

However, one problem arising from a decision to use a lower discount rate for radiological protection than for road safety measures, for example, or for radiological protection than for income earning investment in electricity generation, is that it could lead to sub-optimal use of resources and to higher overall levels of social detriment. (Brown et al, 1986)

So long as the costs considered are limited to monetary costs of reactors, fuel, repositories, etc., and due account is taken of any long term trends in real prices due to resource depletion (of uranium, coal, oil, etc.) there can be no objection to comparison of alternative options using discounting techniques to allow for time preference. If the right discount rate is chosen then a company could, in principle, set aside a sum now, equal to the discounted cost, which with the return on capital would accumulate to meet future cost of, say, decommissioning or a waste repository.

The value of the discount rate is open to debate, but for a private utility the optional return on private sector investment (opportunity cost of capital) appears to be the appropriate choice. Unless there were sound reasons for expecting the opportunity cost to vary up or down with time, most would accept the adoption of a constant rate of discount related to past and present experience.

Government controlled investment can be made on the same basis or can adopt higher or lower returns depending on its policies. There are arguments that an stp rate would be appropriate for government and this, as discussed earlier, would be lower than social opportunity cost. Adoption of an stp rate by government would tend to divert national resources away from private sector investment and would reduce economic growth in the short term. This is clearly a matter of government policy and there is no right answer. In the United Kingdom formal guidelines for public sector investment appraisal are laid down by the treasury (1980). The discount rate to be used for assessing alternative power station investments, new road schemes, hospitals, etc., is currently set at 5 per cent p.a., a figure that is consistent with those used in most other OECD countries (NEA, 1986). Most practical studies would, as a matter of course, explore the sensitivity of cost comparisons to discount rate variations.

If there are real concerns in society about long term deferment of costs then these concerns might be better met by setting time limits within which utilities should decommission plant or dispose of waste, than by arbitrarily adjusting discount rates. Such time limits, which could be based on considerations of technological, economic and social factors, would act like any other constraint and be reflected in cost calculations.

Summary

The practice of discounting is widely accepted in economic assessment and there seems to be no reason for not adopting it in assessing nuclear generation or fuel cycle costs in comparative studies.

The choice of discount rate is a matter of policy and may differ between countries and between government and private utilities.

There is no sound reason for using lower or higher discount rates for energy assessment than for other non-energy projects.

Acknowledgement

This Appendix is an updated version of an article previously published in *Atom*, December 1984 UKAEA, London.

REFERENCES

Brown, M., Blackman, T. F. Jones, P. M. S. and Mc.Keague, R. (1986). *ALARA: A general perspective from the UKAEA*, in Proc. Conf. on Quantitative Optimisation Techniques, Institution of Nuclear Engineers, London.

Collard, D. (1979). 'Faustian projects and the social rate of discount', Working Paper 1179, University of Bath Papers in Political Economy, Bath.

Feldstein, M. S. *Cost benefit analysis*, Penguin Books, Harmondsworth, p.324.

HM Treasury (1980). *Investment appraisal and discounting techniques and the use of test discount rate in the public sector*. HMSO, London.

Jeffery, J. W. (1983). *Evidence to the Sizewell Inquiry*, SSBA/P/1, HMSO, London.

Jones, P. M. S. (1974). 'Cost of environmental quality', in J. T. Coppock and C. B. Wilson (eds.), *Environmental quality*, Scottish Academic Press, pp.132–158.

Jones, P. M. S. (1980). 'Nuclear power economics', in *The fast breeder reactor*, Macmillan, London pp.143–162.

Mishan, E. T. (1975). *Cost Benefit Analysis*, Allen and Unwin, London.

National Radiological Protection Board. (1982). *Cost benefit analysis in the optimization of protection of radiation workers*, NRPB, London.

National Radiological Protection Board. (1982). *Optimization of radiation protection to the public*, NRPB, London.

National Radiological Protection Board. (1986). *Cost benefit analysis in the optimization of radiological protection*, ASP 9, NRPB, London.

Nuclear Energy Agency. (1983). *The costs of generating electricity in nuclear and coal fired stations*, NEA/OECD, Paris, Table 7, p.32.

Nuclear Energy Agency. (1986). *The projected costs of generating electricity*, NEA/OECD, Paris.

Pearce, D. W. (1976). *Environmental economics*, Longmans, London.

Rapport du Groupe de Travail sur la Gestion des Combustibles Irradies (1982). Conseil Superieur de la Surete Nucleaire, Ministere de la Recherche et de l'Industrie, Paris.

Nuclear Power: Policy and Prospects
Edited by P. M. S. Jones

Appendix 2

The Case For Economic Growth

P. M. S. Jones
Department of Economics, Surrey University.

It was noted in Chapter 22 that one aspect of opposition to nuclear power in some developed countries has arisen from a virtual rejection, by a minority, of modern technology and the present structure of society. This appears to centre on an emotional yearning for a simpler life based on self-contained rural communities, which seems to ignore the whole history of economic development and the undoubted benefits this has brought.

Energy, Wealth and Society

Energy

Man's development has been marked by, and depend upon, an expanding use of energy sources of one kind or another. Early hunting communities relied entirely on muscle power to provide the food, tools and materials for their survival; augmented by wood fires for cooking, heating and lighting.

In neolithic times, man harnessed the strength of domesticated livestock and the natural forces of wind and water for transportation, irrigation and other work. Some communities in isolated regions of the world have remained in the early phases of development and a large part of the world's population has not progressed much further.

The creation of 'industries', however, is not new. Many are so old that we can not give them a convincing date of origin. Mining, quarrying, metal working, pottery manufacture and major building were specialized activities well over 3000 years ago. Thermal energy in the form of the wood or charcoal fire was an essential part of the manufacture of pots, the quarrying of stone and the smelting and forming of metals.

Fuels other than wood and charcoal were used where they occurred naturally; pitch or asphalt in Babylonia, in the second millenium BC, and natural gas in China for lighting, cooking and heating before 1000 BC (Ray, 1979); but these were isolated events and not capitalized on widely. The principal source of thermal energy for domestic and industrial purposes remained wood, even in the most advanced nations, until the end of the seventeenth century AD. Mechanical work for hauling, lifting, grinding (corn), cutting, etc., was still provided by a combination of human and animal muscle power augmented from about the tenth century AD by the spread of watermills and, later, windmills.

Successive innovations in the use of energy have from the earliest times enabled man to expand his capabilities and increase his output, and this in turn has increased the supportable population. The use of fire to produce sharp metal tools: the introduction of the horsecollar from China around 1000 AD, which Bernal (1974) has claimed increased the tractive efficiency of the horse five-fold, and shifted the centre of production northwards from the Mediterranean: the advent of wind- and water-mills: all of these were major steps forward.

None, however, was as significant as the invention and exploitation of the steam engine (by Thomas Savary in 1698 and Newcomen in 1705) which enabled man to harness large quantities of energy reliably where and when he wanted it. Industry was no longer confined to the riverside or the hilltop and subject to the vagaries of the weather. From the early eighteenth century onwards man could build reliable, untiring pumps which in turn helped in the improvement of mining productivity by overcoming the problems caused by mine flooding. In the early nineteenth century the advent of the steam locomotive revolutionized bulk transport and heralded the railway age. In parallel with these developments the use of coke as a substitute for wood in iron making (by Darby in 1709) released a major constraint on iron production, which from the sixteenth century onwards had been hampered by the growing shortage and cost of wood charcoal.

These changes laid the foundations of modern industrial society in which plentiful supplies of cheap energy have been used to expand modern man's horizons beyond the wildest imaginings of his eighteenth century predecessors.

Wealth and Society

The mere fact that increased energy usage has been a pre-requisite of industrial expansion does not of itself show that the change is beneficial. Many do yearn for the simple agricultural life, but their view is often idealized and unrelated to the true hardships.

The average Englishman in the seventeenth century lived in a rural community and worked on the land or in related trades. He rented a small one- or two-room cottage, poorly lit by candles, poorly heated with wood or peat and poorly furnished. His wage was low and harvest failure a catastrophe. Starvation was by no means uncommon; travel was difficult and dangerous along unsurfaced rutted roads and tracks; the majority passed their lives within walking distance of their birthplace.

Medicine as we know it was non-existent. The smallest accident could lead to blood poisoning, tetanus and death. There were no anaesthetics or disinfectants. Childbirth was a hazardous event, and infant mortality was frighteningly high. Tuberculosis, smallpox and diphtheria were killers, and epidemics of typhus and cholera not uncommon, even in the nineteenth century.

The industrial revolution, which started in the eighteenth century, set in motion a process of change which has completely revolutionized the way of life and living standards of a large part of the world. At the outset, however, it was not an unmitigated blessing.

The mine and factory owners could offer higher wages than farmers: these enticed men away from the land into the developing industrial townships. At their best, conditions for workers in the towns were probably an improvement on those of their previous rural existence, but trade fluctuated and there was therefore little job security. Working conditions which would appear appalling today seemed adequate to management and workers alike. One cannot judge past times by modern standards; nevertheless, the rapid migration of workers into the new industrial centres led to serious housing and sanitation problems.

Engels wrote of old Manchester, in his *Condition of the Working Class in England* (1899):

> 'Passing along a rough bank . . . one penetrates into this chaos of small one-storied, one-roomed huts, in most of which there are no artificial floor; kitchen, living and sleeping-room all in one . . . Everywhere before the doors residue and offal . . . cattle sheds for human beings.'

Nevertheless, it was the vast increase in productivity afforded by the harnessing of steam power and the newly-invented machines that eventually provided society with the resources to make the major infrastructural developments of the nineteenth century. The linkage was not simple: higher output produced tradable surpluses which could be used to buy produce such as food from other countries (in particular, from the British Empire) and to release domestic resources for further manufacture and construction. Reduced relative costs for steel and coal made economically attractive things that were previously too expensive.

The wealth of Britain and the other nations following the same path permitted the improvement of the transport system, with metalling of the roads, the building of the railways and the development of the merchant navy, which acquired coal-fired, steel-hulled ships. It also permitted major public health measures and the construction of sewers, hospitals and schools. New confidence in man's control of his environment released a flood of inventive genius, and the physical sciences began to flourish. Physics, chemistry and biology developed into the sciences we now know.

Life for the average man in an industrial society has changed beyond recognition. From being a servant working long hours for small reward, with few possessions other than his clothes and a few pieces of simple furniture (Parker, 1976) he has become the master. He works fewer hours, he travels the world, he has his own transport and entertainment systems, he has a wide range of machines to do his work and can afford food and goods brought from all parts of the globe.

Modern technology and the systematic organization of production has increased output and reduced costs, both of raw materials and in manufacturing itself. Fewer resources, including manpower, are needed to produce the essentials, and the released resources can be used to produce yesterday's luxuries—today's necessities.

Industrial growth and new technology produce benefits not merely by enabling us to do new things, or old things in new ways, though this may be important. Their main benefits are felt in the greater freedom of choice and purchasing power they give to the individual.

It is not just a matter of material gain. The individual has benefited from the general improvements in infrastructure and services. It is not generally realized that universal education (in the vernacular) for boys and girls up to the age of 18, followed by university for the best pupils was being urged as early as 1641 (Young, 1932); or a free national health service, with doctors, schoolmasters and lawyers being paid by the state, in 1652 (Nichols). These things have only become possible (and partially realized) with the greater affluence arising from industrial development.

This is not to say that industrial development has occurred without cost. Reference was made earlier to some of the deleterious effects. We have made great strides in overcoming many of these and the problems of overcrowding and consequent health effects have largely disappeared in the industrialized nations. Working conditions have improved beyond recognition in many industries since the early days of the industrial revolution when women and children worked very long hours for a pittance, in the mines and mills. Nevertheless, there are still some industries where working conditions leave a lot to be desired and where automation and mechanization have reduced job satisfaction. The significance of this latter point has been appreciated only in the relatively recent past, and steps have been taken in the development and design of new technologies to match them to the skills and needs of the workforce, rather than solely to optimize economic efficiency. Indeed, such optimization may be less efficient if worker dissatisfaction leads to strikes, or to reduced output rates.

Again, on the environmental side, emissions into rivers and the atmosphere have been greatly reduced, though not eliminated. One can expect that the regulated industrial growth of the future will be accomplished without unacceptable ill-effects and will take due note of the lessons of the past. Indeed, the continued development of industry should materially assist in further improving our environment and doing away with some of the more antiquated and undesirable aspects of existing heavy industry.

Change is a necessary concomitant of economic development. The current European landscape is not the product of natural evolution but is a direct consequence of man's agricultural activity with the forest clearance and land drainage that went with it, stretching back to the bronze age. If man wants manufactured goods he has to accept the associated mines, quarries and factories. If he wants to travel, to have a comfortable home and to have educational and health services he has to accept the back-up industries. There are always trade-offs and conscious or unconscious choices to be made.

The industrialized world has not been alone in damaging its environment. Many poorer countries, with economies based on subsistence agriculture, have done great and lasting damage to their heritage in their efforts to preserve even minimal standards of well being. Arguably this damage is greater and more far reaching than that done even in the less controlled phase of industrial expansion (Maxey, 1979).

Can the World Use Greater Wealth?

A doubling of world population is frequently projected to occur within a generation. There were some signs of a reduction in population growth rates in the first half of the 1970s, declining from a peak of 1.9 per cent p.a. in 1970 to 1.64 per cent p.a. in 1975 due to falling birth rates in the industrial world and China and increasing death rates in many poorer countries (United Nations). Nevertheless, countries such as Algeria and Mexico have sustained population growth in excess of 3 per cent p.a. for over 25 years and there is considerable momentum behind this growth because of the expanding numbers of young adults.

Much of the population of this world is living in conditions no better than those of the United Kingdom in the eighteenth century. If world output (i.e. GDP *per capita*) were to be raised to support consumption globally at the present OECD average level, an increase of some three-fold would be needed with the present world population or, for a doubling of population, some seven-fold. This would require world economic growth rates of the order of 5 per cent p.a. if it was to be achieved by the end of the first quarter of the next century. This is an optimistic but not necessarily too unrealistic an aim when it is appreciated that countries with close to one half of the world's population have experienced growth rates exceeding 6 per cent p.a. for periods of from two to fifteen years in the recent past, and that some non-OPEC countries have achieved 10–12 per cent p.a. (Brown, 1976).

There is little argument that very considerable increases in third world consumption would be desirable and that if the above growth was even approached it would greatly improve the lot of the world's poor. Such improvements could be achieved both directly and indirectly. Increases in the wealth of the industrial world could contribute through aid and investment.

There seems little doubt that the majority of the population in the OECD countries also looks forward to steadily increasing real wealth, and shows little toleration for governments who fail to provide it, even though the extent to which governments themselves can or do control economic growth is debatable. There is a continuing demand for improved welfare services, roads and education, as well as for higher personal incomes and increased leisure time.

Additionally there are many other related and unrelated major challenges facing the world which would demand large scale endeavour: the reclamation of the world's deserts and their return to cultivation; the exploitation of energy supplies; and the conquest of space.

The above arguments would seem to justify the contention that the world both wants and needs continuing economic growth, at least into the foreseeable future. Major challenges exist in the provision of food and mineral and energy resources for the world's population which could not be met without such growth. As a target, figures for world growth in the region of 4 per cent p.a. do not seem excessive in relation to needs, nor, provided the necessary resources can be provided at reasonable cost, do such figures appear intrinsically unrealizable.

Do Energy Shortages Matter?

Economic growth and increases in energy supply necessary to achieve it can be argued to be desirable. The economic consequences of inadequate energy supply reinforce the point. Fuel shortages are not new. The extensive deforestation in the Middle East and Greece posed problems well before Roman times. The later destruction of English woods and forests to feed industrial furnaces and domestic needs, as well as for construction material, gave 'the cottager a cold hearth and a bread-and-cheese diet and sorely restricted the output of the manufacturer' as early as the fourteenth century. These problems continued and fuel crises rose both due to physical shortages and rapidly rising prices of the available fuel wood in the sixteenth century. The solution then was the development of a 'new' energy source, coal, the use of which had been restricted from the fourteenth century because of its pollution which had led 'clergy and nobility to complain of danger of contagion from the stench' (Trevelyan, 1974).

This change led to the movement of the British iron industry from the South of England to the Midlands, the North and Scotland as supplies of charcoal dwindled away.

The phenomenon has continued, though occurring in other countries later. The United States was heavily dependent on wood until the present century yet Benjamin Franklin in the eighteenth century was writing 'wood, our common fewel which within these 100 years might be had at every man's door, must now be fetched near 100 miles to some towns, and makes a very considerable article in the expense of families' (Huekel, 1975).

Today the problem is acute in the Indian subcontinent, in the semi-arid stretches of Africa and even in parts of South America; regions where dependence on wood as a fuel is almost complete. In some of the most remote areas deep in the once heavily forested Himalayan foothills of Nepal the gathering of firewood and fodder are now a day's task where a generation before the same activity occupied an hour or two (Eckholm, 1979). In the Sahel deforestation extends for 100 km and is spreading rapidly as a result of wood consumption for fish drying (Arnold and Jongma,). In the Sudan the ubiquitous Acacia tree, which was common around Khartoum as recently as 1955, has its nearest dense stand some 90 km from the city and the southern boundary of the Sahara is shifting at a rate of 5–6 km per year.

The effects of such shortages are serious. Costs, in terms of effort to gather the fuel, and prices in the market increase and either way the rural poor are obliged to forego consumption of other essential goods—their meagre living standards decline (Eckholm, 1977.

> 'Even in conditions of usually widespread unemployment, fuelwood gathering becomes a significant constraint on other activities in those seasons of the year when agricultural labour is in demand' (Arnold and Jongma,).

Additionally animal dung is diverted from fertilizer use to fuel and smaller vegetation destroyed thus damaging soil fertility and encouraging erosion and desertification (Eckholm and Brown, 1977).

The 'firewood crisis' is important globally and points not only to the urgent need to adopt conservation measures and diversify third world energy supplies, but also to the types of effect that would follow rising energy prices and physical shortages in the industrialized countries. These would be different in kind and scale but worsening living standards and damage to the general economy would be inevitable. The lead times in the development and deployment of new energy sources and major conservation measures is long and if crises are to be avoided action has to be taken well in advance. The costs of overcoming crises and their consequential effects once they have arisen can vastly exceed the costs of intelligent anticipation.

REFERENCES

Arnold, J. E. M., and Jongma, J. ( ). *Fuelwood and charcoal in developing countries*, United Nations Food and Agriculture Organization, Rome.

Bernal, J. D. (1974). *Science in history*, MIT Press, Cambridge, Mass.

Brown, L. R. (1976). *World population*, Worldwatch Institute, Washington, D.C.

Eckholm, E. (1979). *The other energy crisis: firewood*. Worldwatch Institute, Washington, D.C.

Eckholm, E., and Brown, L. R. (1977). *Spreading deserts*. Worldwatch Institute, Washington, D.C.

Huekel, G. (1975). 'A historical approach to future economic growth', *Science*, **187**,

Maxey, M. N. (1979). 'Energy, society and environment: conflict or compromise', in *Ethics and Energy*, Decision Makers Bookshelf,

Nichols, (ed), ( ). *Original letters addressed to Oliver Cromwell*, pp.100, 129.

Parker, R. (1976). *The Common Stream*, Paladin Books, London.

Ray, G. F. (1979). 'Energy economics—a random walk in history', *Energy Econ.*, **July 1979**, 149.

Trevelyan, G. M. (1974). *Illustrated English social history*, Penguin Books, Harmondsworth.

United Nations, (1984). *Statistical Year Book*, United Nations, New York.

Young, R. F. (1932). *Comenius in England*, Oxford University Press, Oxford.

Index